(continued on back inside cover)

SECOND EDITION

Intermediate Algebra

Daniel L. Auvil

KENT STATE UNIVERSITY

 ADDISON-WESLEY PUBLISHING COMPANY

Reading, Massachusetts   Menlo Park, California
Don Mills, Ontario   Wokingham, England   Amsterdam
Sydney   Singapore   Tokyo   Madrid   Bogotá
Santiago   San Juan

Sponsoring Editor	Jeffery Pepper
Production Supervisor	Susanah H. Michener
Copy Editor	Robert Hartwell Fiske
Text Designer	David Ford
Illustrators	Scientific Illustrators
Manufacturing Supervisor	Ann E. DeLacey

Library of Congress Cataloging-in-Publication Data

Auvil, Daniel L.
    Intermediate algebra.

    1. Algebra.   I. Title.
QA1542.A95   1987   512.9   85-18515
ISBN 0-201-11046-6

ABCDEFGHIJ-DO-89876

Preface

This edition of *Intermediate Algebra* differs greatly from the first edition published in 1979. Class testing, rewriting, and redesigning have culminated in a vastly improved text. The book you see represents the best of both worlds in an algebra book. It can be used as a traditional text or as a worktext. However you choose to use this book, you can rely on the following features to provide support for you and your students.

Pedagogical Features

Chapter Introductions   Each chapter starts with a descriptive preview. This preview provides a brief summary of the chapter's topics and features some of the practical applications of the material.

Try Problem Indicators   A unique feature, these margin indicators direct the reader from the examples to the comparable exercises in the problem sets. They provide students with immediate reinforcement of concepts.

Problem Solving Boxes   These applied examples are designed to help students move from words to math and learn to use algebra in daily life.

Floaters   Where warranted, a running commentary explains the steps of an example.

Cautions   Occasional informative sentences help students avoid common mistakes.

Historical Footnotes   Historical material expressed in brief footnotes lets students in on the people and dates behind the algebraic formulas.

Two-Color Format   A second color is used to set off definitions and rules and to highlight important steps within examples.

Index of Applications   To reflect the great variety of applications and to assist the reader in locating them, an Index of Applications is included on the inside covers of the text.

Problems

Problem Sets   These graded exercises are written in pairs to facilitate assigning either the odd or the even problems. Exercises corresponding to the Try Problem Indicators are in second color. Each problem set has a separate section of calculator problems. Some problem sets have a few challenging problems, marked with colored daggers. Answers to all odd problems are found at the end of the text.

Chapter Reviews   These are selections of problems touching all the major concepts of each chapter. The problems are keyed to the appropriate chapter section. Answers to all chapter review problems are in the back of the book.

Chapter Tests   These sample tests allow students to measure their readiness to take an actual classroom test over the chapter. Answers to all chapter test problems are found in the back of the book.

Cumulative Reviews   Four cumulative reviews, each covering the material in the preceding three chapters, provide extra practice for the student. Answers to all cumulative review problems are in the back of the book.

Supplements

The supplements for *Intermediate Algebra* include:

A **Student Supplement** (#11047) containing worked-out solutions to all even-numbered problems in the text, except those in the Chapter Reviews, Chapter Tests, and Cumulative Reviews.

A **Test Bank** (#11049) containing five tests for each chapter and three versions of a final exam.

A **Computerized Test Bank** (contact the publisher) software for the IBM-PC® or Apple II® computers allows instructors to generate and print tests.

Instructional Software (#11048) for the Apple II® Series.

Acknowledgments

I would like to express my sincere appreciation to the reviewers whose comments and suggestions were invaluable in writing this book: Ronald M. Davis (Northern Virginia Community College), Curtis L. Gooden (Cuyahoga Community College), Elaine M. Hubbard (Kennesaw College), Jane Morrison (Thornton Community College), Don Poulson (Mesa Community College), Charles D. Reinauer (San Jacinto College-Central). I would also like to thank Debbie Southworth, who did an excellent job typing the final manuscript.

N. Canton, Ohio.   D.L.A.

Contents

3 First-Degree Equations and Inequalities   96

4 Rational Expressions   139

5 Rational Exponents, Radicals, and Complex Numbers   189

Appendixes   491

Answers   A1

Answers to Odd-Numbered Problems, Chapter Reviews, Chapter Tests, and
Cumulative Reviews

Indexes

This chapter establishes the ground rules for all the algebra we discuss in the subsequent chapters of this book. Some of the material may be review for you, some may be new. All will be vital to you, however, as you learn to use algebra to solve a wide variety of problems.

Early in the chapter we classify the various types of numbers—natural numbers, whole numbers, integers, rational numbers, irrational numbers, and real numbers—and describe the properties of each. We then use these properties to solve simple equations, and in turn apply these equations to solve real-world problems.

1

Real Numbers and Their Properties

1.1 *Order of Operations, Fractions, the Distributive Law*

Algebra is a generalized form of arithmetic. To illustrate what this means, consider the puzzle below.

"Think of a number"	7, for example
"Add eight"	$7 + 8 = 15$
"Multiply by three"	$15 \cdot 3 = 45$
"Subtract nine"	$45 - 9 = 36$
"Multiply by two"	$36 \cdot 2 = 72$
"Divide by six"	$72 \div 6 = 12$
"Subtract the original number"	$12 - 7 = 5$

What's so special about this puzzle? Well, if the instructions are carried out correctly, the result is always 5, regardless of what the original number was. You should try this puzzle with a number of your own choice.

The arithmetic example given above demonstrates that the resulting number is 5 when the original number is 7. At the end of this section, we give an *algebraic* solution to the puzzle. The algebraic solution will make a much more general statement. It will show that the resulting number is 5 when the original number is *any* number.

Before we can prove that our puzzle works for any number, we need to develop a few basic concepts. First, the symbols 7, -2, and $\frac{3}{4}$ are called **constants,** because each represents one and only one value. In contrast, symbols such as x, y, and z are **variables.***

Definition

A symbol that represents one and only one value throughout a particular discussion is called a **constant.** A symbol that may represent any one of a collection of values is called a **variable.**

EXAMPLE 1 Write a mathematical phrase that corresponds to each word phrase.

a) Six more than a number x

$$x + 6$$

b) Three fourths of a number z

$$\frac{3}{4} \cdot z$$

---

* Our present custom of using letters of the alphabet as variables was introduced by the French mathematician François Viète (1540–1603).

c) The value of *t* nickels in cents

Try Problem 7

$$5 \cdot t \blacktriangleleft$$

ORDER OF OPERATIONS

We now discuss rules that indicate the order in which the various arithmetic operations are performed. Let's begin with an example.

EXAMPLE 2   A body builder who weighs 109 pounds wants to gain 2 pounds per week for a competition that is 4 weeks away. If she is successful, what will her weight be for the competition?

The answer is given by the calculation

$$109 + (2 \cdot 4) = 109 + 8 = 117.$$

Try Problem 17

That is, her weight will be 117 pounds. ◄

Note the parentheses placed around the multiplication in Example 2. They indicate that the multiplication must be done *before* the addition. However, if we agree that multiplication will always be done before addition whenever the two operations appear together, we then can write the calculation above more simply as

$$109 + 2 \cdot 4 = 109 + 8 = 117.$$

In those situations where we want to do the addition first, we simply enclose the addition in grouping symbols. For example,

$$(109 + 2) \cdot 4 = 111 \cdot 4 = 444.$$

In general, we agree to the **order of operations** stated below.

Order of Operations

First: perform all operations in grouping symbols. Perform operations above and below a fraction bar separately.

Second: perform all multiplications and divisions from left to right.

Third: perform all additions and subtractions from left to right.

EXAMPLE 3   Compute: **a)** $16 - 6 \cdot 2 + 3$,   **b)** $(8 + 20) \div 2 + 2$.

a) $16 - 6 \cdot 2 + 3 = 16 - 12 + 3$ Multiply before subtracting or adding

$\qquad\qquad\qquad = 4 + 3$ Subtract, since − is on the left

$\qquad\qquad\qquad = 7$ Add

b) $(8 + 20) \div 2 + 2 = 28 \div 2 + 2$ Add first, since $+$ is in parentheses

$$= 14 + 2$$ Divide before adding

Try Problem 25

$$= 16$$ Add ◄

If there is no operation symbol present, we assume that the operation is multiplication. That is,

$$2x \text{ means } 2 \cdot x,$$
$$ab \text{ means } a \cdot b,$$
$$2[26 - 5(4 - 1)] \text{ means } 2 \cdot [26 - 5 \cdot (4 - 1)].$$

EXAMPLE 4   Compute $2[26 - 5(4 - 1)]$.

$2[26 - 5(4 - 1)] = 2[26 - 5 \cdot 3]$ Work with inside grouping symbols first

$$= 2[26 - 15]$$ Multiply before subtracting

$$= 2[11]$$ Subtract, since $-$ is in grouping symbols

Try Problem 39

$$= 22$$ Multiply ◄

Any meaningful collection of constants, variables, and operations, such as

$$3x + 5y + z,$$

is called an **algebraic expression,** or simply an **expression.**

The **terms** of an expression are those quantities that are related by addition. Hence the expression above contains the three terms $3x$, $5y$, and z.

The **factors** of an expression are those quantities that are related by multiplication. The first term, for example, contains the two factors 3 and x.

$$3x + 5y + z \qquad\qquad 3x + 5y + z$$

three terms   two factors

EXAMPLE 5   Determine the value of the expression $3x + 5y + z$ if $x = 6$, $y = 4$, and $z = 2$.

$3 \cdot 6 + 5 \cdot 4 + 2$ Substitute for x, y, and z

$$= 18 + 20 + 2$$ Multiply before adding

Try Problem 55

$$= 40 \ ◄$$

FRACTIONS

The rules for operating with fractions are reviewed on the following page.

Operations on Fractions

Assume no denominator is zero.

Addition: $\dfrac{a}{c} + \dfrac{b}{c} = \dfrac{a+b}{c}$

Subtraction: $\dfrac{a}{c} - \dfrac{b}{c} = \dfrac{a-b}{c}$

Multiplication: $\dfrac{a}{b} \cdot \dfrac{c}{d} = \dfrac{ac}{bd}$

Division: $\dfrac{a}{b} \div \dfrac{c}{d} = \dfrac{a}{b} \cdot \dfrac{d}{c}$

Although it is easy to add and subtract mixed numbers, it is difficult to multiply and divide them. Hence instead of multiplying or dividing mixed numbers, we convert to improper fractions.

EXAMPLE 6   The prices of three different penny stocks are $1\frac{3}{8}$, $2\frac{1}{8}$, and $2\frac{5}{8}$. What is their average price?

$$(1\tfrac{3}{8} + 2\tfrac{1}{8} + 2\tfrac{5}{8}) \div 3 \qquad \text{Add the prices, then divide by 3}$$
$$= 5\tfrac{9}{8} \div 3 \qquad \text{Add the whole numbers, add the fractions}$$
$$= \tfrac{49}{8} \div \tfrac{3}{1} \qquad \text{Convert } 5\tfrac{9}{8} \text{ to an improper fraction}$$
$$= \tfrac{49}{8} \cdot \tfrac{1}{3} \qquad \text{Invert the divisor and multiply}$$
$$= \tfrac{49}{24} \qquad \text{Multiply numerators, multiply denominators}$$

Try Problem 61   The average price is $\frac{49}{24}$, or $2\frac{1}{24}$. ◄

To add or subtract fractions whose denominators are different, we write equivalent fractions having the least common denominator (LCD).

EXAMPLE 7   A carpenter cuts two pieces of lengths $5\frac{3}{4}$ feet and $4\frac{2}{3}$ feet from a 16-foot board, as shown in Figure 1.1. What is the length of the piece that remains?

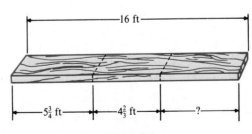

Figure 1.1

$$16 - (5\tfrac{3}{4} + 4\tfrac{2}{3}) \qquad \text{Subtract the sum of } 5\tfrac{3}{4} \text{ and } 4\tfrac{2}{3} \text{ from } 16$$
$$= 16 - (5\tfrac{9}{12} + 4\tfrac{8}{12}) \qquad \text{Write } \tfrac{3}{4} \text{ and } \tfrac{2}{3} \text{ with the LCD } 12$$
$$= 16 - (9\tfrac{17}{12}) \qquad \text{Add the mixed numbers}$$
$$= 15\tfrac{12}{12} - 10\tfrac{5}{12} \qquad \text{Prepare for subtraction}$$
$$= 5\tfrac{7}{12} \qquad \text{Subtract the mixed numbers}$$

Try Problem 65   The length that remains is $5\tfrac{7}{12}$ feet. ◄

THE DISTRIBUTIVE LAW

The next example illustrates one of the most important laws of algebra.

EXAMPLE 8   Tickets to a father-son banquet are priced as follows: $7 for each father and $4 for each son. What is the total cost for 25 father-son pairs?

Method I   Find the cost for 25 father-son pairs.
$$25 \cdot (7 + 4) = 25 \cdot 11 = 275$$

Method II   Find the cost for 25 fathers, then find the cost for 25 sons.
$$25 \cdot 7 + 25 \cdot 4 = 175 + 100 = 275$$

Try Problem 67   Using either method, the total cost is $275. ◄

From Example 8 we see that

$$25 \cdot (7 + 4) = 25 \cdot 7 + 25 \cdot 4.$$

More generally,

$$a \cdot (b + c) = a \cdot b + a \cdot c.$$

We say that "multiplication distributes over addition."

The Distributive Law

$$a(b + c) = ab + ac$$

Actually, multiplication can be distributed over any finite sum. For example,

$$a(b + c + d) = ab + ac + ad.$$

EXAMPLE 9   Use the distributive law to find each product: **a)** $4(x + 5)$, **b)** $2(3y + 15)$, **c)** $12\left(\dfrac{v}{4} + \dfrac{7}{2}\right)$.

a) $4(x + 5)$ $= 4 \cdot x + 4 \cdot 5$ Distribute 4

$= 4x + 20$ Multiply

b) $2(3y + 15)$ $= 2 \cdot 3y + 2 \cdot 15$ Distribute 2

$= 6y + 30$

c) $12\left(\dfrac{v}{4} + \dfrac{7}{2}\right) = 12 \cdot \dfrac{v}{4} + 12 \cdot \dfrac{7}{2}$

$= 3v + 42$ ◄

Try Problem 71

We are now ready to present an algebraic solution to the puzzle given at the beginning of the section.

"Think of a number"	x
"Add eight"	$x + 8$
"Multiply by three"	$3(x + 8) = 3x + 24$
"Subtract nine"	$3x + 24 - 9 = 3x + 15$
"Multiply by two"	$2(3x + 15) = 6x + 30$
"Divide by six"	$\dfrac{6x + 30}{6} = \dfrac{\cancel{6}(x + 5)}{\cancel{6}} = x + 5$
"Subtract the original number"	$x + 5 - x = 5$

Hence we have shown that the resulting number is 5 when the original number is any number x.

PROBLEM SET 1.1

Write a mathematical phrase that corresponds to each word phrase.

1. Eight more than a number x

2. Nine less than a number x

3. Three times a number y

4. Twice a number y

5. Three fifths of a number z

6. Two thirds of a number z

7. The value of t quarters in cents

8. The value of t dimes in cents

9. A person's age four years ago if he is s years old today

10. A person's age in six years if she is s years old today

11. The cost of r pounds of coffee at $2.35 per pound

12. The cost of r pounds of butter at $1.95 per pound

13. Three times a person's age two years from now if he is m years old today

14. Twice a person's age five years ago if she is m years old today

15. The total value of x shares of stock A at $21\frac{7}{8}$ per share and y shares of stock B at $33 per share

16. The total value of x ounces of gold at $379 per ounce and y ounces of silver at $11.80 per ounce

17. Kelly, who weighed 135 pounds, went on a diet and lost 2 pounds per week for 6 weeks. Find her present weight.

18. Charlie added $3 per week for 5 weeks to his savings account. If the account previously contained $127, how much does it contain now?

Compute each of the following.

19. $25 - 5 \cdot 2 + 6$

20. $28 - 8 \cdot 2 + 9$

21. $7 + 8 \cdot 9 - 3$

22. $5 + 6 \cdot 7 - 4$

23. $4 + 18 \div 2 + 4$

24. $3 + 12 \div 3 + 3$

25. $(4 + 18) \div 2 + 4$

26. $3 + 12 \div (3 + 3)$

27. $24 \div 6 - 3$

28. $28 \div 4 - 2$

29. $14 \div 7 \cdot 2$

30. $16 \div 8 \cdot 2$

31. $\frac{1}{2} \cdot \frac{7}{8} + \frac{11}{16}$

32. $\frac{4}{5} \cdot \frac{1}{3} + \frac{14}{15}$

33. $\frac{7}{4} \cdot \left(\frac{5}{4} - \frac{7}{12}\right)$

34. $\frac{2}{3} \cdot \left(\frac{5}{2} - \frac{7}{10}\right)$

35. $\frac{7}{10} + \frac{7}{6} \div 4\frac{2}{3}$

36. $\frac{8}{9} + \frac{8}{15} \div 3\frac{1}{5}$

37. $8[7 - (4 + 3)]$

38. $6[9 - (2 + 7)]$

39. $4[13 - 3(4 - 1)]$

40. $5[14 - 4(3 - 1)]$

41. $12[(\frac{1}{2} - \frac{1}{3}) + (\frac{3}{8} + \frac{5}{8})]$

42. $30[(\frac{1}{2} - \frac{2}{5}) + (\frac{5}{9} + \frac{4}{9})]$

43. $5 + [7 + 3(2 + 9)]$

44. $8 + [4 + 2(3 + 6)]$

45. $2(3[10 - 2(4 - 3)])$

46. $3(2[12 - 2(6 - 5)])$

47. $78 \div [12 - 2(10 - 6)]$

48. $81 \div [14 - 4(10 - 8)]$

49. $91.07 - [6.2(8.6 + 4.8) - 5.5]$

50. $89.09 - [3.4(9.6 + 2.8) - 7.5]$

51. $\dfrac{16 - 8}{4 + 2} \cdot \dfrac{6 + 3}{3 \cdot 2}$

52. $\dfrac{36 - 10}{4 + 2} \cdot \dfrac{12 + 8}{4 \cdot 3}$

53. $\dfrac{19 - 3\left(\dfrac{4 + 6}{5}\right) + 9}{34 + 34 \div 2 - 17}$

54. $\dfrac{17 - 4\left(\dfrac{8 + 6}{7}\right) + 13}{38 + 38 \div 2 - 19}$

Determine the value of each expression if $x = 8$, $y = 3$, and $z = 2$.

55. $3x + 5y - z$

56. $4x - 6y + z$

57. $x + yz$

58. $x - yz$

59. $3(x - y)$

60. $2(x + y)$

61. The prices of three different penny stocks are $3\frac{1}{8}$, $2\frac{3}{8}$, and $1\frac{7}{8}$. What is their average price?

62. The weights of three gears are $3\frac{1}{4}$ pounds, $4\frac{3}{4}$ pounds, and $5\frac{3}{4}$ pounds. Find their average weight.

63. Find the cost of $6\frac{3}{4}$ ounces of silver at $14 per ounce.

64. How high is a stack of 28 sheets of plywood if each sheet is $1\frac{3}{8}$ inches thick?

65. A bolt of material contains 20 yards when full. If sales of $4\frac{2}{3}$ yards and $5\frac{1}{2}$ yards are made from the bolt, how much material remains?

66. A plumber cuts two pieces of lengths $7\frac{3}{4}$ inches and $3\frac{2}{3}$ inches from a 24-inch length of pipe. What is the length of the piece that remains?

67. Twenty-five couples plan to attend a formal dance. Each boy will purchase a bouquet for $18.50, and each girl a boutonniere for $3.50. Use two different methods to calculate the total cost of the 25 bouquets and 25 boutonnieres.

68. A reserved-seat ticket to a concert costs $15.50, and a general-admission ticket costs $8.50. Use two different methods to calculate the total cost of 28 tickets of each kind.

Use the distributive law to find each product.

69. $5(x + 3)$

70. $7(x + 2)$

71. $6(2y + 4)$

72. $4(3y + 8)$

73. $10(\frac{1}{2}z + \frac{2}{5})$

74. $15(\frac{1}{3}z + \frac{3}{5})$

75. $2(a + b + c)$

76. $3(a + b + c)$

77. $24\left(\dfrac{v}{4} + \dfrac{3}{8}\right)$

78. $36\left(\dfrac{v}{4} + \dfrac{2}{9}\right)$

Use the distributive law to complete each statement.

79. $8b + 8c = 8(\underline{\quad ? \quad})$

80. $7b + 7c = 7(\underline{\quad ? \quad})$

81. $2 \cdot x + 2 \cdot 4 = 2(\underline{\quad ? \quad})$

82. $3 \cdot x + 3 \cdot 5 = 3(\underline{\quad ? \quad})$

83. $6x + 18 = 6(\underline{\quad ? \quad})$

84. $6x + 24 = 6(\underline{\quad ? \quad})$

85. $13r + 13s + 13t = \underline{\quad ? \quad}$

86. $11r + 11s + 11t + 11u = \underline{\quad ? \quad}$

87. "Think of a number. Add 5. Multiply by 2. Subtract 4. Multiply by 3. Divide by 6. Subtract the original number. The result is 3." Write an algebraic solution to this puzzle.

88. "Think of a number. Add 7. Multiply by 3. Subtract 9. Multiply by 2. Divide by 6. Subtract the original number. The result is 4." Write an algebraic solution to this puzzle.

Verify each statement. Assume no denominator is zero.

†89. $\dfrac{a}{b} + \dfrac{c}{d} = \dfrac{ad + bc}{bd}$

†90. $\dfrac{a}{b} - \dfrac{c}{d} = \dfrac{ad - bc}{bd}$

CALCULATOR PROBLEMS

91. Compute to the nearest thousandth.

a) $\dfrac{1819.604 - 17.977}{(367.196)(278.593)}$

b) $1.169(2.737)[10.196 - 1.302(8.719 - 2.864)]$

92. Compute $1712(1978 + 51,327)$ using
a) the order of operations.
b) the distributive law.

1.2   *Real Numbers*

Numbers are the building blocks of algebra. The most basic of all the numbers are the **natural numbers.**

Natural numbers = {1, 2, 3, 4, 5, 6, 7, 8, 9, 10, 11, 12, 13, . . .}

When we enclose a collection of objects (in this case the natural numbers) within braces and separate them with commas, we refer to the collection as a **set.** The three dots in the set above tell us to continue the pattern indefinitely.

When zero is included with the set of natural numbers, we obtain the set of **whole numbers.**

Whole numbers = {0, 1, 2, 3, 4, 5, . . .}

The **integers** consist of the whole numbers and the negatives of all the natural numbers.

Integers = {. . . , $-3, -2, -1, 0, 1, 2, 3,$ . . .}

A positive 3 can be denoted $+3$, or simply 3. The integer 0 is neither positive nor negative.

Rational numbers consist of all ratios (that is, quotients) of integers, so long as the divisor is not zero. We cannot list the set of rational numbers, so we write the set using a rule instead.

Rational numbers $= \left\{ \dfrac{a}{b} \,\middle|\, a \text{ and } b \text{ are integers and } b \neq 0 \right\}$

The right side of this equation is read "the set of all numbers of the form $\frac{a}{b}$ such that a and b are integers and b is not equal to zero."

Examples of rational numbers are

$$0, 6, -187, \tfrac{1}{2}, \tfrac{-13}{5}.$$

Notice that the integers 0, 6, and -187 are also rational numbers, since they can be expressed as $\frac{0}{1}$, $\frac{6}{1}$, and $\frac{-187}{1}$. The mixed number $3\frac{2}{5}$ is also a rational number, since it can be expressed as the improper fraction $\frac{17}{5}$.

In our definition of a rational number we excluded zero as a divisor. Let's find out why. Consider the quotient $\frac{6}{2}$:

$$\frac{6}{2} = 3, \quad \text{since } 2 \cdot 3 = 6.$$

Now consider the quotient $6 \div 0$:

$$\text{if } \frac{6}{0} = x, \quad \text{then } 0 \cdot x = 6.$$

However, there is no number x that makes the statement $0 \cdot x = 6$ true. Moreover,

$$\text{if } \frac{0}{0} = x, \quad \text{then } 0 \cdot x = 0.$$

In this case any number x makes the statement $0 \cdot x = 0$ true. Therefore we conclude that *division by zero is undefined.* Division *into* zero is perfectly valid, however. For example,

$$\frac{0}{6} = 0, \quad \text{since } 6 \cdot 0 = 0.$$

A rational number can also be expressed in decimal form.

EXAMPLE 1   Express the rational number $\frac{3}{8}$ as a decimal.

Divide 3 by 8 as follows:

$$
\begin{array}{r}
0.375 \\
8)\overline{3.000} \\
\underline{2\,4} \\
60 \\
\underline{56} \\
40 \\
\underline{40} \\
0
\end{array}
$$

Therefore $\frac{3}{8} = 0.375$. ◄

We can also express the rational number $\frac{3}{8}$ as a percent. Since **percent** means "per hundred," 51% means $\frac{51}{100}$, or 0.51. Hence to change from a percent to a decimal, we divide by 100. To change from a decimal to a percent, we multiply by 100. Therefore

Try Problem 11

$$\frac{3}{8} = 0.375 = 37.5\%.$$

EXAMPLE 2   Express $\frac{1}{3}$ as a decimal and as a percent.

Divide 1 by 3.

$$
\begin{array}{r}
0.33\overline{3} \\
3\overline{\smash{)}1.000} \\
\underline{9} \\
10 \\
\underline{9} \\
10 \\
\underline{9} \\
1
\end{array}
$$

The bar is placed over the 3 to indicate that it repeats indefi....ely. Therefore

Try Problem 13

$$\frac{1}{3} = 0.33\overline{3} = 33.\overline{3}\%. \blacktriangleleft$$

The decimal $0.33\overline{3}$ is called a **repeating decimal.** Other equivalent forms for this decimal are $0.\overline{3}$, $0.3\overline{3}$, etc.

Every rational number has a decimal representation that is either terminating or repeating. Also, every terminating or repeating decimal is a rational number. Hence we have another way to define the set of rational numbers.

Rational numbers $= \{x \mid x$ is a terminating or repeating decimal$\}$

Those decimals that are *neither* terminating nor repeating, such as

$$0.12112111211112 \ldots ,$$

are called **irrational numbers.** An irrational number cannot be expressed as a ratio of two integers.

Irrational numbers $= \{x \mid x$ is a nonterminating, nonrepeating decimal$\}$

Two other examples of irrational numbers are $\sqrt{2}$ and π.* The symbol $\sqrt{2}$ represents the **principal square root** of 2. The symbol π (the Greek letter pi)

---

* The fact that $\sqrt{2}$ is irrational was known to the Greek mathematician Pythagoras over 2000 years ago. However, it wasn't until 1767 that the Swiss mathematician Johann Lambert proved that π was irrational.

represents the ratio of the circumference C of any circle to its diameter d, as shown in Figure 1.2. That is,

$$\pi = \frac{C}{d}$$

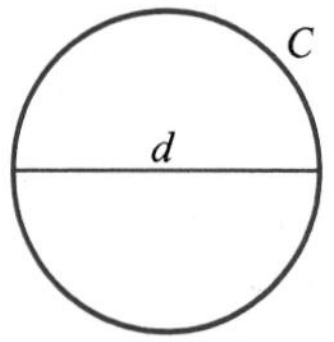

Figure 1.2

Since π is an irrational number, C and d cannot both be integers.

Since $\sqrt{2}$ and π are both irrational numbers, they are represented by nonterminating, nonrepeating decimals. For computational purposes, however, their values are generally approximated. For example, to ten decimal places we write

$$\sqrt{2} \approx 1.4142135624,$$
$$\pi \approx 3.1415926536,$$

where the symbol $\approx$ means "approximately equals."

Numbers such as $\sqrt{3}$, $\sqrt{5}$, and $\sqrt{6}$ are also irrational numbers. Numbers such as $\sqrt{1}$, $\sqrt{4}$, $\sqrt{9}$, $\sqrt{16}$, and $\sqrt{25}$ are all rational numbers, since 1, 4, 9, 16, and 25 are all **perfect squares.**

EXAMPLE 3   Identify each number as being rational or irrational.

a) $\frac{-7}{9}$ is a ratio of the two integers -7 and 9, so it is rational.
b) $\sqrt{15}$ is irrational, since 15 is not a perfect square.
c) $0.6\overline{6}$ is a repeating decimal, so it is rational.
d) $8.575575557 \ldots$ is a nonterminating, nonrepeating decimal, so it is irrational.
e) 0.675 is a terminating decimal, so it is rational.
f) $\sqrt{49} = 7$, which is an integer and hence a rational number.   ◄

Try Problem 35

If we combine the set of rational numbers with the set of irrational numbers, we obtain the set of **real numbers.**

Real numbers $= \{x \mid x \text{ is rational or } x \text{ is irrational}\}$

The relationships between the set of real numbers and its subsets are given in the diagram on the following page.

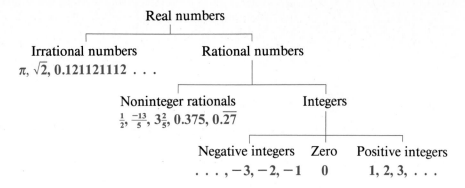

All real numbers obey the basic laws below, as well as the order of operations stated in Section 1.1.

Properties of Real Numbers

Assume a, b, and c represent real numbers.

Commutative Laws

$$a + b = b + a$$
$$ab = ba$$

Example: $5 + 7 = 7 + 5$

Example: $(-3) \cdot 4 = 4 \cdot (-3)$

Associative Laws

$$a + (b + c) = (a + b) + c$$
$$a(bc) = (ab)c$$

Example: $2 + (6 + 8) = (2 + 6) + 8$

Example: $5 \cdot (4 \cdot \frac{1}{2}) = (5 \cdot 4) \cdot \frac{1}{2}$

Identity Laws

$$a + 0 = 0 + a = a$$
$$a \cdot 1 = 1 \cdot a = a$$

Example: $11 + 0 = 0 + 11 = 11$

Example: $9 \cdot 1 = 1 \cdot 9 = 9$

Inverse Laws

$$a + (-a) = (-a) + a = 0$$
$$a \cdot \frac{1}{a} = \frac{1}{a} \cdot a = 1 \ (a \neq 0)$$

Example: $8 + (-8) = (-8) + 8 = 0$

Example: $5 \cdot \frac{1}{5} = \frac{1}{5} \cdot 5 = 1$

Distributive Law

$$a(b + c) = ab + ac$$

Example:
$2[3 + (-4)] = 2 \cdot 3 + 2 \cdot (-4)$

PROBLEM SET 1.2

Use positive and negative numbers to represent each of the following.

1. Four golf strokes over par, six golf strokes under par

2. Ten seconds before blastoff, three seconds after blastoff

3. A rise of $2\frac{7}{8}$, a fall of $4\frac{3}{8}$ in the price of a stock

4. 1700 feet above sea level, 355 feet below sea level

5. A gain of 7 yards, a loss of 3 yards on a football play

6. A temperature of $23°$ above zero, a temperature of $6°$ below zero

Express each rational number as a decimal and as a percent.

7. $\frac{1}{2}$

8. $\frac{1}{4}$

9. $6\frac{3}{4}$

10. $7\frac{4}{5}$

11. $\frac{5}{8}$

12. $\frac{7}{8}$

13. $\frac{2}{9}$

14. $\frac{4}{9}$

15. $\frac{4}{11}$

16. $\frac{2}{11}$

17. $\frac{15}{22}$

18. $\frac{13}{22}$

19. $\frac{-19}{5}$

20. $\frac{-13}{2}$

21. $\frac{0}{8}$

22. $\frac{0}{5}$

23. $\frac{8}{0}$

24. $\frac{5}{0}$

25. Consider the set of numbers below.

$$\{6, 203, +0, 4\tfrac{1}{3}, -10, 0.25, 0.18\overline{18}, \tfrac{25}{2}\}$$

a) Which are natural numbers?
b) Which are whole numbers?
c) Which are integers?
d) Which are rational numbers?

26. Consider the set of numbers below.

$$\{1, 119, -0, 8\tfrac{1}{2}, 0.75, -100, 0.45\overline{45}, \tfrac{25}{3}\}$$

a) Which are natural numbers?
b) Which are whole numbers?
c) Which are integers?
d) Which are rational numbers?

27. Arrange the four numbers below in order from smallest to largest.

$$\tfrac{8}{5}, 1.\overline{6}, 1.06, 1.0\overline{6}$$

28. Arrange the four numbers below in order from smallest to largest.

$$\tfrac{9}{5}, 1.\overline{8}, 1.08, 1.0\overline{8}$$

29. Name a rational number between 8.2 and 8.3. How many such rational numbers are there?

30. Name a rational number between 6.7 and 6.8. How many such rational numbers are there?

31. Write each terminating decimal as a ratio of two integers.
a) 0.017
b) 5.47

32. Write each terminating decimal as a ratio of two integers.
a) 3.1
b) 0.00239

33. Since $0.33\overline{3} = \frac{1}{3}$, what does $0.99\overline{9}$ equal?

34. Since $3.33\overline{3} = \frac{10}{3}$, what does $9.99\overline{9}$ equal?

35. Consider the set of numbers $\{\sqrt{64}, \pi, 3.14, 0.919919991 \ldots, -\frac{15}{7}, \sqrt{10}, 0.\overline{47}\}$
a) Which are rational numbers?   b) Which are irrational numbers?

36. Consider the set of numbers $\{\sqrt{36}, -\pi, -\frac{22}{7}, 0.8181181118 \ldots, 6.17, \sqrt{18}, 0.0\overline{5}\}$
a) Which are rational numbers?   b) Which are irrational numbers?

True or false (Problems 37–46).

37. Every rational number is a real number.

38. There are irrational numbers that are not real numbers.

39. Not every whole number is a natural number.

40. Every whole number is an integer.

41. Every integer is a positive number.

42. Some natural numbers are not integers.

43. Some irrational numbers are negative.

44. Not every rational number is positive.

45. No rational number is also an irrational number.

46. No irrational number is also a rational number.

Use the commutative laws to complete each statement.

47. $6 + 8 = $ _____?_____

48. $9 + 4 = $ _____?_____

49. $(-2) \cdot 7 = $ _____?_____

50. $5 \cdot (-3) = $ _____?_____

Use the associative laws to complete each statement.

51. $3 + (4 + 5) = $ _____?_____

52. $1 + (7 + 9) = $ _____?_____

53. $(x + 4) + (-4) = $ _____?_____

54. $(x + 3) + (-3) = $ _____?_____

55. $9 \cdot (6 \cdot \frac{1}{3}) = $ _____?_____

56. $5 \cdot (8 \cdot \frac{1}{2}) = $ _____?_____

57. $\frac{1}{2} \cdot (2x) = $ _____?_____

58. $\frac{1}{4} \cdot (4x) = $ _____?_____

Use the identity laws to complete each statement.

59. $13 + 0 = $ _____?_____

60. $19 + 0 = $ _____?_____

61. $1 \cdot 5 = $ _____?_____

62. $1 \cdot 8 = $ _____?_____

Use the inverse laws to complete each statement.

63. $(-6) + 6 = $ _____?_____

64. $(-5) + 5 = $ _____?_____

65. $3 \cdot \frac{1}{3} = $ _____?_____

66. $7 \cdot \frac{1}{7} = $ _____?_____

Use the distributive law to complete each statement.

67. $(-2)[5 + (-6)] = $ _____?_____

68. $(-3)[9 + (-2)] = $ _____?_____

69. $3x + 4x = $ _____?_____

70. $5x + 6x = $ _____?_____

Give a specific example to verify each statement.

71. Division is not commutative.

72. Subtraction is not commutative.

73. Subtraction is not associative.

74. Division is not associative.

Write a mathematical phrase that corresponds to each word phrase.

75. The larger of two consecutive integers, given that the smaller integer is n

76. The smaller of two consecutive integers, given that the larger integer is n

77. The smallest of three consecutive integers, given that the largest is m

78. The largest of three consecutive integers, given that the smallest is m

79. Forty-three percent of a number r

80. Fifty-seven percent of a number r

81. Eight and three-fourths percent of a number s

82. Six and one-fourth percent of a number s

83. The number of liters of pure acid in t liters of a solution that is 20% acid

84. The number of liters of pure alcohol in t liters of a solution that is 30% alcohol

85. If n is a whole number, what kind of number is
a) $2n$?　　　　　　　**b)** $2n + 1$?

86. If n is a whole number, what kind of number is
a) $2n + 2$?　　　　　　**b)** $2n + 3$?

Here is a problem you can experiment with in your class. In a random group of n persons, the probability that no two have the same birthday is given by the product below.

$$\frac{365}{365} \cdot \frac{364}{365} \cdot \frac{363}{365} \cdots \cdots \frac{366 - n}{365}$$

†87. What is the probability that no two persons in a random group of five persons have the same birthday? What is the probability that at least two persons in such a group *will* have the same birthday?

†88. What is the probability that at least two persons in a random group of 40 persons will have the same birthday?

CALCULATOR PROBLEMS

89. Use your calculator to express each rational number as a decimal.
a) $\frac{101}{111}$　**b)** $\frac{0}{6}$　**c)** $\frac{6}{0}$　**d)** $\frac{0}{0}$

90. Compute by first converting each common fraction to a decimal fraction.
a) $\frac{2}{3} + \frac{4}{11}$　**b)** $\frac{2}{11} + \frac{4}{27}$　**c)** $\frac{2}{7} + \frac{3}{7}$
Do you see that sometimes it is easier to work with common fractions than decimal fractions?

1.3　　　*Inequalities and Absolute Value*

We can put the set of real numbers in order using a **number line.** In the number-line diagram in Figure 1.3, the number 2 is called the **coordinate** of point P, and point P is called the **graph** of the number 2. The point corresponding to 0 is called the **origin** of the number line.

The graphs of $\frac{1}{2}$, $-2\frac{1}{3}$, and $\sqrt{7}$ appear in the number-line diagram in Figure 1.4.

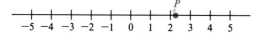

Figure 1.3

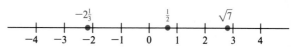

Figure 1.4

To each point on the number line there corresponds a unique real number, and to each real number there corresponds a unique point on the number line. We say that there is a **one-to-one correspondence** between the points on the number line and the set of real numbers.

INEQUALITIES

We now use the number line to define the phrases **is greater than** and **is less than.**

Definition	Suppose that a and b are real numbers. We say that a **is greater than** b, written $a > b$, if the graph of a lies to the *right* of the graph of b on the number line. Equivalently, we say that b **is less than** a, written $b < a$.

This definition is illustrated in Figure 1.5.

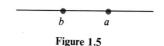

Figure 1.5

$$a > b \quad \text{or} \quad b < a$$

The statements $a > b$ and $b < a$ are called **inequalities,** and the symbols $>$ and $<$ are called **inequality signs.** One way to keep the symbols $>$ and $<$ straight is to remember that *the symbol always points to the lesser number.*

EXAMPLE 1   Which is greater, -5 or -2?

Since the graph of -2 lies to the right of the graph of -5, we say that $-2 > -5$ (see Figure 1.6).

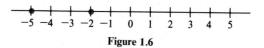

Figure 1.6

Try Problem 3

It may seem strange to say that -2 is greater than -5, but remember that a temperature of $-2°$ is *warmer*, or *higher*, than a temperature of $-5°$. ◄

EXAMPLE 2   Which is greater, $\frac{2}{3}$ or $\frac{3}{5}$?

These fractions are difficult to compare because they don't have the same

denominator. Therefore we write

$$\frac{2}{3} = \frac{2 \cdot 5}{3 \cdot 5} = \frac{10}{15},$$

$$\frac{3}{5} = \frac{3 \cdot 3}{5 \cdot 3} = \frac{9}{15}.$$

Try Problem 15 \   Since $\frac{10}{15} > \frac{9}{15}$, we can say that $\frac{2}{3} > \frac{3}{5}$ (see Figure 1.7).

Figure 1.7

EXAMPLE 3   Use an inequality to express the fact that a U.S. President's age *A* must be at least 35 years.

In this case, *A* can either be *greater* than 35 or *equal* to 35. Therefore we combine the symbols $>$ and $=$ and write

$$A \geq 35 \text{ years,}$$

Try Problem 31 \   which is read "*A* is greater than or equal to thirty-five years." ◄

The various symbols for inequalities are summarized in the table below.

Symbol	Meaning	Examples
$<$	is less than	$2 < 6, -4 < 6$
$>$	is greater than	$14 > -14, 5 > 0$
$\leq$	is less than or equal to	$-1 \leq 0, 4 \leq 7$
$\geq$	is greater than or equal to	$8 \geq 8, -6 \geq -8$

A slash will *negate* any of the symbols above. Thus $\not<$ means "is not less than" (the same as $\geq$), $\not>$ means "is not greater than" (the same as $\leq$), and so on.

EXAMPLE 4   A car with a fuel capacity of 13 gallons gets 27 mpg in the city and 38 mpg on the highway. If *d* denotes the distance the car can travel on a full tank, use inequalities to describe the values that *d* can assume.

If the driving is done entirely in the city, then

$$d = 27 \frac{\text{mi}}{\text{gal}} \cdot 13 \text{ gal} = 351 \text{ mi.}$$

If the driving is done entirely on the highway, then

$$d = 38 \frac{\text{mi}}{\text{gal}} \cdot 13 \text{ gal} = 494 \text{ mi.}$$

Therefore

$$d \geq 351 \text{ mi} \quad \text{and} \quad d \leq 494 \text{ mi.} \blacktriangleleft$$

We can write the two inequalities of Example 4 as

$$351 \text{ mi} \leq d \quad \text{and} \quad d \leq 494 \text{ mi,}$$

and then combine them to form the **double inequality**

$$351 \text{ mi} \leq d \leq 494 \text{ mi.}$$

Try Problem 35

The double inequality above means that "351 miles is less than or equal to d, and d is less than or equal to 494 miles."

The **graph** of an inequality is the graph of all those values of the variable that make the inequality a true statement.

EXAMPLE 5   Graph $\{x | x < 3 \text{ and } x \text{ is a whole number}\}$.

The graph consists of the three points on the number line that correspond to the numbers 0, 1, and 2, as shown in Figure 1.8.

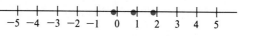

Try Problem 39

Figure 1.8

EXAMPLE 6   Graph $\{x | x < 3 \text{ and } x \text{ is an integer}\}$.

The graph of those integers that are less than 3 consists of the infinite set of points shown in Figure 1.9.

Try Problem 41

Figure 1.9

EXAMPLE 7   Graph $\{x | x < 3\}$.

When it is not stipulated otherwise, we assume that x is a real number. Therefore this graph consists of all points to the left of 3 on the number line, as shown in Figure 1.10.

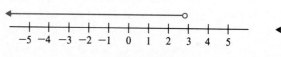

Figure 1.10

The graph of Example 7 is called an **open half line.** The arrow indicates that the graph continues indefinitely to the left. The *open* circle above the number 3 indicates that the point corresponding to 3 is *not* included in the graph. The graph is really *on* the number line but is drawn above it in order to make the diagram easier to read.

Try Problem 43

EXAMPLE 8   Graph $\{x|x \le 3\}$.

This graph consists of all those points to the left of 3, as well as the point corresponding to 3 itself (see Figure 1.11).

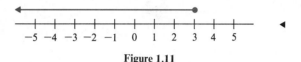

Figure 1.11

The graph of Example 8 is called a **closed half line.** The *closed* circle over the number 3 indicates that the point corresponding to 3 is included in the graph.

Try Problem 45

EXAMPLE 9   Graph $\{x|0 \le x \le 3\}$.

This set consists of all those real numbers between 0 and 3, inclusive (see Figure 1.12).

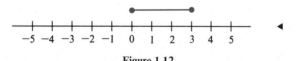

Figure 1.12

The graph of Example 9 is called a **closed interval.** The points corresponding to the numbers 0 and 3 are called the **endpoints** of the interval.

Try Problem 49

EXAMPLE 10   Graph $\{x|-3 < x < 4\}$.

In this case, the endpoints of the interval are not included in the graph (see Figure 1.13).

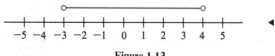

Figure 1.13

Try Problem 51

The graph of Example 10 is called an **open interval.**
Besides being open or closed, an interval may be **half open.** That is, the graph may include one endpoint but not the other.

ABSOLUTE VALUE

We shall now use the number line to define the concept of absolute value. The **absolute value** of a real number is the distance between the graph of that number and the origin. Therefore the absolute value of both $+5$ and -5 is 5, as shown in Figure 1.14. We say that $+5$ and -5 are "equal in absolute value but opposite in sign."

Figure 1.14

The absolute value of a real number is designated by enclosing that number between two vertical bars. That is,

$|x|$ is read "the absolute value of x."

EXAMPLE 11   Find each absolute value.

a) $|-7| = 7$
b) $|7| = 7$
c) $|0| = 0$
d) $|-1\frac{1}{3}| = 1\frac{1}{3}$

Try Problem 67   e) $|-\sqrt{3}| = \sqrt{3}$ ◄

The absolute value of any number x is either a positive number or zero. That is, $|x|$ is a *nonnegative* number.

PROBLEM SET 1.3

Graph each number on a number line.

1. $\{\frac{3}{4}, 4, -4, 5.5, -\frac{7}{3}, \sqrt{3}, -0, \pi\}$ 2. $\{-\frac{3}{4}, 5, -5, 4.5, \frac{10}{3}, \sqrt{2}, +0, -\pi\}$

Replace each comma with the symbol $<$ or the symbol $>$ to make a true statement.

3. $-5, -3$ 4. $-7, -4$ 5. $-101, -100$ 6. $-100, -99$

7. $0, -1$ 8. $0, -9$ 9. $-6, 3$ 10. $-7, 4$

11. $\pi, \frac{22}{7}$ 12. $\pi, 3.14$ 13. $-\sqrt{2}, -1.4$ 14. $-\sqrt{3}, -1.7$

15. $\frac{3}{4}, \frac{5}{7}$ 16. $\frac{4}{5}, \frac{7}{9}$ 17. $-\frac{3}{4}, -\frac{5}{7}$ 18. $-\frac{4}{5}, -\frac{7}{9}$

Use the symbols <, >, ≤, and ≥ to rewrite the following statements (Problems 19–34).

19. x is a positive number.

20. x is a negative number.

21. y is a nonnegative number.

22. y is a nonpositive number.

23. z is at most 7.

24. z is at least 4.

25. $r + 5$ is greater than 14.

26. $r - 3$ is less than 17.

27. s is between 8 and 9.

28. s is between -1 and 0.

29. t is at least -4 but less than 19.

30. t is greater than 6 but at most 23.

31. The voting age v is at least 18 years.

32. The drinking age d is at least 19 years.

33. A girl's weight w fluctuates from 105 pounds to 112 pounds.

34. A bullet's speed s is between 800 ft/sec and 4000 ft/sec.

35. A turnpike has a minimum speed of 45 mph and a maximum speed of 55 mph. A motorist travels for 3 hours on this turnpike, staying within the legal limits. Write a double inequality that describes the distance d the motorist has traveled.

36. Ted lives 5 miles from school and 3 miles from his friend Jerry. Write a double inequality that describes the distance d that Jerry lives from the school.

37. The markup on a toaster oven must be at least 40% of the cost. If the cost is $18.70 and a competitor sells the same appliance for $32, write a double inequality that describes the possible prices p if the competitor is to be undersold.

38. Repeat Problem 37 if
   a) the markup must be at least 60% of the cost.
   b) the markup must be at least 60% of the cost and the competitor's price is $29.

Graph each set on a number line. Then name the graph.

39. $\{x \mid x < 4 \text{ and } x \text{ is a whole number}\}$

40. $\{x \mid x < 2 \text{ and } x \text{ is a whole number}\}$

41. $\{x \mid x < 4 \text{ and } x \text{ is an integer}\}$

42. $\{x \mid x < 2 \text{ and } x \text{ is an integer}\}$

43. $\{x \mid x < 4\}$

44. $\{x \mid x < 2\}$

45. $\{x \mid x \leq 4\}$

46. $\{x \mid x \leq 2\}$

47. $\{y \mid y > 0\}$

48. $\{y \mid y < 0\}$

49. $\{x \mid 0 \leq x \leq 5\}$

50. $\{x \mid 0 \leq x \leq 1\}$

51. $\{x \mid -3 < x < 2\}$

52. $\{x \mid -5 < x < 3\}$

53. $\{r \mid -4 < r \leq -1\}$

54. $\{r \mid -6 \leq r < -2\}$

55. $\{t \mid 3 > t \geq -6\}$

56. $\{t \mid 4 \geq t > -3\}$

Describe each graph algebraically.

57.

```
    +--+--+--•--•--•--•--•--•--•--•   . . .
   -5 -4 -3 -2 -1  0  1  2  3  4  5
```

58.

```
   . . .
    •--•--•--•--•--•--•--•--+--+--+--+
   -5 -4 -3 -2 -1  0  1  2  3  4  5
```

59.

```
              o────────────────────►
    +--+--+--+--+--+--+--+--+--+--+
   -5 -4 -3 -2 -1  0  1  2  3  4  5
```

60.

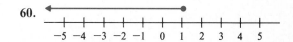

```
   ◄───────────────────•
    +--+--+--+--+--+--+--+--+--+--+
   -5 -4 -3 -2 -1  0  1  2  3  4  5
```

61.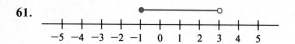

62.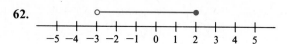

Find each absolute value.

63. $|13|$

64. $|14|$

65. $|-9|$

66. $|-8|$

67. $|-3\frac{1}{3}|$

68. $|-4\frac{2}{3}|$

69. $|-\sqrt{2}|$

70. $|-\sqrt{5}|$

71. $|+0|$

72. $|-0|$

73. $|-\pi|$

74. $|\pi|$

Compute the value of each expression.

75. $-|25|$

76. $-|35|$

77. $-|-99|$

78. $-|-101|$

79. $\dfrac{|-6|+|8|}{|-2|}$

80. $\dfrac{|-6|+|-9|}{|-3|}$

81. $\dfrac{14}{|5|-|-5|}$

82. $\dfrac{15}{|6|-|-6|}$

83. $\left|\dfrac{16-2(1+7)}{8}\right|$

84. $\left|\dfrac{18-6(2+1)}{9}\right|$

85. $\left|\dfrac{12-8}{4+4}\right|$

86. $\left|\dfrac{20-15}{10+5}\right|$

True or false (Problems 87–90).

87. If $|a|=|b|$, then $a=b$.

88. If $a=b$, then $|a|=|b|$.

89. If $a<b$, then $|a|<|b|$.

90. If $|a|>|b|$, then $a>b$.

†91. Imagine a hotel with an infinite number of rooms numbered 1, 2, 3, 4, ... If every room is occupied, explain how the manager might make room in his hotel for a weary traveler.

†92. Imagine two hotels, the Ritz and the Royal, each with an infinite number of rooms numbered 1, 2, 3, 4, ... If every room in both hotels is occupied when the Ritz is suddenly scheduled for demolition, explain how the manager of the Royal might make room for all the occupants of the Ritz.

CALCULATOR PROBLEMS

93. Replace each comma with the symbol $<$ or the symbol $>$ to make a true statement.

   a) $\frac{947}{2221}, \frac{773}{1811}$

   b) $\sqrt{23}+\sqrt{29}, \sqrt{21}+\sqrt{31}$

94. A couple plan on spending anywhere from \$68,847 to \$77,115 to build a new house on a lot they received as a wedding gift. If building costs in their area are \$39.75 per square foot, write a double inequality that describes the size of the house (in terms of the number of square feet n) that they can build.

1.4   *Addition and Subtraction of Signed Numbers*

ADDITION

Suppose that on a particular football play a team gains 4 yards. If it gains 3 yards on the next play, the total yardage gained on the two plays is 7 yards. A computation such as this can be visualized using arrows on a number line. Arrows pointing to the right indicate positive numbers, whereas arrows pointing to the left indicate negative numbers.

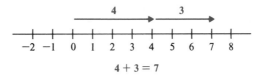

$$4 + 3 = 7$$

A gain of 4 yards followed by a *loss* of 3 yards gives a total yardage of 1 yard.

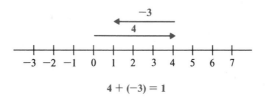

$$4 + (-3) = 1$$

If a gain of 4 yards is followed by a loss of 6 yards, the total yardage is -2 yards.

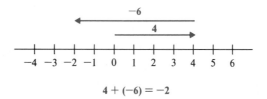

$$4 + (-6) = -2$$

Finally, if a loss of 3 yards is followed by a loss of 5 yards, the total yardage is -8 yards.

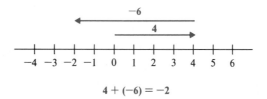

$$(-3) + (-5) = -8$$

We can summarize this discussion by stating the rule following (see page 26).

Rule for Adding Signed Numbers

To add two numbers having

like signs: add the absolute values of the numbers, and then prefix the common sign.

unlike signs: subtract the smaller absolute value from the larger absolute value, and prefix the sign of the number having the larger absolute value.

EXAMPLE 1   Find the sum $(-5) + (-10)$.

The numbers have like signs, so add their absolute values.

$$|-5| + |-10| = 5 + 10 = 15$$

Then prefix their common sign.

Try Problem 1

$$(-5) + (-10) = -15 \blacktriangleleft$$

EXAMPLE 2   Find the sum $3 + (-8)$.

The numbers have unlike signs, so subtract the smaller absolute value from the larger.

$$|-8| - |3| = 8 - 3 = 5$$

Then prefix a negative sign to the answer, since -8 has the larger absolute value.

Try Problem 3

$$3 + (-8) = -5 \blacktriangleleft$$

EXAMPLE 3   Find the sum $7 + (-2)$.

Subtract the smaller absolute value from the larger.

$$|7| - |-2| = 7 - 2 = 5$$

Then prefix a positive sign to the answer, since 7 has the larger absolute value.

Try Problem 5

$$7 + (-2) = +5 \blacktriangleleft$$

These examples illustrate the mechanics of adding signed numbers. *In actual practice, the operations involving absolute values should not be written.* Rather, these operations should be performed mentally. Thus for the last example we would simply write

$$7 + (-2) = 5.$$

When adding several numbers, some of which are positive and some of which are negative, we use the commutative and associative laws of addition to collect all the positive numbers in one group and all the negative numbers in another. This procedure generally produces the simplest calculation.

EXAMPLE 4 The Dow-Jones industrial average posted the following results during one particular week of trading: up 7.76 on Monday, up 3.12 on Tuesday, down 5.43 on Wednesday, up 1.89 on Thursday, down 9.06 on Friday. What was the net change in the Dow that week?

$$7.76 + 3.12 + (-5.43) + 1.89 + (-9.06)$$
$$= [7.76 + 3.12 + 1.89] + [(-5.43) + (-9.06)]$$
$$= 12.77 + (-14.49)$$
$$= -1.72$$

Try Problem 23   The Dow was down 1.72 points. ◄

SUBTRACTION

Before we can define subtraction of signed numbers, we need to discuss the notion of opposites. Consider the number line in Figure 1.15. For every point on one side of the origin, there is a unique point on the other side that is the same distance from the origin as the given point. The coordinates of the two points are called **opposites** of each other.

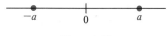

Figure 1.15

We denote the opposite of the real number a by writing $-a$. Therefore the opposite of 3 is denoted -3. Similarly, the opposite of -3 is denoted $-(-3)$. But since the opposite of -3 is 3, it must follow that $-(-3) = 3$. In general, we have the rule below.

Double-Negative Rule

For any real number a,

$$-(-a) = a.$$

EXAMPLE 5 Find the opposite of a, if $a = -8$.

The opposite of a is denoted by $-a$ and is given by

Try Problem 29 $$-a = -(-8) = 8. ◄$$

Note that the number 0 is its own opposite. That is, $-0 = 0$.
We are now ready to define subtraction. First, we note that

$$7 - 3 = 4 \quad \text{and} \quad 7 + (-3) = 4.$$

Generalizing on this example, we have the rule following (see page 28).

Rule for Subtracting Signed Numbers

Subtracting b from a is the same as adding the opposite of b to a. In symbols,
$$a - b = a + (-b).$$

EXAMPLE 6   Find the difference $3 - 7$.

No change
Change subtraction to addition
Opposite of second number

Try Problem 33

$$3 - 7 = 3 + (-7) = -4 \blacktriangleleft$$

EXAMPLE 7   Find the difference $(-11) - 5$.

Try Problem 35

$$(-11) - 5 = (-11) + (-5) = -16 \blacktriangleleft$$

EXAMPLE 8   Find the difference $(-8) - (-9)$.

Try Problem 37

$$(-8) - (-9) = (-8) + 9 = 1 \blacktriangleleft$$

EXAMPLE 9   Find the difference $(-\frac{5}{6}) - (-\frac{7}{10})$.

Try Problem 45

$$(-\tfrac{5}{6}) - (-\tfrac{7}{10}) = -\tfrac{5}{6} + \tfrac{7}{10} = -\tfrac{25}{30} + \tfrac{21}{30} = -\tfrac{4}{30} = -\tfrac{2}{15} \blacktriangleleft$$

EXAMPLE 10   Kristine has a balance of \$37.64 in her checking account when she writes a check for \$51.13. What will her balance be if the check clears?

$$37.64 - 51.13 = 37.64 + (-51.13) = -13.49$$

Try Problem 55

Her balance will be \$$-13.49$. $\blacktriangleleft$

Note that the symbol $-$ has three different, though related, meanings. *First* it denotes a negative number. *Second,* it denotes the opposite of a number. *Third,* it denotes the operation of subtraction.

EXAMPLE 11   Compute $-12 - (-3 + 7) + (11 - 25)$.

$$-12 - (-3 + 7) + (11 - 25)$$
$$= -12 - (4) + (-14) \qquad \text{Operate in parentheses first}$$
$$= -16 + (-14) \qquad \text{Subtract, since subtraction is on the left}$$

Try Problem 63

$$= -30 \qquad \text{Add} \blacktriangleleft$$

In Section 1.3, we defined the absolute value of a number as the distance between the number and 0. This was a *geometric* definition. We can now give an *algebraic* definition of absolute value.

Definition	Let x be a real number. Then

$$|x| = \begin{cases} x, & \text{if } x \geq 0, \\ -x, & \text{if } x < 0. \end{cases}$$

The fact that $|x|$ is equal to $-x$ in the second part of the definition above at first seems contradictory. Most students think that $-x$ is a negative number. But ask yourself: "Is $-x$ a positive or a negative number?" The answer is: "You can't tell until you know the value of x." If x is a negative number, then $-x$ is actually a *positive* number. For example, if $x = -3$, then $-x = -(-3) = 3$.

PROBLEM SET 1.4

Find each sum.

1. $(-5) + (-2)$
2. $(-4) + (-3)$
3. $5 + (-9)$
4. $4 + (-7)$

5. $18 + (-4)$
6. $16 + (-2)$
7. $(+6) + (+7)$
8. $(+8) + (+9)$

9. $(-10) + 0$
10. $(-15) + 0$
11. $(-9) + (-11)$
12. $(-4) + (-12)$

13. $(-4.1) + 3.6$
14. $(-5.3) + 2.8$
15. $-16.9 + 18.27$
16. $-19.1 + 12.36$

17. $(-\frac{11}{5}) + (-\frac{3}{5})$
18. $(-\frac{13}{7}) + (-\frac{2}{7})$
19. $\frac{2}{3} + (-\frac{1}{8})$
20. $\frac{3}{4} + (-\frac{1}{6})$

21. $-134 + (-988)$
22. $-896 + (-547)$

23. The Dow-Jones industrial average posted the following results during one week of trading: up 4.81 on Monday, down 1.78 on Tuesday, down 3.64 on Wednesday, up 0.92 on Thursday, down 2.56 on Friday. What was the net change in the Dow that week?

24. Sally bet on six horse races at a local track with the following results: won $14.40, lost $6, lost $4, won $5.80, won $9.60, lost $23.20. What was the net change in her assets as a result of her betting?

25. The following temperatures were recorded at 6 A.M. in Chicago during one seven-day period: Monday 3°, Tuesday 1°, Wednesday 0°, Thursday −4°, Friday −2°, Saturday 4°, Sunday 5°. Find the average temperature at 6 A.M. that week.

26. A fullback carried the football five times with the following results: 3-yard loss, 15-yard gain, 1-yard loss, 3-yard gain, 4-yard loss. Find his average gain per carry.

For each number a, state its opposite −a in simplest form.

27. $a = 14$
28. $a = 17$

29. $a = -5$
30. $a = -6$

Find each difference.

31. $8 - 4$

32. $5 - 3$

33. $4 - 8$

34. $3 - 5$

35. $(-9) - 3$

36. $(-8) - 5$

37. $(-16) - (-2)$

38. $(-15) - (-4)$

39. $\frac{3}{8} - \frac{5}{8}$

40. $\frac{2}{9} - \frac{8}{9}$

41. $0 - 6$

42. $0 - 8$

43. $(-10) - (-19)$

44. $(-9) - (-22)$

45. $(-\frac{3}{4}) - (-\frac{2}{7})$

46. $(-\frac{5}{6}) - (-\frac{3}{5})$

47. $0 - (-\frac{1}{3})$

48. $0 - (-\frac{1}{2})$

49. $5.4 - 7.51$

50. $4.9 - 9.21$

51. $-6.1 - 2.8$

52. $-8.7 - 5.8$

53. $-439 - 544$

54. $-712 - 629$

55. Renee has a balance of $37.64 in her checking account when she writes two checks, one for $23.08 and one for $19.85. What will her balance be if both checks clear?

56. Even though Sam's checking account is already overdrawn $13, he writes a check for $28.67. What will his balance be if the check clears?

57. The temperature at 8 A.M. is $-3°$. If it rises at the rate of $2°$ per hour, what will the temperature be at noon?

58. The temperature at 5 P.M. is $10°$. If it falls at the rate of $3°$ per hour, what will the temperature be at midnight?

Use the order of operations to compute each of the following.

59. $-4 + 11 + 6$

60. $-7 + 13 + 5$

61. $9 - (-18) - 5$

62. $2 - (-24) - 6$

63. $-14 - (-2 + 8) + (13 - 21)$

64. $-11 - (6 - 4) + (-18 + 7)$

65. $-\frac{5}{6} - (\frac{1}{4} - \frac{2}{9})$

66. $-\frac{2}{3} - (\frac{1}{2} - \frac{7}{9})$

67. $\left| \dfrac{6 - 3(2 + 4)}{27 - 10} \right|$

68. $\left| \dfrac{7 - 2(5 + 1)}{29 - 10} \right|$

69. $|12| - |-4| - (|7| - |-2|)$

70. $|-16| - |-5| - (|-8| - |3|)$

71. $70 - 36 - (14 - 48) - [81 - (-4 - 7 + 3)]$

72. $54 - 28 - (16 - 30) - [63 - (-5 - 8 + 4)]$

If $r = 5$, $s = -8$, and $t = -9$, determine the value of each expression.

73. $r + s$

74. $r + t$

75. $r - s + t$

76. $r + s - t$

77. $r + (s - t) - t$

78. $r - (s - t) + t$

79. Give an example to show that $|a + b|$ and $|a| + |b|$ are not equal for all values of a and b.

80. Give an example to show that $|a - b|$ and $|a| - |b|$ are not equal for all values of a and b.

CALCULATOR PROBLEMS

81. The progress of the Dow-Jones industrial average was charted over a three-week period with the following results: up 4.73, up 2.81, down 1.86, down 3.14, up 1.04, down 0.28, down 7.66, down 5.35, up 2.92, up 6.11, up 0.39, down 2.22, down 8.23, up 1.06, down 5.72. What was the net change in the Dow for the three-week period?

82. Calculate $(-1) + (-2) + (-3) + \cdots + (-98) + (-99) + (-100)$.

1.5   *Multiplication and Division of Signed Numbers*

MULTIPLICATION

We are already familiar with the problem of finding the product of two positive numbers. Now suppose we want to find the product of 4 and -3. One way to look at this problem is in terms of repeated addition. That is, since

$$4 \cdot 3 = 3 + 3 + 3 + 3 = 12,$$

then it should also be true that

$$4 \cdot (-3) = (-3) + (-3) + (-3) + (-3) = -12.$$

Also

$$(-4) \cdot 3 = (-4) + (-4) + (-4) = -12.$$

We conclude that the product of a positive number and a negative number is a negative number.

Now let's turn our attention to finding the product of two negative numbers. First,

$$(-3)[4 + (-4)] = (-3)[0] = 0.$$

But by the distributive law,

$$(-3)[4 + (-4)] = (-3)4 + (-3)(-4)$$
$$= -12 + ?$$

Since $-12 + ?$ must be 0 according to our first calculation, the product $(-3)(-4)$ must be 12. That is, the product of two negative numbers is a positive number. These results are summarized below.

Rule for Multiplying Signed Numbers

To find the product of two numbers, multiply their absolute values. If the numbers have *like* signs, the product is *positive;* if they have *unlike* signs, the product is *negative*.

EXAMPLE 1   Find each product: **a)** $8(-3)$,   **b)** $(-5)6$,   **c)** $(-7)(-9)$.

a) $8(-3) = -24$ Unlike signs, product is negative

b) $(-5)6 = -30$ Unlike signs, product is negative

Try Problem 5   **c)** $(-7)(-9) = 63$ Like signs, product is positive ◄

EXAMPLE 2   A retailer sells two items at a loss of $15 per item, three items at a loss of $12 per item, and eight items at a profit of $24 per item. Determine the net profit on the 13 sales.

We set up our work and follow the order of operations as follows:

$$2(-15) + 3(-12) + 8(24) = (-30) + (-36) + 192$$
$$= 126.$$

Try Problem 23   The net profit is $126. ◄

DIVISION

Using the rule we just developed for multiplying signed numbers, we formulate the rule for dividing signed numbers.

EXAMPLE 3   Find the quotient $(-6) \div (-2)$.

Try Problem 25 $(-6) \div (-2) = 3$,   since $(-2) \cdot 3 = -6$ ◄

EXAMPLE 4   Find the quotient $\frac{-6}{2}$.

Try Problem 29 $$\frac{-6}{2} = -3, \quad \text{since } 2 \cdot (-3) = -6 ◄$$

EXAMPLE 5   Find the quotient $\frac{6}{-2}$.

Try Problem 33 $$\frac{6}{-2} = -3, \quad \text{since } (-2) \cdot (-3) = 6 ◄$$

These results are summarized below.

Rule for Dividing Signed Numbers

To find the quotient of two numbers, divide their absolute values. If the numbers have *like* signs, the quotient is *positive;* if they have *unlike* signs, the quotient is *negative.*

Note that the sign rules for division are the same as the sign rules for multiplication.

EXAMPLE 6   The average January temperatures in Brownsville for the past ten years were $+3°, +1°, -2°, -4°, 0°, +1°, -6°, -5°, -1°$, and $+1°$. What was the average January temperature in Brownsville over the past ten years?

Add the temperatures and divide by 10.

$$\frac{(+3) + (+1) + (-2) + (-4) + 0 + (+1) + (-6) + (-5) + (-1) + (+1)}{10}$$

$$= \frac{-12}{10}$$

$$= -1.2$$

Try Problem 47   The average January temperature was $-1.2°$. ◄

Sometimes we want to *reposition* the sign that occurs in front of a fraction. For example, we can write

$$-\frac{6}{2} = \frac{-6}{2} = \frac{6}{-2},$$

because all three fractions have the same value, namely -3.

Signs of a Fraction

If $b \neq 0$, then

$$-\frac{a}{b} = \frac{-a}{b} = \frac{a}{-b}.$$

Mathematicians generally prefer the forms $-\dfrac{a}{b}$ and $\dfrac{-a}{b}$ to the form $\dfrac{a}{-b}$.

EXAMPLE 7   Simplify each fraction: **a)** $-\dfrac{-5}{8}$,   **b)** $-\dfrac{10}{2}$,   **c)** $\dfrac{-10}{-2}$.

a) $-\dfrac{-5}{8} = \dfrac{-(-5)}{8} = \dfrac{5}{8}$

b) $-\dfrac{10}{2} = -5$

Try Problem 49   **c)** $\dfrac{-10}{-2} = 5$ ◄

Just as every real number has a unique opposite, so does every *nonzero* real number have a unique reciprocal. If $a \neq 0$, the **reciprocal** of a is $\frac{1}{a}$.

EXAMPLE 8   Find the reciprocal of each number: **a)** 5,   **b)** $\frac{1}{5}$,   **c)** $\frac{3}{4}$.

a)   The reciprocal of 5 is $\frac{1}{5}$.

b)   The reciprocal of $\frac{1}{5}$ is

$$\frac{1}{\frac{1}{5}} = 1 \cdot \frac{5}{1} = 5.$$

c)   The reciprocal of $\frac{3}{4}$ is

$$\frac{1}{\frac{3}{4}} = 1 \cdot \frac{4}{3} = \frac{4}{3}. \blacktriangleleft$$

Try Problem 63

The next example illustrates all four arithmetic operations on signed numbers, together with the order of operations.

EXAMPLE 9   Compute $\left[\dfrac{15 + 6(-5)}{8 - 3}\right]\left[\dfrac{10 - 4}{-2 - 4}\right]$.

$$\left[\frac{15 + 6(-5)}{8 - 3}\right]\left[\frac{10 - 4}{-2 - 4}\right] = \left[\frac{15 - 30}{5}\right]\left[\frac{6}{-6}\right]$$

$$= \left[\frac{-15}{5}\right][-1]$$

$$= [-3][-1]$$

$$= 3 \blacktriangleleft$$

Try Problem 83

PROBLEM SET 1.5

Find each product.

1. $3(-7)$

2. $9(-3)$

3. $(-5)4$

4. $(-4)6$

5. $(-5)(-6)$

6. $(-2)(-8)$

7. $(-1)17$

8. $19(-1)$

9. $(-8) \cdot 0 \cdot 35$

10. $44 \cdot 0 \cdot (-9)$

11. $3 \cdot 2 \cdot (-11)$

12. $4 \cdot 5 \cdot (-10)$

13. $(-\frac{1}{2})\frac{1}{4}$

14. $\frac{1}{3}(-\frac{1}{5})$

15. $(-\frac{1}{3})(-\frac{1}{3})$

16. $(-\frac{1}{2})(-\frac{1}{2})$

17. $(-3)(-3)(-3)$

18. $(-4)(-4)(-4)$

19. $(-1)(-1)(-1)(-1)$

20. $(-2)(-2)(-2)(-2)$

21. $84.2(-7.35)$

22. $91.6(-8.15)$

23. A retailer sells two items at a loss of $17 per item, three items at a loss of $8 per item, and seven items at a profit of $32 per item. Determine the net profit on the 12 sales.

24. Allison plays a 50¢ slot machine 16 times. She wins a $5 jackpot twice and a $4 jackpot three times. Find her net profit.

Find each quotient.

25. $(-12) \div (-4)$

26. $(-15) \div (-5)$

27. $(-45) \div (-9)$

28. $(-35) \div (-7)$

29. $\frac{-16}{4}$

30. $\frac{-20}{5}$

31. $\frac{19}{-19}$

32. $\frac{-17}{17}$

33. $\frac{15}{-3}$

34. $\frac{18}{-3}$

35. $\frac{0}{-10}$

36. $\frac{0}{-12}$

37. $\frac{-230}{-10}$

38. $\frac{-240}{-20}$

39. $\frac{-16.65}{3.7}$

40. $\frac{46.72}{-6.4}$

41. $\left(-\frac{2}{3}\right) \div \frac{3}{8}$

42. $\left(-\frac{3}{4}\right) \div \frac{2}{5}$

43. $\left(-\frac{5}{12}\right) \div \left(-\frac{15}{8}\right)$

44. $\left(-\frac{5}{8}\right) \div \left(-\frac{25}{24}\right)$

45. $\frac{-31}{0}$

46. $\frac{-27}{0}$

47. A quarterback carried the football four times with the following results: 3-yard gain, 12-yard loss, 6-yard loss, 7-yard gain. Find his average gain per carry.

48. The following temperatures were recorded in Boston at 4 A.M. during one week: Monday 2°, Tuesday $-4°$, Wednesday $-6°$, Thursday $-7°$, Friday $-10°$, Saturday 0°, Sunday 4°. Find the average temperature at 4 A.M. that week.

Simplify each fraction.

49. $-\frac{-7}{9}$

50. $-\frac{-3}{5}$

51. $-\frac{14}{7}$

52. $-\frac{16}{8}$

53. $\frac{-14}{-7}$

54. $\frac{-16}{-8}$

55. $-\frac{1}{-6}$

56. $-\frac{1}{-4}$

57. $-\frac{-x}{-13}$

58. $-\frac{-x}{-19}$

For each number a, state its reciprocal $\frac{1}{a}$ in simplest form.

59. $a = 7$

60. $a = 8$

61. $a = \frac{1}{7}$

62. $a = \frac{1}{8}$

63. $a = \frac{3}{8}$

64. $a = \frac{5}{6}$

65. $a = -2\frac{4}{5}$

66. $a = -5\frac{1}{5}$

67. $a = 1$

68. $a = -1$

69. $a = -\frac{p}{q}$

70. $a = \frac{p}{q}$

Compute each of the following.

71. $6 \cdot 7 + 4 \cdot (-5)$

72. $(-8) \cdot 9 + 3 \cdot 6$

73. $\frac{-16}{8} + \frac{-12}{9-13}$

74. $\frac{14}{-7} + \frac{-18}{6-15}$

75. $2(3[10 - 2(3 - 1)])$

76. $3(2[12 - 4(8 - 6)])$

77. $6[-19 + 5(6 - 14)]$

78. $8[-13 + 2(5 - 13)]$

79. $32 \div [-6 - 11(3 - 5)]$

80. $21 \div [-3 - 5(7 - 9)]$

81. $\frac{5(-2) + 3(-1) - 9}{-6 - 7 - 2(-1)}$

82. $\frac{4(-1) + 3(-2) - 23}{-1 - 1 - 13(-1)}$

83. $\left[\frac{9 - 5(-3)}{8 - 4}\right]\left[\frac{5 + (-10)}{-2 - 3}\right]$

84. $\left[\frac{(-7) + 3}{-1 + 5}\right]\left[\frac{28 + 4(-10)}{2 - 8}\right]$

85. $\dfrac{12 - 3\left(\dfrac{7+5}{4-8}\right) + 2}{-6 + 4\left(\dfrac{1-4}{6-3}\right) - 9}$

86. $\dfrac{14 - 2\left(\dfrac{8+6}{1-8}\right) + 3}{-8 + 5\left(\dfrac{2-4}{7-5}\right) - 4}$

87. $\frac{-1}{6} \cdot \left(-\frac{9}{5}\right) + \left(\frac{1}{-5}\right) \div \left(-\frac{-2}{7}\right)$

88. $\frac{-1}{8} \cdot \left(-\frac{4}{3}\right) + \left(\frac{-1}{2}\right) \div \left(-\frac{3}{-5}\right)$

If $u = -4$, $v = 8$, and $w = -5$, determine the value of each expression.

89. $\dfrac{2u + 5v}{w}$

90. $\dfrac{4u + 3v}{w}$

91. $\dfrac{9u + 2v - w}{5}$

92. $\dfrac{3u - v + 2w}{6}$

93. $\dfrac{uv - uw + vw}{v}$

94. $\dfrac{uv + uw - vw}{-v}$

95. What can be said about m and n if

    a) $mn > 0?$ **b)** $\dfrac{m}{n} < 0?$

96. What can be said about m and n if

    a) $mn < 0?$ **b)** $\dfrac{m}{n} > 0?$

CALCULATOR PROBLEMS

97. Calculate to the nearest tenth.

    a) $3.61[7.84 - 9.15(6.07 - 4.57)]$

    b) $\dfrac{9.62(-7.45) - 13.8(-19.8)}{-8.73 + 0.199}$

98. A retailer sells 18 items at a loss of $4.37 per item, 29 items at a loss of $2.26 per item, 46 items at a profit of $15.72 per item, and 32 items at a profit of $24.85 per item. What is the retailer's net profit on the 125 items?

1.6   *Solving Simple Equations*

An **equation** is a statement that two quantities are equal. For example,

$$x + 4 = 7$$

is an equation because it states that the quantities $x + 4$ and 7 are equal. We label the parts of this equation as follows:

$$x + 4 = 7.$$

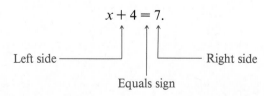

Left side ⎯⎯⎯⎯⎯⎯⎯⎯⎯⎯⎯⎯ Right side

Equals sign

There are three basic types of equations. They are identities, contradictions, and conditional equations.

Identities are equations that are true for all permissible values of the variable. For example, the equation

$$x + 1 = x + 1$$

is an identity because it is true for all values of x.

Contradictions are equations that are false for all values of the variable. The equation

$$x + 1 = x$$

is a contradiction, since it is false for all values of x.

A **conditional equation** is an equation that is true for some values of the variable and false for others. The equation

$$x + 4 = 7$$

is a conditional equation because it is true when x is 3 and false when x has any other value. We call 3 a **solution** of the equation $x + 4 = 7$. We also say that 3 **satisfies** the equation $x + 4 = 7$.

To **solve** an equation means to find its solutions. We shall now develop a systematic approach to solving simple equations. To begin, consider the sequence of three equations below.

$$2x - 3 = 5$$
$$2x = 8$$
$$x = 4$$

Since all three equations have the same solution, namely 4, they are called **equivalent equations**. Note that the solution is most obvious in the last equation, $x = 4$, since the variable x is isolated on one side. Therefore given an equation to solve, we write a sequence of equivalent equations until we have isolated the variable on one side of the equation. To do this, we use the properties of equality stated below.

Addition Property of Equality

The two equations below are equivalent.

$$a = b$$
$$a + c = b + c$$

Multiplication Property of Equality

If $c \neq 0$, the two equations below are equivalent.

$$a = b$$
$$ac = bc$$

The addition property states that adding the same number to both sides of an equation produces an equivalent equation. The multiplication property states that multiplying both sides of an equation by the same nonzero number produces an equivalent equation.

EXAMPLE 1   Solve $x + 3 = 8$.

To isolate x, use the addition property to add -3 to both sides. We choose -3 because it is the opposite of 3, and $3 + (-3) = 0$.

$$x + 3 = 8$$
$$x + 3 + (-3) = 8 + (-3)$$
$$x + 0 = 5$$
$$x = 5$$

The solution is 5. ◄

We can check the solution to Example 1 by substituting 5 for x in the original equation.

CHECK: $5 + 3 \overset{?}{=} 8$

Try Problem 23 $8 \overset{\checkmark}{=} 8$

EXAMPLE 2   Solve $x - 7 = 2$.

Add the opposite of -7, namely 7, to both sides.

$$x - 7 = 2$$
$$x - 7 + 7 = 2 + 7$$
$$x = 9$$

Try Problem 25   The solution is 9. Check in the original equation. ◄

EXAMPLE 3   Solve $4x = 24$.

To isolate x, use the multiplication property to multiply both sides by $\frac{1}{4}$. We choose $\frac{1}{4}$ because it is the reciprocal of 4, and $\frac{1}{4} \cdot 4 = 1$.

$$4x = 24$$
$$\tfrac{1}{4} \cdot 4x = \tfrac{1}{4} \cdot 24$$
$$1 \cdot x = 6$$
$$x = 6$$

Try Problem 27   The solution is 6. Check in the original equation. ◄

EXAMPLE 4 Solve $\frac{2}{3}x = 5$.

Multiply both sides by the reciprocal of $\frac{2}{3}$, namely $\frac{3}{2}$.

$$\frac{2}{3}x = 5$$
$$\frac{3}{2} \cdot \frac{2}{3}x = \frac{3}{2} \cdot 5$$
$$x = \frac{15}{2}$$

Try Problem 29

The solution is $\frac{15}{2}$. Check in the original equation. ◄

Note that subtracting c is the same as adding $-c$. Also, dividing by c is the same as multiplying by $\frac{1}{c}$. Hence we have the two properties below.

Subtraction Property of Equality

The two equations below are equivalent.

$$a = b$$
$$a - c = b - c$$

Division Property of Equality

If $c \neq 0$, the two equations below are equivalent.

$$a = b$$
$$\frac{a}{c} = \frac{b}{c}$$

EXAMPLE 5 Solve $y + 8 = 13$.

To isolate y, use the subtraction property to subtract 8 from both sides.

$$y + 8 = 13$$
$$y + 8 - 8 = 13 - 8 \qquad \text{Same as adding} -8$$
$$y = 5$$

Try Problem 31

The solution is 5. Check in the original equation. ◄

EXAMPLE 6 Solve $3t = 27$.

To isolate t, use the division property to divide each side by 3.

$$3t = 27$$
$$\frac{3t}{3} = \frac{27}{3} \qquad \text{Same as multiplying by } \frac{1}{3}$$
$$t = 9$$

Try Problem 33

The solution is 9. Check in the original equation. ◄

In the following examples, note that we use the addition or the subtraction property before we use the multiplication or the division property.

EXAMPLE 7   Solve $2s + 9 = -11$.

$$2s + 9 - 9 = -11 - 9 \qquad \text{Subtract 9}$$
$$2s = -20$$
$$\frac{2s}{2} = \frac{-20}{2} \qquad \text{Divide by 2}$$
$$s = -10$$

Try Problem 51   The solution is -10. Check in the original equation. ◄

EXAMPLE 8   Solve $-7r - 4 = -4$.

$$-7r - 4 + 4 = -4 + 4 \qquad \text{Add 4}$$
$$-7r = 0$$
$$\frac{-7r}{-7} = \frac{0}{-7} \qquad \text{Divide by } -7$$
$$r = 0$$

Try Problem 59   The solution is 0. Check in the original equation. ◄

EXAMPLE 9   Solve $-p - 6 = 19$.

$$-p - 6 + 6 = 19 + 6 \qquad \text{Add 6}$$
$$-p = 25$$

Then divide by -1, since we want to isolate p, not $-p$.

$$\frac{-p}{-1} = \frac{25}{-1}$$
$$p = -25$$

Try Problem 61   The solution is -25. Check in the original equation. ◄

Problem Solving   **Car Rental**

A salesperson rents a car for one day. If the cost is $30 plus 20¢ per mile, how many miles m can the salesperson travel on $75?

First translate the word equation

$$\$30 \text{ plus } (20¢/\text{mi times no. of miles } m) = \$75$$

into the mathematical equation

$$30 + 0.20 \cdot m = 75.$$

Note that we convert 20¢ to $0.20 so that the units are consistent throughout the equation. Then solve this equation to find the value of m.

$$30 + 0.2m = 75 \qquad \text{Since } 0.20 = 0.2$$
$$30 - 30 + 0.2m = 75 - 30 \qquad \text{Subtract 30}$$
$$0.2m = 45$$
$$\frac{0.2m}{0.2} = \frac{45}{0.2} \qquad \text{Divide by 0.2}$$
$$m = 225$$

The salesperson can travel 225 miles. ◄

PROBLEM SET 1.6

Classify each equation as an identity, a contradiction, or a conditional equation.

1. $a \cdot 1 = a$

2. $a + 0 = a$

3. $a + (-a) = 0$

4. $a \cdot \frac{1}{a} = 1$

5. $x + 2 = x + 2$

6. $x - 1 = x - 1$

7. $x + 2 = x$

8. $x - 1 = x$

9. $x + 2 = 5$

10. $x - 1 = 3$

11. $4(m + 2) = 4m + 8$

12. $5(m + 2) = 5m + 10$

13. $4m = 8$

14. $5m = 10$

15. $ab = ba$

16. $a(bc) = (ab)c$

17. $a + (b + c) = (a + b) + c$

18. $a(b + c) = ab + ac$

19. $0 = 3$

20. $1 = 2$

21. $3 = 3$

22. $2 = 2$

Solve each equation and check your solution.

23. $x + 3 = 7$

24. $x + 4 = 9$

25. $x - 6 = 2$

26. $x - 5 = 3$

27. $5x = 10$

28. $2x = 14$

29. $\frac{3}{4}x = 2$

30. $\frac{3}{4}x = 5$

31. $y + 9 = 14$

32. $y + 6 = 10$

33. $2t = 12$

34. $4t = 12$

35. $r - 1 = -4$

36. $r - 3 = -5$

37. $4s = -8$

38. $6s = -18$

39. $-7w = 28$

40. $-8w = 40$

41. $3z = 0$

42. $5z = 0$

43. $\frac{1}{2}x = -6$

44. $\frac{1}{3}x = -6$

45. $-y = 7$

46. $-y = 9$

47. $-\frac{5}{6}v = -40$

48. $-\frac{3}{8}v = -27$

Solve each equation and check your solution.

49. $2t - 5 = 3$

50. $3t - 4 = 11$

51. $3s + 8 = -10$

52. $2s + 7 = -9$

53. $4x + 1 = 4$

54. $5x + 1 = 5$

55. $5m - 10 = -15$

56. $4m - 6 = -10$

57. $2n + 13 = 13$

58. $3n + 19 = 19$

59. $-6r - 7 = -7$

60. $-8r - 3 = -3$

61. $-p - 9 = 6$

62. $-p - 8 = 12$

63. $-v + 4 = -8$

64. $-v + 8 = -6$

65. $11 - 9q = 2$

66. $13 - 8q = 5$

67. $-12 - 4y = 10$

68. $-17 - 9y = 16$

Write an equation that describes each problem. Then solve the equation to answer the given question.

69. A certain number x minus 17 equals 29. What is x?

70. A certain number x plus 23 equals 38. What is x?

71. Two fifths of a number y equals 7. What is y?

72. Three fifths of a number y equals 8. What is y?

73. Six times a number t plus 13 equals 61. What is t?

74. Four times a number t plus 19 equals 51. What is t?

75. If 12% of a number r is 11.4, what is r?

76. If 16% of a number r is 13.6, what is r?

77. The cost of renting a car is $110 per week plus 30¢ per mile. How many miles m can be traveled in one week on $305?

78. A salesperson earns $175 per week plus a 15% commission on all sales. What must the total sales s in a particular week be for the salesperson to earn $415 that week?

CALCULATOR PROBLEMS

79. Solve each equation to the nearest thousandth.
a) $17.401x - 23.449 = 39.418$
b) $29.667 - 47.045x = 91.323$

80. Given that the perimeter (distance around) of each figure is 117.9748 centimeters, find s to the nearest ten thousandth of a centimeter.

a)
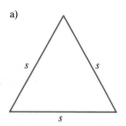

b)
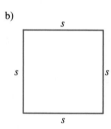

c)
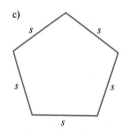

d)
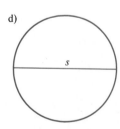

CHAPTER 1 REVIEW

[1.1] *Write a mathematical phrase that corresponds to each word phrase.*

1. Nine less than four times a number x

2. The value of x nickels and y dimes in cents

Compute.

3. $(4 + 20) \div 2 + 2$

4. $3[25 - 5(4 - 2)]$

5. Find the cost of $4\frac{3}{4}$ ounces of silver at \$18 per ounce.

6. Pieces of lengths of $7\frac{1}{2}$ feet and $6\frac{2}{3}$ feet are cut from a 25-foot hose. What is the length of the hose that remains?

Use the distributive law to find each product.

7. $4(3x + 2)$

8. $6\left(\dfrac{y}{2} + \dfrac{4}{3}\right)$

Use the distributive law to complete each statement.

9. $3b + 3c = 3(\underline{\quad ? \quad})$

10. $5x + 15 = 5(\underline{\quad ? \quad})$

[1.2] *Express each rational number as a decimal and as a percent.*

11. $\frac{1}{8}$

12. $\frac{5}{11}$

13. $\frac{0}{4}$

14. $\frac{7}{0}$

List the numbers in the set $\{\sqrt{25}, 2\frac{1}{3}, 0, -4, \sqrt{5}, 0.\overline{37}, 0.8282282228 \ldots, 1.2, -\frac{10}{3}, \pi\}$ that belong to each set below.

15. Natural numbers

16. Integers

17. Rational numbers

18. Irrational numbers

Write each terminating decimal as a ratio of two integers.

19. 5.9

20. 0.007

Use the given law to complete each statement.

21. Commutative law: $1 + 4 = \underline{\quad ? \quad}$

22. Identity law: $6 \cdot 1 = \underline{\quad ? \quad}$

23. Inverse law: $2 + (-2) = \underline{\quad ? \quad}$

24. Associative law: $\frac{1}{3}(3x) = \underline{\quad ? \quad}$

25. Commutative law: $xy = \underline{\quad ? \quad}$

26. Inverse law: $\frac{1}{5} \cdot 5 = \underline{\quad ? \quad}$

27. Identity law: $0 + 9 = \underline{\quad ? \quad}$

28. Associative law: $-7 + (7 + x) = \underline{\quad ? \quad}$

Write a mathematical phrase that corresponds to each word phrase.

29. The largest of four consecutive integers, given that the smallest is n

30. Eight percent of a number r

[1.3]

31. Graph each number on a number line: $\{3, -3, \frac{5}{4}, -\frac{11}{3}, \sqrt{5}\}$.

Replace each comma with the symbol $<$ or the symbol $>$ to make a true statement.

32. $-89, -99$

33. $\frac{9}{16}, \frac{7}{12}$

Use the symbols $<$, $>$, $\leq$, and $\geq$ to rewrite each statement.

34. A driver's speed s must not exceed 55 mph.

35. t is at least 1 but less than 6.

Graph each set on a number line. Then name the graph.

36. $\{x|x > 3\}$

37. $\{y|-2 \leq y < 0\}$

Compute the value of each expression.

38. $\dfrac{|-6| + |10|}{|-2|}$

39. $\dfrac{13}{|8| - |-8|}$

[1.4] *Find each sum.*

40. $13 + (-7)$

41. $(-5.2) + (-7.6)$

42. $(-\frac{8}{9}) + \frac{2}{3}$

State the opposite of each number in simplest form.

43. 4

44. $-[-(-13)]$

Find each difference.

45. $(-6) - 2$

46. $8.71 - 11.5$

47. $(-\frac{3}{8}) - (-\frac{5}{8})$

Compute.

48. $-10 + (-4 + 9) + (13 - 21)$

49. $\left|\dfrac{-18 + 6}{3}\right|$

50. If $r = 3$, $s = 2$, and $t = -6$, determine the value of $r - (s - t) - t$.

[1.5] *Find each product.*

51. $(-2)(-1)(-3)$

52. $\frac{1}{4}(-\frac{2}{5})$

53. $(-9.7)(-1.25)$

Find each quotient.

54. $\dfrac{-24}{6}$

55. $\dfrac{22.1}{-3.4}$

56. $(-\frac{2}{15}) \div (-\frac{1}{4})$

57. Find the average of the numbers 39, -17, -87, 0, 22, and -71.

State the reciprocal of each number in simplest form.

58. 5

59. $-3\frac{2}{7}$

Compute.

60. $3(4[12 - 2(8 - 1)])$

61. $\dfrac{6(-1) + 3(-5) - 1}{-2 - 2 - (15)(-1)}$

62. $\left[\dfrac{8 - 3(-4)}{-1 - 4}\right]\left[\dfrac{(-9) + 5}{7 - 3}\right]$

63. $\left(\dfrac{-2}{9}\right)\left(-\dfrac{3}{4}\right) + \left(\dfrac{1}{-6}\right) \div \left(-\dfrac{5}{6}\right)$

64. If $u = -3$, $v = 5$, and $w = 4$, determine the value of $\dfrac{8u + v - 2w}{3}$.

[1.6] *Classify each equation as an identity, a contradiction, or a conditional equation.*

65. $x + 3 = 7$ **66.** $x + 3 = x - 3$ **67.** $2(x + 3) = 2x + 6$

Solve each equation and check your solution.

68. $\frac{2}{9}x = 18$ **69.** $3y - 5 = 19$ **70.** $2z + 9 = 9$

71. $4m - 3 = -1$ **72.** $-7 - 13r = 6$ **73.** $-p + 8 = -1$

Write an equation that describes each problem. Then solve the equation to answer the given question.

74. The opposite of a number x plus 13 equals negative 2. What is x?

75. Three fourths of a number y equals 12. What is y?

CHAPTER 1 TEST

1. Write a mathematical phrase that corresponds to each word phrase.
   a) Five less than twice a number x
   b) Six percent of a number y

2. Write each decimal as a ratio of two integers.
   a) 1.7
   b) 0.03

3. Use the distributive law to find each product.
   a) $5(2x + 3)$
   b) $8\left(\dfrac{y}{2} + \dfrac{5}{4}\right)$

4. State in simplest form.
   a) The opposite of $-(-8)$
   b) The reciprocal of $2\frac{2}{3}$

5. Express each number as a decimal and as a percent.
   a) $\frac{7}{8}$
   b) $\frac{11}{6}$

6. Use the distributive law to complete each statement.
   a) $9x + 9y = 9 \underline{\quad ? \quad}$
   b) $7x + 28 = \underline{\quad ? \quad}$

7. Use the symbols $<$, $>$, $\leq$, and $\geq$ to rewrite each statement.
   a) The area A is greater than 10.
   b) The cost C is at least 8 but no more than 14.

8. List the numbers in the set $\{-7, 4, \frac{3}{4}, \sqrt{8}, 0.\overline{5}\}$ that belong to each set below.
   a) Natural numbers
   b) Integers
   c) Rational numbers
   d) Irrational numbers

9. Use the given law to complete each statement.
   a) Identity law: $1 \cdot x = $ ___?___
   c) Associative law: $\frac{2}{3}(\frac{3}{2}y) = $ ___?___

   b) Commutative law: $0 + 6 = $ ___?___
   d) Associative law: $-4 + (4 + p) = $ ___?___

Graph each set on a number line.

10. $\{x | x < 4\}$

11. $\{y | -3 < y \le 1\}$

Compute.

12. $-8 + (-2 + 9) + (14 - 19)$

13. $\left| \dfrac{-16 + 4}{2} \right|$

14. $2(5[13 - 3(7 - 3)])$

15. $\dfrac{4(-3) + 5(-1) - 1}{-7 + 2 - (-7)(2)}$

16. $\left[\dfrac{6 - 3(-1)}{-2 + 3} \right] \left[\dfrac{-4 + 8}{-1 - 3} \right]$

17. $\left(\frac{-3}{10}\right) \cdot \left(\frac{5}{-6}\right) + \left(-\frac{1}{-9}\right) \div \left(-\frac{2}{3}\right)$

Solve each equation.

18. $2x + 5 = 13$

19. $6y - 1 = 1$

20. $5z - 8 = -8$

21. $3 - 4p = 11$

22. If $u = -4, v = 2$, and $w = 5$, determine the value of $\dfrac{5u + v - 3w}{3}$.

23. Find the average of the numbers $18, -23, 0, 7$, and -32.

24. A 30-gallon tank is full of kerosene. John siphons off $7\frac{3}{4}$ gallons, and Jim siphons off $10\frac{1}{3}$ gallons. How much kerosene is left in the tank?

25. If three eighths of a number x is 24, find x.

In this chapter we define exponential notation and state the rules that it obeys. You will see how this notation can be applied to topics as diverse as bacterial growth and compound interest.

Using exponential notation we then define a particular type of algebraic expression known as a polynomial. You will see polynomials used in problems from the vertical projection of objects to the construction of a round-robin schedule. You will learn to add, subtract, and multiply polynomials, and you will also learn to perform an operation known as factoring polynomials.

A thorough knowledge of all the topics in this chapter is crucial to your success in solving the equations that appear in Chapter 3, as well as many of the later chapters.

2

Exponents and Polynomials

2.1   *Exponents*

The area A of a square whose side is s is given by the formula

$$A = s \cdot s.$$

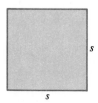

Similarly, the volume V of a cube whose edge is e is given by the formula

$$V = e \cdot e \cdot e.$$

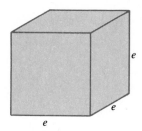

Repeated products such as $s \cdot s$ and $e \cdot e \cdot e$ occur so often in mathematics that we abbreviate them. For example, since $s \cdot s$ has two factors, we write

$$s \cdot s = s^2,$$

where s^2 is read "s to the second power," or "s squared." Similarly since $e \cdot e \cdot e$ has three factors, we write

$$e \cdot e \cdot e = e^3,$$

where e^3 is read "e to the third power," or "e cubed."

Definition   If a is any real number and n is any natural number, then

$$a^n = \underbrace{a \cdot a \cdot a \cdots a.}_{n \text{ factors}}$$

The expression a^n is read "a to the nth power." We say that a^n is written in **exponential form,** whereas $a \cdot a \cdot a \cdots a$ is written in **expanded form.***
Using the definition above, we see that

$$a^1 = a,$$
$$a^2 = a \cdot a,$$
$$a^3 = a \cdot a \cdot a,$$
$$a^4 = a \cdot a \cdot a \cdot a,$$

and so on.

In the expression a^n, the number a is called the **base,** and n is called the **exponent.**

EXAMPLE 1 State the base and the exponent of each exponential expression: **a)** 7^5, **b)** $(-8)^2$, **c)** -8^2, **d)** $(5x)^3$, **e)** $5x^3$.

The solutions are as follows:

a) base $= 7$, exponent $= 5$
b) base $= -8$, exponent $= 2$
c) base $= 8$, exponent $= 2$
d) base $= 5x$, exponent $= 3$

Try Problem 5

e) base $= x$, exponent $= 3$ ◄

CAUTION The base of a power consists only of the symbol that lies immediately to the left of the exponent, unless parentheses indicate otherwise. That is, -8^2 means $-(8^2)$ and $5x^3$ means $5(x^3)$.

EXAMPLE 2 Evaluate: **a)** $(-3)^2$, **b)** -3^2.

a) $(-3)^2 = (-3) \cdot (-3) = 9$

Try Problem 23

b) $-3^2 = -(3 \cdot 3) = -9$ ◄

CAUTION The expressions $(-a)^2$ and $-a^2$ are *not* the same.

EXAMPLE 3 Evaluate: **a)** $(2 \cdot 5)^4$, **b)** $2 \cdot 5^4$.

a) $(2 \cdot 5)^4 = 10^4 = 10 \cdot 10 \cdot 10 \cdot 10 = 10{,}000$

Try Problem 29

b) $2 \cdot 5^4 = 2 \cdot 5 \cdot 5 \cdot 5 \cdot 5 = 1250$ ◄

CAUTION The expressions $(ab)^4$ and ab^4 are *not* the same.

---

* This notation was introduced by the French mathematician and philosopher René Descartes (1596–1650).

Try Problem 33

EXAMPLE 4   Evaluate: **a)** $(4 + 5)^2$,   **b)** $4^2 + 5^2$.

a) $(4 + 5)^2 = 9^2 = 81$

b) $4^2 + 5^2 = 16 + 25 = 41$ ◂

CAUTION An exponent does *not* distribute over a sum. That is, $(a + b)^2$ and $a^2 + b^2$ are *not* the same.

Try Problem 39

EXAMPLE 5   Write $(x + 2)(x + 2)(x + 2)$ in exponential form.

$$(x + 2)(x + 2)(x + 2) = (x + 2)^3 \blacktriangleleft$$

CAUTION You must write parentheses in your answer to Example 5. The expression $x + 2^3$ means $x + 8$.

Try Problem 41

EXAMPLE 6   Write $a \cdot a \cdot b \cdot b \cdot b - 13 \cdot a \cdot b \cdot b \cdot b \cdot b \cdot b$ in exponential form.

$$a \cdot a \cdot b \cdot b \cdot b - 13 \cdot a \cdot b \cdot b \cdot b \cdot b \cdot b = a^2 b^3 - 13ab^5 \blacktriangleleft$$

Problem Solving

Compound Interest

When an interest rate is stated, we shall assume that it is an annual interest rate. Further, unless it is stipulated otherwise, we shall assume that the interest is compounded annually. Under these assumptions, the value V of a savings account after t years is given by the formula

$$V = P(1 + r)^t,$$

where P is the principal, and r is the interest rate expressed as a decimal.

   Suppose that $1000 is deposited in a savings account paying 9% interest. What is the value of the account after two years?

Substitute $P = 1000$, $r = 0.09$, and $t = 2$ into the formula above.

$$V = 1000(1 + 0.09)^2$$
$$= 1000(1.09)^2$$
$$= 1000(1.1881)$$
$$= 1188.1$$

The value of the account is $1188.10.

PROBLEM SET 2.1

State the base and the exponent of each exponential expression.

1. 8^5

2. 9^6

3. $(-4)^2$

4. $(-5)^4$

5. -4^2

6. -5^4

7. $(2x)^3$

8. $(3y)^2$

9. $2x^3$

10. $3y^2$

11. $(-t)^4$

12. $(-t)^2$

13. $-t^4$

14. $-t^2$

15. $(r+2)^7$

16. $(r+3)^8$

17. $(ab)^{n+1}$

18. $(ab)^{n-1}$

19. $\left(\dfrac{a}{b}\right)^{2n}$

20. $\left(\dfrac{a}{b}\right)^{3n}$

Evaluate.

21. 4^2 and 2^4

22. 5^2 and 2^5

23. $(-5)^2$ and -5^2

24. $(-7)^2$ and -7^2

25. $(-10)^3$ and -10^3

26. $(-9)^3$ and -9^3

27. $(-3)^4$ and -3^4

28. $(-2)^4$ and -2^4

29. $(2\cdot4)^3$ and $2\cdot4^3$

30. $(2\cdot5)^3$ and $2\cdot5^3$

31. $8(-\frac{3}{2})^4$ and $-8(\frac{3}{2})^4$

32. $9(-\frac{2}{3})^4$ and $-9(\frac{2}{3})^4$

33. $(2+4)^2$ and 2^2+4^2

34. $(3+5)^2$ and 3^2+5^2

35. $(5-1)^2$ and 5^2-1^2

36. $(4-2)^2$ and 4^2-2^2

37. $(2^3)^2$ and $2^{(3^2)}$

38. $(3^2)^3$ and $3^{(2^3)}$

Write in exponential form.

39. $(x+5)(x+5)(x+5)$

40. $(x+4)(x+4)(x+4)$

41. $19\cdot a\cdot b\cdot b\cdot b\cdot b-a\cdot a\cdot a\cdot b\cdot b$

42. $a\cdot a\cdot a\cdot b-17\cdot a\cdot a\cdot b\cdot b\cdot b\cdot b$

43. $(3r)(3r)(3r)(3r)$

44. $(6s)(6s)(6s)(6s)(6s)$

45. $\dfrac{m}{7}\cdot\dfrac{m}{7}\cdot\dfrac{m}{7}\cdot\dfrac{m}{7}\cdot\dfrac{m}{7}$

46. $\dfrac{n}{9}\cdot\dfrac{n}{9}\cdot\dfrac{n}{9}\cdot\dfrac{n}{9}$

47. Suppose that $20,000 is deposited in an account that pays 10% interest. What is the value of the account after
a) 1 year?   **b)** 2 years?
c) 3 years?   **d)** 4 years?

48. A piece of property is purchased for $50,000. It subsequently appreciates at the rate of 20% per year. What is the value of the property after
a) 1 year?   **b)** 2 years?
c) 3 years?   **d)** 4 years?

49. Construct a table that enumerates all the possible ways that you can complete a true-false test that contains
a) one question.
b) two questions.
c) three questions.
In how many ways can you complete a true-false test containing 10 questions? Containing n questions?

50. Construct a table that enumerates all the possible ways that you can select the winner of
a) one basketball game.
b) two basketball games.
c) three basketball games.
In how many ways can you select the winner of 10 basketball games? Of n basketball games?

CALCULATOR PROBLEMS

51. Calculate to the nearest tenth.
  a) $(0.748)^4$ **b)** π^7
  c) $9 \cdot 514^2$

52. On the day of your birth, your parents deposit $3500 into an account in your name. If the account pays $12\frac{1}{4}\%$ interest, what is its value when you are ready to attend college 18 years later?

2.2	*Laws of Exponents*

Exponential notation simplifies the writing of repeated products. But to take full advantage of exponential notation, we must have laws that allow us to perform operations on powers while keeping them in exponential form.

Before stating the first law of exponents, let's study an example. Consider the product of a^2 and a^3.

$$a^2 \cdot a^3 = \underbrace{a \cdot a}_{2 \text{ factors}} \cdot \underbrace{a \cdot a \cdot a}_{3 \text{ factors}} = \underbrace{a \cdot a \cdot a \cdot a \cdot a}_{5 \text{ factors}} = a^5$$

The exponent of the answer, a^5, is simply the sum of the exponents of a^2 and a^3. That is,

$$a^2 \cdot a^3 = a^{2+3} = a^5.$$

Product Rule for Exponents

$$a^m \cdot a^n = a^{m+n}$$

To verify this law for *any* positive integers m and n, we note that

$$a^m \cdot a^n = \underbrace{a \cdot a \cdots a}_{m \text{ factors}} \cdot \underbrace{a \cdot a \cdots a}_{n \text{ factors}} = \underbrace{a \cdot a \cdot a \cdots a}_{m+n \text{ factors}} = a^{m+n}.$$

EXAMPLE 1   Apply the product rule for exponents.

  a) $x^4 \cdot x^3 = x^{4+3} = x^7$
  b) $2^2 \cdot 2^3 = 2^{2+3} = 2^5$ Do *not* multiply the bases
  c) $4 \cdot 4^3 \cdot 4^6 = 4^{1+3+6} = 4^{10}$ Remember, $4 = 4^1$

Try Problem 3   **d)** $3r^2 \cdot r^7 = 3r^{2+7} = 3r^9$ ◀

CAUTION The product rule does *not* apply when the bases are different. For example, $a^2 b^3$ and $2^{20} \cdot 3^{10}$ cannot be simplified.

Now consider the quotient of a^6 and a^2.

$$\frac{a^6}{a^2} = \frac{\overbrace{a \cdot a \cdot a \cdot a \cdot a \cdot a}^{6 \text{ factors}}}{\underbrace{a \cdot a}_{2 \text{ factors}}} = \overbrace{a \cdot a \cdot a \cdot a}^{4 \text{ factors}} = a^4$$

We can obtain the same result simply by subtracting exponents. That is,

$$\frac{a^6}{a^2} = a^{6-2} = a^4.$$

Quotient Rule for Exponents

If $m > n$ and $a \neq 0$, then

$$\frac{a^m}{a^n} = a^{m-n}.$$

We verify this law by noting that

$$\frac{a^m}{a^n} = \frac{\overbrace{a \cdot a \cdots a}^{m \text{ factors}}}{\underbrace{a \cdot a \cdots a}_{n \text{ factors}}} = \frac{\overbrace{a \cdot a \cdots a}^{n \text{ factors}} \cdot \overbrace{a \cdot a \cdots a}^{m-n \text{ factors}}}{\underbrace{a \cdot a \cdots a}_{n \text{ factors}}} = \overbrace{a \cdot a \cdots a}^{m-n \text{ factors}} = a^{m-n}.$$

EXAMPLE 2 Apply the quotient rule for exponents.

a) $\dfrac{y^9}{y^3} = y^{9-3} = y^6$

b) $\dfrac{5^8}{5^4} = 5^{8-4} = 5^4$ Do *not* divide the bases

c) $\dfrac{2t^6}{t^3} = 2t^{6-3} = 2t^3$

Try Problem 18

d) $\dfrac{a^3 b^5}{ab^2} = \dfrac{a^3}{a} \cdot \dfrac{b^5}{b^2} = a^{3-1} b^{5-2} = a^2 b^3$ ◂

CAUTION The quotient rule does *not* apply when the bases are different. For example, $\dfrac{a^2}{b^3}$ and $\dfrac{2^{20}}{3^{10}}$ cannot be simplified.

We now consider $(a^2)^3$, which is a power to a power.

$$(a^2)^3 = (a^2)(a^2)(a^2) = a^{2+2+2} = a^6$$

In this case, multiplying exponents produces the same result.

Power-to-a-Power Rule

$$(a^m)^n = a^{m \cdot n}$$

This rule is verified in much the same way as the product and quotient rules.

EXAMPLE 3 Apply the power-to-a-power rule.

a) $(x^3)^4 = x^{3 \cdot 4} = x^{12}$

b) $(3^2)^5 = 3^{2 \cdot 5} = 3^{10}$

Try Problem 25 c) $(r^{q+1})^3 = r^{3 \cdot (q+1)} = r^{3q+3}$ ◂

CAUTION Do not confuse the power-to-a-power rule with the product rule. For example, $(a^3)^7 = a^{21}$ but $a^3 \cdot a^7 = a^{10}$.

The next law deals with raising the product ab to a power. Note that

$$(ab)^3 = (ab)(ab)(ab) = a \cdot a \cdot a \cdot b \cdot b \cdot b = a^3 b^3.$$

Product-to-a-Power Rule

$$(ab)^n = a^n b^n$$

EXAMPLE 4 Apply the product-to-a-power rule.

a) $(2p)^3 = 2^3 p^3 = 8p^3$

b) $(-9mn)^2 = (-9)^2 m^2 n^2 = 81 m^2 n^2$

Try Problem 35 c) $(3a)^4 = 3^4 a^4 = 81a^4$ ◂

As an example of a *quotient* to a power, consider

$$\left(\frac{a}{b}\right)^3 = \left(\frac{a}{b}\right)\left(\frac{a}{b}\right)\left(\frac{a}{b}\right) = \frac{a \cdot a \cdot a}{b \cdot b \cdot b} = \frac{a^3}{b^3}.$$

Quotient-to-a-Power Rule

If $b \neq 0$, then

$$\left(\frac{a}{b}\right)^n = \frac{a^n}{b^n}.$$

EXAMPLE 5   Apply the quotient-to-a-power rule.

a) $\left(\dfrac{2}{3}\right)^4 = \dfrac{2^4}{3^4} = \dfrac{16}{81}$

Try Problem 43   b) $\left(\dfrac{-2}{z}\right)^5 = \dfrac{(-2)^5}{z^5} = \dfrac{-32}{z^5}$ ◄

Sometimes we must apply several laws to simplify a given expression.

EXAMPLE 6   Simplify $(3x)(4x^2)$.

Use the commutative and associative laws of multiplication to group the numbers and group the powers of x.

Try Problem 45 $$(3x)(4x^2) = (3 \cdot 4)(x \cdot x^2) = 12x^3 \blacktriangleleft$$

EXAMPLE 7   Simplify $\dfrac{-12(a^3b^4)^2}{6ab^5}$.

$\dfrac{-12(a^3b^4)^2}{6ab^5} = \dfrac{-12(a^3)^2(b^4)^2}{6ab^5}$ Product-to-a-power rule

$= \dfrac{-12a^6b^8}{6ab^5}$ Power-to-a-power rule

Try Problem 55 $= -2a^5b^3$ Quotient rule ◄

EXAMPLE 8   Simplify $(7r^3s^5)(-2r^2s)^3(r^7s^4)^5$.

$(7r^3s^5)(-2r^2s)^3(r^7s^4)^5 = 7r^3s^5(-2)^3(r^2)^3s^3(r^7)^5(s^4)^5$ Product-to-a-power rule

$= 7r^3s^5(-8)r^6s^3r^{35}s^{20}$ Power-to-a-power rule

$= 7(-8)(r^3r^6r^{35})(s^5s^3s^{20})$ Regroup

Try Problem 65 $= -56r^{44}s^{28}$ Product rule ◄

ZERO EXPONENT

If a is nonzero, then

$$\frac{a^n}{a^n} = 1.$$

But if the quotient rule is to apply, then

$$\frac{a^n}{a^n} = a^{n-n} = a^0.$$

This suggests the definition on the following page.

Definition   If a is any number except zero, then
$$a^0 = 1.$$
The expression 0^0 is undefined.

EXAMPLE 9   Evaluate: **a)** 21^0, **b)** $(\frac{2}{3})^0$, **c)** $8x^0$, **d)** $(-5)^0$, **e)** -5^0.

a) $21^0 = 1$
b) $(\frac{2}{3})^0 = 1$
c) $8x^0 = 8 \cdot 1 = 8$ If $x \neq 0$
d) $(-5)^0 = 1$

Try Problem 71   **e)** $-5^0 = -1$ The base is 5, *not* -5 ◄

EXAMPLE 10   A common bacterium known as *Escherichia coli* is capable of doubling its number every hour. If a culture contains 5 cells of *E. coli,* how many cells will it contain in 24 hours?

After 1 hr, the culture contains $5 \cdot 2 = 5 \cdot 2^1$ cells.
After 2 hr, the culture contains $(5 \cdot 2^1) \cdot 2 = 5 \cdot 2^2$ cells.
After 3 hr, the culture contains $(5 \cdot 2^2) \cdot 2 = 5 \cdot 2^3$ cells.

.
.
.

After 24 hr, the culture contains $5 \cdot 2^{24}$ cells.

Try Problem 75   The answer is very large, so we leave it in exponential form. ◄

We summarize the rules of exponents below.

Rules of Exponents

Product rule: $a^m \cdot a^n = a^{m+n}$

Quotient rule: $\dfrac{a^m}{a^n} = a^{m-n}$ $(a \neq 0)$

Power-to-a-power rule: $(a^m)^n = a^{mn}$

Product-to-a-power rule: $(ab)^n = a^n b^n$

Quotient to a power rule: $\left(\dfrac{a}{b}\right)^n = \dfrac{a^n}{b^n}$ $(b \neq 0)$

Zero exponent: $a^0 = 1$ $(a \neq 0)$

PROBLEM SET 2.2

Apply the product rule for exponents.

1. $x^2 \cdot x^5$

2. $x^4 \cdot x^6$

3. $3^2 \cdot 3^3$

4. $4^2 \cdot 4^3$

5. $5 \cdot 5^4 \cdot 5^7$

6. $6^5 \cdot 6^3 \cdot 6$

7. $2r^3 \cdot r^8$

8. $7r^4 \cdot r^9$

9. $t^n t$

10. $t^n t^3$

11. $(y+3)^5(y+3)^{10}$

12. $(y-4)^7(y-4)^3$

Apply the quotient rule for exponents.

13. $\dfrac{y^6}{y^3}$

14. $\dfrac{y^8}{y^2}$

15. $\dfrac{6^4}{6^2}$

16. $\dfrac{8^6}{8^2}$

17. $\dfrac{4t^7}{t^3}$

18. $\dfrac{5t^9}{t^4}$

19. $\dfrac{a^4b^6}{ab^2}$

20. $\dfrac{a^5b^7}{a^3b}$

21. $\dfrac{r^{n+6}}{r^2}$

22. $\dfrac{r^{n+1}}{r}$

23. $\dfrac{(3m-1)^{100}}{(3m-1)^{99}}$

24. $\dfrac{(6m+1)^{201}}{(6m+1)^{200}}$

Apply the power-to-a-power rule.

25. $(x^2)^4$

26. $(x^3)^5$

27. $(2^3)^5$

28. $(3^2)^9$

29. $(b^{10})^{3n}$

30. $(b^{15})^{2n}$

31. $(r^{q+1})^2$

32. $(r^{q-1})^3$

Apply the product-to-a-power rule.

33. $(3p)^3$

34. $(4p)^3$

35. $(-7mn)^2$

36. $(-8mn)^2$

37. $(2a)^4$

38. $(5a)^4$

39. $(9t)^n$

40. $(10t)^n$

Apply the quotient-to-a-power rule.

41. $(\tfrac{3}{5})^2$

42. $(\tfrac{2}{5})^2$

43. $\left(\dfrac{-5}{z}\right)^3$

44. $\left(\dfrac{-3}{z}\right)^3$

Simplify.

45. $(2x)(9x^2)$

46. $(3x^2)(8x)$

47. $\left(\dfrac{3x^2}{y^3}\right)^2$

48. $\left(\dfrac{2x^3}{y^4}\right)^2$

49. $(5p^4)^3$

50. $(4p^5)^3$

51. $(6s^4t)(-9s^2t^3)$

52. $(4st^4)(-7s^2t^6)$

53. $(-3pq^4)^3$

54. $(-2p^6q)^5$

55. $\dfrac{-18(a^3b^5)^2}{6a^2b}$

56. $\dfrac{-16(a^2b^{10})^2}{8ab^5}$

57. $\dfrac{4b^3}{5m}\left(\dfrac{2m^2}{b^7}\right)^5$

58. $\dfrac{3b^4}{5m}\left(\dfrac{3m^3}{b^6}\right)^3$

59. $-x^2(4x^3)^3(-3x^2)^4$

60. $-x^2(-5x^3)^2(2x^4)^3$

61. $\left(\dfrac{2x^7}{5y^2}\right)^4\left(\dfrac{5x^3}{yz^0}\right)^2$

62. $\left(\dfrac{3x^5}{4y^3}\right)^4\left(\dfrac{4x^2}{y^4z^0}\right)^3$

63. $(-5a^2bc^4)(2ab^2c^5)(-3abc^6)$

64. $(-8ab^2c^3)(-3a^2bc^5)(2abc^7)$

65. $(-4r^4s^5)^3(8r^2s^3)(r^6s^6)^5$

66. $(9r^2s)(-3r^4s^4)^3(r^8s^{10})^4$

67. $\dfrac{-m^2}{6n^8}\left(\dfrac{-3n^6}{m^2}\right)^4$ ✓

68. $\dfrac{-m^2}{6n^6}\left(\dfrac{-2n^4}{m^2}\right)^6$

69. $\dfrac{(2w^2)^4(-3w^2)^2}{4w(9w^3)^3}$

70. $\dfrac{(3w^4)^3(-5w^3)^2}{6w(10w^2)^3}$

71. Evaluate.
   a) 101^0
   b) $(\frac{3}{4})^0$
   c) $6x^0$
   d) $(-8)^0$
   e) -8^0

72. Evaluate.
   a) 99^0
   b) $(\frac{2}{5})^0$
   c) $4x^0$
   d) $(-7)^0$
   e) -7^0

73. Write 16^{90} as a power of 2. [*Hint:* $16 = 2^4$.]

74. Write 81^{60} as a power of 3. [*Hint:* $81 = 3^4$.]

75. The bacterium *E. coli* doubles its number every hour. How many cells will a culture that originally contained three cells contain in
   a) 4 hours?   **b)** 24 hours?

76. Under *optimum* conditions for growth, the bacterium *E. coli* can double its number every 20 minutes. If seven *E. coli* cells are placed in a culture maintained at such conditions, how many cells will the culture contain in
   a) 1 hour?   **b)** 6 hours?

77. Verify the product-to-a-power rule for any exponent *n*.

78. Verify the quotient-to-a-power rule for any exponent *n*.

CALCULATOR PROBLEMS

79. Find each power on your calculator (your calculator should register an error for 0^0).
   a) $1{,}234{,}567^0$ **b)** π^0
   c) 0^0 **d)** 1^0

80. Simplify.
   a) $\left(\dfrac{-3a^{555}}{b^{747}}\right)^{15}$
   b) $28r^{3817}s^{1729}(-2r^{987}s^{613})^{18}$

2.3   Polynomials

The expression $3x^2$ is called a **monomial in *x***. The number 3 is called the **numerical coefficient,** or simply the **coefficient,** of the monomial. The **degree** of the monomial is 2, since there are two factors of the variable.

Definition	If $a \neq 0$ and n is a whole number, then

$$ax^n$$

is a **monomial in *x*.** The number a is the **coefficient,** and n is the **degree.**

EXAMPLE 1   State the coefficient and the degree of each monomial: **a)** $9x^3$, **b)** y^5, **c)** $3z$, **d)** $-t^2$, **e)** 8.

The solutions are as follows:

a)   coefficient $= 9$, degree $= 3$
b)   coefficient $= 1$ (since $y^5 = 1 \cdot y^5$), degree $= 5$
c)   coefficient $= 3$, degree $= 1$ (since $3z = 3z^1$)
d)   coefficient $= -1$ (since $-t^2 = (-1) \cdot t^2$), degree $= 2$
e)   coefficient $= 8$, degree $= 0$ (since $8 = 8 \cdot x^0$) ◄

Try Problem 1

Note that nonzero constants, such as 8, are monomials of degree zero. The constant 0, however, is a special case. It is also a monomial, but no degree is assigned to it, since 0 can be written in any of the forms $0 \cdot x^1, 0 \cdot x^2, 0 \cdot x^3$, etc.

Definition

A **polynomial** is either a monomial or a sum of monomials.

The expression

$$5x^3 + 3x^2 - 2x + 8$$

is a polynomial, since it can be written as

$$5x^3 + 3x^2 + (-2x) + 8.$$

This last expression is the sum of the four monomials $5x^3$, $3x^2$, $-2x$, and 8. These monomials are called the **terms** of the polynomial.

Since the term $5x^3$ has the highest degree, namely 3, we say that the **degree of the polynomial** is 3.

Polynomials are easier to read when written in **descending powers** of the variable as follows:

$$5x^3 + 3x^2 - 2x + 8 \qquad \text{Exponents on } x \text{ decrease from left to right}$$

A polynomial of one term is also called a **monomial.** A polynomial of two terms is also called a **binomial,** and a polynomial of three terms a **trinomial.**

Monomials	$3x^2$,	$4y^5$,	6
Binomials	$5x + 2$,	$r + 1$,	$m^2 - 4$
Trinomials	$x^2 - 6x + 5$,	$p^3 + 4p^2 + 8p$,	$t^4 + 9t^2 + 10$

Polynomials of four or more terms do not have a special name.

To **evaluate** a polynomial at $x = a$, substitute a for x wherever x appears in the polynomial.

EXAMPLE 2   Evaluate each polynomial at the given value.

a) $3x + 1$ at $x = -1$
$$3(-1) + 1 = -3 + 1 = -2 \qquad \text{Replace } x \text{ by } -1$$

b) $2r^3 + r^2 - 3r + 5$ at $r = \frac{1}{2}$

$2(\frac{1}{2})^3 + (\frac{1}{2})^2 - 3(\frac{1}{2}) + 5$ Replace r by $\frac{1}{2}$

$= 2(\frac{1}{8}) + \frac{1}{4} - \frac{3}{2} + 5$

$= \frac{1}{4} + \frac{1}{4} - \frac{6}{4} + \frac{20}{4}$ Write with LCD 4

$= \frac{16}{4}$

$= 4$

c) $-t^2 + 4t + 10$ at $t = 3$

$-3^2 + 4(3) + 10$ Replace t by 3

$= -9 + 12 + 10$ Remember, $-3^2 = -9$

Try Problem 49 $= 13$ ◄

Problem Solving

Vertical Projection Problem

Suppose an object is projected vertically upward with an initial velocity of r feet per second from an initial height of s feet. Ignoring air resistance, the height h in feet of the object after t seconds is given by the formula*

$$h = -16t^2 + rt + s.$$

An arrow is shot upward from the top of a building 75 feet high with a bow that gives it an initial velocity of 112 feet per second. Ignoring air resistance, write a formula for the height h of the arrow at any time t. Then find the height of the arrow 4 seconds after it is shot.

Substituting $r = 112$ and $s = 75$ into the formula above, we have

$$h = -16t^2 + 112t + 75$$

This formula gives the height of the arrow at any time t. Therefore the height at $t = 4$ is given by

$h = -16(4)^2 + 112(4) + 75$

$= -16(16) + 448 + 75$

$= -256 + 448 + 75$

$= 267.$

$t = 4$ sec
$h = 267$ ft

75 ft

The height of the arrow at $t = 4$ seconds is 267 feet.

---

* This formula is due largely to the work of the Italian astronomer and mathematician Galileo Galilei (1564–1643) and the English mathematician and physicist Isaac Newton (1642–1727).

PROBLEM SET 2.3

State the coefficient and the degree of each monomial.

1. $6x^5$ 2. $3x^7$ 3. y^2 4. y^3

5. $2z$ 6. $7z$ 7. $-t^4$ 8. $-t^6$

9. 9 10. 16 11. $-\frac{2}{3}r$ 12. $-\frac{4}{5}r$

Write each polynomial in descending powers and state its degree. Identify it as a monomial, a binomial, or a trinomial if appropriate.

13. $7 + 4x + x^2$ 14. $6 + 8x + x^2$ 15. $9 + x^2 - 3x$ 16. $2 + x^2 - 4x$

17. $2 + x$ 18. $3 + x$ 19. $3 + 5y$ 20. $5 + 2y$

21. $-9 + r^2$ 22. $-4 + r^2$ 23. $1 - p^2$ 24. $3 - p^2$

25. $100t^{99}$ 26. $99t^{100}$ 27. $m^2 + 6m^4 - 1$ 28. $m^2 + 8m^4 - 1$

29. $-q^2 + q^5 - 7q + 4$ 30. $-q^2 - 2q^3 + q^5 + q$ 31. $5x$ 32. $3x$

33. 5 34. 3

Evaluate each polynomial at the given value.

35. $x + 6$ at $x = 2$ 36. $x + 4$ at $x = 3$

37. $4x + 2$ at $x = -1$ 38. $5x + 3$ at $x = -1$

39. $y^2 + 1$ at $y = 3$ 40. $y^2 + 6$ at $y = 2$

41. $5y^2 - 8y + 13$ at $y = 5$ 42. $4y^2 - 9y + 12$ at $y = 4$

43. $z^3 - 4z^2 + 7z - 5$ at $z = 0$ 44. $z^3 - 2z^2 + 5z - 8$ at $z = 0$

45. $z^5 - z^3$ at $z = -2$ 46. $z^5 - z^3$ at $z = -3$

47. $2r^3 - 3r^2 + 5r + 1$ at $r = \frac{1}{2}$ 48. $2r^3 - 5r^2 + 4r + 2$ at $r = \frac{1}{2}$

49. $-t^2 + 6t + 11$ at $t = 4$ 50. $-t^2 + 7t + 13$ at $t = 5$

51. $-t^4 - t^2 + t + 7$ at $t = -1$ 52. $-t^4 + t^2 - t + 3$ at $t = -2$

53. A pellet is shot upward with a slingshot from the top of a building 64 feet high. Its initial velocity is 48 ft/sec. Ignoring air resistance, write a formula for the height h of the pellet at any time t. Then find its height at
a) $t = 1$ sec.   b) $t = 2$ sec.
c) $t = 3$ sec.   d) $t = 4$ sec.
Draw a picture of your results. What is the pellet's maximum height?

54. An arrow is shot upward with an initial velocity of 80 ft/sec from a height of 96 ft. Ignoring air resistance, write a formula for the height h of the arrow at any time t. Then find the arrow's height at
a) $t = 1$ sec.   b) $t = 2$ sec.
c) $t = 3$ sec.   d) $t = 4$ sec.
e) $t = 5$ sec.   f) $t = 6$ sec.
Draw a picture of your results. What is the arrow's maximum height?

55. In Problem 53 write a formula for the height h of the pellet at any time t if
  a) the pellet is shot from ground level.
  b) the pellet is simply dropped from the top of the building.

56. In Problem 54 write a formula for the height h of the arrow at any time t if
  a) the arrow is shot from ground level.
  b) the arrow is simply dropped from the top of the building.

CALCULATOR PROBLEMS

57. Evaluate $2x^3 + 7x^2 - 4x + 19$ at $x = 1.7$.

58. Under ideal conditions, the stopping distance d in feet (including reaction time) of a car traveling r mph is given by the formula

$$d = 0.044r^2 + 1.1r.$$

Determine the stopping distance of a car traveling 55 mph.

2.4   *Addition and Subtraction of Polynomials*

Remember that the terms of a polynomial are those quantities that are separated by addition.

$$4x^2 + 5x + 1$$

3 terms

Like terms are terms that have the same variables raised to the same powers. In other words, *like terms may differ only in their coefficients.*

Like terms	Unlike terms
$3x$ and $5x$	$5x$ and $6y$
$8y^2$, $9y^2$, and y^2	$13t$ and $13t^2$
$4rs$ and $-7rs$	$2p^2q^3$ and $3p^3q^2$

COMBINING LIKE TERMS

The distributive law, written in the form

$$ba + ca = (b + c)a,$$

allows us to add like terms.

EXAMPLE 1   Add $3x$ and $5x$.

$$3x + 5x = (3 + 5)x = 8x$$

$$ba + ca = (b + c)a \blacktriangleleft$$

Try Problem 17

The operation of adding or subtracting like terms is known as **combining like terms.**

EXAMPLE 2   Combine the like terms.

a) $8y^2 + 9y^2 + y^2 = (8 + 9 + 1)y^2 = 18y^2$

b) $4rs - 7rs = 4rs + (-7rs) = [4 + (-7)]rs = -3rs$

c) $5m^3 + 3m^2 + 2m^3$

Try Problem 29

$\quad\quad = (5m^3 + 2m^3) + 3m^2 \quad\quad$ Commutative and associative laws

$\quad\quad = 7m^3 + 3m^2 \quad\quad\quad\quad\quad$ Combine like terms ◄

If you study the examples above, you will see that like terms are combined simply by combining their coefficients. Once this is clear, we can perform the middle steps mentally and simply write the answer. That is, $3x + 5x = 8x$ in the same way that $3¢ + 5¢ = 8¢$ or that 3 apples + 5 apples = 8 apples.

We *cannot* combine unlike terms. That is, we cannot simplify the expression $5x + 6y$ any more than we can simplify the expression 5 apples + 6 cows.

REMOVING PARENTHESES

It is easy to verify that multiplication distributes over subtraction as well as over addition. That is,

$$a(b - c) = ab - ac$$

EXAMPLE 3   Write $5(3x - 4)$ without parentheses.

$\quad\quad 5(3x - 4) = 5 \cdot 3x - 5 \cdot 4 \quad\quad$ Distribute 5

Try Problem 35

$\quad\quad\quad\quad\quad = 15x - 20 \quad\quad\quad\quad$ Find each product ◄

Now consider the problem of removing parentheses from the expression $-(r + 7)$. Since

$$-6 = (-1) \cdot 6 \quad \text{and} \quad -x = (-1) \cdot x,$$

we see that

$$-(r + 7) = (-1) \cdot (r + 7).$$

EXAMPLE 4   Write $-(r + 7)$ without parentheses.

$\quad\quad -(r + 7) = (-1) \cdot (r + 7)$

$\quad\quad\quad\quad\quad = (-1) \cdot r + (-1) \cdot 7 \quad\quad$ Distribute -1

$\quad\quad\quad\quad\quad = -r + (-7) \quad\quad\quad\quad\quad$ Find each product

Try Problem 39

$\quad\quad\quad\quad\quad = -r - 7 \quad\quad\quad\quad\quad\quad$ Convert to subtraction ◄

With practice you can omit most of the steps in Example 4. For example, when a negative sign appears in front of parentheses, simply visualize multiplying each term inside the parentheses by -1. This has the effect of changing the sign of each term.

EXAMPLE 5   Write $-(y^2 - y + 4)$ without parentheses.

Change each sign

Try Problem 45

$$-(y^2 - y + 4) = -y^2 + y - 4 \blacktriangleleft$$

ADDITION OF POLYNOMIALS

We add two polynomials by adding their like terms. This can be done either in *horizontal form* or in *vertical form*.

EXAMPLE 6   Find the sum of $7x^3 + 3x^2 - 4$ and $3x^3 - x^2 - x + 5$ in horizontal form.

Use the commutative and associative laws to group the like terms together.

$$(7x^3 + 3x^2 - 4) + (3x^3 - x^2 - x + 5)$$
$$= (7x^3 + 3x^3) + (3x^2 - x^2) + (-x) + (-4 + 5)$$

Then combine like terms.

$$= 10x^3 + 2x^2 - x + 1 \blacktriangleleft$$

EXAMPLE 7   Find the sum of $7x^3 + 3x^2 - 4$ and $3x^3 - x^2 - x + 5$ in vertical form.

$$7x^3 + 3x^2 \quad\quad - 4$$
$$(+) \quad \underline{3x^3 - \ x^2 - x + 5}$$
$$10x^3 + 2x^2 - x + 1 \blacktriangleleft$$

Try Problem 49

Adding polynomials in vertical form may seem easier, since like terms are arranged in the same column. However, both the vertical and the horizontal forms are used extensively in algebra, so you should become comfortable with both methods.

SUBTRACTION OF POLYNOMIALS

We subtract polynomials by subtracting their like terms.

EXAMPLE 8   Using the previous example, subtract the second polynomial from the first.

$$7x^3 + 3x^2 \quad\quad - 4$$
$$(-) \quad \underline{3x^3 - \ x^2 - x + 5}$$
$$4x^3 + 4x^2 + x - 9 \blacktriangleleft$$

Recall that subtracting b is the same as adding the opposite of b. Therefore the subtraction in Example 8 can be done by using the distributive law to mentally change the sign of each term in the second polynomial, and then adding.

$$
\begin{array}{r}
7x^3 + 3x^2 \quad\;\; - 4 \\
(+)\; -3x^3 + \;\; x^2 + x - 5 \\
\hline
4x^3 + 4x^2 + x - 9
\end{array}
$$
  Change each sign
  Add like terms

EXAMPLE 9   Subtract $3x^3 - x^2 - x + 5$ from $7x^3 + 3x^2 - 4$ in horizontal form.

$$(7x^3 + 3x^2 - 4) - (3x^3 - x^2 - x + 5)$$

Change each sign

$$= 7x^3 + 3x^2 - 4 - 3x^3 + x^2 + x - 5$$
$$= 7x^3 - 3x^3 + 3x^2 + x^2 + x - 4 - 5 \qquad \text{Rearrange terms}$$
$$= 4x^3 + 4x^2 + x - 9 \qquad \text{Combine like terms} \blacktriangleleft$$

Try Problem 55

A polynomial may contain more than one variable. For example, the expression

$$5r^3 + r^2s^2 - 4s^3$$

is a polynomial in the two variables r and s.

EXAMPLE 10   Subtract.

$$(5r^3 + r^2s^2 - 4s^3) - (2r^3 + r^2s^2 + 4s^3)$$

Change each sign

$$= 5r^3 + r^2s^2 - 4s^3 - 2r^3 - r^2s^2 - 4s^3$$
$$= 5r^3 - 2r^3 + r^2s^2 - r^2s^2 - 4s^3 - 4s^3 \qquad \text{Rearrange terms}$$
$$= 3r^3 - 8s^3 \qquad \text{Combine like terms} \blacktriangleleft$$

Try Problem 57

SIMPLIFYING POLYNOMIAL EXPRESSIONS

We can use the techniques discussed in this section to simplify expressions involving polynomials.

EXAMPLE 11   Simplify $6u - [3u - 4(2u + 5)]$.

$$
\begin{aligned}
6u - [3u - 4(2u + 5)] &= 6u - [3u - 8u - 20] &\quad& \text{Distribute } -4 \\
&= 6u - [-5u - 20] &\quad& \text{Combine like terms} \\
&= 6u + 5u + 20 &\quad& \text{Distribute } -1 \\
&= 11u + 20 &\quad& \text{Combine like terms} \blacktriangleleft
\end{aligned}
$$

Try Problem 65

EXAMPLE 12   Simplify $(-1)(-11a^3) - (-a^3)(-7) + (12a^3)(-2) - (-4)(-2a^3)$.

Multiply before adding or subtracting.

$$(-1)(-11a^3) - (-a^3)(-7) + (12a^3)(-2) - (-4)(-2a^3)$$
$$= 11a^3 - 7a^3 + (-24a^3) - 8a^3$$
$$= 11a^3 - 39a^3$$

Try Problem '77
$$= -28a^3 \blacktriangleleft$$

PROBLEM SET 2.4

Identify as like terms or unlike terms.

1. $4x, 6x$ **2.** $3x, 2x$ **3.** $5y^2, 7y^2, y^2$ **4.** $8y^2, 6y^2, y^2$

5. $2x, 7y$ **6.** $5x, 8y$ **7.** $12t, 12t^2$ **8.** $15t^2, 15t$

9. $3rs, -6rs$ **10.** $rs, -7rs$ **11.** $p^2q^3, 4p^3q^2$ **12.** $4p^3q^4, 6p^4q^3$

13. m^4, m^3 **14.** m^3, m^2 **15.** $5p, 5$ **16.** $7, 7p$

Combine like terms.

17. $4x + 6x$ **18.** $3x + 2x$

19. $5y^2 + 7y^2 + y^2$ **20.** $8y^2 + 6y^2 + y^2$

21. $-3z + 8z$ **22.** $-4z + 7z$

23. $3rs - 6rs$ **24.** $rs - 7rs$

25. $-13p + 13p + 9q$ **26.** $-11p + 11p + 6q$

27. $-x^2y^2 - 2x^2y^2$ **28.** $-3x^2y^2 - 4x^2y^2$

29. $6m^3 + 2m^2 + 9m^3$ **30.** $5m^3 + m^2 + 3m^3$

31. $-2c^2d - (-c^2d) + 7cd^2$ **32.** $-5cd^2 - (-cd^2) + 4c^2d$

Write without parentheses.

33. $3(x - 2)$ **34.** $4(x - 5)$

35. $6(2x - 4)$ **36.** $7(3x - 2)$

37. $-2(5y - 1)$ **38.** $-3(4y - 1)$

39. $-(r + 8)$ **40.** $-(r + 6)$

41. $-(-t + 2)$ **42.** $-(-t + 3)$

43. $-3(-s^2 - 3s - 1)$ **44.** $-2(-s^2 - 5s - 2)$

45. $-(y^2 - y + 5)$ **46.** $-(y^2 - 3y - 9)$

Add each pair of polynomials, first in horizontal form and then in vertical form.

47. $(3x^2 - 2x + 7) + (5x^2 + 9x + 4)$

48. $(8x^2 - 3x + 2) + (2x^2 + 11x + 1)$

49. $(6x^3 + 4x^2 - 5) + (2x^3 - x^2 - x + 3)$

50. $(8x^3 + 3x^2 - 2) + (5x^3 - 2x^2 - x + 4)$

51. $(7m^2 - 4mn + 2n^2) + (8m^2 + mn + 4n^2)$

52. $(9m^2 - 6mn + n^2) + (5m^2 + 3mn + 7n^2)$

Subtract each pair of polynomials, first in horizontal form and then in vertical form.

53. $(11x^2 + 9x + 6) - (5x^2 + x - 4)$

54. $(10x^2 + 8x + 2) - (4x^2 + 6x - 3)$

55. $(6x^3 + 4x^2 - 5) - (2x^3 - x^2 - x + 3)$

56. $(8x^3 + 3x^2 - 2) - (5x^3 - 2x^2 - x + 4)$

57. $(8r^3 - r^2s^2 + 6s^3) - (3r^3 - r^2s^2 - 6s^3)$

58. $(2r^3 + r^2s^2 - 7s^3) - (r^3 + r^2s^2 + 7s^3)$

Simplify.

59. $3(8p^3 - p + 9) - 6(4p^2 + 3p - 8)$

60. $5(2p^3 + 3p^2 - p) - 4(5p^2 - 4p + 7)$

61. $-5(a - 3b + c) - (4a + b - c) + 3(3a + 4b - c)$

62. $-7(a - 2b + c) + 5(2a - b + c) - (a + b - 2c)$

63. $3 + 7[(r + 1) - (r - 9)]$

64. $2 + 9[(r + 5) - (r - 7)]$

65. $4u - [2u - 5(3u + 4)]$

66. $7u - [2u - 3(5u + 2)]$

67. $2[t - (3t - t - 4) + 5]$

68. $3[t - (2t + t - 6) + 5]$

69. $m - (m - [m - (2m + n)] - 2n)$

70. $n - (n - [m - (3m + n)] - 4m)$

71. $(3x^2y^2)(8xy) - (4x^2y)(5xy^2)$

72. $(7x^3y)(2xy^2) - (x^2y^2)(5x^2y)$

73. $15 - 5(a - 6a) + 7 - (15 + 9a) - (-a)(4)$

74. $14 - 4(a - 8a) + 3 - (12 + 7a) - (a)(-5)$

75. $(6)(5a^2) - (12a^2)(-3) + (-4a^2)(-8) + (10b)(2)$

76. $(9)(4a^2) - (11a^2)(-5) + (-6a^2)(-3) + (15b)(3)$

77. $(-1)(-13a^3) - (-a^3)(-9) + (14a^3)(-3) - (-5)(-2a^3)$

78. $(-1)(-17a^3) - (-a^3)(-8) + (19a^3)(-2) - (-6)(-3a^3)$

79. Find the sum of three consecutive integers given that the smallest integer is n.

80. Find the sum of three consecutive even integers given that the smallest integer is n.

81. If you have m nickels, and six more dimes than nickels, and twice as many quarters as dimes, how much money in cents do you have?

82. If you have m dimes, and four fewer nickels than dimes, and three times as many quarters as nickels, how much money in cents do you have?

83. A new house is priced at p dollars. What will the price of such a house be next year if new-house prices experience an inflation of 10%?

84. A new car is priced at p dollars. What will the price of such a car be next year if new-car prices experience an inflation of 20%?

CALCULATOR PROBLEMS

Simplify.

85. $7.4(8.35k^2 + 0.95k - 17.2) + 0.55(7.68k^2 - 1.19k + 28.6)$

86. $(-0.67)(4.7a) - (-3.5a)(16) + (7.8a)(-5.6) - (-8.2a)(-2.9)$

2.5   *Multiplication of Polynomials*

We find the product of two monomials simply by applying the commutative and associative laws of multiplication along with the laws of exponents.

EXAMPLE 1   Find each product.

a) $3x^2 \cdot 2 = (3 \cdot 2)x^2$ Group the numbers

$\qquad\quad = 6x^2$ Multiply

b) $(5y^3)(6y^4) = (5 \cdot 6)(y^3 \cdot y^4)$ Group the numbers, group the powers of y

$\qquad\qquad\quad = 30y^{3+4}$ Product rule for exponents

$\qquad\qquad\quad = 30y^7$

c) $(-r^2s^3)(4r^3s^5) = (-1 \cdot r^2s^3)(4r^3s^5)$

$\qquad\qquad\qquad = (-1) \cdot 4(r^2r^3)(s^3s^5)$

Try Problem 7 $\qquad\qquad = -4r^5s^8$ ◄

Using the technique for multiplying monomials in conjunction with the distributive law, we can find the product of any two polynomials.

EXAMPLE 2   Find each product.

a) $2x^2(3x + 4) = 2x^2 \cdot 3x + 2x^2 \cdot 4$ Distribute $2x^2$

$\qquad\qquad\quad = 6x^3 + 8x^2$ Multiply the monomials

b) $5r^2(r^2 + 2rs - 3s^2)$

$\qquad = 5r^2 \cdot r^2 + 5r^2 \cdot 2rs - 5r^2 \cdot 3s^2$ Distribute $5r^2$

Try Problem 17 $\quad = 5r^4 + 10r^3s - 15r^2s^2$ Multiply the monomials ◄

EXAMPLE 3   Find the product $(x + 3)(x + 2)$.

Distribute $x + 3$ over the sum $x + 2$.

$$(x + 3)(x + 2) = (x + 3)x + (x + 3)2$$

Then use the distributive law again to find both products on the right side.

$$(x + 3)x + (x + 3)2 = (x \cdot x + 3 \cdot x) + (x \cdot 2 + 3 \cdot 2)$$
$$= x^2 + 3x + 2x + 6$$

Finally, combine like terms.

Try Problem 27 $\qquad\qquad\qquad\qquad = x^2 + 5x + 6$ ◄

If you study Example 3 closely, you see that the product is simply the result of multiplying each term of the first polynomial by each term of the second polynomial. This product is then simplified by combining like terms.

We use this observation to calculate the product in Example 4.

EXAMPLE 4   Find the product $(4x + 2)(x^2 - 5x + 3)$.

First distribute $4x$. Then distribute 2.

$$(4x + 2)(x^2 - 5x + 3) = 4x \cdot x^2 - 4x \cdot 5x + 4x \cdot 3 + 2 \cdot x^2 - 2 \cdot 5x + 2 \cdot 3$$
$$= 4x^3 - 20x^2 + 12x + 2x^2 - 10x + 6$$

Then combine like terms.

Try Problem 29

$$= 4x^3 - 18x^2 + 2x + 6 \blacktriangleleft$$

Sometimes it is more convenient to compute products in vertical form.

EXAMPLE 5   Find the product of Example 4 in vertical form.

Set up your work as shown below.

$$
\begin{array}{ll}
x^2 - 5x + 3 & \\
 4x + 2 & \\
\hline
4x^3 - 20x^2 + 12x & \longleftarrow 4x \text{ times } x^2 - 5x + 3 \\
 2x^2 - 10x + 6 & \longleftarrow 2 \text{ times } x^2 - 5x + 3 \\
\hline
4x^3 - 18x^2 + 2x + 6 & \longleftarrow \text{Add like terms } \blacktriangleleft
\end{array}
$$

Try Problem 39

When multiplying polynomials in vertical form, you should arrange like terms so that they are in the same column. This makes adding them easier. To do this, you must leave spaces for any missing powers of the variable.

EXAMPLE 6   Multiply $2m^3 - 4m^2 + 1$ by $3m - 5$ in vertical form.

$$
\begin{array}{ll}
2m^3 - 4m^2 + 1 & \\
 3m - 5 & \\
\hline
6m^4 - 12m^3 + 3m & \longleftarrow \text{Space for missing } m^2 \text{ term} \\
 - 10m^3 + 20m^2 - 5 & \longleftarrow \text{Space for missing } m \text{ term} \\
\hline
6m^4 - 22m^3 + 20m^2 + 3m - 5 & \longleftarrow \text{Add like terms } \blacktriangleleft
\end{array}
$$

Try Problem 41

MULTIPLYING BINOMIALS

There is a shortcut to finding the product of two binomials. Consider the product of $x + 3$ and $x + 5$.

$$(x + 3)(x + 5) = x \cdot x + x \cdot 5 + 3 \cdot x + 3 \cdot 5$$
$$= x^2 + 5x + 3x + 15$$
$$= x^2 + 8x + 15$$

Let's examine the step immediately before the answer.

The term x^2 is the product of the two *first* terms.

$$(x + 3)(x + 5) \qquad x \cdot x = x^2$$

The term $5x$ is the product of the two *outer* terms.

$$(x + 3)(x + 5) \qquad x \cdot 5 = 5x$$

The term $3x$ is the product of the two *inner* terms.

$$(x + 3)(x + 5) \qquad 3 \cdot x = 3x$$

The term 15 is the product of the two *last* terms.

$$(x + 3)(x + 5) \qquad 3 \cdot 5 = 15$$

Therefore we multiply two binomials by multiplying their first terms **F**, their outer terms **O**, their inner terms **I**, and their last terms **L**. By remembering the key word **FOIL** and by adding the outer product $5x$ and inner product $3x$ mentally, you can perform the multiplication in one step. That is, you can simply write

$$(x + 3)(x + 5) = x^2 + 8x + 15.$$

FOIL Method for Multiplying Binomials

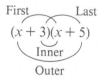

Remember that multiplication is commutative, so

$$(x + 5)(x + 3) = (x + 3)(x + 5).$$

EXAMPLE 7 Use the FOIL method to find each product.

a) $(x + 7)(x - 10) = x^2 - 3x - 70$

b) $(2r - 3)(r + 4) = 2r^2 + 5r - 12$

c) $(3a^2 + b)(6a^2 - 5b) = 18a^4 - 9a^2b - 5b^2$ ◄

Try Problem 51

Remember that an exponent does *not* distribute over a *sum*. Thus $(a + b)^2$ and $a^2 + b^2$ are *not* the same. Instead,

$$(a + b)^2 = (a + b)(a + b) = a^2 + 2ab + b^2.$$

Similarly,

$$(a - b)^2 = (a - b)(a - b) = a^2 - 2ab + b^2.$$

That is, the **square of a binomial** is the square of the first term plus twice the product of the two terms plus the square of the last term.

EXAMPLE 8 Find the square of each binomial.

a) $(x + 5)^2 = x^2 + 10x + 25$

| | Square the first term | Twice the product of the terms | Square the last term |

Try Problem 65

b) $(2y - 3)^2 = 4y^2 - 12y + 9$ ◄

Another special product, which results in a **difference of two squares,** is shown below.

$$(a + b)(a - b) = a^2 - ab + ab - b^2$$
$$= a^2 - b^2$$

EXAMPLE 9   Find each product.

a) $(p + 8)(p - 8) = p^2 - 8^2 = p^2 - 64$

$(a + b)(a - b) = a^2 - b^2$

Try Problem 71   **b)** $(4r + 7s)(4r - 7s) = (4r)^2 - (7s)^2 = 16r^2 - 49s^2$ ◄

These special products will be helpful in factoring polynomials, so we summarize them below.

Special Products

Square of a binomial: $(a + b)^2 = a^2 + 2ab + b^2$
$(a - b)^2 = a^2 - 2ab + b^2$

Difference of two squares: $(a + b)(a - b) = a^2 - b^2$

The formula for the number of games played in a round-robin schedule involves the product of two polynomials. To see this, suppose that each team in a softball league consisting of four teams plays every other team once. That is, each of the four teams plays three games. This is called a round-robin schedule, and the total number of league games is

$$\frac{4 \cdot 3}{2} = 6.$$

The product $4 \cdot 3$ is divided by 2 so that no single game is counted twice, once by each team involved.

Generalizing on this example, we see that if each of the n teams in a league plays every other team once, then the total number of league games is given by

$$\frac{n \cdot (n - 1)}{2}$$

PROBLEM SET 2.5

Find each product of monomials.

1. $2 \cdot 6x$

2. $3 \cdot 8x$

3. $4x^2 \cdot 3$

4. $5x^3 \cdot 2$

5. $5y \cdot y$

6. $9y \cdot y$

7. $(7y^3)(4y^6)$

8. $(6y^2)(7y^5)$

9. $(-2r^2)(-3r^4s^2)$

10. $(-3r^3)(-4r^2s^6)$

11. $(-r^3s^5)(9r^2s^2)$

12. $(2r^2s^5)(-r^3s^3)$

13. $(5x)(8y)$

14. $(8x)(9y)$

Find each product.

15. $x(x + 6)$

16. $x(x - 4)$

17. $4x^2(2x + 5)$

18. $3x^2(4x + 5)$

19. $5r(r^3 - 3r^2 + 1)$

20. $6r(r^3 + 2r^2 - 1)$

21. $3r^2(r^2 + 2rs - 6s^2)$

22. $2r^2(r^2 - 5rs + 8s^2)$

23. $-m(m^2 - mn + n^2)$

24. $-m(m^2 + mn - n^2)$

25. $s^2t(s^3t - s^2t^2 - t^3)$

26. $st^3(s^3 - s^2t^3 - t^4)$

Find each product in horizontal form.

27. $(x + 4)(x + 2)$

28. $(x + 3)(x + 5)$

29. $(5x + 3)(x^2 - 4x + 6)$

30. $(4x - 3)(x^2 + 5x - 7)$

31. $(6r - 7)(2r^3 - 3r^2 + 5r - 1)$

32. $(8r - 3)(5r^3 + 2r^2 - 7r - 4)$

33. $(b + 2)(b - 3)(b + 4)$

34. $(b - 5)(b + 3)(b - 1)$

35. $(2p + 3)^3$

36. $(3p - 2)^3$

Find each product in vertical form.

37. $y + 6$
$\underline{y - 2}$

38. $y - 8$
$\underline{y + 1}$

39. $x^2 - 3x + 5$
$\underline{4x + 1}$

40. $x^2 + 6x - 2$
$\underline{3x - 4}$

41. $3m^3 - 7m^2 + 2$
$\underline{2m - 5}$

42. $4m^3 + 5m - 2$
$\underline{2m - 3}$

43. $a^2 - 8ab - 2b^2$
$\underline{2a + 3b}$

44. $3a^2 - 5ab + b^2$
$\underline{3a - 4b}$

45. $m + 2n - 1$
$\underline{4m - 3n + 5}$

46. $6m - n + 3$
$\underline{2m + 4n - 5}$

Use the FOIL method to find each product.

47. $(x + 3)(x + 4)$

48. $(x + 4)(x + 7)$

49. $(x + 6)(x - 11)$

50. $(x + 8)(x - 3)$

51. $(3r - 1)(r + 2)$

52. $(5r - 4)(r + 2)$

53. $(5s - 2)(4s - 3)$

54. $(2s - 3)(6s - 11)$

55. $(6p + 5q)(2p - 3q)$

56. $(5p + 6q)(4p - 3q)$

57. $(t^2 - 4)(t^2 - 9)$

58. $(t^2 - 1)(t^2 - 16)$

59. $(2a^2 + b)(5a^2 - 3b)$ **60.** $(4a^2 - 5b)(3a^2 + b)$ **61.** $(y^3 + 8z)(y^3 - z)$ **62.** $(y^3 - 9z)(y^3 + z)$

63. $(x^n + 11)(x^n + 17)$ **64.** $(x^n + 13)(x^n + 15)$

Find each special product.

65. $(x + 4)^2$ **66.** $(x + 3)^2$ **67.** $(3y - 2)^2$ **68.** $(2y - 5)^2$

69. $(p + 7)(p - 7)$ **70.** $(p + 4)(p - 4)$ **71.** $(5r + 6s)(5r - 6s)$ **72.** $(7r + 2s)(7r - 2s)$

73. $(4m + 3n)^2$ **74.** $(5m - 3n)^2$ **75.** $(t^{2k} - 1)(t^{2k} + 1)$ **76.** $(t^{3k} - 6)(t^{3k} + 6)$

Simplify.

77. $m^2[m - (m - 2)(m + 3)]$ **78.** $m^2[m - (m + 8)(m - 1)]$

79. $-t(t - 2[3t - 5(t + 1)] + 7)$ **80.** $-t(t - 4[2t - 3(t + 2)] + 8)$

81. $5p^2q - 6[p - p(pq - q)] - pq$ **82.** $6pq^2 - 2[q - q(pq - p)] - pq$

83. Find the product of three consecutive odd integers given that the smallest integer is n.

84. Find the product of three consecutive integers given that the smallest integer is n.

85. If each of the 12 teams in a girls' softball league plays every other team once, how many league games will there be?

86. If each of the 15 teams in a bowling league plays every other team once, how many league matches will there be?

87. If each of the n teams in a touch football league plays every other team twice, how many league games will there be?

88. If each of the n teams in a dart league plays every other team three times, how many league matches will there be?

CALCULATOR PROBLEMS

89. Find the product $(3.85x + 4.72y)(6.4x - 9.5y)$.

90. Just as $(x^6 + 2)(x^6 - 2) = x^{12} - 4$, so does $(10^6 + 2)(10^6 - 2) = 10^{12} - 4$. Because of rounding off, however, your calculator will probably compute $(10^6 + 2) \cdot (10^6 - 2)$ as just 10^{12}. Try it.

2.6   *Greatest Common Factor, Difference of Two Squares*

Consider the equation

$$15 = 3 \cdot 5$$

The natural numbers 3 and 5 are called **factors,** or **divisors,** of 30. To write 15 as the product of 3 and 5 is to **factor** 15.

Factor

Product $\longrightarrow 15 = 3 \cdot 5 \longleftarrow$ Factors

Multiply

Normally when we factor a natural number, we want to express the number as a product of **prime numbers.**

Definition   A natural number greater than 1 that has no natural-number divisors other than 1 and itself is called a **prime number.**

There are an infinite number of prime numbers, the first ten of which are listed below.

2, 3, 5, 7, 11, 13, 17, 19, 23, 29, . . .

When a number is written as a product of prime numbers, it is said to be in **prime factored form.**

EXAMPLE 1   Write 360 in prime factored form.

Divide 360 by the smallest prime 2. Then divide that quotient by 2. Continuing to divide each new quotient by its smallest prime divisor, we obtain the sequence of divisions below.

$$
\begin{array}{r}
2\,)\overline{360} \\
2\,)\overline{180} \\
2\,)\overline{90} \\
3\,)\overline{45} \\
3\,)\overline{15} \\
5
\end{array}
$$

Try Problem 7   Therefore $360 = 2 \cdot 2 \cdot 2 \cdot 3 \cdot 3 \cdot 5 = 2^3 \cdot 3^2 \cdot 5.$ ◀

GREATEST COMMON FACTOR

Definition   The **greatest common factor (GCF)** of a collection of natural numbers is the largest natural number that is a factor (divisor) of each number in the set.

EXAMPLE 2   Find the GCF of 8 and 12.

The GCF is 4, since 4 is the largest natural number that is a factor (divisor) of

Try Problem 13   both 8 and 12. ◀

EXAMPLE 3   Find the GCF of 36, 18, and 90.

First write each number in prime factored form.

$$
36 = 2^2 \cdot 3^2
$$
$$
18 = 2 \cdot 3^2
$$
$$
90 = 2 \cdot 3^2 \cdot 5
$$

The GCF contains each prime factor to the *lowest* power that it appears in any one of the factorizations. That is, the GCF contains 2^1, 3^2, and $5^0 = 1$. Therefore the GCF is $2 \cdot 3^2 = 18$. ◄

Try Problem 15

These ideas can be generalized to apply to expressions involving variables.

EXAMPLE 4   Find the GCF of the terms $18x^3$ and $12x^2$.

The GCF of the coefficients 18 and 12 is 6. The GCF of the powers of x is the smaller of the two powers, that is, x^2. Therefore the GCF of $18x^3$ and $12x^2$ is $6x^2$. ◄

Try Problem 17

EXAMPLE 5   Find the GCF of each collection of terms.

a) $7a$ and $7b$; GCF $= 7$
b) $5x$ and 10; GCF $= 5$
c) $5x$ and 3; GCF $= 1$
Don't say that $5x$ and 3 don't have a GCF. There is always a GCF.
d) $12m^4n^3$, $24m^2n^4$, and $18mn^5$; GCF $= 6mn^3$ ◄

Try Problem 25

To **factor** a polynomial means to write the polynomial as a product of two or more simpler polynomials. One factoring technique involves finding the GCF of the terms of the polynomial. The distributive law is then used to "factor out" this GCF.

EXAMPLE 6   Factor out the GCF from $4x + 12$.

The GCF of $4x$ and 12 is 4. Using the distributive law, we factor out 4 as follows:
$$4x + 12 = 4 \cdot x + 4 \cdot 3 = 4(x + 3).$$
Remember that you can always check a factorization by multiplying. ◄

Try Problem 27

Normally we do not factor monomials. That's why we did not write the answer to Example 6 as $2 \cdot 2(x + 3)$.

EXAMPLE 7   Factor out the GCF.

a) $5a + 5b = 5 \cdot a + 5 \cdot b$ GCF is 5
       $\qquad\quad = 5(a + b)$ Distributive law
b) $21m - 35n = 7 \cdot 3m - 7 \cdot 5n$ GCF is 7
       $\qquad\qquad = 7(3m - 5n)$ Distributive law
c) $3y - 8$ GCF is 1, which we don't factor out ◄

Try Problem 31

Since we cannot write $3y - 8$ as a product of simpler polynomials, we say that $3y - 8$ is a **prime polynomial.**

The technique of factoring out the GCF also applies when the GCF contains variables.

EXAMPLE 8   Factor out the GCF.

a) $6x^3 + 15x^2 = 3x^2 \cdot 2x + 3x^2 \cdot 5$ GCF is $3x^2$

$= 3x^2(2x + 5)$

b) $a^3 + a = a \cdot a^2 + a \cdot 1$ GCF is a

$= a(a^2 + 1)$

c) $-5p^5 + 10p^4 - 15p^3 = 5p^3(-p^2) + 5p^3(2p) - 5p^3(3)$ GCF is $5p^3$

$= 5p^3(-p^2 + 2p - 3)$ ◄

Try Problem 43 \\

We could have used $-5p^3$ as the GCF in Example 8(c). This gives the factorization

$$-5p^3(p^2 - 2p + 3).$$

Both factorizations are correct.

EXAMPLE 9   Factor out the GCF.

a) $8r^4s - 32r^3s^2 + 12r^2s^2$

$= 4r^2s \cdot 2r^2 - 4r^2s \cdot 8rs + 4r^2s \cdot 3s$ GCF is $4r^2s$

$= 4r^2s(2r^2 - 8rs + 3s)$

b) $3x(m + n) - 7y(m + n)$ GCF is $m + n$

$= (m + n)(3x - 7y)$

c) $z^{n+5} + z^{n+3} = z^{n+3} \cdot z^2 + z^{n+3} \cdot 1$ GCF is z^{n+3}

Try Problem 53 \\

$= z^{n+3}(z^2 + 1)$ ◄

DIFFERENCE OF TWO SQUARES

In Section 2.5 we observed that

$$(a - b)(a + b) = a^2 - b^2.$$

We simply turn this equation around to get our next factoring formula.

Difference of Two Squares

$$a^2 - b^2 = (a - b)(a + b)$$

Since multiplication is commutative, we can write $a^2 - b^2$ as $(a - b)(a + b)$ or as $(a + b)(a - b)$.

EXAMPLE 10 Factor each difference of two squares.

a) $x^2 - 25 = x^2 - 5^2 = (x - 5)(x + 5)$

$$a^2 - b^2 = (a - b)(a + b)$$

Try Problem 59

b) $16m^2 - 9n^2 = (4m)^2 - (3n)^2 = (4m - 3n)(4m + 3n)$ ◄

The polynomial $x^2 + 4$ is not a difference of two squares but rather a *sum* of two squares. It cannot be factored using real numbers. You can verify that plausible factorizations such as $(x + 2)(x + 2)$, $(x - 2)(x - 2)$, and $(x + 2)(x - 2)$ do not work. Therefore $x^2 + 4$ is a prime polynomial.

EXAMPLE 11 Factor $(v + 7)^2 - 25$.

Use the difference-of-two-squares formula with $a = v + 7$ and $b = 5$.

$$(v + 7)^2 - 25 = (v + 7)^2 - 5^2 = [(v + 7) - 5][(v + 7) + 5]$$

$$a^2 - b^2 = [a - b][a + b]$$

and

Try Problem 69

$$[(v + 7) - 5][(v + 7) + 5] = (v + 2)(v + 12) ◄$$

CAUTION When factoring any polynomial, always look for the GCF before attempting any other type of factoring.

EXAMPLE 12 Factor $7x^2 - 28$.

First factor out the GCF 7.

$$7x^2 - 28 = 7(x^2 - 4)$$

Then factor the difference of two squares $x^2 - 4$.

$$= 7(x - 2)(x + 2) ◄$$

Try Problem 75

Since 7, $x - 2$, and $x + 2$ are all prime polynomials, the answer to Example 12 is said to be in **prime factored form.**

EXAMPLE 13 Write $r^3 - 100r$ in prime factored form.

$$r^3 - 100r = r(r^2 - 100) \qquad \text{Factor out the GCF } r$$

Try Problem 77

$$= r(r - 10)(r + 10) \qquad \text{Difference of two squares} ◄$$

EXAMPLE 14 Write $36p^5q - 36pq$ in prime factored form.

First factor out the GCF $36pq$.

$$36p^5q - 36pq = 36pq(p^4 - 1)$$

Then factor $p^4 - 1$ as the difference of two squares, since $p^4 - 1 = (p^2)^2 - 1^2$.

$$= 36pq(p^2 - 1)(p^2 + 1)$$

Finally, factor the difference of two squares $p^2 - 1$. Remember that the sum of two squares $p^2 + 1$ is a prime polynomial.

Try Problem 91

$$= 36pq(p - 1)(p + 1)(p^2 + 1) \triangleleft$$

PROBLEM SET 2.6

Write each number in prime factored form.

1. 21   **2.** 35   **3.** 12   **4.** 18

5. 90   **6.** 60   **7.** 504   **8.** 540

Find the GCF of each collection of numbers.

9. 4, 6   **10.** 6, 9   **11.** 9, 18   **12.** 4, 12

13. 16, 12   **14.** 18, 12   **15.** 45, 30, 60   **16.** 24, 36, 48

Find the GCF of each collection of terms.

17. $6x^3, 9x^2$ **18.** $8x^3, 20x^4$ **19.** $9a, 9b$ **20.** $11a, 11b$

21. $6x, 12$ **22.** $4x, 8$ **23.** $7x, 2$ **24.** $5x, 4$

25. $10m^5n^3, 25m^3n^4$ **26.** $15m^4n^2, 25m^5n^3$

Factor out the GCF from each polynomial.

27. $6x + 18$ **28.** $4x + 20$

29. $13a + 13b$ **30.** $11a + 11b$

31. $28m - 35n$ **32.** $35m - 45n$

33. $4y - 9$ **34.** $6y - 25$

35. $5x^2 + 5$ **36.** $8x^2 + 8$

37. $8x^3 + 12x^2$ **38.** $9x^3 + 12x^2$

39. $a^5 + a$ **40.** $a^6 + a$

41. $16a^2 - 12ab$ **42.** $24a^2 - 18ab$

43. $-4p^5 - 12p^3 + 16p^2$ **44.** $-3p^7 + 12p^5 - 9p^3$

45. $45r^3s^4 - 30r^2s^3 - 60rs$ **46.** $24r^6s^3 - 36r^4s^4 - 48rs$

47. $99z^{55} + 11z^{53} - 33z^{51}$ **48.** $100z^{75} - 10z^{50} + 40z^{25}$

49. $-10r^3s^2t^4 - 25r^2s^2t^2$ **50.** $-15r^2s^2t^2 - 25r^4s^2t^3$

51. $\pi a^2 bc + \pi a b^2 c - \pi a b c^2$

52. $\pi x^2 y^2 z - \pi x y^2 z - \pi x y^2 z^2$

53. $5x(3m - 2n) + y(3m - 2n)$

54. $7x(4m + n) - y(4m + n)$

55. $x^2(x^2 + 9) + 9(x^2 + 9)$

56. $x^2(x^2 + 4) + 4(x^2 + 4)$

57. $z^{n+3} + z^{n+1}$

58. $z^{n+4} + z^{n+2}$

Factor each difference of two squares.

59. $x^2 - 9$

60. $x^2 - 16$

61. $4x^2 - 9$

62. $9x^2 - 16$

63. $25m^2 - 16n^2$

64. $25m^2 - 49n^2$

65. $x^2 + 9$

66. $x^2 + 16$

67. $36r^2 s^2 - 49$

68. $100r^2 s^2 - 81$

69. $(v + 2)^2 - 9$

70. $(v - 2)^2 - 4$

71. $a^2 - (b - c)^2$

72. $a^2 - (b + c)^2$

73. $(2x + 5)^2 - (3x - 1)^2$

74. $(4x - 3)^2 - (2x + 1)^2$

Write in prime factored form.

75. $5x^2 - 20$

76. $6x^2 - 54$

77. $r^3 - 25r$

78. $r^3 - 49r$

79. $t^4 - 16$

80. $t^4 - 81$

81. $10x^4 - 10y^4$

82. $15x^4 - 15y^4$

83. $x^4 y^4 - 256$

84. $x^4 y^4 - 625$

85. $77rs - 77r^3 s^3$

86. $66rs - 66r^3 s^3$

87. $m^6 - 4$

88. $m^6 - 9$

89. $1 - 81x^4$

90. $1 - 16x^4$

91. $25p^5 q^2 - 25pq^2$

92. $4p^6 q - 64p^2 q$

93. Calculate $85^2 - 15^2$ by first factoring this expression.

94. Calculate $75^2 - 25^2$ by first factoring this expression.

95. The area of the ring-shaped region shown below is given by $\pi R^2 - \pi r^2$. Write this expression in prime factored form.

96. The volume of the cylindrical shell shown below is given by $\pi R^2 h - \pi r^2 h$. Write this expression in prime factored form.

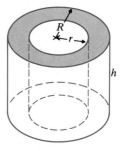

CALCULATOR PROBLEMS

Write in prime factored form.

97. $283t^5 - 1981t^3 + 5377t$

98. $7921m^2 - 9409n^2$

2.7   *Factoring Trinomials*

In Section 2.5 we used the FOIL method to multiply two binomials. For example,

$$F \quad O \quad I \quad L$$
$$(x + 2)(x + 3) = x^2 + 3x + 2x + 6$$
$$= x^2 + 5x + 6.$$

In this section we learn to reverse the procedure. That is, given the trinomial $x^2 + 5x + 6$ we learn to write it as the product of the two binomials $x + 2$ and $x + 3$.

$$\text{Factor}$$
$$x^2 + 5x + 6 = (x + 2)(x + 3)$$
$$\text{Multiply}$$

EXAMPLE 1 Factor $x^2 + 5x + 6$.

We begin by writing

$$\text{Factors of 6}$$
$$x^2 + 5x + 6 = (x + \underline{\ \ })(x + \underline{\ \ }).$$

Next we list all pairs of integers whose product is 6. Beside this list we write the sum of each pair.

Factors of 6	Sum of factors
$1 \cdot 6$	$1 + 6 = 7$
$2 \cdot 3$	$2 + 3 = 5$
$(-1)(-6)$	$(-1) + (-6) = -7$
$(-2)(-3)$	$(-2) + (-3) = -5$

Then we fill in the blanks with 2 and 3, since $2 \cdot 3 = 6$ (the last term) and $2 + 3 = 5$ (the middle coefficient).

Try Problem 1

$$x^2 + 5x + 6 = (x + 2)(x + 3) \blacktriangleleft$$

To Factor $x^2 + bx + c$

1. Write $x^2 + bx + c = (x + \underline{\ \ })(x + \underline{\ \ })$.
2. List all pairs of integers whose product is c.
3. Fill in the blanks with the pair whose sum is b.

EXAMPLE 2   Factor $x^2 - 16x + 15$.

$$x^2 - 16 + 15 = (x + \underline{})(x + \underline{})$$

Factors of 15	Sum of factors
$1 \cdot 15$	16
$3 \cdot 5$	8
$(-1)(-15)$	-16
$(-3)(-5)$	-8

Fill in the blanks with -1 and -15, since their sum is -16, the middle coefficient.

$$x^2 - 16x + 15 = [x + (-1)][x + (-15)]$$
$$= (x - 1)(x - 15)$$

You can check this factorization by computing the product on the right.

CHECK:

Try Problem 3

$$(x - 1)(x - 15) = x^2 - 16x + 15 \blacktriangleleft$$

$x^2 \quad 15$

$-x$

$-15x$

EXAMPLE 3   Factor $y^2 + 2y - 15$.

$$y^2 + 2y - 15 = (y + \underline{})(y + \underline{})$$

Factors of -15	Sum of factors
$1(-15)$	-14
$3(-5)$	-2
$(-1)15$	14
$(-3)5$	2

Try Problem 7   Therefore $y^2 + 2y - 15 = (y - 3)(y + 5)$. $\blacktriangleleft$

EXAMPLE 4   Factor $p^2 - 3p + 4$.

$$p^2 - 3p + 4 = (p + \underline{})(p + \underline{})$$

Factors of 4	Sum of factors
$1 \cdot 4$	5
$2 \cdot 2$	4
$(-1)(-4)$	-5
$(-2)(-2)$	-4

Try Problem 13   Since no pair of factors has a sum of -3, we conclude that $p^2 - 3p + 4$ is a prime polynomial. $\blacktriangleleft$

This technique extends to polynomials of more than one variable.

EXAMPLE 5 Factor $x^2 - 4xy - 12y^2$.

$$x^2 - 4xy - 12y^2 = (x + \underline{\quad}y)(x + \underline{\quad}y)$$

Factors of -12	Sum of factors
$1(-12)$	-11
$2(-6)$	-4
$3(-4)$	-1
$(-1)12$	11
$(-2)6$	4
$(-3)4$	1

Try Problem 23

Therefore $x^2 - 4xy - 12y^2 = (x + 2y)(x - 6y)$. ◄

To factor $ax^2 + bx + c$ when $a \neq 1$ requires some trial and error.

EXAMPLE 6 Factor $2x^2 + 9x + 7$.

First write

Positive factors of 2

$$2x^2 + 9x + 7 = (\underline{\quad}x + \underline{\quad})(\underline{\quad}x + \underline{\quad}).$$

Factors of 7

Next list all pairs of integers whose product is 2, and all pairs whose product is 7. We prefer that the first coefficient in each term be positive, so list only positive factors of 2.

Factors of 2 (positive only)	Factors of 7
$1 \cdot 2$	$1 \cdot 7$
	$(-1)(-7)$

Then fill in the first blanks in each binomial with the factors of 2, and the second blanks with the factors of 7. *You must consider all possible combinations.*

$(x + 1)(2x + 7)$	Right, since $7x + 2x = 9x$
$(x + 7)(2x + 1)$	Wrong, since $x + 14x = 15x$
$(x - 1)(2x - 7)$	Wrong, since $-7x - 2x = -9x$
$(x - 7)(2x - 1)$	Wrong, since $-x - 14x = -15x$

Try Problem 29

Therefore $2x^2 + 9x + 7 = (x + 1)(2x + 7)$. ◄

Note that $2x^2 + 9x + 7 = (-x - 1)(-2x - 7)$ is also a correct factorization, but as we stated previously, we prefer that the first coefficient in each term be positive. With this agreement, there is only one way to factor this trinomial.

To Factor $ax^2 + bx + c$ $(a \neq 1)$

1. Write $ax^2 + bx + c = (\underline{}x + \underline{})(\underline{}x + \underline{})$.
2. List all pairs of integers whose product is a. Use only positive integers. If a is negative, begin by factoring out -1.
3. List all pairs of integers whose product is c.
4. Fill in the first blanks in each binomial with the factors of a, and the second blanks with the factors of c. Consider all possible combinations.
5. Use trial and error to determine the combination in step 4 that produces the correct middle term.

EXAMPLE 7   Factor $6x^2 - 7x - 5$.

$$6x^2 - 7x - 5 = (\underline{}x + \underline{})(\underline{}x + \underline{})$$

Factors of 6	Factors of -5
$1 \cdot 6$	$1(-5)$
$2 \cdot 3$	$(-1)5$

Write all possible combinations.

$(x + 1)(6x - 5)$	Wrong, since $-5x + 6x = x$
$(x - 5)(6x + 1)$	Wrong, since $x - 30x = -29x$
$(x - 1)(6x + 5)$	Wrong, since $5x - 6x = -x$
$(x + 5)(6x - 1)$	Wrong, since $-x + 30x = 29x$
$\mathbf{(2x + 1)(3x - 5)}$	Right, since $-10x + 3x = -7x$
$(2x - 5)(3x + 1)$	Wrong, since $2x - 15x = -13x$
$(2x - 1)(3x + 5)$	Wrong, since $10x - 3x = 7x$
$(2x + 5)(3x - 1)$	Wrong, since $-2x + 15x = 13x$

Try Problem 39   Therefore $6x^2 - 7x - 5 = (2x + 1)(3x - 5)$. ◄

As you become more familiar with the procedure for factoring trinomials, you will find that you can accomplish much of the trial-and-error process mentally. One important rule that will help is this: *If the last term is positive, then both of its factors must have the same sign. Consider only those pairs that have the same sign as the middle term.*

EXAMPLE 8   Factor $4y^2 - 12y + 9$.

$$4y^2 - 12y + 9 = (\underline{}y + \underline{})(\underline{}y + \underline{})$$

Factors of 4	Factors of 9
$1 \cdot 4$	$1 \cdot 9$
$2 \cdot 2$	$3 \cdot 3$
	$(-1)(-9)$
	$(-3)(-3)$

Since the last term is positive, its factors must have the same sign. In addition, the middle term is negative, so consider only the negative factors of 9.

$(y - 1)(4y - 9)$	Wrong, since $-9y - 4y = -13y$
$(y - 9)(4y - 1)$	Wrong, since $-y - 36y = -37y$
$(y - 3)(4y - 3)$	Wrong, since $-3y - 12y = -15y$
$(2y - 1)(2y - 9)$	Wrong, since $-18y - 2y = -20y$
$\mathbf{(2y - 3)(2y - 3)}$	Right, since $-6y - 6y = -12y$

Try Problem 67

Therefore $4y^2 - 12y + 9 = (2y - 3)(2y - 3) = (2y - 3)^2$. ◄

If the coefficient of the middle term in Example 8 had been positive, we would have considered only the positive factors of 9. This would give the factorization

$$4y^2 + 12y + 9 = (2y + 3)(2y + 3) = (2y + 3)^2.$$

Each of the trinomials $4y^2 - 12y + 9$ and $4y^2 + 12y + 9$ is called a **perfect-square trinomial,** because each can be expressed as the square of a binomial. If we reverse the special products given in Section 2.5 for the square of a binomial, we obtain the factoring formulas below.

Perfect-Square Trinomials

$$a^2 + 2ab + b^2 = (a + b)^2$$
$$a^2 - 2ab + b^2 = (a - b)^2$$

We can factor the trinomial of Example 8 using the second perfect-square trinomial formula as follows:

$$4y^2 - 12y + 9 = (2y)^2 - 2(2y)(3) + 3^2 = (2y - 3)^2.$$

$$a^2 - 2ab + b^2 = (a - b)^2$$

We recognize a perfect-square trinomial by the fact that its first and last terms are perfect squares, and its middle term is twice the product of the square roots of the first and last terms.

EXAMPLE 9   Factor $25 + 10r + r^2$.

First use the commutative law of addition to write the trinomial in descending powers.

$$25 + 10r + r^2 = r^2 + 10r + 25$$

Since the first and last terms are perfect squares, and since

$$2 \cdot r \cdot 5 = 10r \quad \text{(the middle term)},$$

this is a perfect-square trinomial. The middle term is positive, so the answer must take the form $(a + b)^2$. Therefore $r^2 + 10r + 25 = (r + 5)^2$. ◄

Try Problem 69

Remember that we always factor out the GCF before attempting other types of factoring.

EXAMPLE 10   Write in prime factored form.

$$144m^3n^2 - 102m^2n^3 - 120mn^4$$
$$= 6mn^2(24m^2 - 17mn - 20n^2) \quad \text{Factor out the GCF } 6mn^2$$
$$= 6mn^2(3m - 4n)(8m + 5n) \quad \text{Factor the trinomial} ◄$$

Try Problem 83

Sometimes we can convert a complicated factorization problem into a simpler one using the **method of substitution.**

EXAMPLE 11   Write $12t^4 - 17t^2 - 5$ in prime factored form.

Make the substitution $u = t^2$. Then $u^2 = (t^2)^2 = t^4$, and the given trinomial becomes

$$12u^2 - 17u - 5.$$

Then factor the trinomial in u.

$$12u^2 - 17u - 5 = (3u - 5)(4u + 1)$$

Substituting $u^2 = t^4$ and $u = t^2$, we have

$$12t^4 - 17t^2 - 5 = (3t^2 - 5)(4t^2 + 1). ◄$$

Try Problem 87

PROBLEM SET 2.7

Factor each trinomial.

1. $x^2 + 8x + 15$ **2.** $x^2 + 7x + 10$ **3.** $x^2 - 7x + 6$ **4.** $x^2 - 10x + 9$

5. $y^2 + 2y - 3$ **6.** $y^2 + 4y - 5$ **7.** $y^2 + 3y - 10$ **8.** $y^2 + 4y - 21$

9. $r^2 - 3r - 18$ **10.** $r^2 - 2r - 24$ **11.** $r^2 + 7r - 18$ **12.** $r^2 + 5r - 24$

13. $p^2 + 4p - 6$ **14.** $p^2 - 8p - 10$ **15.** $35 - 12p + p^2$ **16.** $12 - 8p + p^2$

17. $10t + t^2 - 11$ **18.** $12t + t^2 - 13$ **19.** $16 - 6t - t^2$ **20.** $16 + 6t - t^2$

21. $x^2 + 4xy + 3y^2$ **22.** $x^2 + 6xy + 5y^2$ **23.** $x^2 - xy - 12y^2$ **24.** $x^2 + 11xy - 12y^2$

25. $r^2 + 4rs - 21s^2$ **26.** $r^2 - 2rs - 35s^2$ **27.** $p^2 - 9pq + 14q^2$ **28.** $p^2 - 9pq + 20q^2$

Factor each trinomial.

29. $2x^2 + 7x + 5$ **30.** $2x^2 + 7x + 3$ **31.** $3x^2 - 5x + 2$ **32.** $3x^2 - 8x + 5$

33. $2y^2 + 5y - 3$ **34.** $2y^2 - 5y - 3$ **35.** $11y^2 + 6y - 17$ **36.** $13y^2 + 2y - 11$

37. $3m^2 + 13m - 10$ **38.** $3m^2 - 13m - 10$ **39.** $6x^2 - 7x - 3$ **40.** $6x^2 - 13x - 5$

41. $15r^2 - 14r - 8$ **42.** $12r^2 - 17r - 5$ **43.** $4t^2 - 5t - 12$ **44.** $6t^2 + 5t - 10$

45. $9s^2 - 34s - 8$ **46.** $9s^2 - 14s - 8$ **47.** $6 + 17x + 5x^2$ **48.** $12 + 11x + 2x^2$

49. $12 - 2x^2 - 5x$ **50.** $15 - 2x^2 - 7x$ **51.** $18p^2 - 19p - 12$ **52.** $8p^2 - 27p - 20$

53. $2x^2 - 3xy + y^2$ **54.** $3x^2 + 4xy + y^2$ **55.** $3x^2 + 5xy + 2y^2$ **56.** $2x^2 - 5xy + 3y^2$

57. $3m^2 - mn - 10n^2$ **58.** $4m^2 - 11mn - 3n^2$ **59.** $36p^2 - 13pq - 40q^2$ **60.** $36p^2 + 5pq - 24q^2$

Factor each perfect-square trinomial.

61. $x^2 + 6x + 9$ **62.** $x^2 + 4x + 4$ **63.** $z^2 - 2z + 1$ **64.** $z^2 - 8z + 16$

65. $4t^2 + 4t + 1$ **66.** $9t^2 + 6t + 1$ **67.** $4y^2 - 20y + 25$ **68.** $9y^2 - 12y + 4$

69. $36 + 12r + r^2$ **70.** $49 + 14r + r^2$ **71.** $m^2 + 8mn + 16n^2$ **72.** $m^2 + 2mn + n^2$

73. $9r^2 - 24rs + 16s^2$ **74.** $4r^2 - 20rs + 25s^2$ **75.** $4s^2t^2 + 12st + 9$ **76.** $81s^2t^2 + 18st + 1$

Write in prime factored form.

77. $2x^2 + 6x - 20$ **78.** $3x^2 - 6x - 24$

79. $3x^2 + 3x + 3$ **80.** $7x^2 - 7x + 7$

81. $10x^2y + 140xy + 130y$ **82.** $15x^2y + 180xy + 165y$

83. $18m^3n^2 - 24m^2n^3 - 42mn^4$ **84.** $12m^3n^2 - 42m^2n^3 - 54mn^4$

85. $z^4 + 5z^2 + 6$ **86.** $z^4 + 7z^2 + 12$

87. $6t^4 + 7t^2 - 3$ **88.** $6t^4 + 5t^2 - 25$

89. $w^5 - 3w^3 - 4w$ **90.** $w^5 - 8w^3 - 9w$

91. $4r^4 - 33r^2s^2 - 27s^4$ **92.** $6r^4 + 25r^2s^2 - 25s^4$

93. Factor $(3x - 5)^2 - 8(3x - 5) + 12$ **94.** Factor $(2x + 4)^2 - 7(2x + 4) + 12$
  a) by first multiplying and simplifying.   **a)** by first multiplying and simplifying.
  b) by first making the substitution $u = 3x - 5$. **b)** by first making the substitution $u = 2x + 4$.

CALCULATOR PROBLEMS

95. The product of two binomials is $8633x^2 - 342x - 14{,}351$. If one of the binomials is $97x - 127$, find the other.

96. By direct substitution, verify that $2x^2 - 5x - 3 = (2x + 1)(x - 3)$ for
  a) $x = 0.198$.   **b)** $x = 677$.

| 2.8 | *Sum and Difference of Cubes, Grouping, Factoring Strategy* |

In Section 2.6 we factored the difference of two squares using the formula

$$a^2 - b^2 = (a - b)(a + b).$$

We also noted that the sum of two squares, $a^2 + b^2$, was a prime polynomial.

SUM AND DIFFERENCE OF CUBES

We can factor both a **sum of two cubes** and a **difference of two cubes** using the formulas below.

Sum of two cubes: $a^3 + b^3 = (a + b)(a^2 - ab + b^2)$

Difference of two cubes: $a^3 - b^3 = (a - b)(a^2 + ab + b^2)$

You can verify these formulas by computing the products on the right side.

Notice that the first factor on the right side of each formula is the same as the original expression, except that a and b each have an exponent of 1 rather than 3.

$$a^3 + b^3 = (a + b)(\qquad)$$

↑
Same as original expression
except a and b each have an
exponent of 1
↓

$$a^3 - b^3 = (a - b)(\qquad)$$

The second factor in both formulas can be obtained from the first factor as follows:

1. Square the first term of the first factor to get the first term of the second factor.
2. Find the product of the two terms in the first factor and change the sign to get the second term of the second factor.
3. Square the last term of the first factor to get the last term of the second factor.

CAUTION Do not confuse the second factors with the perfect-square trinomials $a^2 - 2ab + b^2$ and $a^2 + 2ab + b^2$. The expressions $a^2 - ab + b^2$ and $a^2 + ab + b^2$ are prime polynomials.

EXAMPLE 1 Factor $x^3 - 8$.

Use the difference-of-two-cubes formula with $a = x$ and $b = 2$.

$$x^3 - 8 = x^3 - 2^3$$
$$= (x - 2)(x^2 + 2x + 4)$$

1. Square x to get x^2
2. Product of x and -2 with sign change
3. Square -2 to get 4 ◄

Try Problem 1

The second-degree trinomial $x^2 + 2x + 4$ is a prime polynomial. Therefore the answer to Example 1 is in prime factored form.

EXAMPLE 2 Factor $64x^3 + y^3$.

Use the sum-of-two-cubes formula with $a = 4x$ and $b = y$.

$$64x^3 + y^3 = (4x)^3 + y^3$$
$$= (4x + y)(16x^2 - 4xy + y^2)$$

1. Square $4x$ to get $16x^2$
2. Product of $4x$ and y with sign change
3. Square y to get y^2 ◄

Try Problem 9

EXAMPLE 3 Factor each difference of two cubes.

$$a^3 - b^3 = (a - b)(a^2 + ab + b^2)$$

a) $x^6 - 125y^3 = (x^2)^3 - (5y)^3 = (x^2 - 5y)(x^4 + 5x^2y + 25y^2)$

$$a^3 - b^3$$

b) $(m + 5)^3 - 27 = (m + 5)^3 - 3^3$

$$(a - b)(a^2 + ab + b^2)$$

$$= [(m + 5) - 3][(m + 5)^2 + 3(m + 5) + 9]$$
$$= (m + 2)(m^2 + 10m + 25 + 3m + 15 + 9)$$
$$= (m + 2)(m^2 + 13m + 49) ◄$$

Try Problem 17

GROUPING

Sometimes a polynomial can be factored by **grouping.**

EXAMPLE 4   Factor $x^3 + 5x^2 + 3x + 15$.

Note that there is no common factor. However, we can obtain a common factor. First, group the terms.

$$x^3 + 5x^2 + 3x + 15 = (x^3 + 5x^2) + (3x + 15)$$

Then factor x^2 out of the first group, and factor 3 out of the second group.

$$= x^2(x + 5) + 3(x + 5)$$

This procedure exposes the common binomial factor $x + 5$, which we factor out.

Try Problem 25

$$= (x + 5)(x^2 + 3) \blacktriangleleft$$

EXAMPLE 5   Factor $rs - 7r + 2s - 14$.

$$\begin{aligned} rs - 7r + 2s - 14 &= (rs - 7r) + (2s - 14) &&\text{Group the terms} \\ &= r(s - 7) + 2(s - 7) &&\text{Factor out } r, \text{ factor out 2} \\ &= (s - 7)(r + 2) &&\text{Factor out } s - 7 \blacktriangleleft \end{aligned}$$

Try Problem 29

EXAMPLE 6   Factor $a^2 + 10a + 25 - b^2$.

To factor this polynomial, group the first three terms together.

$$a^2 + 10a + 25 - b^2 = (a^2 + 10a + 25) - b^2$$

Then factor the perfect-square trinomial in parentheses.

$$= (a + 5)^2 - b^2$$

Finally, factor this last expression as a difference of two squares.

$$= [(a + 5) - b][(a + 5) + b]$$
$$= (a + 5 - b)(a + 5 + b) \blacktriangleleft$$

Try Problem 37

EXAMPLE 7   Factor $x^2y + x^2 - 4y - 4$.

Group. Note the sign change

$$\begin{aligned} x^2y + x^2 - 4y - 4 &= (x^2y + x^2) - (4y + 4) \\ &= x^2(y + 1) - 4(y + 1) &&\text{Factor out } x^2, \text{ factor out 4} \\ &= (y + 1)(x^2 - 4) &&\text{Factor out } y + 1 \\ &= (y + 1)(x - 2)(x + 2) &&\text{Factor } x^2 - 4 \blacktriangleleft \end{aligned}$$

Try Problem 45

Remember that we always factor out the GCF before attempting other types of factoring.

EXAMPLE 8   Write in prime factored form.

a) $24r^3s - 81s$

$\quad\quad = 3s(8r^3 - 27)$ $\qquad\qquad\qquad$ Factor out $3s$

$\quad\quad = 3s((2r)^3 - 3^3)$

$\quad\quad = 3s(2r - 3)(4r^2 + 6r + 9)$ $\qquad$ Factor the difference of two cubes

b) $10v^6 + 6v^3 - 4$

$\quad\quad = 2(5v^6 + 3v^3 - 2)$ $\qquad\qquad$ Factor out 2

$\quad\quad = 2(5v^3 - 2)(v^3 + 1)$ $\qquad\qquad$ Factor the trinomial

$\quad\quad = 2(5v^3 - 2)(v + 1)(v^2 - v + 1)$ $\quad$ Factor the sum of two cubes ◄

Try Problem 59

FACTORING STRATEGY

We now summarize the various factoring techniques and the strategy involved in their use.

Factoring Strategy

1. First factor out the GCF (unless it is 1). You may want to factor out -1.
2. To factor a binomial, use

$\quad\quad a^2 - b^2 = (a - b)(a + b),$ $\qquad$ Difference of two squares

$\quad\quad a^3 + b^3 = (a + b)(a^2 - ab + b^2),$ $\qquad$ Sum of two cubes

$\quad\quad a^3 - b^3 = (a - b)(a^2 + ab + b^2).$ $\qquad$ Difference of two cubes

Note that $a^2 + b^2$ is a prime polynomial.

3. To factor a perfect-square trinomial, use

$$a^2 + 2ab + b^2 = (a + b)^2,$$
$$a^2 - 2ab + b^2 = (a - b)^2.$$

To factor other trinomials, use the methods of Section 2.7.

4. To factor a polynomial with four terms, try grouping.
5. Continue to factor each result until you reach prime factored form. There is only one prime factored form.

To be successful at factoring, you must be able to recognize which factoring technique to use in a given situation. The factoring problems in the chapter test are scrambled to allow you to test your skill at recognizing the different factoring types.

PROBLEM SET 2.8

Factor each sum or difference of two cubes.

1. $x^3 - 27$
2. $x^3 - 64$
3. $y^3 + 8$
4. $y^3 + 125$
5. $z^3 + 1$
6. $z^3 - 1$
7. $8 - r^3$
8. $27 - r^3$
9. $125x^3 + y^3$
10. $8x^3 + y^3$
11. $27s^3 - 125t^3$
12. $125s^3 + 8t^3$
13. $p^3q^3 + 1$
14. $p^3q^3 - 1$
15. $343u^3 + 216v^3$
16. $125u^3 - 216v^3$
17. $x^6 - 64y^3$
18. $x^6 + 27y^3$
19. $x^9 + 8$
20. $x^9 - 27$
21. $(m - 4)^3 - 8$
22. $(m - 4)^3 + 8$
23. $(p + q)^3 + (p - q)^3$
24. $(p - q)^3 - (p + q)^3$

Factor each polynomial by grouping.

25. $x^3 + 7x^2 + 3x + 21$
26. $x^3 + 4x^2 + 5x + 20$
27. $ab + 13a + 2b + 26$
28. $ab + 11a + 3b + 33$
29. $rs - r + 3s - 3$
30. $rs - r + 6s - 6$
31. $5x^2 + 5xy + 2x + 2y$
32. $3x^2 + 3xy + 7x + 7y$
33. $uv - v - u + 1$
34. $uv + v - u - 1$
35. $s^2 + 8st^2 + 6st + 48t^3$
36. $s^3 + 9s^2t + 7st + 63t^2$
37. $a^2 + 6a + 9 - b^2$
38. $a^2 + 4a + 4 - b^2$
39. $x^2 + 2x + 1 - 4y^2$
40. $x^2 + 10x + 25 - 9y^2$
41. $a^2 - b^2 - 8b - 16$
42. $a^2 - b^2 + 8b - 16$
43. $4r^2 - 20rs + 25s^2 - 9t^2$
44. $9r^2 - 12rs + 4s^2 - 25t^2$
45. $x^2y + x^2 - 9y - 9$
46. $x^2y - x^2 - y + 1$
47. $2y^3 + y^2 - 10y - 5$
48. $3y^3 + y^2 - 18y - 6$
49. $p^2q^2 - 16q^2 + 9p^2 - 144$
50. $p^2q^2 - 9q^2 + 4p^2 - 36$
51. $t^4 + 5t^3 - 8t - 40$
52. $t^4 - 2t^3 + 27t - 54$
53. $t^4 + s^3t + st^3 + s^4$
54. $t^4 - s^3t - st^3 + s^4$

Write in prime factored form.

55. $2x^3 + 16$
56. $3x^3 + 24$
57. $3000x^3 - 81$
58. $4000x^3 - 32$
59. $40r^3s + 135s$
60. $54r^3s^2 + 16s^2$
61. $12v^6 + 3v^3 - 9$
62. $15v^6 - 5v^3 - 10$
63. $27x^{12n} + 64y^{6m}$
64. $27x^{12n} - 64y^{6m}$
65. $64k^6 - 1$
66. $k^6 - 729$

67. Verify that $a^3 + b^3 = (a + b)(a^2 - ab + b^2)$ by computing the product on the right side.

68. Verify that $a^3 - b^3 = (a - b)(a^2 + ab + b^2)$ by computing the product on the right side.

CALCULATOR PROBLEMS

Factor each polynomial.

69. $2197y^3 - 4913$

70. $19rs + 1007r + 29s + 1537$

CHAPTER 2 REVIEW

[2.1]   *State the base and the exponent of each expression.*

1. 8^4

2. $(-6)^2$

3. -6^2

4. $(7x)^n$

Evaluate.

5. 5^4

6. $2 \cdot 3^3$

7. $(-4)^2$

8. -4^2

Write in exponential form.

9. $(x+2)(x+2)(x+2)$

10. $5 \cdot a \cdot a \cdot b - a \cdot b \cdot b \cdot b$

11. You deposit $5000 in an account that pays 8% interest. What is the value of the account after two years? Use $V = P(1+r)^t$.

[2.2]   *Apply the appropriate rule of exponents to simplify each expression.*

12. $x \cdot x^3 \cdot x^5$

13. $5^8 \cdot 5^4$

14. $\dfrac{t^{10}}{t^2}$

15. $(z^3)^3$

16. $\dfrac{a^3 b^5}{a^2 b}$

17. 10^0

18. $(2p)^5$

19. $\dfrac{4^{20}}{4^2}$

20. $(a^n)^2$

21. $7x^0$

22. $(-5ab)^2$

23. $(6y)(2y^3)$

Simplify.

24. $(-2a^4 b^5 c)^3 (3a^3 b c^2)^2$

25. $m^2 (5m^7)^2 (-2m)^4$

26. $\dfrac{(4x^3 y)^2}{x^2 y}$

27. $\dfrac{r^6}{10s^8} \left(\dfrac{4s^5}{r^6} \right)^2$

[2.3]   *State the coefficient and the degree of each monomial.*

28. $5x^4$

29. y^3

30. $-z$

31. 6

Write each polynomial in descending powers and state its degree. Identify it as a monomial, a binomial, or a trinomial, if appropriate.

32. $3 + 4x + 6x^2$

33. $-5 + 2x$

34. $p^2 + 7p^3 - p + 8$

Evaluate each polynomial at the given value.

35. $x^2 - 2x + 7$ at $x = -1$

36. $-y^2 + y - 9$ at $y = 3$

37. $4r^3 + 6r^2 - 3r + 1$ at $r = \frac{1}{2}$

[2.4]   *Write without parentheses.*

38. $3(4x - 5)$

39. $-(t^2 - 4t + 2)$

40. Add.

$$4x^2 - 6x + 9$$
$$2x^2 + 3x - 5$$

41. Subtract.

$$6y^3 - 2y^2 \quad\;\; + 5$$
$$3y^3 + 7y^2 - y - 5$$

Simplify.

42. $2(3p^3 - p + 8) - 6(2p^2 + 3p + 1)$

43. $8t - [4t - 3(2t + 7)]$

44. $(-2)(10a) - (-a)(4) + (3a)(6) - (-a)(-1)$

[2.5]   *Find each product and simplify.*

45. $3y^2(4y + 5)$

46. $(a + 2)(a + 6)$

47. $(4r - 5)(2r^3 + r^2 - 3r + 6)$

48. $(2x + 3y)^2$

49. $2m^3 - 8m^2 + 1$
$$\underline{\qquad\quad 3m\; - 5}$$

50. $a^2 - 2ab + 3b^2$
$$\underline{\qquad\qquad a + 4b}$$

51. Simplify $t[t^2 - (t + 4)(t - 4)]$.

[2.6]   *Write in prime factored form.*

52. $18r^5 - 36r^3 + 27r^2$

53. $3x(5a + b) - y(5a + b)$

54. $9t^2 - 25$

55. $4a^2 - 121b^2$

56. $(x - 1)^2 - 4$

57. $x^4 - 625$

58. $7p^4 - 7q^4$

59. $x^3y^3 - 16xy$

[2.7]   *Write in prime factored form.*

60. $z^2 - 2z - 8$

61. $r^2 + 3rs - 28s^2$

62. $16 - 8p + p^2$

63. $y^2 - 8y + 7$

64. $3t^2 + 14t - 5$

65. $9x^2 - 12xy + 4y^2$

66. $2m^2 + 5m - 12$

67. $5x^2 - 5x + 10$

68. $6x^2y^2 - 17xy^2 + 12y^2$

69. $t^4 + 5t^2 - 36$

[2.8]   *Write in prime factored form.*

70. $p^3 + 27$

71. $rs + r + 2s + 2$

72. $8a^3 + 125b^3$

73. $m^4 + 7m^3 - m - 7$

74. $x^2 - 2x + 1 - y^2$

75. $x^3 + 3x^2 - 4x - 12$

76. $2t^3 - 128$

77. $3z^6 + 21z^3 - 24$

CHAPTER 2 TEST

1. Evaluate.
   a) 3^4 b) $2 \cdot 5^2$ c) $(-8)^2$ d) -8^2

2. Simplify.
   a) $7^5 \cdot 7^9$ b) $(4x)(3x^2)$ c) $\dfrac{a^4b^3}{ab^2}$ d) 6^0

3. State the coefficient and the degree of each monomial.
   a) $2x$ b) $-y^3$

4. Add.
   $$5m^2 + 3m + 10$$
   $$\underline{m^2 - 4m - 6}$$

5. Subtract.
   $$8p^3 \qquad - p + 9$$
   $$\underline{2p^3 + 6p^2 - 7p - 9}$$

Simplify.

6. $(-3a^2b^5c)^3(2a^4b^3c^2)$

7. $\dfrac{r^2}{6s^5}\left(\dfrac{4s^3}{r^2}\right)^2$

8. Evaluate $t^3 + 3t^2 - 4t - 13$ at $t = -2$.

Simplify.

9. $3(5x^3 - x^2 + 4) - 4(x^2 + 2x - 1)$

10. $(-1)(8a) - (a)(-6) + (2a)(4) - (-a)(-3)$

Find each product and simplify.

11. $(p + 8)(p - 6)$

12. $(5x + y)^2$

13. $(3r - 4)(r^3 + 2r^2 - 5r + 7)$

Write in prime factored form.

14. $9x^2 - 49 = (3x+7)(3x-7)$

15. $y^2 - 4y - 12$

16. $r^2 + 6rs + 9s^2$

17. $4z^2 + 4z - 15$

18. $a^3 - 27$

19. $12p^5 + 18p^3 - 24p^2$

20. $rs + 5r + 2s + 10$

21. $t^4 - 16$

22. $6x^2y - 7xy - 10y$

23. $5m^3 + 40$

24. $x^2 - 8x + 16 - y^2$

25. You deposit $4000 in an account that pays 10% interest. What is the value of the account after two years? Use $V = P(1 + r)^t$.

In this chapter you will learn to solve equations more complicated than those presented in Section 1.6. You will also learn to solve inequalities. You will then apply equations and inequalities to solve many types of applied problems, from locating the center of an earthquake to finding the break-even point of a business. You will learn to work with formulas, and you will see how formulas can be used to answer many everyday questions.

The skills that you develop in this chapter are among the most important in algebra. You will use them again and again throughout the rest of the text.

3

First-Degree Equations
and Inequalities

3.1	*First-Degree Equations*

A **first-degree equation** is so named because the highest power of its variable is one. Hence the equations

$$3x + 5x + 1 = 17 \quad \text{and} \quad 4(y + 2) = y + 11$$

are first-degree equations.

Definition	A **first-degree equation,** or **linear equation,** in the variable x is an equation that can be expressed in the form
	$$ax + b = 0,$$
	where a $\neq 0$.

For example, the equation $3x + 5x + 1 = 17$ can be expressed in the form $ax + b = 0$ as follows:

$$8x + (-16) = 0.$$

We solve a first-degree equation using the four properties of equality discussed in Section 1.6 to isolate the variable on one side of the equation.

EXAMPLE 1   Solve $3x + 5x + 1 = 17$.

$$
\begin{aligned}
3x + 5x + 1 &= 17 \\
8x + 1 &= 17 & \text{Combine like terms} \\
8x + 1 - 1 &= 17 - 1 & \text{Subtract 1} \\
8x &= 16 \\
\frac{8x}{8} &= \frac{16}{8} & \text{Divide by 8} \\
x &= 2
\end{aligned}
$$

The solution is 2.

CHECK:

$$
\begin{aligned}
3 \cdot 2 + 5 \cdot 2 + 1 &\overset{?}{=} 17 \\
6 + 10 + 1 &\overset{?}{=} 17 \\
17 &\overset{\checkmark}{=} 17 \blacktriangleleft
\end{aligned}
$$

Try Problem 1

EXAMPLE 2   Solve $4(y + 2) = y + 11$.

First, remove parentheses by distributing 4.

$$
\begin{aligned}
4(y + 2) &= y + 11 \\
4y + 8 &= y + 11
\end{aligned}
$$

Then collect all variable terms on one side and all constant terms on the other.

$$4y + 8 - 8 = y + 11 - 8 \qquad \text{Subtract 8}$$
$$4y = y + 3$$
$$4y - y = y + 3 - y \qquad \text{Subtract } y$$
$$3y = 3$$
$$\frac{3y}{3} = \frac{3}{3} \qquad\qquad \text{Divide by 3}$$
$$y = 1$$

Try Problem 11

The number 1 checks, so the solution is 1. ◄

EXAMPLE 3   Solve $x - (3x - 1) = 3x + 21$.

$$x - (3x - 1) = 3x + 21$$
$$x - 3x + 1 = 3x + 21 \qquad \text{Remove parentheses by distributing } -1$$
$$-2x + 1 = 3x + 21 \qquad \text{Combine like terms}$$
$$-2x + 1 - 1 = 3x + 21 - 1 \qquad \text{Subtract 1}$$
$$-2x = 3x + 20$$
$$-2x - 3x = 3x + 20 - 3x \qquad \text{Subtract } 3x$$
$$-5x = 20$$
$$\frac{-5x}{-5} = \frac{20}{-5} \qquad\qquad \text{Divide by } -5$$
$$x = -4$$

Try Problem 19

The number -4 checks, so the solution is -4. ◄

To solve an equation with fractions, multiply each side by the least common denominator (LCD). This will clear the equation of fractions.

EXAMPLE 4   Solve $\dfrac{r}{3} - 2 = \dfrac{5r}{12} + \dfrac{3}{4}$.

$$12\left(\frac{r}{3} - 2\right) = 12\left(\frac{5r}{12} + \frac{3}{4}\right) \qquad \text{Multiply by the LCD 12}$$

$$12 \cdot \frac{r}{3} - 12 \cdot 2 = 12 \cdot \frac{5r}{12} + 12 \cdot \frac{3}{4} \qquad \text{Distribute 12 over every term}$$

$$4r - 24 = 5r + 9 \qquad \text{Find each product}$$
$$4r - 24 + 24 = 5r + 9 + 24 \qquad \text{Add 24}$$
$$4r = 5r + 33$$
$$4r - 5r = 5r + 33 - 5r \qquad \text{Subtract } 5r$$
$$-r = 33$$
$$\frac{-r}{-1} = \frac{33}{-1} \qquad\qquad \text{Divide by } -1$$
$$r = -33$$

Try Problem 37 The number -33 checks, so the solution is -33. ◄

CAUTION Be sure to distribute the LCD over every term in the equation, whether that term contains a fraction or not.

EXAMPLE 5 Solve $\dfrac{3m-2}{6} + 2 = 3 - \dfrac{m-1}{9}$.

$$18\left[\dfrac{3m-2}{6} + 2\right] = 18\left[3 - \dfrac{m-1}{9}\right] \qquad \text{Multiply by the LCD 18}$$

$$18 \cdot \dfrac{3m-2}{6} + 18 \cdot 2 = 18 \cdot 3 - 18 \cdot \dfrac{m-1}{9} \qquad \text{Distribute 18 over every term}$$

$$3(3m-2) + 36 = 54 - 2(m-1) \qquad \text{Careful! Write parentheses here}$$
$$9m - 6 + 36 = 54 - 2m + 2 \qquad \text{Remove parentheses}$$
$$9m + 30 = 56 - 2m \qquad \text{Combine like terms}$$
$$2m + 9m + 30 = 56 - 2m + 2m \qquad \text{Add } 2m$$
$$11m + 30 = 56$$
$$11m + 30 - 30 = 56 - 30 \qquad \text{Subtract 30}$$
$$11m = 26$$
$$\dfrac{11m}{11} = \dfrac{26}{11} \qquad \text{Divide by 11}$$
$$m = \dfrac{26}{11}$$

Try Problem 43 The number $\frac{26}{11}$ checks, so the solution is $\frac{26}{11}$. ◄

So far all the equations in this section have been conditional equations, as opposed to identities or contradictions. The next example shows how to recognize each type of equation during the solving process.

EXAMPLE 6 Classify as a conditional equation, an identity, or a contradiction.

a) $3(x+1) = 2x + 8$
$$3x + 3 = 2x + 8 \qquad \text{Remove parentheses}$$
$$3x = 2x + 5 \qquad \text{Subtract 3}$$
$$x = 5 \qquad \text{Subtract } 2x$$

Since $x = 5$ is true when x is 5 and false when x is any other number, the original equation is a conditional equation. The solution is 5.

b) $3(x+1) = 3x + 3$
$$3x + 3 = 3x + 3 \qquad \text{Remove parentheses}$$

Since $3x + 3 = 3x + 3$ is true for all values of x, the original equation is an identity. Any real number is a solution.

c) $3(x + 1) = 3(x - 1)$

$$3x + 3 = 3x - 3 \qquad \text{Remove parentheses}$$
$$3 = -3 \qquad \text{Subtract } 3x$$

Since $3 = -3$ is false for all values of x, the original equation is a contradiction. There is no solution. ◄

Try Problem 53

We can summarize the steps for solving a first-degree equation as follows:

To Solve a First-Degree Equation

1.   Remove grouping symbols and combine like terms.
2.   Clear fractions by multiplying each side by the LCD. Distribute the LCD over every term.
3.   Collect all variable terms on one side of the equation and all constant terms on the other side. Combine like terms in the process.
4.   Write the variable term with a coefficient of 1.
5.   Check your solution by substituting it into the original equation.
6.   If a contradiction results, then the equation has no solution. If an identity results, then every real number is a solution.

We now use a first-degree equation to solve an applied problem.

EXAMPLE 7   A furniture store has a standard markup of 60%. What was the store's cost c for a lamp that it has priced at $72?

Translate the given word equation into a mathematical equation.

Store's cost + 60% of store's cost = 72

$$c + 0.60c = 72$$

Then solve this equation.

$$1.6c = 72 \qquad \text{Since } 1c + 0.6c = 1.6c$$
$$\frac{1.6c}{1.6} = \frac{72}{1.6} \qquad \text{Divide by 1.6}$$
$$c = 45$$

Try Problem 69

The store's cost was $45. ◄

PROBLEM SET 3.1

Solve each equation.

1. $4x + 5x + 1 = 28$

2. $2x + 3x + 1 = 21$

3. $4(x - 2) + 5 = 9$

4. $3(x - 2) + 4 = 10$

5. $6z + 4 = 5z + 11$

6. $3z + 5 = 2z + 12$

7. $t + 8 = -t + 2$

8. $t + 9 = -t + 3$

9. $7m - 5 = 4m - 5$

10. $8m - 2 = 3m - 2$

11. $3(y + 2) = y + 8$

12. $5(y + 2) = y + 14$

13. $10 - (p - 1) = 4p + 21$

14. $5 - (p - 1) = 2p + 15$

15. $8 + 2(3q + 5) = 11 - 14$

16. $6 + 4(2q + 5) = 13 - 15$

17. $r - (2r + 1) = 4$

18. $r - (2r + 3) = 5$

19. $x - (4x - 1) = 4x + 22$

20. $x - (5x - 1) = 5x + 28$

21. $5(r - 2) - 4(3r - 5) = 3$

22. $3(r - 5) - 2(6r - 7) = -2$

23. $8 - 3(4s - 2) + 5s = 35$

24. $9 - 7(2s - 3) + 6s = 62$

25. $t - (6t - 10) = -8 - 2(4t - 9)$

26. $t - (10t - 3) = -7 - 5(3t - 2)$

27. $-(-2 + 9n) - (8 - 8n) + n = -(5 + 6n) + n - 1$

28. $-(-4 + 6n) - (7 - 7n) - n = -(4 + 9n) + n + 1$

29. $7 - 2\,[r - 5(r - 1)] = 1 + 3r - 6[9 - (1 - r)]$

30. $11 - 3[r - 3(r - 1)] = 3 + 5r - 6[7 - (1 - r)]$

Solve each equation.

31. $\dfrac{x}{2} + \dfrac{x}{4} = 6$

32. $\dfrac{x}{3} + \dfrac{x}{6} = 5$

33. $\dfrac{a}{5} - \dfrac{a}{15} = -4$

34. $\dfrac{a}{2} - \dfrac{a}{10} = -2$

35. $d - \dfrac{13}{3} = \dfrac{7d}{3} - \dfrac{9}{2}$

36. $d - \dfrac{27}{8} = \dfrac{7d}{4} - \dfrac{7}{2}$

37. $\dfrac{r}{3} - 1 = \dfrac{5r}{12} + \dfrac{3}{4}$

38. $\dfrac{r}{2} - 2 = \dfrac{7r}{12} + \dfrac{2}{3}$

39. $\dfrac{h + 1}{6} - \dfrac{h}{9} - \dfrac{1}{2} = 0$

40. $\dfrac{h + 1}{3} - \dfrac{h}{4} - \dfrac{1}{2} = 0$

41. $\dfrac{k + 6}{5} - \dfrac{13}{2} = \dfrac{7}{10} - \dfrac{k - 3}{4}$

42. $\dfrac{k + 7}{2} - \dfrac{1}{6} = \dfrac{2}{3} - \dfrac{k + 9}{9}$

43. $\dfrac{3m - 2}{9} + 1 = 2 - \dfrac{m - 1}{6}$

44. $\dfrac{2m - 3}{6} + 2 = 3 - \dfrac{m - 1}{9}$

45. $3.1y - 6 = 4.5y + 2.4$

46. $6.2y - 1.1 = 8.9y + 7$

47. $0.05q + 0.25(8 - q) = 1.2(4)$

48. $0.35q + 0.05(6 - q) = 2.4(2)$

Classify as a conditional equation, an identity, or a contradiction.

49. $5(x + 2) = 4x + 25$

50. $4(x - 2) = 3x + 4$

51. $5(x + 2) = 5x + 10$

52. $4(x - 2) = 4x - 8$

53. $5(x + 2) = 5(x + 3)$

54. $4(x - 2) = 4(x + 1)$

55. $3x - (5x - 3) = 2(4 - x) + 1$

56. $6x - (9x - 3) = 4 + 3(1 - x)$

57. $\dfrac{y + 1}{6} - \dfrac{y}{4} = \dfrac{1}{6} - \dfrac{y}{12}$

58. $\dfrac{y + 3}{9} - \dfrac{y}{6} = \dfrac{1}{3} - \dfrac{y}{18}$

Solve each equation.

59. $x(x + 4) - x(x + 2) = 14$

60. $x(x - 6) - x(x - 3) = 12$

61. $2u(3u - 4) + 13u = 6u^2 + 90$

62. $2u(2u - 3) + 3u = 4u^2 + 42$

63. $(w - 5)^2 - (w - 1)^2 = 4$

64. $(w + 5)^2 - (w + 8)^2 = 1$

65. $v - (v + 1)^2 = -v^2 + 4$

66. $v - (v + 1)^2 = -v^2 - 9$

67. Three times a number n is 7 more than five times the number. Find n.

68. Twice a number n is 5 more than six times the number. Find n.

69. A jewelry store has a standard markup of 40%. What was the store's cost c for a bracelet that it has priced at $49?

70. A vase is marked down 20% to $68. What was the original price p of the vase?

71. The grades on Amy's first three accounting tests were 85, 89, and 97. What score x must she get on the final exam, which is counted as two tests, to have a final average of 93?

72. The grades on Todd's first four biology tests were 79, 86, 90, and 82. What score x must he get on the final exam, which is counted as three tests, to have a final average of 85?

CALCULATOR PROBLEMS

Solve each equation to the nearest hundredth.

73. $6.189p - (2.718p + 13.488) = 0.199(5.861 - 0.765p)$

74. $q = \dfrac{q}{353} - \dfrac{839}{797}$

3.2   *Formulas*

We can use the equation-solving skills from Section 3.1 to work with formulas. A **formula** is a rule that relates two or more variable quantities. For example, the formula

$$A = lw$$

relates the area A of a rectangle with the length l and the width w (see Figure 3.1).

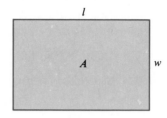

Figure 3.1

EXAMPLE 1   Given that $A = lw$, $A = 12$, and $w = 3$, find l.

Replace A by 12 and w by 3 in the formula $A = lw$.

$$12 = l \cdot 3$$

Then solve this equation for l.

$$\frac{12}{3} = \frac{l \cdot 3}{3} \qquad \text{Divide by 3}$$

$$4 = l$$

Try Problem 1

or $l = 4$ ◄

The value V of a principal P invested for one year at an interest rate r is given by the formula

$$V = P + Pr$$

EXAMPLE 2   Given that $V = P + Pr$, $V = 1650$, and $P = 1500$, find r.

$$1650 = 1500 + 1500r \qquad \text{Replace } V \text{ by 1650 and } P \text{ by 1500}$$

$$1650 - 1500 = 1500 + 1500r - 1500 \qquad \text{Subtract 1500}$$

$$150 = 1500r$$

$$\frac{150}{1500} = \frac{1500r}{1500} \qquad \text{Divide by 1500}$$

Try Problem 5

$$\frac{1}{10} = r \qquad \text{Reduce} ◄$$

Another commonly used formula is $d = rt$, which relates the quantities distance d, rate r, and time t.

EXAMPLE 3   At what rate must you travel to cover 230 miles in 5 hours?

Method I   Substitute $d = 230$ mi and $t = 5$ hr into the formula $d = rt$, and then solve for r.

$$d = rt$$

$$230 \text{ mi} = r(5 \text{ hr})$$

$$\frac{230 \text{ mi}}{5 \text{ hr}} = \frac{r(5 \text{ hr})}{5 \text{ hr}} \qquad \text{Divide by 5 hr}$$

$$46\frac{\text{mi}}{\text{hr}} = r$$

Method II   Solve the formula $d = rt$ for r, and then substitute.

$$d = rt$$

$$\frac{d}{t} = \frac{rt}{t} \qquad \text{Divide by } t$$

$$\frac{d}{t} = r$$

Substituting $d = 230$ mi and $t = 5$ hr into the formula $r = \frac{d}{t}$, we have

$$r = \frac{230 \text{ mi}}{5 \text{ hr}} = 46\frac{\text{mi}}{\text{hr}}.$$

Try Problem 11   Using either method, we see that the required rate is 46 mph. ◀

In Example 3 we substituted $d = 230$ mi and $t = 5$ hr, and the solution turned out with the correct units, namely mi/hr. Normally, however, we simply substitute $d = 230$ and $t = 5$, and then attach the correct units after the numerical part of the answer is obtained.

The perimeter P of a rectangle is given by the formula

$$P = 2w + 2l,$$

where w is the width and l is the length (see Figure 3.2).

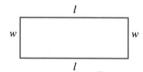

Figure 3.2

EXAMPLE 4   Solve $P = 2w + 2l$ for w.

$$P - 2l = 2w + 2l - 2l \qquad \text{Subtract } 2l$$

$$P - 2l = 2w$$

$$\frac{P - 2l}{2} = \frac{2w}{2} \qquad \text{Divide by 2}$$

Try Problem 13 $$\frac{P - 2l}{2} = w \blacktriangleleft$$

The formula

$$C = \tfrac{5}{9}(F - 32)$$

relates Celsius temperature C with Fahrenheit temperature F.*

EXAMPLE 5   Solve $C = \tfrac{5}{9}(F - 32)$ for F.

$$9 \cdot C = 9 \cdot \frac{5}{9}(F - 32) \qquad \text{Multiply by the LCD 9}$$

$$9C = 5(F - 32)$$

$$\frac{9C}{5} = \frac{5(F - 32)}{5} \qquad \text{Divide by 5}$$

$$\frac{9C}{5} = F - 32$$

$$\frac{9}{5}C + 32 = F - 32 + 32 \qquad \text{Add 32}$$

Try Problem 15 $\qquad \dfrac{9}{5}C + 32 = F$ ◄

EXAMPLE 6   Solve $V = P + Pr$:   **a)** for r,   **b)** for P.

a) $\qquad V = P + Pr$

$\qquad V - P = P + Pr - P \qquad \text{Subtract } P$

$\qquad V - P = Pr$

$\qquad \dfrac{V - P}{P} = \dfrac{Pr}{P} \qquad \text{Divide by } P$

$\qquad \dfrac{V - P}{P} = r$

b) $\qquad V = P + Pr$

$\qquad V = P(1 + r) \qquad \text{Factor out } P$

$\qquad \dfrac{V}{1 + r} = \dfrac{P(1 + r)}{1 + r} \qquad \text{Divide by } 1 + r$

Try Problem 27 $\qquad \dfrac{V}{1 + r} = P$ ◄

Using a formula to solve an applied problem generally involves three steps.

Using a Formula to Solve an Applied Problem

1. Choose the formula that fits the problem.
2. Substitute the known values into the formula.
3. Solve the formula for the unknown value.

---

* These scales are named after their inventors Gabriel Fahrenheit (1686–1736) and Anders Celsius (1701–1744).

Appendix 2 contains a list of commonly used formulas from geometry. You may need to refer to them to solve some of the problems in Problem Set 3.2. You do not have to memorize them. There are literally thousands of mathematical formulas, and no one could remember them all. What you should learn from this section is how to use formulas.

Problem Solving

Estimating the Distance of Lightning

The speed of light is approximately 186,300 miles per second. Therefore light travels over short distances almost instantaneously. Sound, however, travels at the much lower speed of about 1100 feet per second.

You see a flash of lightning 3.5 seconds before you hear the thunder. How far away was the lightning?

$$d = rt \qquad \text{Choose the appropriate formula}$$
$$d = (1100)(3.5) \qquad \text{Substitute the known values}$$
$$d = 3850 \qquad \text{Solve for the unknown value}$$

The flash of lightning was 3850 feet away.

PROBLEM SET 3.2

Use the formula and the known values to determine the unknown value.

1. $A = lw$; $A = 20$, $w = 4$

2. $A = lw$; $A = 30$, $w = 5$

3. $A = \frac{1}{2}bh$; $A = 14$, $b = 7$

4. $A = \frac{1}{2}bh$; $A = 18$, $b = 9$

5. $V = P + Pr$; $V = 1050$, $P = 1000$

6. $V = P + Pr$; $V = 1250$, $P = 1100$

7. $S = \dfrac{a}{1-r}$; $S = 60$, $r = -\dfrac{1}{2}$

8. $S = \dfrac{a}{1-r}$; $S = 48$, $r = -\dfrac{1}{3}$

9. $T = 2\pi r(h + r)$; $T = 94.2$, $\pi \approx 3.14$, $r = 2.5$

10. $T = 2\pi r(h + r)$; $T = 109.9$, $\pi \approx 3.14$, $r = 3.5$

11. Use two methods to determine the rate at which you must travel to cover 248 miles in 4 hours.

12. Use two methods to determine the time it would take to travel 165 miles at 55 miles per hour.

13. Solve $P = 2w + 2l$ for l. Then find the length of a rectangle whose perimeter is 36 meters and whose width is 7 meters.

14. Solve $P = 2w + 2l$ for w. Then find the width of a rectangle whose perimeter is 46 meters and whose length is 15 meters.

15. Solve $C = \frac{5}{9}(F - 32)$ for F. Then find the Fahrenheit temperature that corresponds to a Celsius temperature of $-10°$.

16. Solve $F = \frac{9}{5}C + 32$ for C. Then find the Celsius temperature that corresponds to a Fahrenheit temperature of $5°$.

Solve each equation for the specified variable.

17. $y - 3x + 12 = 0$ for y

18. $y - 4x + 16 = 0$ for y

19. $A = \frac{1}{3}Bh$ for h

20. $V = \frac{1}{2}bh$ for b

21. $s = \frac{1}{2}(a + b + c)$ for c

22. $s = \frac{1}{2}(a + b + c)$ for b

23. $2y - x + 4 = 0$ for y

24. $5y - x + 10 = 0$ for y

25. $V = P - Pr$ for r

26. $V = P + Prt$ for t

27. $V = P - Pr$ for P

28. $V = P + Prt$ for P

29. $s = vt + \frac{1}{2}at^2$ for a

30. $s = vt + \frac{1}{3}at^2$ for a

31. $7x - 2y + 1 = 0$ for y

32. $4x - 3y + 1 = 0$ for y

33. $c^2 = a^2 + b^2$ for b^2

34. $c^2 = a^2 + b^2$ for a^2

35. $E = mc^2$ for c^2

36. $S = 4\pi r^2$ for r^2

37. $A = a + (n - 1)d$ for n

38. $B = 2b + (n - 2)d$ for n

39. $T = \pi r(l + h)$ for l

40. $T = 2\pi r(h + r)$ for h

41. $a(x - b) = 3 - 2x$ for x

42. $a(x + b) = 5 + 2x$ for x

43. Determine the area of the shaded region in Figure 3.3.

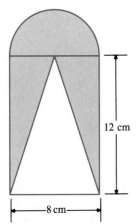

44. Determine the area of the shaded region in Figure 3.4.

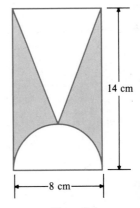

14 cm

8 cm

Figure 3.4

12 cm

8 cm

Figure 3.3

45. How many square feet of paint would it take to cover a spherical water tank of diameter 50 feet?

46. How many square feet of padding would it take to cover the floor, walls, and ceiling of a cubical padded cell whose width, length, and height are each 12 feet?

47. How much water would it take to fill the inverted cone in Figure 3.5?

48. How much water would it take to fill the cylinder in Figure 3.6?

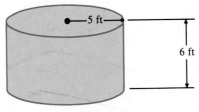

5 ft

6 ft

Figure 3.6

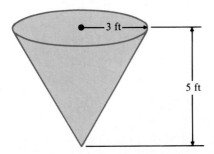

3 ft

5 ft

Figure 3.5

49. You see fireworks 2.5 seconds before you hear them. How far away are the fireworks?

50. You see the flash from a starter's gun 0.3 second before you hear the shot. How far away is the starter?

51. Two bicyclists 25 miles apart start toward each other, one traveling at 10 mph and the other at 15 mph. At the same instant, a fly leaves the handlebars of the first bike and travels toward the second bike. Upon reaching the second bike, the fly immediately turns and heads back toward the first bike. If the fly continues in this back-and-forth fashion at 30 mph until the two bikes meet, what is the total distance traveled by the fly?

52. A car averages 30 mph over the first mile of a 2-mile track. What must it average over the second mile in order to have an average speed of 60 mph for the entire 2-mile lap?

53. Determine the grazing area of a horse that is tied with a 40-foot rope to one corner of a shed that is 20 feet by 20 feet.

54. Do Problem 53 if the rope is tied to a point halfway along one side of the shed.

†**55.** The worker on the trestle is surprised to look up and see a train approaching at a constant speed (Figure 3.7). Which way should he run?

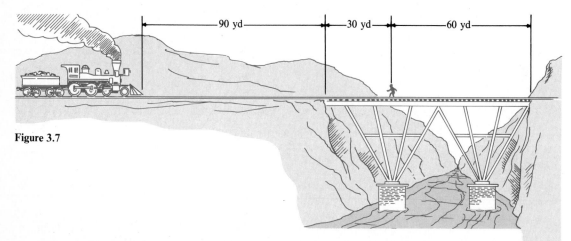

Figure 3.7

†**56.** What is the only temperature that reads the same on both the Celsius and the Fahrenheit scale?

†**57.** What is the difference in temperature between 30° Celsius and 30° Fahrenheit in terms of Fahrenheit degrees?

†**58.** Find the radius x of the circle inscribed in the triangle in Figure 3.8.

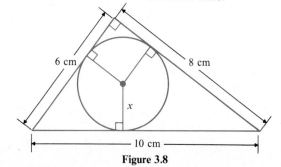

6 cm   8 cm

x

10 cm

Figure 3.8

CALCULATOR PROBLEMS

59. What principal must be deposited into an account that pays 11.32% interest so that the value of the account will be $12,650 in 8 years. [*Hint:* Use $V = P(1 + r)^t$.]

60. In 1985, Carlos Lopes of Portugal ran a marathon (26 miles, 385 yards) in the record time of 2 hours, 7 minutes, and 11 seconds. What was his average speed in miles per hour?

3.3   *Word Problems I*

Now that we have discussed how to solve equations and work with formulas, we shall use these skills to solve certain types of word problems. We can solve each of the word problems in this section and in Section 3.4 by carefully following each of the five steps below.

To Solve a Word Problem

1. Read the problem carefully, and then write down the precise quantities to be determined.
2. Represent one of these quantities by a variable, say x, and then represent the other quantities in terms of x.
3. Translate a statement in the problem into an equation involving x. Sometimes a diagram can be helpful here.
4. Solve the equation to determine the value of x, as well as the values of the other quantities.
5. Check your solution with the statement of the original problem.

These steps are illustrated in Example 1.

EXAMPLE 1   The sum of two numbers is 91, and the larger number is six times the smaller. Find the numbers.

Step 1   Read the problem carefully, and then write down the precise quantities to be determined.

smaller number

larger number

Step 2   Represent one of these quantities by a variable, say x, and then represent the other quantity in terms of x.

x = smaller number

$6x$ = larger number

Step 3 Translate a statement in the problem into an equation involving x.

The sum of the two numbers is 91

$$\underbrace{x + 6x} \qquad = 91$$

Step 4 Solve the equation to determine the value of x, as well as the value of the other quantity.

$$
\begin{aligned}
x + 6x &= 91 \\
7x &= 91 \qquad \text{Combine like terms} \\
x &= 13 \qquad \text{Divide by 7} \\
\text{and} \quad 6x &= 78
\end{aligned}
$$

The numbers are 13 and 78.

Step 5 Check your solution with the statement of the original problem.

Try Problem 1 We note that the sum of 13 and 78 is 91, and 78 is six times 13. ◄

Observe how we use these same steps to solve the next four examples.

EXAMPLE 2 Find three consecutive integers such that three times the first minus the third is 12 more than the second.

$$
\begin{aligned}
x &= \text{first integer} \\
x + 1 &= \text{second integer} \\
x + 2 &= \text{third integer}
\end{aligned}
$$

Three times the first minus the third is 12 more than the second

$$3 \cdot x \quad - \quad (x + 2) \quad = 12 \quad + \quad (x + 1)$$

$$
\begin{aligned}
3x - x - 2 &= 12 + x + 1 \qquad \text{Remove parentheses} \\
2x - 2 &= 13 + x \qquad \text{Combine like terms} \\
2x - x &= 13 + 2 \qquad \text{Subtract } x \text{ and add 2} \\
x &= 15 \\
\text{and} \quad x + 1 &= 16 \\
\text{and} \quad x + 2 &= 17
\end{aligned}
$$

The three integers are 15, 16, and 17. Check these numbers with the statement

Try Problem 7 of the original problem. ◄

EXAMPLE 3 Determine the measures of the three interior angles of the triangle shown in Figure 3.9 if the second angle is 7° more than the first, and the third angle is 41° less than four times the second. (The sum of the interior angles of any triangle is 180°.)

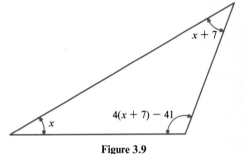

$x =$ first angle
$x + 7 =$ second angle
$4(x + 7) - 41 =$ third angle

Figure 3.9

$$\underbrace{\text{First angle}} + \underbrace{\text{second angle}} + \underbrace{\text{third angle}} = 180$$

$$x + (x + 7) + 4(x + 7) - 41 = 180$$

$$
\begin{aligned}
x + x + 7 + 4x + 28 - 41 &= 180 && \text{Remove parentheses} \\
6x - 6 &= 180 && \text{Combine like terms} \\
6x &= 186 && \text{Add 6} \\
x &= 31 && \text{Divide by 6}
\end{aligned}
$$

$$\text{and}\quad x + 7 = 38$$
$$\text{and}\quad 4(x + 7) - 41 = 111$$

Try Problem 13

The measures of the angles are 31°, 38°, and 111°. ◄

EXAMPLE 4   Kelly is 4 years older than Kim. Twelve years ago she was twice as old as Kim. Find their present ages.

$$y = \text{Kim's age}$$
$$y + 4 = \text{Kelly's age}$$

$$\underbrace{\text{Kelly's age 12 years ago}} = \underbrace{\text{twice Kim's age 12 years ago}}$$

$$(y + 4) - 12 \quad = \quad 2(y - 12)$$

$$
\begin{aligned}
y - 8 &= 2y - 24 && \text{Remove parentheses, combine like terms} \\
-y &= -16 && \text{Subtract } 2y, \text{ add 8} \\
y &= 16 && \text{Divide by } -1
\end{aligned}
$$

$$\text{and}\quad y + 4 = 20$$

Try Problem 17

Kelly is 20 years old, and Kim is 16 years old. ◄

EXAMPLE 5   A collection of nickels, dimes, and quarters has a total value of $3.80. If there are 2 more dimes than nickels, and three times as many quarters as nickels, how many of each kind of coin are there?

$$n = \text{number of nickels}$$
$$n + 2 = \text{number of dimes}$$
$$3n = \text{number of quarters}$$

Value of nickels + value of dimes + value of quarters = total value

$$5n \quad + \quad 10(n + 2) \quad + \quad 25(3n) \quad = 380$$

$$5n + 10n + 20 + 75n = 380$$
$$20 + 90n = 380$$
$$90n = 360$$
$$n = 4$$
$$\text{and} \quad n + 2 = 6$$
$$\text{and} \quad 3n = 12$$

Try Problem 21   There are 4 nickels, 6 dimes, and 12 quarters. ◄

Carefully follow the five steps outlined in this section to solve each of the word problems below. Don't use trial and error. You must practice these five steps so that you will be able to solve those problems where trial and error doesn't work.

PROBLEM SET 3.3

Solve the following number problems.

1. The sum of two numbers is 102, and the larger number is five times the smaller. Find the numbers.

2. The sum of two numbers is 65, and the larger is four times the smaller. Find the numbers.

3. Two consecutive even integers have a sum of 234. Find the integers.

4. Two consecutive odd integers have a sum of 256. Find the integers.

5. One fourth the sum of three consecutive integers is 18. Find the integers.

6. One fifth the sum of four consecutive integers is 18. Find the integers.

7. Find three consecutive integers such that three times the first minus the third is 8 more than the second.

8. Find four consecutive integers such that three times the first minus the third is 12 more than the fourth.

9. Find three consecutive odd integers such that the smallest subtracted from twice the largest is 26.

10. Find three consecutive even integers such that the smallest subtracted from three times the largest is 50.

Solve the following geometry problems.

11. The length of a soccer field is 30 yards less than twice the width. If the perimeter is 390 yards, find the dimensions.

12. The length of a tennis court for singles play is 24 feet more than twice the width. If the perimeter is 210 feet, find the dimensions.

13. Determine the measures of the three interior angles of a triangle if the second angle is 5° more than twice the first and the third angle is 17° less than twice the second.

14. Determine the measures of the three interior angles of a triangle if the second angle is 4° less than three times the first and the third angle is 38° less than three times the second.

15. The length of a rectangle is 3 meters more than the width. If each dimension is increased by 1 meter, the area is increased by 14 square meters. Find the dimensions of the original rectangle.

16. The length of a rectangle is 4 meters more than the width. If each dimension is increased by 2 meters, the area is increased by 40 square meters. Find the dimensions of the original rectangle.

Solve the following age problems.

17. Karen is 5 years older than Kathy. Two years ago she was twice as old as Kathy. Find their present ages.

18. Jerry is 8 years older than Jimmy. In two years he will be twice as old as Jimmy. Find their present ages.

19. Stacey is half as old as her sister. In 7 years she will be the same age her sister was 5 years ago. Find the present ages of Stacey and her sister.

20. Eric is one third as old as his brother. In 9 years he will be the same age his brother was 13 years ago. Find the present ages of Eric and his brother.

Solve the following coin problems.

21. A piggy bank contains $3.35 in nickels, dimes, and quarters. If it contains two more dimes than quarters, and twice as many nickels as quarters, how many of each kind of coin are in the bank?

22. A collection of nickels, dimes, and quarters has a total value of $4.35. If there are three more quarters than dimes, and twice as many nickels as dimes, how many of each kind of coin are in the collection?

23. A collection of 40 nickels and dimes is worth $3.05. How many of each kind of coin are in the collection?

24. Sixty dimes and quarters are worth $10.65. How many dimes and how many quarters are there?

Solve the following word problems.

25. Each multiple-choice question on the SAT exam has five choices. To discourage guessing, a student's final score is determined by subtracting one fourth of a point for each incorrect answer from the number of correct answers. If a student who answered all 75 questions on a particular exam received a final score of 45, how many answers were correct? How many incorrect?

26. Do Problem 25 if the student's final score was 65.

†27. In 1987 Deidre was 16 and her mom was 40. In what year
a) will her mom be twice as old as Deidre?
b) was her mom three times as old as Deidre?

†28. There is an interesting 5-digit number. When the number is followed by a 1, the resulting number is three times as large as when a 1 is placed in front of the number. Find the number.

CALCULATOR PROBLEMS

29. The sum of 10.113 and a number is 4.8663 less than 3.45 times the number. Find the number.

30. Mr. Smith, who is married but files separately, has a federal income tax bill for 1983 of $18,384.76. Given the partial 1983 tax table below, determine his taxable income for the year.

Taxable income I	Tax on I
$30,000 \leq I < $42,800	$ 8,007 + 44% of amount over $30,000
$42,800 \leq I < $54,700	$13,639 + 48% of amount over $42,800
$54,700 \leq I	$19,351 + 50% of amount over $54,700

3.4	*Word Problems II*

In this section we discuss money, mixture, motion, and business word problems.

Before we get to our first example, consider the following question: If you have $6300 to invest, and you invest $4300 at a low-risk 7%, how much would you have left to invest at a higher-risk 11%? The answer of course is given by the subtraction

$$\$6300 - \$4300 = \$2000.$$

That is, you would have $2000 to invest at 11%. Now, if instead of $4300 you invest x dollars at 7%, how much is left to invest at 11%? The answer is

$$6300 - x.$$

CAUTION Remember that 6300 is greater than x, so subtract x from 6300. Do *not* subtract 6300 from x.

Notice how we use the five steps given in Section 3.3 to solve each of the following examples.

EXAMPLE 1 How can $6300 be invested, part at a low-risk 7% and the rest at a higher-risk 11%, so that the interest will be the same on each investment?

$$x = \text{amount invested at 7\%}$$
$$6300 - x = \text{amount invested at 11\%}$$

Interest on 7% investment $=$ interest on 11% investment

$$7\% \text{ of } x = 11\% \text{ of } (6300 - x)$$
$$0.07x = 0.11(6300 - x)$$
$$0.07x = 693 - 0.11x \qquad \text{Distribute } 0.11$$
$$0.18x = 693 \qquad \text{Add } 0.11x$$
$$x = \frac{693}{0.18} \qquad \text{Divide by } 0.18$$
$$x = 3850$$
$$\text{and} \quad 6300 - x = 2450$$

Therefore $3850 should be invested at 7%, and $2450 at 11%. Check this answer in the statement of the original problem. ◄

Try Problem 1

EXAMPLE 2   One solution is 12% acid, and another is 30% acid. How many liters of each should be used to make 72 liters of a solution that is 20% acid?

$$x = \text{number of liters of 12\% solution}$$
$$72 - x = \text{number of liters of 30\% solution}$$

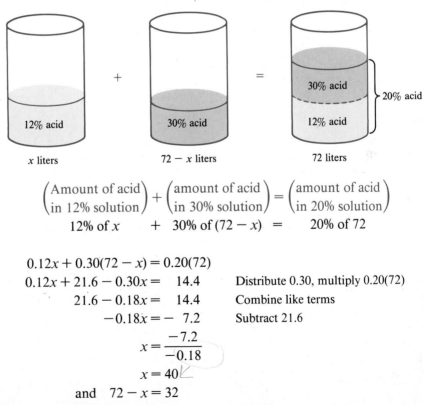

12% acid

x liters

30% acid

$72 - x$ liters

30% acid
12% acid

72 liters

20% acid

$$\left(\begin{array}{c}\text{Amount of acid} \\ \text{in 12\% solution}\end{array}\right) + \left(\begin{array}{c}\text{amount of acid} \\ \text{in 30\% solution}\end{array}\right) = \left(\begin{array}{c}\text{amount of acid} \\ \text{in 20\% solution}\end{array}\right)$$
$$12\% \text{ of } x \quad + \quad 30\% \text{ of } (72 - x) \quad = \quad 20\% \text{ of } 72$$

$$0.12x + 0.30(72 - x) = 0.20(72)$$
$$0.12x + 21.6 - 0.30x = \quad 14.4 \qquad \text{Distribute } 0.30, \text{ multiply } 0.20(72)$$
$$21.6 - 0.18x = \quad 14.4 \qquad \text{Combine like terms}$$
$$-0.18x = - \ 7.2 \qquad \text{Subtract } 21.6$$
$$x = \frac{-7.2}{-0.18}$$
$$x = 40$$
$$\text{and} \quad 72 - x = 32$$

Try Problem 7

Forty liters of 12% solution and 32 liters of 30% solution should be used. ◄

An earthquake sets off two major types of vibrations, or waves, in the earth: *P*-waves (primary waves) and *S*-waves (secondary waves). *P*-waves travel at approximately 5 mi/sec, and cause the rock to move in the same direction as the wave. *S*-waves travel at approximately 2.75 mi/sec, and cause the rock to move at right angles to the direction of the wave.

EXAMPLE 3 A seismograph measures a time interval of 36 seconds between the *P*-wave and the *S*-wave of an earthquake. How long did it take each wave to arrive at the seismic station? Use this information to determine the distance from the seismic station to the *epicenter* (the center of the earthquake).

$$t = \text{time } P\text{-wave travels}$$
$$t + 36 = \text{time } S\text{-wave travels}$$

Distance traveled by *P*-wave = distance traveled by *S*-wave

$5t = 2.75(t + 36)$	Since distance = rate × time
$5t = 2.75t + 99$	Distribute 2.75
$2.25t = 99$	Subtract $2.75t$
$t = 44$	Divide by 2.25

$$\text{and} \quad t + 32 = 80$$

Try Problem 13

The *P*-wave took 44 seconds and the *S*-wave took 80 seconds to arrive at the seismic station. Therefore the epicenter is (5 mi/sec) · (44 sec) = 220 miles away. ◄

EXAMPLE 4 How long will it take a runner who travels at 10 mph to lap a runner who travels at 8 mph on a 1-mile track?

$$t = \text{time faster runner travels before lapping slower runner}$$

$$\begin{pmatrix} \text{Distance traveled} \\ \text{by faster runner} \end{pmatrix} = \begin{pmatrix} \text{distance traveled} \\ \text{by slower runner} \end{pmatrix} + 1 \text{ mile}$$

$10t = 8t + 1$	Since $d = rt$
$2t = 1$	Subtract $8t$
$t = \frac{1}{2}$	Divide by 2

Try Problem 17

It will take the faster runner $\frac{1}{2}$ hour to lap the slower runner. ◄

Problem Solving

Break-Even Point

The total cost of manufacturing an item is the sum of two types of costs: fixed cost and variable cost. **Fixed cost,** also called overhead, consists of those costs that do not depend on the number of items produced. They may include such costs as rent and utilities. **Variable cost** consists of those costs that do depend on the number of items produced. They may include such costs as materials and labor. The total income, or **total revenue,** derived from selling an item is determined by multiplying the price of the item by the number of items sold. A **break-even point** occurs when money coming in equals money going out; that is, when revenue equals cost.

The fixed cost of producing a record album is determined to be $60,000. The variable cost is $4.50 per album. How many albums must be sold at $6.00 per album to break even?

$$x = \text{number of albums}$$

Use the fact that break-even occurs when revenue equals cost.

$$\text{Revenue} = \text{cost}$$
$$\text{Price} \cdot \text{quantity} = \text{cost per album} \cdot \text{quantity} + \text{fixed cost}$$

$$6.00x = 4.50x + 60,000$$
$$1.5x = 60,000$$
$$x = 40,000$$

Therefore 40,000 albums must be sold.

PROBLEM SET 3.4

Solve the following money problems.

1. How can $7500 be invested, part at a low-risk 8% and the rest at a higher-risk 12%, so that the interest will be the same on each investment?

2. How can $8800 be invested, part at a low-risk 9% and the rest at a higher-risk 13%, so that the interest will be the same on each investment?

3. If $5000 is invested at 6%, how much additional money must be invested at 7% so that the total yearly interest from both investments is $510?

4. If $4000 is invested at 8%, how much additional money must be invested at 9% so that the total yearly interest from both investments is $500?

5. The total receipts for a football game were $1030 for 500 tickets sold. If adult tickets were $3.50 and student tickets were $1.50, how many of each were sold?

6. The total receipts for a neighborhood play were $2390 for 700 seats sold. If reserved seats were $4.50 and general admission seats were $2.50, how many of each were sold?

Solve the following mixture problems.

7. One alloy contains 20% gold, and another contains 28% gold. How many grams of each should be melted together to produce 72 grams of an alloy that is 22% gold?

8. 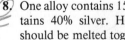 One alloy contains 15% silver, and another contains 40% silver. How many grams of each should be melted together to produce 25 grams of an alloy that is 20% silver?

9. How many liters of pure alcohol should be added to 5 liters of a 40% alcohol solution to obtain a 50% alcohol solution?

10. How many liters of formaldehyde should be added to 15 liters of a 40% formaldehyde solution to obtain a 50% formaldehyde solution?

11. A radiator with a capacity of 12 quarts is filled with a solution that is 40% antifreeze. How many quarts of this solution should be drained and replaced with pure antifreeze to make the final solution 50% antifreeze?

12. Do Problem 11 if the original solution is only 20% antifreeze.

Solve the following motion problems.

13. A seismograph measures a time interval of 27 seconds between the *P*-wave and the *S*-wave of an earthquake. How long did it take each wave to arrive at the seismic station? Use this information to determine the distance from the seismic station to the epicenter.

14. Do Problem 13 if the time interval is 45 seconds.

15. A motorist starts out on a trip traveling 45 mph. A half hour later a second motorist leaves from the same point and travels the same route at 50 mph. How long does it take the second motorist to overtake the first?

16. A jogger leaves the starting point of a trail traveling 10 mph. One-quarter hour later, another jogger begins the same trail traveling 12 mph. How long does it take the second jogger to overtake the first?

17. How long will it take a runner who travels at 11 mph to lap a runner who travels at 8 mph on a 1-mile track?

18. Do Problem 17 if the faster runner travels at 12 mph.

19. One runner finishes a race in 1 hour. A second runner finishes 20 minutes later. If the rate of the faster runner is 2 mph more than the rate of the slower, find the rate of each.

20. Do Problem 19 if the faster runner is 3 mph faster than the slower runner.

21. A motorboat can travel 8 mph upstream and 12 mph downstream. How far upstream could you travel in this boat and still return in a total time of 4 hours?

22. Do Problem 21 if the total time is 6 hours.

Solve the following business problems.

23. The fixed cost of publishing a book is determined to be $75,000. The variable cost is $6.50 per book. How many books must be sold at $12.50 per book to break even?

24. A manufacturer can produce calculators at a cost of $6 per calculator plus a daily overhead of $1400. How many calculators must be sold at $18.50 each day to break even?

25. A theater normally seats 4000 with tickets selling for $8.50. How many seats would have to be added so that tickets could be sold for $8 without changing the total revenue realized from a sold-out theater?

26. A basketball arena normally seats 15,000 with tickets selling for $6.50. How many seats would have to be added so that tickets could be sold for $6 without changing the total revenue realized from a sold-out arena?

Solve the following miscellaneous problems.

27. Tom can stuff 32 envelopes per minute, and Lisa can stuff 40. How long would it take Lisa and Tom working together to stuff 1296 envelopes?

28. In Problem 27, how long would it take Lisa and Tom working together to stuff 1656 envelopes?

CALCULATOR PROBLEMS

29. How can $37,450 be invested, part at a low-risk 8.31% and the rest at a higher-risk 13.78%, so that the total yearly interest will be $4093.96?

30. A policeman fires a gun and 1.5 seconds later hears the bullet hit the target. If the bullet travels at 773 ft/sec and sound travels at 1100 ft/sec, approximately how far away is the target?

3.5   *First-Degree Inequalities*

In Section 3.1 we defined a first-degree equation as an equation of the form

$$ax + b = 0,$$

where $a \neq 0$. A **first-degree inequality** is defined similarly.

Definition

A **first-degree inequality,** or **linear inequality,** in the variable x is an inequality that can be expressed in the form

$$ax + b < 0,$$

where $a \neq 0$.

The inequalities $ax + b \leq 0$, $ax + b > 0$, and $ax + b \geq 0$ are also first-degree inequalities. In general, the definitions and rules we state for $<$ are also valid for $\leq$, $>$, and $\geq$.

A **solution** to an inequality is a number that makes the inequality a true statement when it is substituted for the variable. To **solve** an inequality means to find its solutions. We solve an inequality by writing a sequence of **equivalent inequalities** (inequalities that have the same solution) until we obtain an inequality with the variable isolated on one side.

The **addition-subtraction property of inequality** gives us one way of writing equivalent inequalities. Assume a, b, and c are real numbers.

Addition-Subtraction Property of Inequality

The three inequalities below are equivalent.

$$a < b, \qquad a + c < b + c, \qquad a - c < b - c$$

Note that the addition-subtraction property of inequality is essentially the same as the addition and subtraction properties of equality. The **multiplication-division property of inequality,** however, has one dramatic difference. To see this, consider the true statement

$$-3 < 5.$$

If we multiply each side by a *positive* number, say 4, we get

$$(-3)4 < 5 \cdot 4$$
$$-12 < 20. \qquad \text{True}$$

However, if we multiply each side by a *negative* number, say -4, we get

$$(-3)(-4) < 5(-4)$$
$$12 < -20. \qquad \text{False}$$

To make this last inequality a true statement, we must *reverse* the direction of the inequality sign and write

$$12 > -20.$$

We must also reverse the direction of the inequality sign when we *divide* each side of an inequality by a negative number. This suggests the following rule for the real numbers a, b, and c.

Multiplication-Division Property of Inequality

If c is a positive number, then the inequalities below are equivalent.

$$a < b, \qquad ac < bc, \qquad \frac{a}{c} < \frac{b}{c}$$

If c is a negative number, then the inequalities below are equivalent.

$$a < b, \qquad ac > bc, \qquad \frac{a}{c} > \frac{b}{c}$$

CAUTION We solve a first-degree inequality in the same way that we solve a first-degree equation, except that when we multiply or divide each side of the inequality by a negative number, the direction of the inequality sign must be reversed.

EXAMPLE 1   Solve $3x + 5 < 17$. Then graph the solution.

First, subtract 5 from each side.

$$3x + 5 - 5 < 17 - 5$$
$$3x < 12$$

Then divide each side by 3. Since you are dividing by a positive number, do *not* reverse the inequality sign.

$$\frac{3x}{3} < \frac{12}{3}$$
$$x < 4$$

The solution consists of all numbers less than 4 (see Figure 3.10).

Try Problem 11

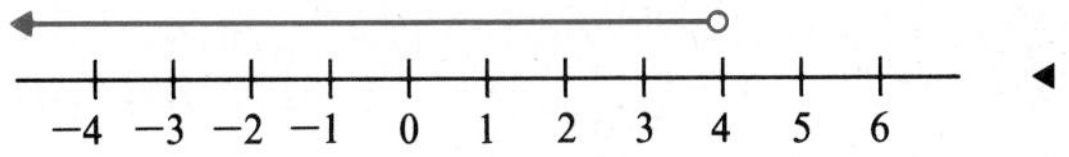

Figure 3.10

Note that the equation $3x + 5 = 17$ has only the one solution 4, but the inequality $3x + 5 < 17$ has an infinite number of solutions.

EXAMPLE 2 Solve $-4y - 7 \le 1$. Then graph the solution.

First, add 7 to each side.

$$-4y - 7 + 7 \le 1 + 7$$
$$-4y \le 8$$

Then divide each side by -4. Since you are dividing each side by a negative number, you must reverse the inequality sign.

$$\frac{-4y}{-4} \ge \frac{8}{-4}$$
$$y \ge -2$$

The solution consists of -2 and all numbers greater than -2 (see Figure 3.11).

Try Problem 13

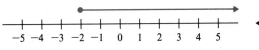

Figure 3.11

EXAMPLE 3 Solve $8 - 4(r + 1) > 2r + 14$.

Remove parentheses and combine like terms.

$$8 - 4r - 4 > 2r + 14$$
$$-4r + 4 > 2r + 14$$

To isolate the variable on the right, add $4r$ to each side.

$$-4r + 4r + 4 > 2r + 4r + 14$$
$$4 > 6r + 14$$

To get the constant terms on the left, subtract 14 from each side.

$$4 - 14 > 6r + 14 - 14$$
$$-10 > 6r$$
$$\frac{-10}{6} > \frac{6r}{6} \qquad \text{Divide by 6, do not}$$
$$\qquad\qquad\qquad\qquad \text{reverse inequality sign}$$
$$\frac{-5}{3} > r$$

$$\text{or} \quad r < -\frac{5}{3}$$

Try Problem 27   The solution consists of all numbers less than $-\frac{5}{3}$. ◄

When the inequality involves fractions, multiply each side by the LCD to clear the fractions.

EXAMPLE 4 Solve $\dfrac{m}{3} - \dfrac{m-6}{4} \geq \dfrac{m}{6} + 1$.

$$12 \cdot \left(\frac{m}{3} - \frac{m-6}{4} \right) \geq 12 \cdot \left(\frac{m}{6} + 1 \right) \qquad \text{Multiply by the LCD 12}$$

$$12 \cdot \frac{m}{3} - 12 \cdot \frac{m-6}{4} \geq 12 \cdot \frac{m}{6} + 12 \cdot 1 \qquad \text{Distribute 12 over every term}$$

$$4m - 3(m-6) \geq 2m + 12 \qquad \text{Careful! Write parentheses here}$$
$$4m - 3m + 18 \geq 2m + 12 \qquad \text{Distribute} -3$$
$$m + 18 \geq 2m + 12$$
$$m + 18 - 18 \geq 2m + 12 - 18 \qquad \text{Subtract 18}$$
$$m \geq 2m - 6$$
$$m - 2m \geq 2m - 6 - 2m \qquad \text{Subtract } 2m$$
$$-m \geq -6$$
$$\frac{-m}{-1} \leq \frac{-6}{-1} \qquad \text{Divide by} -1, \text{reverse inequality sign}$$
$$m \leq 6$$

Try Problem 43   The solution consists of 6 and all numbers less than 6. ◄

To Solve a First-Degree Inequality

1. Remove grouping symbols and combine like terms.
2. Clear fractions by multiplying each side by the LCD. Distribute the LCD over every term.
3. Collect all variable terms on one side of the inequality and all constant terms on the other side. Combine like terms in the process.
4. Write the variable term with a coefficient of 1. Reverse the inequality sign if you multiply or divide by a negative number.
5. If a contradiction results, then the inequality has no solution. If an identity results, then every real number is a solution.

EXAMPLE 5 Solve and graph $0 < x - 2 < 1$.

To solve this double inequality, we isolate x in the middle. We do this by adding

2 to each of the three parts.

$$0 + 2 < x - 2 + 2 < 1 + 2$$
$$2 < x < 3$$

The solution consists of all numbers between 2 and 3 (see Figure 3.12).

Try Problem 45

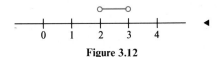

Figure 3.12

EXAMPLE 6   Solve and graph $3 < 1 - 2y \le 10$.

$$3 - 1 < 1 - 2y - 1 \le 10 - 1 \qquad \text{Subtract 1 from each part}$$
$$2 < -2y \le 9$$
$$\frac{2}{-2} > \frac{-2y}{-2} \ge \frac{9}{-2} \qquad \text{Divide by } -2, \text{ reverse } both \text{ inequality signs}$$
$$-1 > y \ge -\frac{9}{2}$$
$$\text{or} \quad -\frac{9}{2} \le y < -1$$

The solution consists of all numbers between $-\frac{9}{2}$ and -1, including $-\frac{9}{2}$ but not including -1 (see Figure 3.13).

Try Problem 51

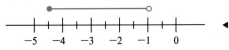

Figure 3.13

The **profit** P of a company is determined by subtracting cost C from revenue R. That is,

$$P = R - C.$$

EXAMPLE 7   A graphics company can print posters at a cost of $4 per poster plus a daily overhead of $600. How many posters must be printed and sold each day at $5.20 per poster if the company's daily profit is to be greater than 10% of its daily costs?

$$x = \text{the number of posters printed and sold each day}$$

The daily cost C is variable cost plus fixed cost.

$$C = 4x + 600$$

The daily revenue R is price times quantity.

$$R = 5.20x$$

The daily profit P is revenue minus cost.

$$P = R - C = 5.20x - (4x + 600) = 1.2x - 600$$

Translate the given word statement into a mathematical inequality as follows:

Daily profit $>$ 10% of daily costs

$$1.2x - 600 > 0.10(4x + 600).$$

Then solve this inequality.

$1.2x - 600 > 0.4x + 60$	Distribute 0.10
$0.8x > 660$	Add 600, subtract $0.4x$
$x > 825$	Divide by 0.8

Try Problem 69   Therefore in excess of 825 posters must be printed and sold each day. ◄

PROBLEM SET 3.5

Solve each inequality. Then graph your solution.

1. $x - 1 < 3$

2. $x - 3 < 2$

3. $x + 4 \le 9$

4. $x + 5 \le 9$

5. $4x > 12$

6. $6x > 12$

7. $-4x > 12$

8. $-6x > 12$

9. $-x < 5$

10. $-x < 4$

11. $3x + 7 < 13$

12. $2x + 5 < 11$

13. $-5y - 1 \le 4$

14. $-3y - 4 \le 2$

15. $-t + 10 \ge 8$

16. $-t + 17 \ge 16$

17. $-2m \le 0$

18. $-4m \le 0$

Solve each inequality.

19. $8x - 5 < 4x + 11$

20. $6x - 8 < x + 2$

21. $15 - 6y \le 9y - 15$

22. $10 - 3y \le 7y - 10$

23. $7z + 9 \ge -3z + 9$

24. $5z + 1 \ge -2z + 1$

25. $4(m + 1) - 2(3m - 4) < 0$

26. $2(m + 2) - 5(2m - 4) < 0$

27. $7 - 3(r + 1) > 3r + 18$

28. $5 - 2(r + 1) > 2r + 13$

29. $5p - (7p + 13) < p + 2$

30. $3p - (5p + 14) < p + 4$

31. $2(2q - 1) \le 6 - (q + 8)$

32. $2(2q - 1) \le 2 - (q + 4)$

Solve each inequality.

33. $\frac{2}{3}x \le -2$

34. $\frac{3}{4}x \le -3$

35. $\frac{y}{5} - 1 \ge y + \frac{3}{5}$

36. $\frac{y}{3} - 2 \ge y + \frac{2}{3}$

37. $\frac{1}{3}t + \frac{7}{12} > \frac{1}{2}t + \frac{3}{4}$

38. $\frac{1}{9}t + \frac{5}{18} > \frac{1}{6}t + \frac{2}{3}$

39. $\frac{4v - 3}{5} + 8 \le 6 + \frac{3v}{2}$

40. $\frac{3v - 1}{4} + 6 \le 2 + \frac{v}{3}$ $v \le -9$

41. $1 < 11 - \frac{3u + 2}{5}$

42. $1 < 10 - \frac{4u + 5}{5}$

43. $\dfrac{m}{3} - \dfrac{m-2}{2} \geq \dfrac{m}{4} - 4$

44. $\dfrac{3m}{7} - \dfrac{m-4}{3} \geq 4 + \dfrac{2m}{7}$

Solve each double inequality. Then graph your solution.

45. $1 < x - 2 < 4$

46. $3 < x - 1 < 5$

47. $-6 \leq 3r \leq 9$

48. $-4 \leq 2r \leq 10$

49. $0 \leq 4s + 12 < 12$

50. $-10 \leq 5s - 10 < 0$

51. $4 < 1 - 3y \leq 8$

52. $5 < 3 - 2y \leq 10$

53. $-\dfrac{5}{2} \leq \dfrac{1-2q}{6} \leq -\dfrac{5}{3}$

54. $-\dfrac{3}{2} \leq \dfrac{1-3q}{6} \leq -\dfrac{4}{3}$

55. $t - 10 < 5t - 6 < t + 10$

56. $t - 6 < 4t - 3 < t + 9$ $-1 < t < 4$

Classify as a conditional inequality, an identity, or a contradiction.

57. $3(x + 2) - 4 < 8 + 3x$

58. $4(x + 1) - 5 < 11 + 4x$

59. $5x - (x + 2) < 3 - (2x - 1)$

60. $6x - 3(x - 1) < 9 - (x - 2)$

61. $\dfrac{1}{3} + \dfrac{9x - 1}{6} < \dfrac{3x}{2}$

62. $\dfrac{3}{4} + \dfrac{12x - 1}{8} < \dfrac{3x}{2}$

63. Five more than six times a number x is between -7 and 17. What are the possible values for x?

64. Six more than five times a number x is between -9 and 21. What are the possible values for x?

65. Within what range must the temperature stay in Fahrenheit degrees if it is to stay between $-10°$ and $25°$ in Celsius degrees? [*Hint:* Solve the double inequality $-10 < \frac{5}{9}(F - 32) < 25$.]

66. Within what range must the temperature stay in Celsius degrees if it is to stay between $23°$ and $68°$ in Fahrenheit degrees? [*Hint:* Solve the double inequality $23 < \frac{9}{5}C + 32 < 68$.]

67. Sally scores 96, 83, 79, and 94 on her first four chemistry tests. If 80 is a B and 90 is an A, what must she score on the fifth and final test to get a B in the course?

68. Repeat Problem 67 for Andy, whose test scores are 95, 87, 80, and 99.

69. A textile plant can produce T-shirts at a cost of $4 per T-shirt plus a daily overhead of $1600. How many T-shirts must be produced and sold each day at $7.20 per T-shirt if the company's daily profit is to be
a) greater than zero?
b) greater than $400?
c) greater than 20% of its costs?

70. Do Problem 69 if the production cost per T-shirt is $5.60.

CALCULATOR PROBLEMS

71. Solve the inequality $17.876x - 53.925 > 21.048x - 33.7035$.

72. The price-demand equation $p = 500 - 1.25x$ gives the quantity x demanded at price p. How many units must be demanded to get the price below 40?

3.6   *Absolute-Value Equations and Inequalities*

ABSOLUTE-VALUE EQUATIONS

Recall from Section 1.3 that the absolute value of x is the distance between x and 0. Therefore the absolute-value equation

$$|x| = 3$$

means that the distance between x and 0 is 3 (see Figure 3.14). That is,

$$x = 3 \quad \text{or} \quad x = -3.$$

```
|←—3 units—→|←—3 units—→|
 +  +  ●  +  +  +  +  +  ●  +  +
-5 -4 -3 -2 -1  0  1  2  3  4  5
```

Figure 3.14

This suggests the following law, where Q denotes any quantity that represents a number.

First Law of Absolute Value

If c is a positive number, then

$$|Q| = c \quad \text{is equivalent to} \quad Q = c \quad \text{or} \quad Q = -c.$$

EXAMPLE 1   Use the first law of absolute value to solve $|2x - 3| = 7$. In this case $Q = 2x - 3$ and $c = 7$.

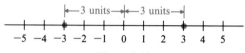

$$|Q| = c \text{ is equivalent to } Q = c \text{ or } Q = -c$$

$$|2x - 3| = 7 \text{ is equivalent to } 2x - 3 = 7 \quad \text{or} \quad 2x - 3 = -7$$

Solve each equation separately.

$$
\begin{array}{rclcrcl}
2x - 3 &=& 7 & \text{or} & 2x - 3 &=& -7 \\
2x &=& 10 & \text{or} & 2x &=& -4 \\
x &=& 5 & \text{or} & x &=& -2
\end{array}
$$

Try Problem 7   The solution is 5 or -2. ◄

CAUTION A common mistake is to ignore the absolute-value symbol and simply solve the equation $2x - 3 = 7$. This incorrectly yields only the solution 5.

EXAMPLE 2   Solve $|6y + 1| = -5$.

We cannot use the first law of absolute value because -5 is not a positive number. Instead, we determine by inspection that the equation is a contradiction, since the absolute value of a quantity is never negative. There is no solution. ◄

Try Problem 13

We now consider absolute-value equations of the form

$$|P| = |Q|.$$

The quantities P and Q will have the same absolute value if they are *equal* or if they are *opposites*. Hence we have the law below.

Second Law of Absolute Value

$|P| = |Q|$ is equivalent to $P = Q$ or $P = -Q$.

EXAMPLE 3   Use the second law of absolute value to solve $|2x - 1| = |x + 4|$.

$|P| = |Q|$ is equivalent to $P = Q$ or $P = -Q$

$|2x - 1| = |x + 4|$ is equivalent to $2x - 1 = x + 4$ or $2x - 1 = -(x + 4)$

Solve each equation separately.

$$2x - 1 = x + 4 \quad \text{or} \quad 2x - 1 = -(x + 4)$$
$$x = 5 \quad\quad \text{or} \quad 2x - 1 = -x - 4$$
$$3x = -3$$
$$x = -1$$

Try Problem 15

The solution is 5 or -1. ◄

EXAMPLE 4   Solve $|4y + 5| = |8 - 4y|$.

Apply the second law of absolute value.

$$4y + 5 = 8 - 4y \quad \text{or} \quad 4y + 5 = -(8 - 4y)$$
$$8y = 3 \quad\quad \text{or} \quad 4y + 5 = -8 + 4y$$
$$y = \tfrac{3}{8} \quad\quad \text{or} \quad\quad 5 = -8$$

Try Problem 19

Since $5 = -8$ is a contradiction, the only solution is $\tfrac{3}{8}$. ◄

ABSOLUTE-VALUE INEQUALITIES

As we noted earlier, the solution to the equation $|x| = 3$ consists of numbers whose distance from the origin is equal to 3. However, the solution to the absolute-value *inequality* $|x| < 3$ consists of numbers whose distance from the origin is *less* than 3 (see Figure 3.15). Therefore the solution to $|x| < 3$ is given by the double inequality $-3 < x < 3$.

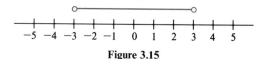

Figure 3.15

Third Law of Absolute Value

If c is a positive number, then

$$|Q| < c \quad \text{is equivalent to} \quad -c < Q < c.$$

EXAMPLE 5 Solve $2|3r + 6| - 35 < -5$.

First, isolate the absolute-value expression.

$$2|3r + 6| - 35 < -5$$
$$2|3r + 6| < 30 \qquad \text{Add 35}$$
$$|3r + 6| < 15 \qquad \text{Divide by 2}$$

Apply the third law of absolute value

$$|Q| < c \text{ is equivalent to } -c < Q < c$$

$$|3r + 6| < 15 \quad \text{is equivalent to} \quad -15 < 3r + 6 < 15$$

Solve the double inequality.

$$-15 < 3r + 6 < 15$$
$$-21 < 3r < 9 \qquad \text{Subtract 6}$$
$$-7 < r < 3 \qquad \text{Divide by 3}$$

The solution consists of all those numbers between -7 and 3 (see Figure 3.16).

Try Problem 33

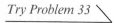
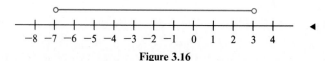

Figure 3.16

EXAMPLE 6   Solve $\left|\dfrac{8s+2}{3}\right| \le 6$.

$$\left|\frac{8s+2}{3}\right| \le 6$$

$$-6 \le \frac{8s+2}{3} \le 6 \qquad \text{Third law of absolute value}$$

$$-18 \le 8s+2 \le 18 \qquad \text{Multiply by 3}$$

$$-20 \le 8s \le 16 \qquad \text{Subtract 2}$$

$$-\frac{5}{2} \le s \le 2 \qquad \text{Divide by 8}$$

The solution consists of all those numbers between $-\frac{5}{2}$ and 2, inclusive (see Figure 3.17).

Try Problem 35

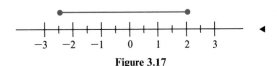

Figure 3.17

The solution to the inequality $|x| < 3$ consists of those numbers *less* than 3 units from the origin, and the solution to the inequality $|x| > 3$ consists of those numbers *greater* than 3 units from the origin (see Figure 3.18). Therefore the solution to $|x| > 3$ is given by $x < -3$ or $x > 3$.

Figure 3.18

Fourth Law of Absolute Value

If c is a positive number, then

$$|Q| > c \quad \text{is equivalent to} \quad Q < -c \quad \text{or} \quad Q > c.$$

EXAMPLE 7   Solve $11 \le |5 - 4t|$.

First, write the inequality as $|5 - 4t| \ge 11$. Then apply the fourth law of absolute value.

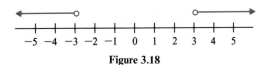

$$|5 - 4t| \ge 11 \quad \text{is equivalent to} \quad 5 - 4t \le -11 \quad \text{or} \quad 5 - 4t \ge 11$$

Solve each inequality separately.

$$5 - 4t \le -11 \quad \text{or} \quad 5 - 4t \ge 11$$
$$-4t \le -16 \quad \text{or} \quad -4t \ge 6 \qquad \text{Subtract 5}$$
$$t \ge 4 \qquad \text{or} \qquad t \le -\frac{3}{2} \qquad \text{Divide by } -4, \text{ reverse the inequality sign}$$

The solution consists of all those numbers less than or equal to $-\frac{3}{2}$ *or greater than or equal to 4 (see Figure 3.19).*

Try Problem 45

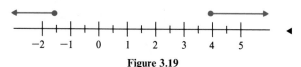

Figure 3.19

CAUTION The solution to Example 7 corresponds to two separate half lines. Therefore the solution *cannot* be written as a double inequality.

EXAMPLE 8   Solve $|3w + 10| > -2$.

We cannot use the fourth law of absolute value because -2 is not a positive number. You should notice, however, that $|3w + 10| \ge 0$ for *any* value of w. The original inequality, then, is an identity. The solution consists of all real numbers (see Figure 3.20).

Try Problem 49

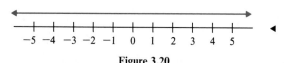

Figure 3.20

EXAMPLE 9   The greens fees at a local golf club are $10 for members and $15 for nonmembers. If it costs $50 to become a member, for what number of rounds played will the difference in the two costs be less than $25?

$$x = \text{the number of rounds of golf played}$$
$$15x = \text{nonmembers cost for } x \text{ rounds}$$
$$10x + 50 = \text{members cost for } x \text{ rounds}$$
$$\text{Difference in cost (in either order)} < 25$$
$$|15x - (10x + 50)| < 25$$
$$|5x - 50| < 25$$
$$-25 < 5x - 50 < 25 \qquad \text{Third law of absolute value}$$
$$25 < 5x < 75$$
$$5 < x < 15$$

The difference in cost will be less than $25 if between 5 and 15 rounds of golf are played. ◄

Try Problem 87

PROBLEM SET 3.6

Use the first law of absolute value to solve each equation.

1. $|x| = 4$

2. $|x| = 2$

3. $|3x| = 6$

4. $|2x| = 8$

5. $|y - 6| = 7$

6. $|y - 3| = 5$

7. $|2x - 5| = 9$

8. $|2x - 7| = 3$

9. $|5r| + 3 = 18$

10. $|4r| + 5 = 17$

11. $|4p + 3| - 7 = 20$

12. $|5p + 6| - 3 = 18$

13. $|7y + 2| = -3$

14. $|8y + 3| = -5$

Use the second law of absolute value to solve each equation.

15. $|2x + 3| = |x + 9|$

16. $|3x + 4| = |2x + 11|$

17. $|6t - 1| = |3t + 2|$

18. $|5t - 4| = |2t + 5|$

19. $|5y - 9| = |6 - 5y|$

20. $|4y - 9| = |3 - 4y|$

21. $|4v + 7| = |4v - 7|$

22. $|6v + 1| = |6v - 1|$ —

Use the third law of absolute value to solve each inequality.

23. $|x| < 4$

24. $|x| < 2$

25. $|2x| < 6$

26. $|3x| < 6$

27. $|y - 4| < 1$

28. $|y - 2| < 5$

29. $|3z + 9| \le 6$

30. $|2z + 5| \le 3$

31. $|8k| - 24 \le 0$

32. $|6k| - 12 \le 0$

33. $3|2r + 1| - 25 < -4$

34. $4|2r + 5| - 33 < -5$

35. $\left| \dfrac{4s + 1}{3} \right| \le 5$

36. $\left| \dfrac{8s + 1}{3} \right| \le 5$

Use the fourth law of absolute value to solve each inequality.

37. $|x| > 4$

38. $|x| > 2$

39. $|5x| > 10$

40. $|4x| > 12$

41. $|r + 2| > 3$

42. $|r + 1| > 4$

43. $|6s - 1| \ge 11$

44. $|5s - 3| \ge 7$

45. $13 \le |5 - 4t|$

46. $11 \le |3 - 4t|$

47. $|\tfrac{3}{4}t| - 6 > 0$

48. $|\tfrac{4}{5}t| - 8 > 0$

49. $|3w - 4| \ge -1$

50. $|2w - 1| \ge -3$

Solve each equation or inequality.

51. $|3x - 5| = 7$

52. $|3x - 4| = 8$

53. $|2y - 3| < 5$

54. $|2y - 7| < 3$

55. $|7 + 2r| > 1$

56. $|1 + 2r| > 5$

57. $|3t - 4| = |2t - 1|$ **58.** $|2t - 3| = |t - 5|$ **59.** $|m + 1| = |m|$

60. $|m + 3| = |m|$ **61.** $|8 - q| \geq 3$ **62.** $|6 - q| \geq 1$

63. $\left|\dfrac{3s + 1}{2}\right| = 4$ **64.** $\left|\dfrac{2s + 1}{3}\right| = 5$ **65.** $|3 - 4x| \leq 5$

66. $|1 - 4x| \leq 9$ **67.** $|z| = |2z|$ **68.** $|3z| = |z|$

Solve each equation or inequality.

69. $|p - 7| + 19 = 10$ **70.** $|p - 8| + 16 = 12$ **71.** $|5 - w| = |w - 5|$

72. $|3 - w| = |w - 3|$ **73.** $|r - 2| > 0$ **74.** $|r - 4| > 0$

75. $|r - 2| < 0$ **76.** $|r - 4| < 0$ **77.** $|r - 2| \leq 0$

78. $|r - 4| \leq 0$

Write an absolute-value equation or inequality that describes each graph.

79. a)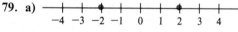

b)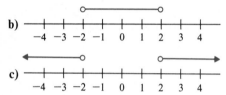

c)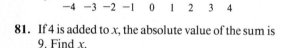

80. a)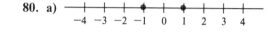

b)

c)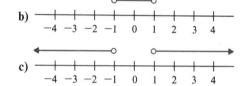

81. If 4 is added to x, the absolute value of the sum is 9. Find x.

82. If 3 is subtracted from x, the absolute value of the difference is 8. Find x.

83. The absolute value of x is no greater than 6. Find all possible values of x.

84. The absolute value of x is no less than 5. Find all possible values of x.

85. If 5 is added to twice a number x, the absolute value of the result is greater than 7. What are the possible values for x?

86. If 6 is added to three times a number x, the absolute value of the result is less than 9. What are the possible values for x?

87. It costs $13 per hour for members to play at a local tennis club and $15 per hour for non-members. If it costs $25 to become a member, for what number of hours played will the difference in the two costs be less than $15?

88. It costs $15 to join a record club. If members can purchase records for $4 per record, and each record costs $7 for nonmembers, for what number of record purchases will the difference in the two costs be less than $12?

CALCULATOR PROBLEMS

Solve.

89. $|0.16x - 0.33| = 0.005$

90. $\left|\dfrac{2.5x + 19.284}{0.815}\right| < 22.06$

CHAPTER 3 REVIEW

[3.1]   *Solve each equation.*

1. $7x - 3x - 5 = 23$

2. $3(t + 3) = t + 5$

3. $4(r - 1) - 3(2r - 5) = 7$

4. $-(3z - 4) + 2 - z = -(-3 + 2z) - (z - 5)$

5. $\dfrac{x}{2} + \dfrac{x}{6} = 2$

6. $\dfrac{2r}{3} - \dfrac{1}{12} = \dfrac{3r}{4} + \dfrac{1}{6}$

7. $\dfrac{h - 3}{3} - \dfrac{1}{5} = \dfrac{2}{3} - \dfrac{h - 2}{15}$

8. $\dfrac{3m - 2}{4} = 2 - \dfrac{m - 1}{6}$

9. $(v + 4)^2 = v^2 + 4$

10. $x(x + 4) - x(x - 3) = 14$

Classify as a conditional equation, an identity, or a contradiction.

11. $5(x - 2) = 5x - 2$

12. $2y - (3y + 4) = 1 - (y + 5)$

13. Three times a number n is 8 less than five times the number. Find n.

14. A television set is marked down 25% to $330. What was the original price p of the set?

[3.2]   *Use the formula and the known values to determine the unknown value.*

15. $P = 2w + 2l$; $P = 68, w = 7$

16. $V = P + Pr$; $V = 1380, P = 1200$

17. $F = \frac{9}{5}C + 32$; $C = -25$

Solve each equation for the specified variable.

18. $2y - 8x + 6 = 0$ for y

19. $V = \pi r^2 h$ for h

20. $A = a + (n - 1)d$ for d

21. $s = \frac{1}{2}(a + b + c)$ for a

22. At what rate must you travel to cover 270 miles in 5 hours?

23. How much water would it take to fill a cylinder of height 5 cm and radius 2 cm?

24. You see an auto accident 1.5 seconds before you hear the crash. How far away is the accident?

[3.3]

25. The sum of two numbers is 85, and the larger number is four times the smaller. Find the numbers.

26. Find three consecutive integers such that four times the first minus twice the third is 14 more than the second.

27. The length of a badminton court for singles play is 10 feet more than twice the width. If the perimeter is 122 feet, find the dimensions.

28. Adam is twice as old as his brother. Three years ago he was three times as old as his brother. Find their present ages.

29. A collection of dimes and quarters is worth $1.80. If there are twice as many dimes as quarters, how many dimes and how many quarters are there?

[3.4]

30. The total receipts for a school play were $1378 for 400 seats sold. If adult tickets were $4.50 and student tickets were $2.50, how many of each were sold?

31. One alloy contains 22% gold and another contains 30% gold. How many grams of each should be melted together to produce 52 grams of an alloy that is 28% gold?

32. A ship leaves port traveling 15 mph. Two and one-half hours later a second ship leaves the same port traveling the same course at 20 mph. How long does it take the second ship to overtake the first?

33. A manufacturer can produce tennis rackets at a cost of $17 per racket plus a daily overhead of $5760. How many rackets must be sold at $49 each day to break even?

[3.5] *Solve each inequality. Then graph your solution.*

34. $\frac{3}{4}x > 3$

35. $-3x \leq 6$

36. $2t - 4 < 6$

37. $-y + 11 \geq 8$

38. $-2 \leq x + 2 \leq 5$

39. $-8 < 4t - 8 \leq 0$

40. $-\frac{2}{5} \leq \frac{1 - 2q}{10} < \frac{1}{2}$

41. $t - 8 < 3t - 4 < t + 8$

Solve each inequality.

42. $7m - 3 > 3m + 5$

43. $5 - (p + 8) < 3(p - 1)$

44. $3(2x - 3) \geq 9 - (x + 4)$

45. $\frac{3}{4}y - 1 \leq \frac{1}{2}y + \frac{1}{4}$

46. $\frac{m - 4}{3} \leq -1 - \frac{m + 1}{3}$

47. $\frac{2t}{3} - \frac{t - 4}{12} \geq \frac{t + 3}{6}$

48. Eight more than four times a number x is between -4 and 28. What are the possible values for x?

[3.6] *Solve each equation or inequality.*

49. $|4x| = 8$

50. $|2z - 5| < 3$

51. $|4t - 3| > 7$

52. $|2p - 3| = 9$

53. $|3y - 2| = |4y - 5|$

54. $|4 - m| \geq 1$

55. $\left|\frac{2x - 1}{3}\right| \leq 5$

56. $|2v + 3| = |2v - 3|$

57. $|x - 2| = |2 - x|$

58. $|y - 3| < 0$

59. $|r - 6| \geq 0$

60. $|p - 5| + 11 = 8$

61. If 7 is added to x, the absolute value of the sum is 4. Find x.

62. If 5 is added to two times a number y, the absolute value of the result is less than 7. What are the possible values for y?

CHAPTER 3 TEST

Use the formula and the known values to determine the unknown value.

1. $P = 2w + 2l$; $P = 56$, $l = 5$

2. $V = P + Pr$; $V = 1000$, $P = 800$

Solve each equation for the specified variable.

3. $I = Prt$ for r

4. $2y - 6x + 4 = 0$ for y

Classify as a conditional equation, an identity, or a contradiction.

5. $2(x + 3) = 3x - (x - 6)$

6. $3(x - 1) = 3x - 1$

Solve each equation.

7. $x + 4 = -x + 1$

8. $t - (2t - 1) = 3$

9. $\dfrac{x}{3} - \dfrac{x}{4} = 2$

10. $\dfrac{h + 2}{6} - \dfrac{h}{2} + \dfrac{1}{3} = 0$

11. $(y - 2)^2 = (y + 3)(y - 3)$

12. $5x - (x + 6) = 2 - (3x - 8)$

Solve each inequality. Then graph your solution.

13. $x - 2 > 2$

14. $\dfrac{2}{3}x \le 6$

15. $2 < y + 3 < 4$

16. $-1 \le \dfrac{1 - m}{2} < 3$

Solve each inequality.

17. $7 - 3r \le 2(r + 1)$

18. $\dfrac{3}{4}y - 2 > \dfrac{1}{2}y + \dfrac{1}{4}$

Solve each equation or inequality.

19. $|5x| = 15$

20. $|2x + 3| = |3x + 7|$

21. $|y + 4| > 3$

22. $\left| \dfrac{2r - 3}{5} \right| \le 1$

Solve the following word problems.

23. The sum of two numbers is 76, and the larger number is three times the smaller. Find the numbers.

24. The length of a rectangle is 5 feet more than three times the width. If the perimeter is 88 feet, find the dimensions.

25. One alloy contains 25% gold and another contains 45% gold. How many grams of each should be melted together to produce 100 grams of an alloy that is 40% gold?

CUMULATIVE REVIEW FOR CHAPTERS 1-3

[1]   *Compute.*

1. $(25 - 7) \div 3 + 6$

2. $4[17 - 2(6 - 1)]$

3. $\dfrac{|+4| - |-4|}{|-2|}$

4. $4(2[8 - 2(8 - 2)])$

Use the given law to complete each statement.

5. Commutative law: $ab = $ ___?___

6. Inverse law: $\frac{1}{3} \cdot 3 = $ ___?___

7. Distributive law: $2(a + b) = $ ___?___

8. Associative law: $-3 + (3 + t) = $ ___?___

Replace each comma with the symbol $<$ or the symbol $>$ to make a true statement.

9. $\frac{9}{13}, \frac{11}{16}$

10. $-63, -73$

Graph each set on a number line. Then name the graph.

11. $\{t | -3 < t \le -1\}$

12. $\{x | x > -1\}$

Perform the indicated operations.

13. $\left(-\dfrac{2}{3}\right) + \dfrac{1}{3}$

14. $(-3.7) + (-2.4)$

15. $(-4) - 4.1$

16. $\left(-\dfrac{2}{7}\right) - \left(-\dfrac{5}{7}\right)$

17. $(-2)(-2)(-2)$

18. $\dfrac{2}{3}\left(-\dfrac{3}{4}\right)$

19. $\left(-\dfrac{3}{8}\right) \div \left(-\dfrac{2}{3}\right)$

20. $\dfrac{12}{-2.4}$

Solve each equation and check your solution.

21. $\frac{2}{7}x = 14$

22. $3m + 5 = 17$

23. Pieces of lengths $4\frac{3}{4}$ feet and $5\frac{1}{6}$ feet are cut from a 20-foot wire. What is the length of the wire that remains?

[2]   *Evaluate.*

24. $3 \cdot 2^3$

25. -2^4

Apply the appropriate rule of exponents to simplify each expression.

26. $x^3 \cdot x^6$

27. 5^0

28. $\dfrac{a^7}{a^2}$

29. $(2m^3)(5m^2)$

30. $(t^4)^4$

31. $(2xy)^3$

32. $\left(\dfrac{2}{b}\right)^3$

33. $\dfrac{(3x^2y^3)^2}{3xy}$

34. Evaluate the polynomial $-2y^2 - y + 4$ at $y = -2$.

35. Add.

$$2x^2 + 3x - 4$$
$$4x^2 - x - 3$$

36. Subtract.

$$3t^3 - 4t + 2$$
$$2t^3 - t^2 + 5t - 2$$

Find each product and simplify.

37. $(x + 5)(x + 4)$

38. $(3a - 2b)^2$

39. $3a^2 - 2ab - b^2$
$$\underline{ 2a - b}$$

40. $(2r + 3)(3r^3 - r^2 + 6r - 4)$

Write in prime factored form.

41. $12x^4 - 18x^2 - 30x$

42. $16t^2 - 25$

43. $m^2 - 6mn + 5n^2$

44. $4a^2 - 12ab + 9b^2$

45. $27r^3 + 8s^3$

46. $a^2 + 4a + 4 - b^2$

[3] *Solve each equation.*

47. $6x - 2x - 3 = 13$

48. $3(y + 2) - 2(3y - 2) = 1$

49. $\dfrac{x}{6} + \dfrac{x}{12} = 1$

50. $\dfrac{h-3}{3} - 3 = \dfrac{2}{3} - \dfrac{h+4}{15}$

51. $t^2 + 2 = (t + 2)^2$

52. $x(x - 6) - x(x - 2) = 16$

Solve each equation for the specified variable.

53. $3x - 4y = 12$ for y

54. $A = 2\pi pw$ for w

55. Use the formula and the known values to determine the unknown value.

$$P = \dfrac{V}{1+r}; \quad P = 1400, \ r = 0.12$$

56. A VCR is marked down 40% to $402. What was the original price p of the VCR?

57. The sum of two numbers is 63, and the larger is twice the smaller. Find the numbers.

58. A collection of nickels, dimes, and quarters has a total value of $12.95. If there are three times as many dimes as nickels, and twice as many quarters as dimes, how many of each kind of coin are in the collection?

59. One alloy contains 18% gold and another contains 26% gold. How many grams of each should be melted together to produce 48 grams of an alloy that is 23.5% gold?

60. A motorboat can travel 8 mph upstream and 10 mph downstream. How far downstream could you travel in this boat and still return in a total time of 3 hours?

Solve each inequality. Then graph your solution.

61. $x - 3 \le 2$

62. $9 - 2y > 5$

63. $3 - y < 3y + 7$

64. $2(3p - 4) \ge 3 - (p - 3)$

65. $-\dfrac{2}{7}m \le 2$

66. $\dfrac{3t}{2} - \dfrac{t+3}{12} > \dfrac{t-4}{6}$

Solve each double inequality. Then graph your solution.

67. $-5 < 2x - 3 < -1$

68. $-\dfrac{4}{5} \le \dfrac{1 - 3q}{10} < \dfrac{1}{2}$

Solve each equation or inequality.

69. $|2t + 3| > 3$

70. $|4 - 3x| = 5$

71. $|r - 6| < 0$

72. $|3 - s| = |s - 3|$

73. If 3 is added to two times a number y, the absolute value of the result is less than 9. What are the possible values for y?

Rational expressions are to algebra what rational numbers are to arithmetic. Just as many real-world problems in arithmetic involve operating with rational numbers, so do many real-world problems in algebra involve operating with rational expressions. Therefore your first task will be to learn how to add, subtract, multiply, and divide rational expressions. You will find it easier to learn these operations if you observe how they parallel the same operations with rational numbers. Once these operations are learned, you will use them to solve equations and work with formulas that involve rational expressions.

Throughout the chapter you will apply rational expressions to solve applied problems involving the Doppler effect, the resistance in an electrical circuit, the focal length of a lens, and the time it would take several people working together to complete a task.

4

Rational Expressions

4.1   *Simplifying and Building Expressions*

We defined rational numbers in Section 1.2 as ratios of integers. We define **rational expressions** as ratios of polynomials.

Definition

A **rational expression** is an algebraic expression that can be written as a ratio (quotient) of two polynomials, where the divisor is not zero.

All the expressions

$$\frac{x+3}{x-5}, \quad \frac{3m-2}{m^2-5m+6}, \quad \text{and} \quad y^3+8$$

are rational expressions. The third is a rational expression because it can be written as

$$\frac{y^3+8}{1},$$

where y^3+8 and 1 are polynomials.

The variable in a rational expression can be replaced by any number, so long as that number does not make the denominator equal zero.

EXAMPLE 1   For what value is the rational expression $\dfrac{4y}{y-3}$ undefined?

Set the denominator equal to zero and solve.

$$y-3=0$$
$$y=3$$

Try Problem 3

The expression $\dfrac{4y}{y-3}$ is undefined when $y=3$. ◄

Sometimes for emphasis we shall specify that a statement involving rational expressions is valid only for those values that do not make any denominator equal zero. Often, however, we shall let that go without saying.

SIMPLIFYING RATIONAL EXPRESSIONS

The rules for operating with rational expressions are the same rules we used in Section 1.1 to operate with rational numbers. We state them again for review as we discuss each topic.

Fundamental Law of Rational Expressions

If $b \neq 0$ and $c \neq 0$, then

$$\frac{a}{b} = \frac{a \cdot c}{b \cdot c}.$$

In words, the fundamental law states that if the numerator and denominator of a rational expression are multiplied or divided by the same nonzero quantity, the result is an equivalent rational expression.

EXAMPLE 2   Simplify each rational expression.

a) $\dfrac{5x}{8x} = \dfrac{5 \cdot x}{8 \cdot x} = \dfrac{5}{8}$ Divide numerator and denominator by x

Try Problem 9   **b)** $\dfrac{5 + x}{8 + x}$ cannot be simplified ◄

Remember that *factors* are separated by multiplication, whereas *terms* are separated by addition. In Example 2(a) the numerator and denominator contained the common factor x, which we divided out. In Example 2(b) the numerator and denominator contained the common term x, which we cannot divide out.

CAUTION You cannot divide out common terms from the numerator and denominator of a rational expression. You can divide out only common factors.

EXAMPLE 3   Simplify each rational expression.

a) $\dfrac{7xy^3}{21x^4y^2} = \dfrac{7xy^2 \cdot y}{7xy^2 \cdot 3x^3} = \dfrac{y}{3x^3}$

b) $\dfrac{x + 5}{4x + 20} = \dfrac{1(x + 5)}{4(x + 5)}$ Factor

Try Problem 17 $= \dfrac{1}{4}$ Divide numerator and denominator by $x + 5$ ◄

To Simplify a Rational Expression

1.   Write the numerator and the denominator in prime factored form.
2.   Divide numerator and denominator by their greatest common factor (GCF).

EXAMPLE 4   Simplify each rational expression.

a) $\dfrac{r+s}{r^2-s^2} = \dfrac{r+s}{(r-s)(r+s)}$ Factor

$= \dfrac{1}{r-s}$ Divide numerator and denominator by $r+s$

b) $\dfrac{5t^2+10t-75}{10t^2-30t} = \dfrac{5(t^2+2t-15)}{10t(t-3)}$

$\left. \vphantom{\dfrac{5(t+5)(t-3)}{5\cdot 2t(t-3)}} \right\}$ Factor

$= \dfrac{5(t+5)(t-3)}{5\cdot 2t(t-3)}$

Try Problem 23

$= \dfrac{t+5}{2t}$ Divide out $5(t-3)$ ◄

EXAMPLE 5   Simplify.

a) $\dfrac{x+4}{4+x} = \dfrac{x+4}{x+4} = 1$

Try Problem 37

b) $\dfrac{x-4}{4-x} = \dfrac{x-4}{-(-4+x)} = \dfrac{1(x-4)}{-1(x-4)} = \dfrac{1}{-1} = -1$ ◄

Did you notice in Example 5(b) that $x-4$ and $4-x$ are opposites? Whenever the numerator and the denominator of a rational expression are opposites, the expression always simplifies to -1.

EXAMPLE 6   Simplify $\dfrac{c^3-8d^3}{2d^2-cd-8d+4c}$.

The numerator is a difference of two cubes. The denominator is factored by grouping.

$\dfrac{c^3-8d^3}{2d^2-cd-8d+4c} = \dfrac{c^3-(2d)^3}{d(2d-c)-4(2d-c)}$

$= \dfrac{(c-2d)(c^2+2cd+4d^2)}{(2d-c)(d-4)}$

$= \dfrac{c-2d}{2d-c}\cdot\dfrac{c^2+2cd+4d^2}{d-4}$

$= (-1)\cdot\dfrac{c^2+2cd+4d^2}{d-4}$ Since $\dfrac{c-2d}{2d-c} = -1$

Try Problem 43

$= -\dfrac{c^2+2cd+4d^2}{d-4}$ ◄

BUILDING RATIONAL EXPRESSIONS

When adding or subtracting rational expressions having different denominators, we are faced with the task of building a rational expression. Building is the reverse of simplifying.

Building

$$\frac{3a}{5} = \frac{3a \cdot 4a}{5 \cdot 4a} = \frac{12a^2}{20a}$$

Simplifying

EXAMPLE 7 Supply the missing numerator $\dfrac{3}{4x} = \dfrac{?}{20x}$.

To write $\dfrac{3}{4x}$ as an equivalent expression having a denominator of $20x$, multiply numerator and denominator by 5.

Try Problem 49

$$\frac{3}{4x} = \frac{3 \cdot 5}{4x \cdot 5} = \frac{15}{20x} \blacktriangleleft$$

EXAMPLE 8 Supply the missing numerator $\dfrac{2m}{m+3} = \dfrac{?}{m(m+3)}$.

Multiply numerator and denominator by m.

Try Problem 55

$$\frac{2m}{m+3} = \frac{2m \cdot m}{(m+3) \cdot m} = \frac{2m^2}{m(m+3)} \blacktriangleleft$$

When building a fraction it is generally to your advantage to multiply out the numerator, but to leave the denominator in factored form.

EXAMPLE 9 Supply the missing numerator $\dfrac{q+2}{q-1} = \dfrac{?}{6(q-1)(q+1)}$.

Multiply numerator and denominator by $6(q+1)$.

Try Problem 59

$$\frac{q+2}{q-1} = \frac{(q+2) \cdot 6(q+1)}{(q-1) \cdot 6(q+1)} = \frac{6(q^2+3q+2)}{6(q-1)(q+1)} = \frac{6q^2+18q+12}{6(q-1)(q+1)} \blacktriangleleft$$

EXAMPLE 10 Supply the missing numerator $\dfrac{p+2}{p-4} = \dfrac{?}{p^2-16}$.

Factor the second denominator. The problem then becomes

$$\frac{p+2}{p-4} = \frac{?}{(p-4)(p+4)}$$

Multiply numerator and denominator by $p + 4$.

Try Problem 61

$$\frac{p+2}{p-4} = \frac{(p+2)(p+4)}{(p-4)(p+4)} = \frac{p^2+6p+8}{(p-4)(p+4)}$$ ◀

Problem Solving

The Doppler Effect

Have you ever noticed the sharp reduction in pitch of an automobile horn just as it passes by? This is known as the Doppler effect.* As the car approaches, each successive vibration produced by the horn travels a shorter distance than the last to reach you. Therefore the frequency (number of vibrations per second) of the sound reaching you is *greater* than the normal frequency of the horn. As the car moves away, the opposite occurs. This frequency is *less* than the normal frequency of the horn. The sudden change of frequency just as the car passes accounts for the sharp reduction in pitch (see Figure 4.1).

The formula that describes the Doppler effect is

$$f = \frac{1100F}{1100 - v},$$

where 1100 ft/sec is the speed of sound, f is the observed frequency, F is the actual frequency, and v is the velocity (directed speed) of the car in ft/sec relative to the observer. If a horn produces a sound with a frequency F of 483 vibrations per second, what is the observed frequency f **a)** as the car approaches at 50 mph? **b)** as the car moves away at 50 mph?

a) Substitute $F = 483$ and $v = 50$ into the Doppler-effect formula.

$$f = \frac{1100\,(483)}{1100 - 50} = \frac{531{,}300}{1050} = 506$$

The observed frequency f is 506 vibrations per second.

b) Substitute $F = 483$ and $v = -50$.

$$f = \frac{1100\,(483)}{1100 - (-50)} = \frac{531{,}300}{1150} = 462$$

The observed frequency f is 462 vibrations per second.

a)   b)

Figure 4.1

---

* The Austrian physicist Christian Johann Doppler (1803–1853) first explained this phenomenon, and so it bears his name.

PROBLEM SET 4.1

For what value is each rational expression undefined?

1. $\dfrac{1}{4x}$

2. $\dfrac{1}{6x}$

3. $\dfrac{3y}{y-1}$

4. $\dfrac{5y}{y-2}$

5. $\dfrac{p-1}{p+4}$

6. $\dfrac{p-2}{p+3}$

7. $\dfrac{3r+7}{5}$

8. $\dfrac{2r+9}{7}$

Simplify each rational expression.

9. $\dfrac{5x}{9x}$

10. $\dfrac{7x}{8x}$

11. $\dfrac{5+x}{9+x}$

12. $\dfrac{7+x}{8+x}$

13. $\dfrac{8b^2}{14}$

14. $\dfrac{9b^2}{15}$

15. $\dfrac{-42c}{14c^3}$

16. $\dfrac{-30c}{12c^2}$

17. $\dfrac{4xy^2}{12x^4y^3}$

18. $\dfrac{7xy^4}{14x^3y^2}$

19. $\dfrac{x+3}{6x+18}$

20. $\dfrac{x-2}{7x-14}$

21. $\dfrac{py-4pz}{qy-4qz}$

22. $\dfrac{3py+pz}{3qy+qz}$

23. $\dfrac{2r+s}{4r^2-s^2}$

24. $\dfrac{r-s}{r^2-s^2}$

25. $\dfrac{20t+15}{16t^2-9}$

26. $\dfrac{12t+20}{9t^2-25}$

27. $\dfrac{(w+8)^2}{(w+8)^6}$

28. $\dfrac{(w+6)^3}{(w+6)^6}$

29. $\dfrac{x^2-y^2}{(x-y)^2}$

30. $\dfrac{(x-y)^2}{x^2-y^2}$

31. $\dfrac{6t^2-30t+36}{4t^2-8t}$

32. $\dfrac{10t^2-10t-60}{6t^2-18t}$

33. $\dfrac{12p^2+17pq-40q^2}{4p^2+3pq-10q^2}$

34. $\dfrac{15p^2+2pq-24q^2}{3p^2-5pq-12q^2}$

35. $\dfrac{x+6}{6+x}$

36. $\dfrac{a+b}{b+a}$

37. $\dfrac{x-6}{6-x}$

38. $\dfrac{a-b}{b-a}$

39. $\dfrac{m+2}{m^3+8}$

40. $\dfrac{m-1}{m^3-1}$

41. $\dfrac{ab^2-b+abc-c}{ab^2-b-abc+c}$

42. $\dfrac{t-rst+s^2-rs^3}{t-rst-s^2+rs^3}$

43. $\dfrac{c^3-27d^3}{3d^2-cd-12d+4c}$

44. $\dfrac{c^3-64d^3}{4d^2-cd-8d+2c}$

45. $\dfrac{(r-5)2-r}{(r-5)r-50}$

46. $\dfrac{(r-3)2-r}{(r-3)r-18}$

47. $\dfrac{(q^2 + 2)4 - 8}{(q^2 + 2)q}$

48. $\dfrac{(q^2 + 3)2 - 6}{(q^2 + 3)q}$

Supply the missing numerator.

49. $\dfrac{5}{6x} = \dfrac{?}{24x}$

50. $\dfrac{3}{4x} = \dfrac{?}{36x}$

51. $\dfrac{1}{3xy^2} = \dfrac{?}{21x^8y^2}$

52. $\dfrac{1}{7x^3y^2} = \dfrac{?}{28x^3y^9}$

53. $r = \dfrac{?}{r + 9}$

54. $r = \dfrac{?}{r - 8}$

55. $\dfrac{5m}{m + 5} = \dfrac{?}{m(m + 5)}$

56. $\dfrac{2m}{m - 2} = \dfrac{?}{m(m - 2)}$

57. $\dfrac{s + 1}{s - 1} = \dfrac{?}{6s(s - 1)}$

58. $\dfrac{s - 3}{s + 3} = \dfrac{?}{4s(s + 3)}$

59. $\dfrac{q + 2}{q - 5} = \dfrac{?}{3(q - 5)(q + 4)}$

60. $\dfrac{q + 1}{q - 6} = \dfrac{?}{2(q - 6)(q + 3)}$

61. $\dfrac{p + 2}{p - 7} = \dfrac{?}{p^2 - 49}$

62. $\dfrac{p - 5}{p + 8} = \dfrac{?}{p^2 - 64}$

63. $\dfrac{15x^2}{x + 2} = \dfrac{?}{4x + 8}$

64. $\dfrac{17x^2}{x + 3} = \dfrac{?}{2x + 6}$

65. $\dfrac{2t + 1}{t - 3} = \dfrac{?}{(t - 3)^2}$

66. $\dfrac{3t - 1}{t + 5} = \dfrac{?}{(t + 5)^2}$

67. $\dfrac{w}{w + 1} = \dfrac{?}{w^3 + 1}$

68. $\dfrac{w}{w - 2} = \dfrac{?}{w^3 - 8}$

The measure m in degrees of each interior angle of a regular polygon having s sides is given by the formula*

$$m = \dfrac{180s - 360}{s}.$$

69. Find the measure of each interior angle of a regular pentagon (5 sides).

70. Find the measure of each interior angle of a regular hexagon (6 sides).

CALCULATOR PROBLEMS

71. If a train whistle produces a sound with a frequency of 635 vibrations per second, what is the observed frequency
    a) as the train approaches at 55 mph?
    b) as the train moves away at 55 mph?

72. The cube law of politics states that if x represents the ratio of votes for a given political party, then the ratio of seats won by that party will by y, where

$$y = \dfrac{x^3}{3x^2 - 3x + 1}.$$

If the ratio of votes for the Democrats in an election is 0.57, find the ratio of seats won by that party.

---

* This formula was known to the early Greeks.

4.2 *Multiplying and Dividing Expressions*

MULTIPLYING EXPRESSIONS

To multiply two rational expressions, multiply their numerators and multiply their denominators.

Multiplying Rational Expressions

If $b \neq 0$ and $d \neq 0$, then

$$\frac{a}{b} \cdot \frac{c}{d} = \frac{a \cdot c}{b \cdot d}.$$

EXAMPLE 1 Multiply and simplify.

a) $\dfrac{x^3}{8} \cdot \dfrac{2}{x^2} = \dfrac{2x^3}{8x^2} = \dfrac{2x^2 \cdot x}{2x^2 \cdot 4} = \dfrac{x}{4}$

b) $\dfrac{4x^2}{5y} \cdot \dfrac{35y^4}{6x^3} = \dfrac{140x^2y^4}{30x^3y}$ Multiply numerators, multiply denominators

$\qquad\qquad = \dfrac{10x^2y \cdot 14y^3}{10x^2y \cdot 3x}$ Factor out the GCF $10x^2y$

Try Problem 1 $\qquad\qquad = \dfrac{14y^3}{3x}$ Divide out the GCF ◄

CAUTION Usually it is best to divide out common factors *before* we multiply. This is particularly true when a numerator or denominator is not a monomial.

EXAMPLE 2 Multiply and simplify.

a) $\dfrac{5q + 15}{8} \cdot \dfrac{64q}{3q + 9} = \dfrac{5(q + 3)}{8} \cdot \dfrac{8 \cdot 8q}{3(q + 3)}$ Factor

$\qquad\qquad = \dfrac{5}{1} \cdot \dfrac{8q}{3}$ Divide out common factors

$\qquad\qquad = \dfrac{40q}{3}$ Multiply

b) $\dfrac{p^2 - 4}{p^2} \cdot \dfrac{8p^2 - 48p}{p^2 - 4p - 12} = \dfrac{(p - 2)(p + 2)}{p \cdot p} \cdot \dfrac{8p(p - 6)}{(p - 6)(p + 2)}$

Try Problem 13 $\qquad\qquad = \dfrac{8(p - 2)}{p}$ ◄

To Multiply Rational Expressions

1. Write each numerator and each denominator in prime factored form.
2. Divide out all common factors *(not terms)* that any numerator has with any denominator.
3. Multiply the remaining numerators, and multiply the remaining denominators.

DIVIDING EXPRESSIONS

To divide two rational expressions, invert the second expression and multiply.

Dividing Rational Expressions

If $b \neq 0$, $c \neq 0$, and $d \neq 0$, then

$$\frac{a}{b} \div \frac{c}{d} = \frac{a}{b} \cdot \frac{d}{c}.$$

EXAMPLE 3   Divide and simplify $\dfrac{5m}{n} \div \dfrac{75}{2m}$.

Invert the second expression and change to multiplication.

$$\frac{5m}{n} \div \frac{75}{2m} = \frac{5m}{n} \cdot \frac{2m}{75}$$

$$= \frac{m}{n} \cdot \frac{2m}{15} \qquad \text{Divide out 5}$$

$$= \frac{2m^2}{15n} \qquad \text{Multiply} \blacktriangleleft$$

Try Problem 17

CAUTION You *cannot* divide out the *m*'s in Example 3. You can divide out common factors only when the operation is multiplication.

EXAMPLE 4   Divide out and simplify $\dfrac{q-3}{2} \div \dfrac{q^2+4q+4}{5q-15}$.

$$\frac{q-3}{2} \div \frac{q^2+4q+4}{5q-15} = \frac{q-3}{2} \cdot \frac{5q-15}{q^2+4q+4} \qquad \text{Invert and change to multiplication}$$

$$= \frac{q-3}{2} \cdot \frac{5(q-3)}{(q+2)^2} \qquad \text{Factor}$$

$$= \frac{5(q-3)^2}{2(q+2)^2} \qquad \text{Multiply}$$

Try Problem 31   Note that there are no common factors to be divided out. ◄

EXAMPLE 5   Divide and simplify $\dfrac{ms - mt - ns + nt}{m - n} \div (t^3 - s^3)$.

Since $t^3 - s^3 = \dfrac{t^3 - s^3}{1}$, invert and multiply as follows:

$$\frac{ms - mt - ns + nt}{m - n} \div (t^3 - s^3) = \frac{ms - mt - ns + nt}{m - n} \cdot \frac{1}{t^3 - s^3}.$$

Factor the first numerator by grouping, and the second denominator as a difference of two cubes.

$$= \frac{m(s - t) - n(s - t)}{m - n} \cdot \frac{1}{(t - s)(t^2 + st + s^2)}$$

$$= \frac{(s - t)(m - n)}{m - n} \cdot \frac{1}{(t - s)(t^2 + st + s^2)}$$

Since $\dfrac{s - t}{t - s} = -1$ and $\dfrac{m - n}{m - n} = 1$, we have

Try Problem 39

$$= \frac{-1}{t^2 + st + s^2}. \blacktriangleleft$$

CAUTION Do not find the LCD when multiplying or dividing rational expressions.

PROBLEM SET 4.2

Multiply or divide as indicated. Write your answer in simplest form.

1. $\dfrac{x^2}{10} \cdot \dfrac{2}{x}$

2. $\dfrac{x^3}{15} \cdot \dfrac{3}{x}$

3. $\dfrac{5x^2}{10y^3} \cdot \dfrac{4y^4}{35x^5}$

4. $\dfrac{8x^3}{20y} \cdot \dfrac{15y^2}{18x^5}$

5. $\dfrac{2r}{3} \div \dfrac{5}{r}$

6. $\dfrac{4r}{5} \div \dfrac{3}{r}$

7. $\dfrac{a^2b^3}{4b^2} \div \dfrac{a^3b^2}{4a^3}$

8. $\dfrac{a^3b^2}{6b^3} \div \dfrac{a^2b^3}{6a^2}$

9. $-\dfrac{4r}{5s^2} \cdot \dfrac{-3rs}{8} \cdot \dfrac{2s}{-9r^3}$

10. $-\dfrac{4r}{15s^4} \cdot \dfrac{3r^3}{-14} \cdot \dfrac{-7s^2}{10r^4}$

11. $\dfrac{-6ab^5c}{5(bc)^2} \div \dfrac{3(ac)^3}{-b^3}$

12. $\dfrac{14ac^3}{-3(bc)^3} \div \dfrac{-7(ab)^2}{a^4b^2c^5}$

13. $\dfrac{6q + 18}{5} \cdot \dfrac{25q}{7q + 21}$

14. $\dfrac{11q}{3q - 9} \cdot \dfrac{4q - 12}{22}$

15. $\dfrac{p^2 - 9}{p^3} \cdot \dfrac{9p^2 - 45p}{p^2 - 2p - 15}$

16. $\dfrac{p^2 - 16}{p^3} \cdot \dfrac{8p^2 + 16p}{p^2 - 2p - 8}$

17. $\dfrac{10m}{n} \div \dfrac{100}{3m}$

18. $\dfrac{15n}{m} \div \dfrac{75}{n}$

19. $\dfrac{xy}{xy + 1} \div \dfrac{x}{y}$

20. $\dfrac{xy}{xy - 1} \div \dfrac{y}{x}$

21. $\dfrac{r^3 - 5r^2}{r(3r - 15)} \div \dfrac{3r - 15}{5r + 15}$

22. $\dfrac{2r + 6}{r^3 + 8r^2} \div \dfrac{3r + 9}{r(2r + 6)}$

23. $\dfrac{s^2 + s - 12}{s^2 - 2s - 3} \cdot \dfrac{s + 1}{8s^3 + 32s^2}$

24. $\dfrac{s^2 + 3s - 4}{s^2 - 6s + 5} \cdot \dfrac{s - 5}{7s^3 + 28s^2}$

25. $\dfrac{m^2 + 5m + 6}{m^2 + 4m + 3} \div \dfrac{m^2 - 4}{m^2 - 3m - 4}$

26. $\dfrac{m^2 + 4m - 12}{m^2 + 2m - 8} \div \dfrac{m^2 + 7m + 6}{m^2 - 1}$

27. $\dfrac{x - 3}{x + 3} \div (x^3 - 27)$

28. $\dfrac{x + 2}{x - 2} \div (x^3 + 8)$

29. $\dfrac{c^2 - d^2}{3c^2 + 11cd - 4d^2} \cdot \dfrac{c^2 + 3cd - 4d^2}{(c - d)^2}$

30. $\dfrac{2c^2 - 9cd - 5d^2}{c^2 - d^2} \cdot \dfrac{(c + d)^2}{c^2 - 4cd - 5d^2}$

31. $\dfrac{q - 2}{5} \div \dfrac{q^2 + 4q + 4}{3q - 6}$

32. $\dfrac{q + 3}{2} \div \dfrac{q^2 - 6q + 9}{5q + 15}$

33. $\dfrac{8w^2 + 14w - 15}{6w^2 + 7w - 20} \cdot \dfrac{9w^2 - 16}{16w^2 - 9}$

34. $\dfrac{20w^2 - 23w + 6}{8w^2 - 26w + 15} \cdot \dfrac{4w^2 - 25}{25w^2 - 4}$

35. $\dfrac{2u^2 - 21uv - 36v^2}{u^2 - 11uv - 12v^2} \div \dfrac{10u + 15v}{u^3 - uv^2}$

36. $\dfrac{3u^2 + 7uv - 20v^2}{u^2 + 6uv + 8v^2} \div \dfrac{9u - 15v}{u^2v - 4v^3}$

37. $\dfrac{s^3 + 8t^3}{s + 2t} \div \dfrac{s^2 - 2st + 4t^2}{s + t}$

38. $\dfrac{s - t}{s - 3t} \div \dfrac{s^2 + 3st + 9t^2}{s^3 - 27t^3}$

39. $\dfrac{ms + mt + ns + nt}{m + n} \div (t^3 + s^3)$

40. $\dfrac{ms + mt - ns - nt}{m - n} \div (t^3 + s^3)$

41. $\dfrac{ax - ay - bx + by}{x^3} \div \dfrac{y^2 - x^2}{x^2}$

42. $\dfrac{ax + ay - bx - by}{b^5} \div \dfrac{b^2 - a^2}{b^2}$

43. $\dfrac{1}{2u^2 - 6uv + 3u - 9v} \div \dfrac{1}{8u^3 + 27}$

44. $\dfrac{1}{2u^2 - 6uv - 5u + 15v} \div \dfrac{1}{8u^3 - 125}$

45. $\left(\dfrac{w^3 - w^2}{1 - w^2} \div \dfrac{1 - w^2}{(w + 1)^2} \right) \cdot \dfrac{z^2 - z^2w}{w^2 - w^2z}$

46. $\left(\dfrac{w^4 - 2w^3}{4 - w^2} \div \dfrac{4 - w^2}{(w + 2)^2} \right) \cdot \dfrac{2z^3 - z^3w}{2w^3 - w^3z}$

47. The price p of a certain item is related to the quantity demanded x by the equation

$$p = \dfrac{675 - 0.5x}{x + 4}.$$

Given that revenue equals price times quantity, write the revenue R in terms of x.

48. The total cost C of producing x units of a particular product is given by

$$C = \dfrac{0.5x^2 + 29x + 480}{x + 3}.$$

Given that the average cost per unit of producing x units equals the total cost divided by the number of units, write the average cost AC in terms of x.

CALCULATOR PROBLEMS

49. Find the product

$$\frac{649r + 803s}{113} \cdot \frac{1017}{3481r^2 - 5329s^2}.$$

50. Evaluate

$$\frac{m^2 + m - 2}{m^2 - 4} \div \frac{1 - m^2}{m^2 - m - 2}$$

at $m = 2.37$. Then divide the two rational expressions and simplify.

4.3   *Adding and Subtracting Expressions*

To add or subtract two rational expressions having the same denominator, add or subtract their numerators and place that sum or difference over their common denominator.

Adding and Subtracting Rational Expressions

If $c \neq 0$, then

$$\frac{a}{c} + \frac{b}{c} = \frac{a + b}{c} \quad \text{and} \quad \frac{a}{c} - \frac{b}{c} = \frac{a - b}{c}.$$

EXAMPLE 1   Perform the indicated operations $\dfrac{5}{9r^2} + \dfrac{2}{9r^2} - \dfrac{1}{9r^2}$.

Since the denominators are the same, simply perform the indicated operations on the numerators.

Try Problem 1

$$\frac{5}{9r^2} + \frac{2}{9r^2} - \frac{1}{9r^2} = \frac{5 + 2 - 1}{9r^2} = \frac{6}{9r^2} = \frac{2}{3r^2} \blacktriangleleft$$

EXAMPLE 2   Add and simplify.

$$\frac{3x}{x^2 + 2x - 15} + \frac{15}{x^2 + 2x - 15}$$

$$= \frac{3x + 15}{x^2 + 2x - 15} \qquad \text{Add numerators}$$

$$= \frac{(x + 5)}{(x - 3)(x + 5)} \qquad \text{Factor numerator and denominator}$$

Try Problem 9

$$= \frac{3}{x - 3} \qquad \text{Divide out } x + 5 \blacktriangleleft$$

EXAMPLE 3   Subtract and simplify.

$$\frac{m^2 + 4n^2}{m - 2n} - \frac{4mn}{m - 2n} = \frac{m^2 + 4n^2 - 4mn}{m - 2n} \qquad \text{Subtract numerators}$$

$$= \frac{m^2 - 4mn + 4n^2}{m - 2n} \qquad \text{Rearrange terms}$$

$$= \frac{(m - 2n)^2}{m - 2n} \qquad \text{Factor the numerator}$$

Try Problem 11

$$= m - 2n \qquad \text{Simplify} \blacktriangleleft$$

EXAMPLE 4   Add and simplify $\dfrac{s^2 + t^2}{s - t} + \dfrac{t^2}{t - s}$.

In this case the denominators are opposites. Therefore multiply numerator and denominator of the second expression by -1 to get like denominators. Then add numerators.

$$\frac{s^2 + t^2}{s - t} + \frac{t^2}{t - s} = \frac{s^2 + t^2}{s - t} + \frac{(-1) \cdot t^2}{(-1) \cdot (t - s)}$$

$$= \frac{s^2 + t^2}{s - t} + \frac{-t^2}{s - t}$$

$$= \frac{s^2 + t^2 + (-t^2)}{s - t}$$

Try Problem 15

$$= \frac{s^2}{s - t} \blacktriangleleft$$

   To add or subtract rational expressions having different denominators, we first build each fraction to an equivalent fraction having the least common denominator (LCD). The **LCD** of a collection of rational expressions is the simplest polynomial that is a multiple of each denominator in the collection. Sometimes the LCD is obvious by inspection. When it is not obvious, you can find the LCD as follows:

To Find the LCD

1.   Write each denominator in prime factored form.
2.   The LCD contains each prime factor to the highest power to which it appears in any one of the factorizations.

EXAMPLE 5   Find the LCD of $\dfrac{3}{8c}$ and $\dfrac{5}{2c^2}$.

Step 1 Write each denominator in prime factored form.

$$8c = 2^3 \cdot c$$
$$2c^2 = 2 \cdot c^2$$

Step 2 The LCD contains each prime factor to the highest power to which it appears in any one of the factorizations.

The highest power of 2 is 2^3, the highest power of c is c^2. Therefore

Try Problem 23a
$$\text{LCD} = 2^3 \cdot c^2 = 8c^2. \blacktriangleleft$$

EXAMPLE 6 Find the LCD of $\dfrac{1}{24x^2y}$ and $\dfrac{7}{90xy^3}$.

$$24x^2y = 2^3 \cdot 3 \cdot x^2 \cdot y$$
$$90xy^3 = 2 \cdot 3^2 \cdot 5 \cdot x \cdot y^3$$

Try Problem 25a
$$\text{LCD} = 2^3 \cdot 3^2 \cdot 5 \cdot x^2 \cdot y^3 = 360x^2y^3 \blacktriangleleft$$

EXAMPLE 7 Find the LCD of p and $\dfrac{p^2}{p+4}$.

Since $p = \dfrac{p}{1}$, neither denominator can be factored.

$$1 = 1$$
$$p + 4 = p + 4$$

Try Problem 29a
$$\text{LCD} = 1 \cdot (p + 4) = p + 4 \blacktriangleleft$$

Note that we could have determined the LCD in Example 7 simply by inspection.

EXAMPLE 8 Find the LCD of $\dfrac{r}{r-2}$ and $\dfrac{r-1}{r+3}$.

$$\left. \begin{array}{l} r - 2 = r - 2 \\ r + 3 = r + 3 \end{array} \right\} \qquad \text{Both denominators are prime}$$

Try Problem 35a
$$\text{LCD} = (r - 2)(r + 3) \blacktriangleleft$$

CAUTION It is generally best to leave the LCD in factored form.

EXAMPLE 9 Find the LCD of $\dfrac{2}{m^2 - m}$ and $\dfrac{2}{m^2 - 2m + 1}$.

$$m^2 - m = m(m - 1)$$
$$m^2 - 2m + 1 = (m - 1)^2$$

Try Problem 43a
$$\text{LCD} = m(m - 1)^2 \blacktriangleleft$$

EXAMPLE 10 Find the LCD of $\dfrac{x}{4x+8}$, $\dfrac{x+10}{x^2-4}$, and $\dfrac{3}{x-2}$.

$$4x + 8 = 4(x + 2)$$
$$x^2 - 4 = (x - 2)(x + 2)$$
$$x - 2 = x - 2$$

Try Problem 49a

$$\text{LCD} = 4(x + 2)(x - 2) \blacktriangleleft$$

Now let's try adding and subtracting rational expressions having different denominators.

EXAMPLE 11 Add $\dfrac{3}{8c} + \dfrac{5}{2c^2}$.

In Example 5 we found that the LCD of these two rational expressions was $8c^2$. To find the sum, build each expression into an equivalent expression having a denominator of $8c^2$.

$$\frac{3}{8c} = \frac{3 \cdot c}{8c \cdot c} + \frac{3c}{8c^2}$$

$$\frac{5}{2c^2} = \frac{5 \cdot 4}{2c^2 \cdot 4} = \frac{20}{8c^2}$$

Now add the numerators.

Try Problem 23b

$$\frac{3}{8c} + \frac{5}{2c^2} = \frac{3c}{8c^2} + \frac{20}{8c^2} = \frac{3c + 20}{8c^2} \blacktriangleleft$$

CAUTION Do not add the terms $3c$ and 20. They are not like terms.

EXAMPLE 12 Add $\dfrac{1}{24x^2y} + \dfrac{7}{90xy^3}$.

From Example 6 we know that the LCD is $360x^2y^3$.

$$\frac{1}{24x^2y} + \frac{7}{90xy^3} = \frac{1 \cdot 15y^2}{24x^2y \cdot 15y^2} + \frac{7 \cdot 4x}{90xy^3 \cdot 4x} \qquad \text{Supply the missing factors}$$

$$= \frac{15y^2}{360x^2y^3} + \frac{28x}{360x^2y^3}$$

Try Problem 25b

$$= \frac{15y^2 + 28x}{360x^2y^3} \qquad \text{Add numerators} \blacktriangleleft$$

EXAMPLE 13   Subtract $p - \dfrac{p^2}{p + 4}$.

From Example 7 we know that the LCD is $p + 4$.

$$p - \frac{p^2}{p + 4} = \frac{p}{1} - \frac{p^2}{p + 4}$$

$$= \frac{p \cdot (p + 4)}{1 \cdot (p + 4)} - \frac{p^2}{p + 4} \qquad \text{Supply the missing factors}$$

$$= \frac{p^2 + 4p}{p + 4} - \frac{p^2}{p + 4} \qquad \text{Multiply out the first numerator}$$

$$= \frac{p^2 + 4p - p^2}{p + 4} \qquad \text{Subtract numerators}$$

Try Problem 29b
$$= \frac{4p}{p + 4} \blacktriangleleft$$

CAUTION Do not divide out the p's or the 4's in Example 13. Both are terms (not factors) of the denominator.

EXAMPLE 14   Subtract $\dfrac{r}{r - 2} - \dfrac{r - 1}{r + 3}$.

From Example 8 we know that the LCD is $(r - 2)(r + 3)$.

$$\frac{r}{r - 2} - \frac{r - 1}{r + 3}$$

$$= \frac{r \cdot (r + 3)}{(r - 2) \cdot (r + 3)} - \frac{(r - 1) \cdot (r - 2)}{(r + 3) \cdot (r - 2)}$$

$$= \frac{r^2 + 3r}{(r - 2)(r + 3)} - \frac{r^2 - 3r + 2}{(r - 2)(r + 3)} \qquad \text{Multiply out numerators}$$

Careful! Write parentheses here

$$= \frac{r^2 + 3r - (r^2 - 3r + 2)}{(r - 2)(r + 3)}$$

$$= \frac{r^2 + 3r - r^2 + 3r - 2}{(r - 2)(r + 3)} \qquad \text{Remove parentheses}$$

$$= \frac{6r - 2}{(r - 2)(r + 3)} \qquad \text{Combine terms in the numerator}$$

Factor the numerator to see if the answer can be simplified.

Try Problem 35b
$$= \frac{2(3r - 1)}{(r - 2)(r + 3)} \qquad \text{Cannot be simplified} \blacktriangleleft$$

EXAMPLE 15 Subtract $\dfrac{2}{m^2 - m} - \dfrac{2}{m^2 - 2m + 1}$.

From Example 9 we know that the LCD is $m(m - 1)^2$.

$$\frac{2}{m(m - 1)} - \frac{2}{(m - 1)^2} = \frac{2 \cdot (m - 1)}{m(m - 1) \cdot (m - 1)} - \frac{2 \cdot m}{(m - 1)^2 \cdot m}$$

$$= \frac{2m - 2 - 2m}{m(m - 1)^2}$$

Try Problem 43b

$$= \frac{-2}{m(m - 1)^2} \blacktriangleleft$$

EXAMPLE 16 Perform the indicated operations.

$$\frac{x}{4x + 8} - \frac{x + 10}{x^2 - 4} + \frac{3}{x - 2}$$

$$= \frac{x}{4(x + 2)} - \frac{x + 10}{(x - 2)(x + 2)} + \frac{3}{x - 2} \qquad \text{Factor denominators}$$

$$= \frac{x \cdot (x - 2)}{4(x + 2) \cdot (x - 2)} - \frac{(x + 10) \cdot 4}{(x - 2)(x + 2) \cdot 4}$$

$$+ \frac{3 \cdot 4(x + 2)}{(x - 2) \cdot 4(x + 2)} \qquad \text{Use LCD from Example 10}$$

$$= \frac{x^2 - 2x}{4(x + 2)(x - 2)} - \frac{4x + 40}{4(x + 2)(x - 2)}$$

$$+ \frac{12x + 24}{4(x + 2)(x - 2)}$$

$$\text{Write parentheses}$$

$$= \frac{x^2 - 2x - (4x + 40) + 12x + 24}{4(x + 2)(x - 2)}$$

$$= \frac{x^2 - 2x - 4x - 40 + 12x + 24}{4(x + 2)(x - 2)}$$

$$= \frac{x^2 + 6x - 16}{4(x + 2)(x - 2)}$$

$$= \frac{(x + 8)(x - 2)}{4(x + 2)(x - 2)} \qquad \text{Factor the numerator}$$

Try Problem 49b

$$= \frac{x + 8}{4(x + 2)} \qquad \text{Simplify} \blacktriangleleft$$

PROBLEM SET 4.3

Perform the indicated operations. Simplify your answer.

1. $\dfrac{4}{15r^2} + \dfrac{7}{15r^2} - \dfrac{1}{15r^2}$

2. $\dfrac{9}{25r^2} + \dfrac{4}{25r^2} - \dfrac{3}{25r^2}$

3. $\dfrac{k^2}{k^3} - \dfrac{k}{k^3} + \dfrac{1}{k^3}$

4. $\dfrac{k^2}{k^4} - \dfrac{k}{k^4} + \dfrac{7}{k^4}$

5. $\dfrac{p+4}{p+1} - \dfrac{3}{p+1}$

6. $\dfrac{p-6}{p+2} + \dfrac{8}{p+2}$

7. $\dfrac{6}{q} - \dfrac{q+6}{q}$

8. $\dfrac{7}{q} - \dfrac{q+7}{q}$

9. $\dfrac{3x}{x^2+x-12} + \dfrac{12}{x^2+x-12}$

10. $\dfrac{5x}{x^2-3x-10} + \dfrac{10}{x^2-3x-10}$

11. $\dfrac{m^2+9n^2}{m-3n} - \dfrac{6mn}{m-3n}$

12. $\dfrac{m^2+25n^2}{m-5n} - \dfrac{10mn}{m-5n}$

13. $\dfrac{x(x+3)}{(x-3)(x+3)} - \dfrac{3(x-3)}{(x+3)(x-3)}$

14. $\dfrac{x(x+4)}{(x-4)(x+4)} - \dfrac{4(x-4)}{(x+4)(x-4)}$

15. $\dfrac{s^2+36}{s-6} + \dfrac{36}{6-s}$

16. $\dfrac{s^2-64}{s-8} - \dfrac{64}{8-s}$

17. $\dfrac{c}{c^2-d^2} - \dfrac{d}{d^2-c^2}$

18. $\dfrac{c}{c^2-d^2} + \dfrac{d}{d^2-c^2}$

*For each problem **a)** find the LCD, **b)** use the LCD to perform the indicated operations. Simplify your answer.*

19. $\dfrac{1}{3a} + \dfrac{1}{5a}$

20. $\dfrac{1}{2a} + \dfrac{1}{7a}$

21. $\dfrac{2}{b} - \dfrac{1}{b^2}$

22. $\dfrac{3}{b} - \dfrac{1}{b^3}$

23. $\dfrac{4}{9c} + \dfrac{7}{3c^2}$

24. $\dfrac{5}{16c} + \dfrac{3}{4c^2}$

25. $\dfrac{1}{18xy^2} + \dfrac{3}{40x^3y}$

26. $\dfrac{1}{27x^2y^2} + \dfrac{7}{60xy^4}$

27. $\dfrac{1}{k} + k$

28. $\dfrac{1}{k} - k$

29. $p - \dfrac{p^2}{p+3}$

30. $p - \dfrac{p^2}{p+2}$

31. $\dfrac{8}{q(q+2)} + \dfrac{4}{q+2}$

32. $\dfrac{18}{q(q+3)} + \dfrac{6}{q+3}$

33. $\dfrac{10}{h-10} - \dfrac{5}{h-5}$

34. $\dfrac{12}{h-12} - \dfrac{4}{h-4}$

35. $\dfrac{r}{r-4} - \dfrac{r-3}{r+1}$

36. $\dfrac{r}{r-2} - \dfrac{r-5}{r+1}$

37. $\dfrac{-4}{s+10} - \dfrac{2s}{5s+50}$

38. $\dfrac{-4}{s+6} - \dfrac{2s}{3s+18}$

39. $\dfrac{30}{x^2-25} + \dfrac{2}{x+5}$

40. $\dfrac{20}{x^2 - 49} + \dfrac{3}{x + 7}$

41. $\dfrac{1}{2y + 18} - \dfrac{y}{y^2 - 81}$

42. $\dfrac{1}{2y + 16} - \dfrac{y}{y^2 - 64}$

43. $\dfrac{6}{m^2 - 2m} - \dfrac{6}{m^2 - 4m + 4}$

44. $\dfrac{5}{m^2 + m} - \dfrac{5}{m^2 + 2m + 1}$

45. $\dfrac{a^2}{a^2 - b^2} + \dfrac{b}{a^2 + ab}$

46. $\dfrac{a}{ab - b^2} - \dfrac{b^2}{a^2 - b^2}$

47. $\dfrac{1}{k^2 + k - 12} + \dfrac{1}{k^2 - k - 6}$

48. $\dfrac{1}{k^2 + 3k - 10} + \dfrac{1}{k^2 - k - 2}$

49. $\dfrac{x}{2x + 6} - \dfrac{x + 15}{x^2 - 9} + \dfrac{3}{x - 3}$

50. $\dfrac{x}{5x + 10} - \dfrac{x + 6}{x^2 - 4} + \dfrac{2}{x - 2}$

51. $\dfrac{t + 2}{t - 5} + \dfrac{t + 1}{t + 3} - \dfrac{t^2 + 7}{t^2 - 2t - 15}$

52. $\dfrac{t - 2}{t + 3} + \dfrac{t - 5}{t - 4} - \dfrac{t^2 - 23}{t^2 - t - 12}$

53. $1 - \dfrac{1}{a + 1} - \dfrac{a}{(a + 1)^2}$

54. $1 + \dfrac{1}{a - 1} + \dfrac{a}{(a - 1)^2}$

55. $\dfrac{4r}{r - 1} - r - 4$

56. $\dfrac{5r}{r + 1} - r - 3$

57. $\dfrac{s}{(s - 8)^2} + \dfrac{s}{s^2 - 64}$

58. $\dfrac{s}{(s - 6)^2} + \dfrac{s}{s^2 - 36}$

59. $\dfrac{xy}{x^3 + y^3} + \dfrac{1}{x + y}$

60. $\dfrac{1}{x - y} - \dfrac{xy}{x^3 - y^3}$

61. $\left(\dfrac{c}{2c - d} - \dfrac{c}{2c + d} \right) \div \dfrac{cd}{8c^2 + 2cd - 3d^2}$

62. $\left(\dfrac{d}{c + 3d} + \dfrac{d}{c - 3d} \right) \div \dfrac{cd}{6c^2 - 13cd - 15d^2}$

63. $\dfrac{m}{m^2 + 5m + 4} - \dfrac{m - 1}{m^2 + 2m + 1} - \dfrac{1}{m^2 + 4m}$

64. $\dfrac{m}{m^2 - m - 2} - \dfrac{m + 2}{m^2 - 4m + 4} - \dfrac{1}{m^2 + m}$

65. Find the sum of the reciprocals of two consecutive integers if the smaller integer is n.

66. Find the sum of the reciprocals of two consecutive odd integers if the smaller integer is n.

67. The velocity of a meteor from inside our solar system is given by the formula

$$v^2 = Mg\left(\dfrac{2}{R} - \dfrac{1}{a} \right).$$

Express this formula's right side as one fraction.

68. The velocity of a meteor from outside our solar system is given by the formula

$$v^2 = Mg\left(\dfrac{2}{R} + \dfrac{1}{a} \right).$$

Express this formula's right side as one fraction.

CALCULATOR PROBLEMS

69. Add $\dfrac{83}{137x^5} + \dfrac{89}{139x^5} + \dfrac{97}{149x^5}$.

70. Calculate the value of

$$\left(1 - \dfrac{1}{n} \right)\left(1 - \dfrac{1}{n - 1} \right)\left(1 - \dfrac{1}{n - 2} \right)$$

at $n = 3.75$. Then simplify this expression and perform the calculation again.

4.4   *Complex Fractions*

Suppose that John runs to his summer job at the beach each morning at the rate of 10 mph (see Figure 4.2). If he walks home each evening at the rate of 4 mph, what is his average rate for the round trip?

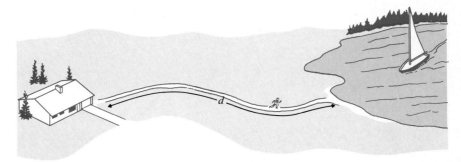

Figure 4.2

The answer to the question posed above is *not* 7 mph, which is the arithmetic mean of the two rates. Instead, John's average rate is closer to 4 mph than it is to 10 mph because he walks at 4 mph for a longer period of time than he runs at 10 mph. If we denote the distance to the beach as d, we can solve this problem.

$$\text{Time to beach} = \frac{\text{distance to beach}}{\text{rate to beach}} = \frac{d}{10}$$

$$\text{Time home} = \frac{\text{distance home}}{\text{rate home}} = \frac{d}{4}$$

$$\text{Average rate} = \frac{\text{total distance}}{\text{total time}} = \frac{2d}{\dfrac{d}{10} + \dfrac{d}{4}}$$

This last expression is called a complex fraction. A **complex fraction** is a fraction that contains other fractions in its numerator or denominator (or both). A fraction that is not a complex fraction is called a **simple fraction.**

To find John's average rate, we must simplify the complex fraction above. This may be done using either of two methods.

To Simplify a Complex Fraction

Method I   Write the numerator as one fraction and the denominator as one fraction. Then invert the denominator and multiply.

Method II   Multiply the numerator and the denominator of the complex fraction by the LCD of the fractions within the numerator and the denominator. Then simplify.

We use the first method in Examples 1 and 2.

EXAMPLE 1   Find John's average rate by simplifying the complex fraction

$$\frac{2d}{\dfrac{d}{10}+\dfrac{d}{4}}.$$

Write the numerator as one fraction, and write the denominator as one fraction.

$$\frac{2d}{\dfrac{d}{10}+\dfrac{d}{4}}=\frac{\dfrac{2d}{1}}{\dfrac{2d}{20}+\dfrac{5d}{20}}=\frac{\dfrac{2d}{1}}{\dfrac{7d}{20}}$$

Invert the denominator and multiply.

$$=\frac{2d}{1}\cdot\frac{20}{7d}$$

$$=\frac{40}{7}$$

Try Problem 5
John's average rate for his daily round trip is $\frac{40}{7}=5\frac{5}{7}$ mph. It is interesting to note that John's average rate does *not* depend on d, the distance to the beach. ◄

EXAMPLE 2   Simplify $\dfrac{\dfrac{x+3}{x}}{\dfrac{x-3}{2x}}.$

The numerator is one fraction, and the denominator is one fraction. Simply invert the denominator and multiply.

Try Problem 7
$$\frac{\dfrac{x+3}{x}}{\dfrac{x-3}{2x}}=\frac{x+3}{x}\cdot\frac{2x}{x-3}=\frac{2(x+3)}{x-3}$$ ◄

We now illustrate the second method for simplifying a complex fraction.

EXAMPLE 3   Simplify $\dfrac{\dfrac{1}{xy}+1}{\dfrac{1}{x^2}+\dfrac{y}{x}}.$

The LCD of $\dfrac{1}{xy}$, $\dfrac{1}{x^2}$, and $\dfrac{y}{x}$ is x^2y. Therefore multiply numerator and denominator by x^2y.

$$\frac{\dfrac{1}{xy}+1}{\dfrac{1}{x^2}+\dfrac{y}{x}}=\frac{\left(\dfrac{1}{xy}+1\right)\cdot x^2y}{\left(\dfrac{1}{x^2}+\dfrac{y}{x}\right)\cdot x^2y}$$

$$=\frac{\dfrac{1}{xy}\cdot x^2y+1\cdot x^2y}{\dfrac{1}{x^2}\cdot x^2y+\dfrac{y}{x}\cdot x^2y}\qquad \text{Distribute } x^2y \text{ over every term}$$

$$=\frac{x+x^2y}{y+xy^2}\qquad\qquad \text{Find each product}$$

$$=\frac{x(1+xy)}{y(1+xy)}\qquad\qquad \text{Factor}$$

Try Problem 13

$$=\frac{x}{y}\qquad\qquad\qquad \text{Divide out } 1+xy \blacktriangleleft$$

EXAMPLE 4   Simplify $\dfrac{12+\dfrac{11}{p}-\dfrac{5}{p^2}}{4-\dfrac{3}{p}-\dfrac{10}{p^2}}.$

Multiply numerator and denominator by the LCD p^2.

$$\frac{\left(12+\dfrac{11}{p}-\dfrac{5}{p^2}\right)\cdot p^2}{\left(4-\dfrac{3}{p}-\dfrac{10}{p^2}\right)\cdot p^2}=\frac{12\cdot p^2+\dfrac{11}{p}\cdot p^2-\dfrac{5}{p^2}\cdot p^2}{4\cdot p^2-\dfrac{3}{p}\cdot p^2-\dfrac{10}{p^2}\cdot p^2}$$

$$=\frac{12p^2+11p-5}{4p^2-3p-10}$$

$$=\frac{(3p-1)(4p+5)}{(p-2)(4p+5)}\left.\begin{matrix}\\\\\\\end{matrix}\right\} \text{Simplify}$$

Try Problem 21

$$=\frac{3p-1}{p-2}\blacktriangleleft$$

EXAMPLE 5   Simplify $\dfrac{\dfrac{2}{m}+\dfrac{1}{n}}{\dfrac{1}{n^3}+\dfrac{8}{m^3}}$.

Multiply numerator and denominator by the LCD m^3n^3.

$$\frac{\left(\dfrac{2}{m}+\dfrac{1}{n}\right)\cdot m^3n^3}{\left(\dfrac{1}{n^3}+\dfrac{8}{m^3}\right)\cdot m^3n^3}=\frac{\dfrac{2}{m}\cdot m^3n^3+\dfrac{1}{n}\cdot m^3n^3}{\dfrac{1}{n^3}\cdot m^3n^3+\dfrac{8}{m^3}\cdot m^3n^3}$$

$$=\frac{2m^2n^3+m^3n^2}{m^3+8n^3}$$

Factor out m^2n^2 from the numerator. The denominator is the sum of two cubes $m^3+(2n)^3$.

$$=\frac{m^2n^2(2n+m)}{(m+2n)(m^2-2mn+4n^2)}$$

$$=\frac{m^2n^2}{m^2-2mn+4n^2}\blacktriangleleft$$

Try Problem 29

PROBLEM SET 4.4

Simplify the two complex fractions in each problem below and note that they are not equal.

1. a) $\dfrac{2}{\dfrac{3}{\dfrac{3}{4}}}$ **b)** $\dfrac{\dfrac{2}{3}}{4}$

2. a) $\dfrac{u}{\dfrac{v}{w}}$ **b)** $\dfrac{\dfrac{u}{v}}{w}$

Simplify each complex fraction.

3. $\dfrac{\frac{3}{4}+\frac{5}{6}}{\frac{1}{2}+\frac{2}{3}}$

4. $\dfrac{\frac{5}{8}+\frac{2}{3}}{\frac{5}{6}+\frac{1}{2}}$

5. $\dfrac{3d}{\dfrac{d}{10}+\dfrac{d}{6}}$

6. $\dfrac{2d}{\dfrac{d}{15}+\dfrac{d}{6}}$

7. $\dfrac{\dfrac{x+1}{x}}{\dfrac{x-1}{3x}}$

8. $\dfrac{\dfrac{x+2}{x}}{\dfrac{x-2}{5x}}$

9. $\dfrac{\dfrac{p^2-4}{p^2}}{\dfrac{p+2}{p}}$

10. $\dfrac{\dfrac{p^3}{p^2-9}}{\dfrac{p^2}{p-3}}$

11. $\dfrac{\dfrac{6}{r}-\dfrac{6}{s}}{\dfrac{6}{r}+\dfrac{6}{s}}$

12. $\dfrac{\dfrac{2}{r}+\dfrac{2}{s}}{\dfrac{2}{r}-\dfrac{2}{s}}$

13. $\dfrac{\dfrac{1}{xy}-1}{\dfrac{1}{x^2}-\dfrac{y}{x}}$

14. $\dfrac{1+\dfrac{1}{xy}}{\dfrac{x^2}{y^2}+\dfrac{x}{y^3}}$

15. $\dfrac{16 - \dfrac{1}{m^2}}{4 + \dfrac{1}{m}}$

16. $\dfrac{25 - \dfrac{1}{m^2}}{5 - \dfrac{1}{m}}$

17. $\dfrac{\dfrac{b^2}{a^2} - 1}{\dfrac{1}{7a} - \dfrac{b}{7a^2}}$

18. $\dfrac{1 - \dfrac{a^2}{b^2}}{\dfrac{a}{8b^2} - \dfrac{1}{8b}}$

19. $\dfrac{\dfrac{c}{9} - \dfrac{1}{c}}{1 + \dfrac{c+6}{c}}$

20. $\dfrac{\dfrac{c}{4} - \dfrac{1}{c}}{1 + \dfrac{c+4}{c}}$

21. $\dfrac{6 + \dfrac{5}{p} - \dfrac{4}{p^2}}{3 - \dfrac{11}{p} - \dfrac{20}{p^2}}$

22. $\dfrac{15 - \dfrac{17}{p} - \dfrac{4}{p^2}}{3 + \dfrac{5}{p} - \dfrac{12}{p^2}}$

23. $\dfrac{\dfrac{5}{d+3} + 1}{1 - \dfrac{2}{3+d}}$

24. $\dfrac{1 - \dfrac{7}{d+1}}{\dfrac{4}{1+d} + 1}$

25. $\dfrac{x+y}{\dfrac{1}{x} + \dfrac{1}{y}}$

26. $\dfrac{x-y}{\dfrac{1}{y} - \dfrac{1}{x}}$

27. $\dfrac{1 + \dfrac{1}{t}}{t^3 + 1}$

28. $\dfrac{1 + \dfrac{2}{t}}{t^3 + 8}$

29. $\dfrac{\dfrac{3}{m} + \dfrac{1}{n}}{\dfrac{1}{n^3} + \dfrac{27}{m^3}}$

30. $\dfrac{\dfrac{5}{m} + \dfrac{1}{n}}{\dfrac{1}{n^3} + \dfrac{125}{m^3}}$

31. $\dfrac{\dfrac{1}{u^2} - \dfrac{1}{v^2}}{\dfrac{1}{u^3} - \dfrac{1}{v^3}}$

32. $\dfrac{\dfrac{1}{u^2} - \dfrac{1}{v^2}}{\dfrac{1}{u^3} + \dfrac{1}{v^3}}$

33. $\dfrac{1 - \dfrac{2}{q}}{1 - \dfrac{4q-4}{q^2}}$

34. $\dfrac{1 - \dfrac{1}{q}}{1 - \dfrac{2q-1}{q^2}}$

35. $\dfrac{k+3 + \dfrac{12}{k-5}}{k+5 + \dfrac{16}{k-5}}$

36. $\dfrac{k+5 + \dfrac{7}{k-3}}{k+3 - \dfrac{7}{k-3}}$

37. $\dfrac{\dfrac{w+1}{w-1} - \dfrac{w-1}{w+1}}{\dfrac{w+1}{w-1} + \dfrac{w-1}{w+1}}$

38. $\dfrac{\dfrac{w-3}{w+3} + \dfrac{w+3}{w-3}}{\dfrac{w-3}{w+3} - \dfrac{w+3}{w-3}}$

39. $x - \dfrac{1}{1 + \dfrac{1}{x-1}}$

40. $x + \dfrac{1}{1 + \dfrac{1}{x+1}}$

41. $1 + \dfrac{2}{1 + \dfrac{2}{1 + \dfrac{2}{1+1}}}$

42. $2 - \dfrac{1}{2 - \dfrac{1}{2 - \dfrac{1}{2-1}}}$

43. A lumberjack travels 3 mph going up a tree and 12 mph coming down. What is the lumberjack's average rate?

44. Melissa drives to work at 30 mph and drives home at 50 mph. What is her average rate?

CALCULATOR PROBLEMS

45. Compute to the nearest hundredth.

$$\dfrac{\dfrac{1}{3.08}}{\dfrac{1}{14.99} + \dfrac{1}{24.77}}$$

46. Allison rides the ski lift up the slope at 2 mph and skis down the slope at 35 mph. Find her average rate.

4.5 *Equations with Rational Expressions*

To solve an equation with rational expressions, it is generally best to begin by multiplying each side by the LCD. This will clear the equation of fractions.

EXAMPLE 1 Solve $\dfrac{x}{2} + \dfrac{x}{3} = 10$.

$$6 \cdot \left(\frac{x}{2} + \frac{x}{3}\right) = 6 \cdot 10 \qquad \text{Multiply each side by the LCD 6}$$

$$6 \cdot \frac{x}{2} + 6 \cdot \frac{x}{3} = 6 \cdot 10 \qquad \text{Distribute 6 over every term}$$

$$3x + 2x = 60 \qquad \text{Find each product}$$

$$5x = 60 \qquad \text{Combine like terms}$$

$$x = 12 \qquad \text{Divide by 5}$$

Try Problem 1

The solution is 12. Check in the original equation. ◄

CAUTION Do not confuse the two problems below.

$$\textbf{Solve} \quad \frac{x}{2} + \frac{x}{3} = 10 \qquad \textbf{Add} \quad \frac{x}{2} + \frac{x}{3}$$

We *solve* the equation and get a value for *x*. We *add* the two rational expressions and get a third rational expression.

$$\textbf{Solve} \qquad \frac{x}{2} + \frac{x}{3} = 10$$

$$6 \cdot \left(\frac{x}{2} + \frac{x}{3}\right) = 6 \cdot 10$$

$$3x + 2x = 60$$

$$5x = 60$$

$$x = 12$$

$$\textbf{Add} \qquad \frac{x}{2} + \frac{x}{3} = \frac{x \cdot 3}{2 \cdot 3} + \frac{x \cdot 2}{3 \cdot 2}$$

$$= \frac{3x}{6} + \frac{2x}{6}$$

$$= \frac{5x}{6}$$

EXAMPLE 2   Solve $\dfrac{1}{6x} - \dfrac{1}{4x} = \dfrac{1}{8}$.

$$24x \cdot \left(\frac{1}{6x} - \frac{1}{4x} \right) = 24x \cdot \frac{1}{8} \qquad \text{Multiply by the LCD } 24x$$

$$24x \cdot \frac{1}{6x} - 24x \cdot \frac{1}{4x} = 24x \cdot \frac{1}{8} \qquad \text{Distribute } 24x$$

$$4 - 6 = 3x \qquad \text{Find each product}$$

$$-2 = 3x \qquad \text{Combine like terms}$$

$$-\frac{2}{3} = x \qquad \text{Divide by 3}$$

Try Problem 5   The number $-\frac{2}{3}$ checks, so the solution is $-\frac{2}{3}$. ◄

CAUTION A common error is to write the equation of Example 2 as $6x - 4x = 8$. This produces the *incorrect* solution $x = 4$. When writing the equation in this way, you might think that you are performing the same operation on each side. But actually you are performing the same operation on each *term.*

Consider the equation

$$\frac{a}{b} = \frac{c}{d}.$$

If we multiply each side by the LCD bd, we have

$$bd \cdot \frac{a}{b} = bd \cdot \frac{c}{d}$$

$$ad = bc.$$

That is, if an equation simply equates two fractions, we can solve by cross-multiplying.

Cross-Multiplying

If $b \neq 0$ and $d \neq 0$, the equations below are equivalent.

$$\frac{a}{b} = \frac{c}{d} \quad \text{and} \quad ad = bc$$

EXAMPLE 3   Solve $\dfrac{5}{u+8} = \dfrac{1}{u-8}$.

$$5 \cdot (u - 8) = (u + 8) \cdot 1 \qquad \text{Cross-multiply}$$

$$5u - 40 = u + 8 \qquad \text{Find each product}$$

$$4u = 48 \qquad \text{Subtract } u, \text{ add } 40$$

$$u = 12 \qquad \text{Divide by 4}$$

Try Problem 15

The number 12 checks, so the solution is 12. ◄

EXAMPLE 4 Solve $\dfrac{4}{r} + 2 = \dfrac{2r}{r-4}$.

We cannot cross-multiply, because the left side is not a single fraction. Instead, we multiply every term by the LCD $r(r-4)$.

$$r(r-4) \cdot \frac{4}{r} + r(r-4) \cdot 2 = r(r-4) \cdot \frac{2r}{r-4}$$
$$4(r-4) + 2r(r-4) = 2r^2$$
$$4r - 16 + 2r^2 - 8r = 2r^2$$
$$2r^2 - 4r - 16 = 2r^2$$
$$-4r - 16 = 0 \qquad \text{Subtract } 2r^2$$
$$-4r = 16$$
$$r = -4$$

Try Problem 25

The number -4 checks, so the solution is -4. ◄

EXAMPLE 5 Solve $\dfrac{5}{p-3} - \dfrac{6}{p+3} = \dfrac{30}{p^2-9}$.

Multiply every term by the LCD $p^2 - 9 = (p-3)(p+3)$.

$$(p-3)(p+3) \cdot \frac{5}{p-3} - (p-3)(p+3) \cdot \frac{6}{p+3} = (p-3)(p+3) \cdot \frac{30}{p^2-9}$$
$$5(p+3) - 6(p-3) = 30$$
$$5p + 15 - 6p + 18 = 30$$
$$-p + 33 = 30$$
$$-p = -3$$
$$p = 3$$

However, the number 3 does not check in the original equation since division by zero is undefined.

CHECK:
$$\frac{5}{3-3} - \frac{6}{3+3} \overset{?}{=} \frac{30}{3^2-9}$$
$$\frac{5}{0} - \frac{6}{6} \neq \frac{30}{0}$$

Try Problem 35

Therefore there is no solution. ◄

CAUTION When an equation involves variables in any denominator, the solution must be checked in the original equation to ensure it does not make a denominator equal zero.

EXAMPLE 6   Solve $\dfrac{m+1}{m^2+m-20}+\dfrac{3}{2m+10}=0$.

Factor the denominators to determine the LCD.

$$\frac{m+1}{(m+5)(m-4)}+\frac{3}{2(m+5)}=0$$

Multiply every term by the LCD $2(m+5)(m-4)$.

$$2(m+5)(m-4)\cdot\frac{m+1}{(m+5)(m-4)}+$$

$$2(m+5)(m-4)\cdot\frac{3}{2(m+5)}=2(m+5)(m-4)\cdot 0$$

$$2(m+1)+3(m-4)=0$$

$$2m+2+3m-12=0$$

$$5m-10=0$$

$$5m=10$$

$$m=2$$

Try Problem 41

The number 2 checks, so the solution is 2. ◄

To Solve an Equation with Rational Expressions

1. Factor the denominators to determine the LCD.
2. Multiply every term on each side by the LCD.
3. Solve the equation resulting from step 2.
4. Check all solutions in the original equation.

If the equation is of the form $\dfrac{a}{b}=\dfrac{c}{d}$, you can cross-multiply.

We can use the techniques developed in this section to solve a formula for one of its variables.

EXAMPLE 7   Solve $S=\dfrac{a}{1-r}$ for r.

$$S\cdot(1-r)=\frac{a}{1-r}\cdot(1-r) \qquad \text{Multiply each side by the LCD } 1-r$$

$$S-Sr=a \qquad\qquad\qquad \text{Find each product}$$

$$-Sr=a-S \qquad\qquad\qquad \text{Subtract } S$$

$$r=\frac{a-S}{-S} \qquad\qquad\qquad \text{Divide by } -S$$

Try Problem 47

$$r=\frac{S-a}{S} \qquad\qquad\qquad \text{Multiply numerator and denominator by } -1 \text{ ◄}$$

Problem Solving

Resistance in an Electrical Circuit

Suppose two resistors are connected in parallel as shown in Figure 4.3. The first resistor has a resistance of R_1 ohms, and the second has a resistance of R_2 ohms. (The numbers 1 and 2 are called **subscripts,** and R_1 and R_2 are read "R sub 1" and "R sub 2" respectively. Subscripts allow us to use the same letter to represent different quantities.) The total resistance R of the circuit is given by the formula

$$\frac{1}{R} = \frac{1}{R_1} + \frac{1}{R_2}.$$

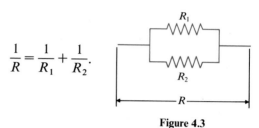

Figure 4.3

We can solve this formula for R_1.

$$RR_1R_2 \cdot \frac{1}{R} = RR_1R_2 \cdot \frac{1}{R_1} + RR_1R_2 \cdot \frac{1}{R_2} \qquad \text{Multiply by the LCD } RR_1R_2$$

$$R_1R_2 = RR_2 + RR_1 \qquad \text{Find each product}$$

$$R_1R_2 - RR_1 = RR_2 \qquad \text{Collect all terms with } R_1 \text{ on one side}$$

$$R_1(R_2 - R) = RR_2 \qquad \text{Factor out } R_1$$

$$R_1 = \frac{RR_2}{R_2 - R} \qquad \text{Divide by } R_2 - R$$

PROBLEM SET 4.5

Solve each equation.

1. $\dfrac{x}{2} + \dfrac{x}{5} = 14$

2. $\dfrac{x}{4} + \dfrac{x}{5} = 18$

3. $\dfrac{x}{3} + \dfrac{x}{2} = -5$

4. $\dfrac{x}{5} + \dfrac{x}{2} = -7$

5. $\dfrac{1}{8x} - \dfrac{1}{4x} = \dfrac{1}{12}$

6. $\dfrac{1}{4x} - \dfrac{1}{3x} = \dfrac{1}{6}$

7. $\dfrac{1}{y} = \dfrac{1}{2} - \dfrac{1}{3}$

8. $\dfrac{2}{y} = \dfrac{1}{3} - \dfrac{1}{5}$

9. $3 - \dfrac{5}{2k} = \dfrac{3}{k} - \dfrac{5}{2}$

10. $2 - \dfrac{4}{3k} = \dfrac{2}{k} - \dfrac{4}{3}$

11. $\dfrac{5x - 3}{x + 4} = 0$

12. $\dfrac{7x - 2}{x + 6} = 0$

13. $\dfrac{3t + 13}{t - 1} = 1$

14. $\dfrac{3t + 19}{t - 3} = 1$

15. $\dfrac{3}{u + 3} = \dfrac{1}{u - 3}$

16. $\dfrac{5}{u + 2} = \dfrac{3}{u - 2}$

17. $\dfrac{6}{4v - 1} = \dfrac{4}{2v + 3}$

18. $\dfrac{3}{5v - 2} = \dfrac{5}{3v + 4}$

19. $\dfrac{w}{w + 1} = \dfrac{w}{w - 1}$

20. $\dfrac{w}{w - 4} = \dfrac{w}{w + 4}$

21. $\dfrac{4s}{s + 7} = \dfrac{12}{s + 7}$

22. $\dfrac{2s}{s + 8} = \dfrac{10}{s + 8}$

23. $\dfrac{x}{x + 10} + 1 = \dfrac{x - 8}{x + 10}$

24. $\dfrac{x}{x - 7} + 1 = \dfrac{x + 11}{x - 7}$

25. $\dfrac{8}{r} + 5 = \dfrac{5r}{r - 2}$

26. $\dfrac{6}{r} + 3 = \dfrac{3r}{r - 3}$

27. $\dfrac{5}{s} + \dfrac{4}{s - 2} = \dfrac{10}{s}$

28. $\dfrac{3}{s} + \dfrac{2}{s - 1} = \dfrac{4}{s}$

29. $\dfrac{t + 1}{3t + 15} + \dfrac{t}{t + 5} = \dfrac{5}{3}$

30. $\dfrac{t + 2}{2t + 8} + \dfrac{t}{t + 4} = \dfrac{1}{2}$

31. $\dfrac{x}{x - 2} + 4 = \dfrac{2}{x - 2}$

32. $\dfrac{x}{x - 3} + 3 = \dfrac{x}{x - 3}$

33. $\dfrac{2}{y + 1} - \dfrac{4}{y - 1} = \dfrac{y}{y^2 - 1}$

34. $\dfrac{4}{y + 2} - \dfrac{5}{y - 2} = \dfrac{y}{y^2 - 4}$

35. $\dfrac{6}{p - 4} - \dfrac{7}{p + 4} = \dfrac{48}{p^2 - 16}$

36. $\dfrac{3}{p - 6} - \dfrac{4}{p + 6} = \dfrac{48}{p^2 - 36}$

37. $\dfrac{2}{q^2 + q - 2} = \dfrac{1}{q + 2} + \dfrac{1}{q - 1}$

38. $\dfrac{2}{q^2 + q - 12} = \dfrac{1}{q + 4} + \dfrac{1}{q - 3}$

39. $\dfrac{3}{4r - 12} - \dfrac{1}{r - 3} = \dfrac{1}{6r + 12}$

40. $\dfrac{5}{6r - 12} - \dfrac{1}{r - 2} = \dfrac{1}{4r + 12}$

41. $\dfrac{m + 2}{m^2 - 7m - 8} + \dfrac{2}{3m + 3} = 0$

42. $\dfrac{m + 7}{m^2 - 3m - 10} + \dfrac{1}{5m + 10} = 0$

43. $\dfrac{u - 4}{u + 5} - \dfrac{u + 1}{u} = \dfrac{25}{u^2 + 5u}$

44. $\dfrac{u - 3}{u + 6} - \dfrac{u + 1}{u} = \dfrac{14}{u^2 + 6u}$

Solve each equation for the specified variable.

45. $\dfrac{PV}{T} = \dfrac{pv}{t}$ for P

46. $\dfrac{PV}{T} = \dfrac{pv}{t}$ for V

47. $S = \dfrac{a}{2 - r}$ for r

48. $S = \dfrac{a}{1 + r}$ for r

49. $\dfrac{1}{R} = \dfrac{1}{R_1} + \dfrac{1}{R_2}$ for R_2

50. $\dfrac{1}{f} = \dfrac{1}{d_1} + \dfrac{1}{d_2}$ for d_1

51. $x = \dfrac{1}{y} - 2$ for y

52. $x = \dfrac{1}{y} - 5$ for y

53. $d = \dfrac{r}{1 + rt}$ for r

54. $d = \dfrac{r}{1 - rt}$ for r

55. $y = \dfrac{x + 3}{x - 3}$ for x

56. $y = \dfrac{x - 1}{x + 1}$ for x

57. Five plus the reciprocal of x is 2. Find x.

58. Seven minus the reciprocal of x is 5. Find x.

CALCULATOR PROBLEMS

59. Solve

$$\frac{1187}{x + 1999} = \frac{1933}{x - 1999}$$

to the nearest integer.

60. Solve

$$\frac{8.77}{4.73x} + 0.281 = \frac{9.63}{5.09x}$$

to the nearest thousandth.

4.6   *Formulas and Word Problems*

FORMULAS

Sometimes we need to use a formula that involves rational expressions in order to solve an applied problem. For example, the focal length f of a simple convex lens is given by the formula

$$\frac{1}{f} = \frac{1}{d_o} + \frac{1}{d_i},$$

where d_o is the distance from the lens to the object, and d_i is the distance from the lens to the image (see Figure 4.4).

EXAMPLE 1   Find the focal length of a simple convex lens if the distances from the lens to the object and to the image are, respectively, 15 cm and 5 cm.

Substitute $d_o = 15$ and $d_i = 5$ into the lens formula above.

$$\frac{1}{f} = \frac{1}{15} + \frac{1}{5}$$

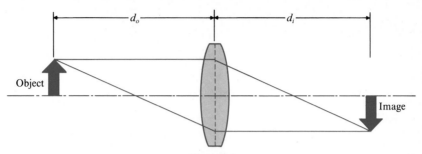

Figure 4.4

Then solve this equation for f.

$$15f \cdot \frac{1}{f} = 15f \cdot \left(\frac{1}{15} + \frac{1}{5} \right) \qquad \text{Multiply by the LCD } 15f$$

$$15f \cdot \frac{1}{f} = 15f \cdot \frac{1}{15} + 15f \cdot \frac{1}{5} \qquad \text{Distribute } 15f \text{ over every term}$$

$$15 = f + 3f \qquad \text{Find each product}$$

$$15 = 4f \qquad \text{Combine like terms}$$

$$f = \frac{15}{4} \qquad \text{Divide by 4, reverse sides}$$

Try Problem 1   The focal length is $\frac{15}{4} = 3.75$ cm. ◀

WORD PROBLEMS

We solve the word problems in this section using the same five steps given in Section 3.3. The only difference here is that the equation we write in step 3 involves rational expressions.

EXAMPLE 2   The sum of the reciprocals of two consecutive odd integers is equal to 8 divided by the product of the two integers. Find the integers.

Step 1   Write down the quantities to be determined.

smaller odd integer
larger odd integer

Step 2   Represent one quantity by x, and then represent the other quantity in terms of x.

$x =$ smaller odd integer
$x + 2 =$ larger odd integer

Step 3 Write an equation involving x.

$$\left(\begin{array}{c}\text{Reciprocal of}\\\text{smaller integer}\end{array}\right) + \left(\begin{array}{c}\text{reciprocal of}\\\text{larger integer}\end{array}\right) = \frac{8}{\text{product of integers}}$$

$$\frac{1}{x} \quad + \quad \frac{1}{x+2} \quad = \quad \frac{8}{x \cdot (x+2)}$$

Step 4 Solve the equation.

We distribute the LCD $x(x+2)$ over every term.

$$x(x+2) \cdot \frac{1}{x} + x(x+2) \cdot \frac{1}{x+2} = x(x+2) \cdot \frac{8}{x(x+2)}$$
$$(x+2) + x = 8$$
$$2x + 2 = 8$$
$$2x = 6$$
$$x = 3$$
$$x + 2 = 5$$

The integers are 3 and 5.

Step 5 Check in the original problem.

Try Problem 9

The sum of the reciprocals of 3 and 5 is $\frac{1}{3} + \frac{1}{5} = \frac{5}{15} + \frac{3}{15} = \frac{8}{15}$, which is 8 divided by the product of the two integers. ◄

An important type of word problem that involves rational expressions is the work problem. To solve work problems, we use the following assumption: If it takes 4 hours to do a job, then the portion of the job completed in 1 hour is $\frac{1}{4}$. We can extend this example to the general rule below.

Rule for Solving Work Problems

If it takes t hours to do a job, then the portion of the job completed in 1 hour is $\frac{1}{t}$.

EXAMPLE 3 It takes Lynn 4 hours to mow the lawn, and Elaine just 3 hours. How long will it take them to mow the lawn if they work together?

$$t = \text{time to mow lawn working together}$$

$$\begin{pmatrix} \text{Portion of job} \\ \text{done by Lynn} \\ \text{in 1 hour} \end{pmatrix} + \begin{pmatrix} \text{portion of job} \\ \text{done by Elaine} \\ \text{in 1 hour} \end{pmatrix} = \begin{pmatrix} \text{portion of job} \\ \text{done by both} \\ \text{in 1 hour} \end{pmatrix}$$

$$\frac{1}{4} \qquad + \qquad \frac{1}{3} \qquad = \qquad \frac{1}{t}$$

$$12t \cdot \frac{1}{4} + 12t \cdot \frac{1}{3} = 12t \cdot \frac{1}{t} \qquad \text{Distribute the LCD } 12t$$

$$3t + 4t = 12$$
$$7t = 12$$
$$t = \frac{12}{7}$$

It will take $\frac{12}{7} = 1\frac{5}{7}$ hours for Lynn and Elaine to mow the lawn working together. ◄

Try Problem 11

EXAMPLE 4   A cold-water faucet can fill an empty tub in 8 minutes, whereas the hot-water faucet takes 12 minutes. A pump can empty a full tub in 6 minutes. If by mistake both faucets are open and the pump is working, how long will it take to fill the empty tub?

$$t = \text{time to fill tub}$$

$$\begin{pmatrix} \text{Portion of job} \\ \text{done by cold} \\ \text{faucet in 1} \\ \text{minute} \end{pmatrix} + \begin{pmatrix} \text{portion of job} \\ \text{done by hot} \\ \text{faucet in 1} \\ \text{minute} \end{pmatrix} + \begin{pmatrix} \text{portion of job} \\ \text{done by pump} \\ \text{in 1 minute} \end{pmatrix} = \begin{pmatrix} \text{portion of job} \\ \text{done by all pipes} \\ \text{in 1 minute} \end{pmatrix}$$

$$\frac{1}{8} \qquad + \qquad \frac{1}{12} \qquad + \qquad \left(-\frac{1}{6}\right) \qquad = \qquad \frac{1}{t}$$

$$24t \cdot \frac{1}{8} + 24t \cdot \frac{1}{12} + 24t \cdot \left(-\frac{1}{6}\right) = 24t \cdot \frac{1}{t} \qquad \text{Distribute the LCD } 24t$$

$$3t + 2t - 4t = 24$$
$$t = 24$$

Try Problem 13   It will take 24 minutes to fill the tub. ◄

To solve uniform-motion problems we must recall the formula $d = rt$, which can also be written as $r = \dfrac{d}{t}$ or as $t = \dfrac{d}{r}$.

EXAMPLE 5   A motorboat travels 15 mph in still water. If the boat takes the same amount of time to travel 2 miles upstream as it does to travel 3 miles downstream, find the rate of the current.

$$x = \text{rate of current}$$

You may find it helpful to construct a chart like the one below.

	d	r	t
Upstream	2	$15 - x$	$\dfrac{2}{15 - x}$
Downstream	3	$15 + x$	$\dfrac{3}{15 + x}$

Since $t = \dfrac{d}{r}$

Time upstream = time downstream

$$\frac{2}{15 - x} = \frac{3}{15 + x}$$

Since this equation equates two fractions, we can cross-multiply.

$$2(15 + x) = 3(15 - x)$$
$$30 + 2x = 45 - 3x$$
$$5x = 15$$
$$x = 3$$

Try Problem 15 The rate of the current is 3 mph. ◄

PROBLEM SET 4.6

Solve the following formula problems.

1. Find the focal length of a simple convex lens if the distances from the lens to the object and to the image are, respectively, 20 cm and 5 cm. (See Example 1.)

2. Find the focal length of a simple convex lens if the distances from the lens to the object and to the image are, respectively, 15 cm and 10 cm. (See Example 1.)

3. Find the resistance in one branch of a parallel-wired circuit if the resistance in the other branch is 5 ohms and the total resistance is 3 ohms. (See the Problem Solving example in Section 4.5.)

4. Find the resistance in one branch of a parallel-wired circuit if the resistance in the other branch is 7 ohms and the total resistance is 2 ohms. (See the Problem Solving example in Section 4.5.)

Solve the following number problems.

5. What number must be added to the numerator and the denominator of the fraction $\frac{10}{13}$ to make the result equal $\frac{4}{5}$?

6. What number must be subtracted from the numerator and the denominator of the fraction $\frac{9}{17}$ to make the result equal $\frac{3}{7}$?

7. One number is 3 times another. If the sum of their reciprocals is $\frac{2}{9}$, find the numbers.

8. One number is 5 times another. If the sum of their reciprocals is $\frac{2}{15}$, find the numbers.

9. The sum of the reciprocals of two consecutive odd integers is equal to 12 divided by the product of the two integers. Find the integers.

10. The sum of the reciprocals of two consecutive even integers is equal to 10 divided by the product of the two integers. Find the integers.

Solve the following work problems.

11. It takes Karen 3 hours to type the company payroll checks, whereas Kathy can type them in just 2 hours. How long will it take them to type the checks if they work together?

12. John can plow the family's field in 7 hours using the old tractor; his father can plow the field with the new tractor in just 5 hours. How long will it take them to plow the field if they work together?

13. One inlet pipe can fill an empty tank in 6 minutes; another takes 18 minutes. A pump can empty a full tank in 9 minutes. How long will it take to fill the empty tank if both pipes are open and the pump is on?

14. One inlet pipe can fill an empty swimming pool in 3 days; another takes 6 days. A pump can empty a full pool in 4 days. How long will it take to fill the empty pool if both pipes are open and the pump is on?

Solve the following motion problems.

15. A motorboat travels 10 mph in still water. If the boat takes the same amount of time to travel 2 miles upstream as it does to travel 3 miles downstream, find the rate of the current.

16. A helicopter can travel 80 mph in still air. If the helicopter takes the same amount of time to travel 2 miles against the wind as it does to travel 3 miles with the wind, find the speed of the wind.

17. An express train travels 160 miles in the same time that a freight train travels 100 miles. If the express travels 30 mph faster than the freight, find the speed of each.

18. A plane travels 840 miles in the same time that a car travels 240 miles. If the plane travels 150 mph faster than the car, find the speed of each.

CALCULATOR PROBLEMS

19. Find the focal length of a simple convex lens if the distances from the lens to the object and to the image are, respectively, 16.8 cm and 12.5 cm. (See Example 1.)

20. The relationship between the sidereal period I and the synodic period Y of Mars is given by the formula

$$\frac{1}{I} = \frac{1}{E} - \frac{1}{Y},$$

where E is the sidereal period of the Earth (365.25 days). Find I if $Y = 779.87$ days.

4.7   *Dividing Polynomials*

DIVIDING BY A MONOMIAL

The simplest case of dividing two polynomials occurs when the divisor is a monomial. Recall that fractions are added and subtracted according to the rule

$$\frac{a}{d} + \frac{b}{d} - \frac{c}{d} = \frac{a+b-c}{d}.$$

If we turn this rule around, we have

$$\frac{a + b - c}{d} = \frac{a}{d} + \frac{b}{d} - \frac{c}{d}.$$

Therefore we have the rule below.

Dividing by a Monomial

To divide a polynomial by a monomial we divide each term of the polynomial by the monomial.

EXAMPLE 1   Divide $9x^2 + 6x - 12$ by 3.

$$\frac{9x^2 + 6x - 12}{3} = \frac{9x^2}{3} + \frac{6x}{3} - \frac{12}{3} \qquad \text{Divide each term by 3}$$

$$= 3x^2 + 2x - 4 \qquad \text{Simplify} \blacktriangleleft$$

Try Problem 1

We can check the answer to Example 1 as follows:

$$\text{Divisor} \cdot \text{Quotient} = \text{Dividend}$$

$$3 \cdot (3x^2 + 2x - 4) = 9x^2 + 6x - 12.$$

EXAMPLE 2   Divide.

a) $\dfrac{6x^2 - 14x + 8}{2x} = \dfrac{6x^2}{2x} - \dfrac{14x}{2x} + \dfrac{8}{2x} = 3x - 7 + \dfrac{4}{x}$

b) $\dfrac{64r^5s^5 + 10r^3s^2 - 4r^2s^2}{4r^2s^2} = \dfrac{64r^5s^5}{4r^2s^2} + \dfrac{10r^3s^2}{4r^2s^2} - \dfrac{4r^2s^2}{4r^2s^2}$

Try Problem 3

$$= 16r^3s^3 + \frac{5}{2}r - 1 \blacktriangleleft$$

Now let's consider the case where the divisor is *not* a monomial.

EXAMPLE 3   Divide $3m^2 + 13m - 10$ by $m + 5$.

$$\frac{3m^2 + 13m - 10}{m + 5} = \frac{(3m - 2)(m + 5)}{m + 5} \qquad \text{Factor the numerator}$$

Try Problem 17a

$$= 3m - 2 \qquad \text{Divide out } m + 5 \blacktriangleleft$$

LONG DIVISION OF POLYNOMIALS

We were able to find the quotient in Example 3 because the numerator was factorable and one of its factors was identical to the denominator. Oftentimes, however, this is not the case. In that event we use a procedure called **long division of polynomials.**

Long Division of Polynomials

1. Write the divisor and the dividend in descending powers. Write any missing power with a coefficient of 0.
2. Divide the first term of the divisor into the first term of the dividend to obtain the first term of the quotient.
3. Multiply the term of the quotient obtained in step 2 by every term in the divisor.
4. Subtract the product obtained in step 3 from the dividend to obtain the new dividend.
5. Repeat steps 2, 3, and 4 with the new dividend. Continue repeating these steps until the degree of the new dividend is less than the degree of the divisor. This last dividend is the remainder.

We illustrate these steps in Example 4.

EXAMPLE 4   Use long division of polynomials to find the quotient in Example 3.

$$m + 5 \overline{)3m^2 + 13m - 10} \leftarrow \text{Write divisor and dividend in descending powers}$$

$$
\begin{array}{r}
3m \qquad\qquad \leftarrow \text{Divide } m \text{ into } 3m^2 \text{ to get } 3m \\
m + 5 \overline{)3m^2 + 13m - 10}
\end{array}
$$

$$
\begin{array}{r}
\overparen{3m} \\
(m + 5) \overline{)3m^2 + 13m - 10} \\
3m^2 + 15m \qquad \leftarrow \text{Multiply } 3m \text{ by } m + 5
\end{array}
$$

$$
\begin{array}{r}
3m \qquad\qquad\quad \\
m + 5 \overline{)3m^2 + 13m - 10} \\
\underline{3m^2 + 15m \qquad\quad} \\
-\ 2m - 10 \leftarrow \text{Subtract to get new dividend}
\end{array}
$$

$$
\begin{array}{r}
3m \;-\; 2 \qquad \leftarrow \text{Divide } m \text{ into } -2m \text{ to get } -2 \\
m + 5 \,\overline{)\,3m^2 + 13m - 10} \\
3m^2 + 15m \\
\hline
-\;2m - 10
\end{array}
$$

$$
\begin{array}{r}
3m \;\;(-2) \\
(m + 5)\,\overline{)\,3m^2 + 13m - 10} \\
3m^2 + 15m \\
\hline
-\;2m - 10 \\
-\;2m - 10 \quad \leftarrow \text{Multiply } -2 \text{ by } m + 5
\end{array}
$$

$$
\begin{array}{r}
3m - 2 \\
m + 5 \,\overline{)\,3m^2 + 13m - 10} \\
3m^2 + 15m \\
\hline
-\;2m - 10 \\
-\;2m - 10 \\
\hline
0 \quad \leftarrow \text{Subtract to get remainder}
\end{array}
$$

Try Problem 17b

Since the remainder is 0, the quotient is $3m - 2$. This is the same quotient we obtained in Example 3. ◄

EXAMPLE 5   Divide $10y^3 - y^2 + 12y - 9$ by $5y - 3$.

$$
\begin{array}{r}
\qquad\qquad\qquad\qquad\;\; 5y \text{ into } 10y^3 \\
\qquad\qquad\qquad\qquad\;\; 5y \text{ into } 5y^2 \\
2y^2 + \;\; y \;\; + \;\; 3 \;\;\longleftarrow\;\; 5y \text{ into } 15y \\
5y - 3 \,\overline{)\,10y^3 - \;\; y^2 + 12y - 9} \\
10y^3 - 6y^2 \qquad\qquad \leftarrow 2y^2 \text{ times } 5y - 3 \\
\hline
5y^2 + 12y - 9 \;\;\leftarrow \text{Subtract} \\
5y^2 - \;\; 3y \qquad\;\; \leftarrow y \text{ times } 5y - 3 \\
\hline
15y - 9 \;\;\leftarrow \text{Subtract} \\
15y - 9 \;\;\leftarrow 3 \text{ times } 5y - 3 \\
\hline
0 \;\;\leftarrow \text{Subtract}
\end{array}
$$

Try Problem 23

The quotient is $2y^2 + y + 3$. The remainder is 0. ◄

EXAMPLE 6   Divide $4p + 8p^3 + 10$ by $1 + 2p$.

Write both polynomials in descending powers, and write $0p^2$ for the missing second-degree term in the dividend.

$$
\begin{array}{r}
4p^2 - 2p\ + 3 \\
2p + 1\overline{)\ 8p^3 + 0p^2 + 4p + 10} \\
\underline{8p^3 + 4p^2} \\
-4p^2 + 4p + 10 \\
\underline{-4p^2 - 2p} \\
6p + 10 \\
\underline{6p +\ \ 3} \\
7
\end{array}
$$

The quotient is $4p^2 - 2p + 3$ and the remainder is 7. ◄

The answer to Example 6 can be checked as follows:

Divisor ·   Quotient   + Remainder =   Dividend

$$
\underbrace{(2p+1)}\cdot\underbrace{(4p^2-2p+3)}+\ \ \ \underbrace{7}\ \ \ =\underbrace{8p^3+4p+10}.
$$

If we turn this equation around and divide both sides by $2p + 1$, we have

$$
\frac{8p^3 + 4p + 10}{2p + 1} = \frac{(2p+1)(4p^2 - 2p + 3)}{2p + 1} + \frac{7}{2p + 1}
$$

Dividend → $\dfrac{8p^3 + 4p + 10}{2p + 1}$ = $\underbrace{4p^2 - 2p + 3}_{\text{Quotient}} + \dfrac{7}{2p + 1}\cdot$ ← Remainder

Divisor →   ← Divisor

Try Problem 27 \   We shall write our answers in this form from now on.

EXAMPLE 7   Divide $\dfrac{6w^4 + w^3 - 5}{3w^2 - 4w}$.

We arrange our work as shown below.

$$
\begin{array}{r}
2w^2 + 3w\ + \ 4 \\
3w^2 - 4w\overline{)\ 6w^4 +\ \ w^3 +\ 0w^2 +\ \ 0w - 5}\ \ \leftarrow\text{Write }0w^2\text{ and }0w \\
\underline{6w^4 - 8w^3} \\
9w^3 +\ \ 0w^2 +\ \ 0w - 5 \\
\underline{9w^3 - 12w^2} \\
12w^2 +\ \ 0w - 5 \\
\underline{12w^2 - 16w} \\
16w - 5\ \ \leftarrow\text{Stop, since degree is less}
\end{array}
$$

                              than degree of divisor

Try Problem 35 Therefore the answer is $2w^2 + 3w + 4 + \dfrac{16w - 5}{3w^2 - 4w}$. ◄

Long division of polynomials can be used to divide two polynomials whenever the degree of the dividend is equal to or greater than the degree of the divisor.

PROBLEM SET 4.7

Divide.

1. $\dfrac{8x^2 + 4x - 6}{2}$

2. $\dfrac{10x^2 + 15x - 20}{5}$

3. $\dfrac{9x^2 - 15x + 6}{3x}$

4. $\dfrac{8x^2 - 16x + 12}{4x}$

5. $(y^3 + y^2 - y) \div y$

6. $(y^5 + y^3 - y^2) \div y^2$

7. $\dfrac{6a^5 - 12a^4 + 10a^3 - 18a}{6a^3}$

8. $\dfrac{16a^5 - 8a^4 + 10a^3 - 24a}{8a^3}$

9. $\dfrac{27p^2q^2 + 18pq - 3p}{9p^2q}$

10. $\dfrac{50p^2q^2 - 75pq + 5q}{25pq^2}$

11. $\dfrac{45x^3y^3z^5 + 15x^3y^2z^4 + 30x^2y^3z^3}{15xyz}$

12. $\dfrac{80x^3y^4z^6 + 16x^3y^3z^5 + 48x^2y^4z^4}{16xyz}$

13. $\dfrac{64r^5s^5 + 10r^2s^3 - 8r^2s^2}{8r^2s^2}$

14. $\dfrac{48r^6s^6 + 20r^3s^4 - 12r^3s^3}{12r^3s^3}$

*Find each quotient **a)** by factoring and simplifying, **b)** by long division of polynomials.*

15. $\dfrac{4x - 12}{x - 3}$

16. $\dfrac{3x - 15}{x - 5}$

17. $\dfrac{3m^2 + 10m - 8}{m + 4}$

18. $\dfrac{3m^2 + m - 10}{m + 2}$

19. $\dfrac{6r^2 + 11r - 10}{3r - 2}$

20. $\dfrac{8r^2 + 14r - 15}{4r - 3}$

Use long division of polynomials to find the quotient and the remainder for each division.

21. $\dfrac{x^3 + 3x^2 + 5x + 3}{x + 1}$

22. $\dfrac{x^3 + 2x^2 + 4x + 3}{x + 1}$

23. $\dfrac{10y^3 - 23y^2 + y - 15}{2y - 5}$

24. $\dfrac{9y^3 - 3y^2 + 4y + 4}{3y - 2}$

25. $\dfrac{4x^3 - x^2 + 17x - 15}{4x + 3}$

26. $\dfrac{5x^3 - 3x^2 - 22x + 8}{5x + 2}$

27. $(11p + 9p^3 + 10) \div (1 + 3p)$

28. $(7p + 12p^3 + 13) \div (1 + 2p)$

29. $\dfrac{6t^3 - 5t^2 + 18t - 15}{6t - 5}$

30. $\dfrac{8t^3 - 7t^2 + 16t - 14}{8t - 7}$

31. $\dfrac{2r^3 + 5r^2 - 3r - 21}{r^2 - r - 3}$

32. $\dfrac{3r^3 + 7r^2 - r - 8}{r^2 + r - 2}$

33. $\dfrac{6s^4 - 10s^3 - s^2 + 15s - 9}{2s^2 - 4s + 3}$

34. $\dfrac{4s^4 + 6s^3 + 22s^2 - 15s + 75}{2s^2 - 3s + 5}$

35. $\dfrac{20w^4 + 13w^3 - 8}{5w^2 - 3w}$

36. $\dfrac{8w^4 + 10w^3 - 7}{4w^2 - 3w}$

37. $\dfrac{8m^3 - 26m^2 + 28m - 91}{2m^2 + 7}$

38. $\dfrac{12m^3 - 39m^2 + 32m - 104}{3m^2 + 8}$

39. $(x^4 - 81) \div (3 + x)$

40. $(x^4 - 16) \div (2 + x)$

41. The average cost AC per unit when x units are produced is determined by dividing the total cost C of producing x units by the number of units x. Write a formula for AC if $C = 7.5x + 1500$.

42. Do Problem 41 if $C = 0.01x^2 + 4x + 2500$.

†43. Determine the value of k that makes $2x - 1$ a factor of $2x^3 + 5x^2 + 9x + k$.

†44. Determine the value of k that makes $3x - 1$ a factor of $3x^3 + 5x^2 + 13x + k$.

CALCULATOR PROBLEMS

Divide.

45. $\dfrac{68.242p^5 - 11.908p^3 + 5.038}{9.16p}$

46. $\dfrac{35.568r^3 + 21.594r^2 - 8.508r - 7.143}{6.24r + 1.38}$

4.8   *Synthetic Division*

We can streamline the process of long division of polynomials if the divisor is of the form $x - k$ or $x + k$, where k is a constant. Consider the long division of $2x^3 - x^2 - 8x - 10$ by $x - 3$, shown on the left below.

$$
\begin{array}{r}
2x^2 + 5x + 7 \\
x - 3\overline{)\,2x^3 - x^2 - 8x - 10} \\
\underline{2x^3 - 6x^2} \\
5x^2 - 8x - 10 \\
\underline{5x^2 - 15x} \\
7x - 10 \\
\underline{7x - 21} \\
11
\end{array}
$$

$$
\begin{array}{r}
2 \quad 5 \quad 7 \\
1 \;\; -3\overline{)\,2 \;\; -1 \;\; -8 \;\; -10} \\
\underline{(2) \;\; -6} \\
5 \;\; (-8) \;\; (-10) \\
\underline{(5) \;\; -15} \\
7 \;\; (-10) \\
\underline{(7) \;\; -21} \\
11
\end{array}
$$

Now look at this same division with the variables removed, as shown on the right above. The numbers in parentheses are simply repetitions of the numbers directly above them. Omit these numbers and obtain the arrangement shown on the left below.

$$
\begin{array}{r}
\,2 \quad\;\; 5 \qquad 7 \\
\hline
1 \;\; -3\,)\overline{2 \quad -1 \quad -8 \quad -10} \\
\downarrow\,-6 \\
\hline
2 \qquad 5 \\
-15 \\
\hline
7 \\
-21 \\
\hline
11
\end{array}
\qquad
\begin{array}{r}
1 \;\; -3\,)\overline{2 \quad -1 \quad -8 \quad -10} \\
-6 \quad -15 \quad -21 \\
\hline
2 \qquad 5 \qquad 7 \qquad 11
\end{array}
$$

If we bring down the first number 2 in the dividend, then the numbers in color duplicate the numbers on the top row (the coefficients of the quotient). Therefore omit the top row and condense the remaining numbers as shown on the right above.

The primary cause of the repetitions is the number **1** in the divisor **1 − 3**. These repetitions occurred because $\dfrac{a}{1} = a$ and $a \cdot \mathbf{1} = a$. Therefore we omit the 1 in 1 − 3. Also, we write the opposite of the number −3 in 1 − 3. Now when each term in the third row is multiplied by +3, the resulting term in the second row has the opposite sign to what it had in the previous arrangement (see below). This means that instead of *subtracting* we can *add*, which is easier.

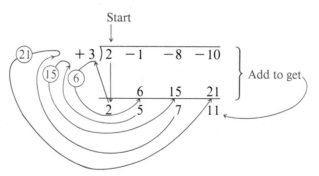

Finally, we box the remainder 11 to facilitate writing the answer, which we obtain by inserting the appropriate powers of x.

$$
\begin{array}{r}
\underline{3|}\;\; 2 \qquad -1 \qquad -8 \qquad -10 \\
 6 \qquad 15 \qquad 21 \\
\hline
2 \qquad\;\; 5 \qquad\;\; 7 \qquad\;\; \boxed{11} \leftarrow \text{Remainder}
\end{array}
$$

$$\text{Answer} \rightarrow 2x^2 + \;\; 5x + \;\; 7 + \frac{11}{x-3} \leftarrow \text{Divisor}$$

The shortcut procedure outlined above is called **synthetic division.** It is used only when the divisor can be written in the form $x - k$ or $x + k$.

EXAMPLE 1   Use synthetic division to divide $3x^3 + 2x^2 - 9x - 10$ by $x - 2$.

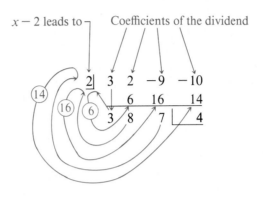

Try Problem 7   The answer is $3x^2 + 8x + 7 + \dfrac{4}{x - 2}$. ◄

CAUTION You must write the dividend in descending powers and insert a coefficient of 0 for each missing power.

EXAMPLE 2   Use synthetic division to divide $\dfrac{2x^3 + x^4 - 5x^2 - 58}{x + 4}$.

Write the dividend in descending powers as $x^4 + 2x^3 - 5x^2 + 0x - 58$.

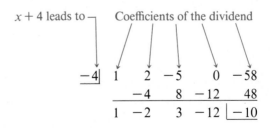

Try Problem 25   The answer is $x^3 - 2x^2 + 3x - 12 - \dfrac{10}{x + 4}$. ◄

With practice, the process of synthetic division soon becomes quite mechanical.

PROBLEM SET 4.8

Use synthetic division to divide.

1. $(3x^2 + 4x + 1) \div (x - 2)$

2. $(2x^2 + 4x + 1) \div (x - 3)$

3. $(3x^2 + 4x + 1) \div (x + 2)$

4. $(2x^2 + 4x + 1) \div (x + 3)$

5. $(x^2 + 3x - 10) \div (x - 2)$

6. $(x^2 + 2x - 15) \div (x - 3)$

7. $(2x^3 - x^2 - 7x - 11) \div (x - 3)$

8. $(3x^3 - x^2 - 2x - 7) \div (x - 2)$

9. $(2x^3 - 5x^2 - 6x - 10) \div (x - 4)$

10. $(2x^3 - 7x^2 - 6x - 25) \div (x - 5)$

11. $\dfrac{x^3 + 5x^2 - x + 4}{x + 3}$

12. $\dfrac{x^3 + 4x^2 - x + 7}{x + 2}$

13. $\dfrac{y^3 + y^2 - 4}{y - 1}$

14. $\dfrac{y^3 + y^2 - 5}{y - 1}$

15. $\dfrac{4m^3 + 2m^2 - 7}{m + 1}$

16. $\dfrac{5m^3 + 3m^2 - 1}{m + 1}$

17. $\dfrac{5x^3 + 10x^2 + 4x + 11}{x + 2}$

18. $\dfrac{4x^3 + 12x + 2x + 11}{x + 3}$

19. $\dfrac{r^3 - r^2 - 10r}{r - 5}$

20. $\dfrac{r^3 - r^2 - 15r}{r - 5}$

21. $\dfrac{p^4 - 2p^3 + 2p^2 - 10p - 8}{p - 3}$

22. $\dfrac{p^4 - 3p^3 + 6p^2 - 4p - 1}{p - 2}$

23. $\dfrac{2t^4 + 15t^3 - 44t^2 + 61t + 10}{t + 10}$

24. $\dfrac{3t^4 + 25t^3 - 49t^2 + 12t + 20}{t + 10}$

25. $\dfrac{3x^3 + x^4 - 2x^2 - 42}{x + 4}$

26. $\dfrac{6x^3 + 11x^2 + x^4 - 59}{x + 4}$

27. $\dfrac{2y^4 - 5y^2 + y + 1}{y + 1}$

28. $\dfrac{3y^4 - 2y^3 + y + 1}{y + 1}$

29. $\dfrac{3 - 5x - 4x^2 + 4x^3 + x^4}{x + 5}$

30. $\dfrac{2 - 5x + 4x^2 + 6x^3 + x^4}{x + 5}$

31. $\dfrac{m^5 - 2m^4 + 3m^3 - 5m^2 - 4}{m + 2}$

32. $\dfrac{m^5 - 2m^4 - 3m^3 + m^2 - 9}{m + 3}$

33. $(x^3 - 1) \div (x - 1)$

34. $(x^3 - 8) \div (x - 2)$

35. $(x^5 + 32) \div (x + 2)$

36. $(x^5 + 1) \div (x + 1)$

37. $(4y^3 + 5y - 3) \div (y - \frac{1}{2})$

38. $(8y^3 + 2y - 2) \div (y - \frac{1}{2})$

CALCULATOR PROBLEMS

Use synthetic division to divide.

39. $\dfrac{3x^3 + 19x^2 - 1868x - 34{,}853}{x - 29}$

40. $\dfrac{3.4x^3 + 7.3x^2 - 5.6x - 4.8}{x + 2.5}$

CHAPTER 4 REVIEW

[4.1] *For what value is each rational expression undefined?*

1. $\dfrac{1}{3x}$

2. $\dfrac{3y}{2 - y}$

Simplify each rational expression.

3. $\dfrac{6a}{9a}$

4. $\dfrac{x + 6}{x + 9}$

5. $\dfrac{6x - 18}{x - 3}$

6. $\dfrac{15t + 20}{9t^2 - 16}$

7. $\dfrac{-15y}{12y^3}$

8. $\dfrac{(a - b)^2}{b^2 - a^2}$

9. $\dfrac{m^3 + 8n^3}{m + 2n}$

10. $\dfrac{12p^2 + 8pq - 15q^2}{6p^2 + 7pq - 3q^2}$

Supply the missing numerator.

11. $\dfrac{1}{4x^2y} = \dfrac{?}{24x^3y^3}$

12. $t = \dfrac{?}{t + 2}$

13. $\dfrac{s - 1}{s + 1} = \dfrac{?}{3s(s + 1)}$

14. $\dfrac{p + 3}{p + 5} = \dfrac{?}{p^2 - 25}$

[4.2] *Multiply or divide as indicated. Write your answer in simplest form.*

15. $\dfrac{x^4}{12} \cdot \dfrac{4}{x}$

16. $\dfrac{a^3b}{3b^3} \div \dfrac{ab^3}{3a^2}$

17. $\dfrac{5q - 15}{5} \cdot \dfrac{25q}{3q - 9}$

18. $\dfrac{p^2 - 4p - 21}{2p - 14} \cdot \dfrac{p^4}{p^2 - 9}$

19. $\dfrac{x^2 - 3x - 4}{x^2 - 4} \div \dfrac{x^2 + 4x + 3}{x^2 + 5x + 6}$

20. $\dfrac{c^3 - cd^2}{2c + 3d} \cdot \dfrac{2c^2 - 21cd - 36d^2}{c^2 - 11cd - 12d^2}$

21. $(x^3 + 8) \div \dfrac{x + 2}{x - 2}$

[4.3] *Perform the indicated operations. Simplify your answer.*

22. $\dfrac{y + 8}{y + 3} - \dfrac{5}{y + 3}$

23. $\dfrac{3}{20x^2} + \dfrac{4}{20x^2} - \dfrac{2}{20x^2}$

*For each problem **a)** find the LCD, **b)** use the LCD to perform the indicated operations. Simplify your answer.*

24. $\dfrac{5}{8a} + \dfrac{3}{2a^2}$

25. $\dfrac{3}{16x^2y} - \dfrac{1}{20xy^3}$

26. $\dfrac{p^2}{p+1} - p$

27. $\dfrac{8}{h-8} - \dfrac{2}{h-2}$

28. $\dfrac{y}{2y-10} - \dfrac{2}{y^2-25}$

29. $\dfrac{a}{a^2-b^2} + \dfrac{b}{a^2-ab}$

30. $\dfrac{x+y}{x^2-xy} - \dfrac{x-y}{x^2+xy}$

31. $\dfrac{k}{k^2+k-6} + \dfrac{1}{k^2+2k-8}$

32. $\dfrac{1}{r+1} + r - 1$

33. $\dfrac{y-6}{y+2} + \dfrac{16y}{y^2-4}$

[4.4]   *Simplify each complex fraction.*

34. $\dfrac{\dfrac{3}{x} + \dfrac{3}{y}}{\dfrac{3}{x} - \dfrac{3}{y}}$

35. $\dfrac{\dfrac{1}{a} + \dfrac{1}{ab}}{\dfrac{1}{b} + \dfrac{1}{ab}}$

36. $\dfrac{9 - \dfrac{1}{m^2}}{3 + \dfrac{1}{m}}$

37. $\dfrac{8 - \dfrac{10}{p} - \dfrac{3}{p^2}}{4 - \dfrac{4}{p} - \dfrac{3}{p^2}}$

38. $\dfrac{1 - \dfrac{3}{x}}{x^3 - 27}$

39. $\dfrac{\dfrac{1}{m^3} - \dfrac{1}{n^3}}{\dfrac{1}{m^2} - \dfrac{1}{n^2}}$

40. $\dfrac{\dfrac{1}{x+1} + \dfrac{1}{x-1}}{\dfrac{x+1}{x-1} - \dfrac{x-1}{x+1}}$

41. $x - \dfrac{1}{1 - \dfrac{1}{1 + \dfrac{1}{x}}}$

[4.5]   *Solve each equation.*

42. $\dfrac{x}{6} + \dfrac{x}{2} = 4$

43. $\dfrac{3y+2}{y+2} = 0$

44. $\dfrac{w}{w-3} = \dfrac{w}{w+3}$

45. $\dfrac{x+1}{2x-2} - \dfrac{x-2}{3x-3} = \dfrac{1}{6}$

46. $\dfrac{3}{t+3} - \dfrac{2}{t+5} = \dfrac{1}{t+2}$

47. $\dfrac{3x-2}{x+3} - \dfrac{3x+2}{x-3} = \dfrac{40-2x}{x^2-9}$

Solve each equation for the specified variable.

48. $\dfrac{PV}{T} = \dfrac{pv}{t}$ for v

49. $S = \dfrac{rl - a}{r - 1}$ for r

[4.6]

50. What number must be added to the numerator and to the denominator of the fraction $\frac{3}{5}$ to make the result equal $\frac{5}{6}$?

51. One inlet pipe can fill an empty swimming pool in 7 hours, whereas another takes 6 hours. How long will it take both pipes to fill the pool?

52. A plane travels 1330 miles in the same time that a car travels 175 miles. If the plane travels 330 mph faster than the car, find the speed of each.

[4.7]   *Divide.*

53. $\dfrac{3x^3 - 9x^2 + 12x}{3x}$

54. $\dfrac{8a^3bc^2 + 12ab^2c^2 - 36a^2b^3c^3}{4abc^2}$

55. $\dfrac{x^2 - 7x + 10}{x - 2}$

56. $\dfrac{6y^3 + y^2 - 8y + 6}{2y + 3}$

57. $(6x^2 - 5x + 3) \div (2x - 3)$

58. $\dfrac{4y^3 - 3y + 1}{1 + 2y}$

59. $\dfrac{6x^3 - 5x^2 + 5}{2x^2 - 3x + 2}$

60. $\dfrac{x^5 - 1}{x - 1}$

Use synthetic division to divide.

61. $(a^2 - a - 6) \div (a - 3)$

62. $(t^3 - 6t^2 + 11t - 6) \div (t - 2)$

63. $(2x^4 + 13x - 3) \div (x - 2)$

64. $(4y^3 + 10y^2 - 19y + 16) \div (y + 4)$

65. $\dfrac{x^4 - 16}{x + 2}$

66. $(x^3 + 2x^2 - 8x - 24) \div (x - 3)$

CHAPTER 4 TEST

Supply the missing numerator.

1. $\dfrac{1}{4x} = \dfrac{?}{12x^2}$

2. $\dfrac{x}{x + 2} = \dfrac{?}{x^2 + 3x + 2}$

Simplify.

3. $\dfrac{5a - 15}{a - 3}$

4. $\dfrac{t^2 - t - 6}{t^2 - 9}$

Simplify.

5. $\dfrac{x + \dfrac{x}{y}}{1 + \dfrac{1}{y}}$

6. $\dfrac{1 - \dfrac{7}{p} + \dfrac{12}{p^2}}{1 - \dfrac{5}{p} + \dfrac{6}{p^2}}$

Perform the indicated operation. Write your answer in simplest form.

7. $\dfrac{2y}{y + 5} + \dfrac{10}{y + 5}$

8. $\dfrac{3a^3}{14b^3} \cdot \dfrac{18b^5}{27a^6}$

9. $\dfrac{5}{6a} + \dfrac{2}{3a^2}$

10. $\dfrac{x + 2}{x^2 - 3x} \cdot \dfrac{x - 3}{x^2 - 4}$

11. $\dfrac{x + 1}{x - 1} - \dfrac{3}{x + 2}$

12. $\dfrac{6x^2 - 11x - 10}{3x^2 - 4x + 1} \div \dfrac{2x^2 + x - 15}{3x^2 + 8x - 3}$

13. $\dfrac{y}{y + 3} - \dfrac{18}{y^2 - 9}$

Solve each equation.

14. $\dfrac{x}{3} + \dfrac{x}{6} = 1$

15. $\dfrac{x + 2}{x + 9} = 0$

16. $\dfrac{t + 3}{t} = \dfrac{t - 1}{t + 2}$

17. $\dfrac{3}{5m + 5} - \dfrac{1}{10} = \dfrac{1}{2m + 2}$

Solve each equation for the specified variable.

18. $\dfrac{PV}{T} = \dfrac{pv}{t}$ for T

19. $d = \dfrac{r}{1 - rt}$ for t

Divide.

20. $\dfrac{15t^3 - 20t^2 + 10t}{5t}$

21. $\dfrac{x^4 - 3x^3 + x^2 + 8x - 1}{x^2 - 2}$

22. $(y^3 - 7y + 6) \div (y - 1)$

23. Divide $x^3 - 2x^2 - x + 2$ by $x - 3$ using synthetic division.

24. What number must be added to the numerator and to the denominator of the fraction $\frac{5}{7}$ to make the result equal $\frac{5}{6}$?

25. Karen can make a quilt in 10 hours. Kathy can make the same quilt in 6 hours. How long will it take them to make the quilt if they work together?

Many applied problems and formulas involve exponents that are not positive integers. For example, scientists use positive and negative exponents to express very large and very small numbers in "scientific notation." Kepler's third law of planetary motion, $T = d^{3/2}$, uses a rational-number exponent. In this chapter we define negative-integer and rational-number exponents. These exponents obey the rules for exponents introduced in Section 2.2. Since rational exponents are defined in terms of radicals, we shall also examine radical notation and its rules.

The chapter concludes with a discussion of a type of number that may be unfamiliar to you—an imaginary number. You will learn to operate with imaginary numbers, and you will see how imaginary numbers combine with the set of real numbers to form the complex number system.

5

Rational Exponents, Radicals, and Complex Numbers

5.1 *Negative Exponents*

If we apply the rule $a^m \cdot a^n = a^{m+n}$ to find the product of a^3 and a^{-3}, we have

$$a^3 \cdot a^{-3} = a^{3+(-3)} = a^0 = 1.$$

But if $a^3 \cdot a^{-3} = 1$, then a^{-3} must be the reciprocal of a^3. That is,

$$a^{-3} = \frac{1}{a^3}.$$

In general, we have the definition below.*

Definition

If a is any nonzero real number and n is any whole number, then

$$a^{-n} = \frac{1}{a^n}.$$

Based on this definition, we see that

$$a^{-1} = \frac{1}{a},$$

$$a^{-2} = \frac{1}{a^2},$$

$$a^{-3} = \frac{1}{a^3},$$

and so on.

We shall use the word "simplify" in this section to mean "write with positive exponents in simplest form."

EXAMPLE 1 Simplify: **a)** 5^{-1}, **b)** 3^{-2}, **c)** $(-4)^{-3}$, **d)** $(-2)^{-4}$.

a) $\quad 5^{-1} = \dfrac{1}{5}$

b) $\quad 3^{-2} = \dfrac{1}{3^2} = \dfrac{1}{9}$

c) $\quad (-4)^{-3} = \dfrac{1}{(-4)^3} = -\dfrac{1}{64}$

---

* The English mathematician John Wallis (1616–1703) was the first to fully explain the meaning of negative exponents.

Try Problem 3 **d)** $(-2)^{-4} = \dfrac{1}{(-2)^4} = \dfrac{1}{16}$ ◄

CAUTION The negative sign on the exponent has no influence on the sign of the answer. That is, 5^{-1} does *not* equal $5 \cdot (-1)$.

EXAMPLE 2 Simplify: **a)** $5x^{-2}$, **b)** $(5x)^{-2}$.

a) $5x^{-2} = 5 \cdot \dfrac{1}{x^2} = \dfrac{5}{x^2}$

Try Problem 15 **b)** $(5x)^{-2} = \dfrac{1}{(5x)^2} = \dfrac{1}{25x^2}$ ◄

CAUTION The expressions $5x^{-2}$ and $(5x)^{-2}$ are not the same.

Suppose that a^{-3} appears in a denominator. Then

$$\frac{1}{a^{-3}} = \frac{1}{\dfrac{1}{a^3}} = 1 \cdot \frac{a^3}{1} = a^3.$$

In general, if $a \neq 0$ and n is a whole number, then

$$\frac{1}{a^{-n}} = a^n.$$

Based on this rule, we see that

$$\frac{1}{a^{-1}} = a,$$

$$\frac{1}{a^{-2}} = a^2,$$

$$\frac{1}{a^{-3}} = a^3,$$

and so on.

EXAMPLE 3 Simplify: **a)** $\dfrac{1}{10^{-3}}$, **b)** $\dfrac{3}{10^{-3}}$.

a) $\dfrac{1}{10^{-3}} = 10^3 = 1000$

Try Problem 17 **b)** $\dfrac{3}{10^{-3}} = 3 \cdot \dfrac{1}{10^{-3}} = 3 \cdot 10^3 = 3 \cdot 1000 = 3000$ ◄

EXAMPLE 4 Simplify: **a)** $\dfrac{2}{3r^{-2}}$, **b)** $\dfrac{2}{(3r)^{-2}}$.

a) $\dfrac{2}{3r^{-2}} = \dfrac{2}{3} \cdot \dfrac{1}{r^{-2}} = \dfrac{2}{3} r^2 \left(\text{or } \dfrac{2r^2}{3}\right)$

Try Problem 19 **b)** $\dfrac{2}{(3r)^{-2}} = 2 \cdot \dfrac{1}{(3r)^{-2}} = 2 \cdot (3r)^2 = 2 \cdot 9r^2 = 18r^2$ ◄

We can shorten our work in some problems by noting that
$$\left(\dfrac{a}{b}\right)^{-3} = \dfrac{1}{\left(\dfrac{a}{b}\right)^3} = 1 \cdot \left(\dfrac{b}{a}\right)^3 = \left(\dfrac{b}{a}\right)^3.$$

In general, if $a \neq 0$ and $b \neq 0$ and n is a whole number, then

$$\left(\dfrac{a}{b}\right)^{-n} = \left(\dfrac{b}{a}\right)^{n}.$$

EXAMPLE 5 Simplify: **a)** $(\tfrac{3}{4})^{-2}$, **b)** $(-\tfrac{1}{2})^{-3}$.

a) $(\tfrac{3}{4})^{-2} = (\tfrac{4}{3})^2 = \tfrac{16}{9}$

Try Problem 21 **b)** $(-\tfrac{1}{2})^{-3} = (-\tfrac{2}{1})^3 = -8$ ◄

CAUTION The expression $(\tfrac{3}{4})^{-2}$ does *not* mean $\tfrac{3}{4}$ times -2.

We can use the rules for exponents stated in Section 2.2 to simplify expressions with negative exponents.

EXAMPLE 6 Simplify: **a)** $s^4 \cdot s^{-14} \cdot s$, **b)** $10^6 \cdot 10^{-2}$.

Add exponents to multiply powers having the same base.

a) $s^4 \cdot s^{-14} \cdot s = s^{4+(-14)+1} = s^{-9} = \dfrac{1}{s^9}$

Try Problem 29 **b)** $10^6 \cdot 10^{-2} = 10^{6+(-2)} = 10^4$ (or 10,000) ◄

EXAMPLE 7 Simplify: **a)** $\dfrac{m^{-5}}{m}$, **b)** $\dfrac{m^{-6}}{m^{-8}}$.

Subtract the lower exponent from the upper exponent to divide powers having the same base.

a) $\dfrac{m^{-5}}{m} = m^{-5-1} = m^{-6} = \dfrac{1}{m^6}$

Try Problem 31 **b)** $\dfrac{m^{-6}}{m^{-8}} = m^{-6-(-8)} = m^{-6+8} = m^2$ ◄

EXAMPLE 8   Simplify: **a)** $(p^{-5})^4$,   **b)** $(p^{-1}q^3)^{-2}$,   **c)** $\left(\dfrac{p^2}{q^3}\right)^{-20}$.

a)   Multiply exponents when raising a power to a power.
$$(p^{-5})^4 = p^{(-5)\cdot 4} = p^{-20}$$

b)   Since $(ab)^n = a^n b^n$, we have
$$(p^{-1}q^3)^{-2} = (p^{-1})^{-2}(q^3)^{-2} = p^2 q^{-6} = \frac{p^2}{q^6}.$$

c)   Since $\left(\dfrac{a}{b}\right)^n = \dfrac{a^n}{b^n}$, we have

Try Problem 39
$$\left(\frac{p^2}{q^3}\right)^{-20} = \frac{(p^2)^{-20}}{(q^3)^{-20}} = \frac{p^{-40}}{q^{-60}} = \frac{q^{60}}{p^{40}}. \blacktriangleleft$$

EXAMPLE 9   Simplify: **a)** $(-3u^{-2}v^4)^2$,   **b)** $\dfrac{x^{-3}y^3}{x^4 y^{-5}}$.

a) $(-3u^{-2}v^4)^2 = (-3)^2(u^{-2})(v^4)^2 = 9u^{-4}v^8 = \dfrac{9v^8}{u^4}$

Try Problem 49   **b)** $\dfrac{x^{-3}y^3}{x^4 y^{-5}} = \dfrac{x^{-3}}{x^4} \cdot \dfrac{y^3}{y^{-5}} = x^{-3-4} \cdot y^{3-(-5)} = x^{-7}y^8 = \dfrac{y^8}{x^7} \blacktriangleleft$

Sometimes negative exponents give rise to complex fractions.

EXAMPLE 10   Simplify: **a)** $\dfrac{t^{-1}}{t^{-2}+1}$,   **b)** $(x^{-1}+y^{-1})^{-1}$.

a) $\dfrac{t^{-1}}{t^{-2}+1} = \dfrac{\dfrac{1}{t}}{\dfrac{1}{t^2}+1} = \dfrac{\dfrac{1}{t}}{\dfrac{1}{t^2}+1} \cdot \dfrac{t^2}{t^2}$

$$= \dfrac{\dfrac{1}{t}\cdot t^2}{\dfrac{1}{t^2}\cdot t^2 + 1\cdot t^2} = \dfrac{t}{1+t^2}$$

b) $(x^{-1}+y^{-1})^{-1} = \dfrac{1}{x^{-1}+y^{-1}} = \dfrac{1}{\dfrac{1}{x}+\dfrac{1}{y}}$

Try Problem 71
$$= \dfrac{1}{\dfrac{1}{x}+\dfrac{1}{y}}\cdot\dfrac{xy}{xy} = \dfrac{1\cdot xy}{\dfrac{1}{x}\cdot xy + \dfrac{1}{y}\cdot xy} = \dfrac{xy}{y+x} \blacktriangleleft$$

CAUTION The expression $(x^{-1}+y^{-1})^{-1}$ is *not* equal to $x+y$.

Simplify.

1. 8^{-1}
2. 6^{-1}
3. 4^{-2}
4. 5^{-2}
5. $(-2)^{-3}$
6. $(-3)^{-3}$
7. $(-2)^{-4}$
8. $(-3)^{-4}$
9. $(-7)^{-1}$
10. $(-9)^{-1}$
11. $2 \cdot 4^{-1}$
12. $3 \cdot 5^{-1}$
13. $3^{-1} + 4^{-1}$
14. $2^{-1} + 5^{-1}$

Simplify and compare.

15. $6x^{-2}$ and $(6x)^{-2}$
16. $8x^{-2}$ and $(8x)^{-2}$
17. $\dfrac{1}{5^{-3}}$ and $\dfrac{2}{5^{-3}}$
18. $\dfrac{1}{4^{-3}}$ and $\dfrac{3}{4^{-3}}$
19. $\dfrac{3}{4r^{-2}}$ and $\dfrac{3}{(4r)^{-2}}$
20. $\dfrac{2}{5r^{-2}}$ and $\dfrac{2}{(5r)^{-2}}$
21. $\left(\frac{2}{3}\right)^{-2}$ and $\left(-\frac{1}{4}\right)^{-3}$
22. $\left(\frac{2}{5}\right)^{-2}$ and $\left(-\frac{1}{3}\right)^{-3}$
23. $\dfrac{1}{2^{-1}}$ and 2^{-1}
24. $\dfrac{1}{3^{-1}}$ and 3^{-1}

Use the rules for exponents to simplify each expression.

25. $x^{-8} \cdot x^{15}$
26. $x^{13} \cdot x^{-4}$
27. $s^3 \cdot s^{-11} \cdot s$
28. $s^{-17} \cdot s \cdot s^9$
29. $10^7 \cdot 10^{-4}$
30. $10^{-5} \cdot 10^8$
31. $\dfrac{m^{-6}}{m}$
32. $\dfrac{m^{-9}}{m}$
33. $\dfrac{m^{-3}}{m^{-9}}$
34. $\dfrac{m^{-4}}{m^{-8}}$
35. $\dfrac{y^3}{y^{-5}}$
36. $\dfrac{y^4}{y^{-7}}$
37. $\dfrac{10^{-12}}{10^{-2}}$
38. $\dfrac{10^{-14}}{10^{-1}}$
39. $(p^{-3})^5$
40. $(p^{-2})^9$
41. $(p^{-1}q^2)^{-4}$
42. $(p^3q^{-1})^{-6}$
43. $\left(\dfrac{p^3}{q^6}\right)^{-10}$
44. $\left(\dfrac{p^2}{q^4}\right)^{-15}$
45. $(-2u^3v^{-2})^2$
46. $(-3u^{-4}v^2)^2$
47. $(4a^2b^{-4})^{-3}$
48. $(5a^{-3}b^2)^{-3}$
49. $\dfrac{x^{-2}y^3}{x^4y^{-4}}$
50. $\dfrac{x^{-3}y^5}{x^2y^{-4}}$
51. $\left(\dfrac{3x}{y^{-2}}\right)^{-2}$
52. $\left(\dfrac{2x}{y^{-3}}\right)^{-2}$
53. $\dfrac{2a^{-1}}{5b^{-1}}$
54. $\dfrac{3a^{-1}}{4b^{-1}}$

55. $\dfrac{ax^{-2}}{by^{-3}}$

56. $\dfrac{ax^{-3}}{by^{-4}}$

57. $\dfrac{(3a^{-4})^2}{a^{-5}}$

58. $\dfrac{(2a^{-3})^2}{a^{-6}}$

59. $\left(\dfrac{5x^{-3}y^2}{4z^2}\right)^{-2}$

60. $\left(\dfrac{4x^{-5}y^3}{5z^4}\right)^{-2}$

61. $(-2x^{-3}y^4)^5(xy^{-1})^{-2}$

62. $(-3x^{-5}y^2)^4(xy^{-2})^{-1}$

63. $\left(\dfrac{a^{-11}b^4}{c^{-5}}\right)^2\left(\dfrac{5a^{-1}}{8b^{-1}}\right)$

64. $\left(\dfrac{a^6b^{-14}}{c^{-3}}\right)^2\left(\dfrac{3a^{-2}}{7b^{-2}}\right)$

Write as a single fraction in simplest form.

65. $(a+b)^{-2}$

66. $(a-b)^{-2}$

67. $c+c^{-1}$

68. $c^{-2}+c$

69. $\dfrac{1-d^{-2}}{1+d^{-2}}$

70. $\dfrac{d^{-1}+1}{d^{-1}-1}$

71. $\dfrac{t^{-2}}{t^{-1}+1}$

72. $\dfrac{t^{-2}+1}{t^{-1}}$

73. $\dfrac{r^{-1}-s^{-1}}{r-s}$

74. $\dfrac{r-s}{r^{-2}-s^{-2}}$

75. $(x^{-1}-y^{-1})^{-1}$

76. $(x^{-2}+y^{-2})^{-1}$

CALCULATOR PROBLEMS

77. Compute to the nearest hundredth.

    a) $(0.8145)^{-2}$ **b)** $\dfrac{1}{\pi^{-3}}$ **c)** $\left(\dfrac{4.87}{5.66}\right)^{-4}$

78. What principal P must be invested at $13\frac{1}{4}\%$ so that its value V in 8 years is \$25,000? Use $P = V(1+r)^{-t}$.

5.2   *Scientific Notation*

If you square 2,500,000 on your calculator, the display will probably read as follows:

This means that

$$(2{,}500{,}000)^2 = 6.25 \times 10^{12}.$$

Your calculator displays the answer in this form because the answer is too large to be displayed in standard form.

Definition   A positive real number is in **scientific notation** when it is written in the form

$$a \times 10^n,$$

where $1 \le a < 10$ and n is an integer.

We use the symbol $\times$ to denote the multiplication in scientific notation to avoid confusing the dot multiplication symbol with the decimal point.

EXAMPLE 1   The internal temperature of the sun is about 2×10^7 degrees Celsius. Write this number in standard form.

Try Problem 1
$$2 \times 10^7 = 2 \times 10,000,000 = 20,000,000 \blacktriangleleft$$

When we multiply the number 2 by 10^7, we move the decimal point 7 places to the right. Once this is noted, we can omit the middle step in Example 1 and simply write

$$2 \times 10^7 = 20,000,000.$$

7 places

EXAMPLE 2   The greatest recorded eruption of a volcano occurred on the island of Sumbawa in 1815. The energy of this eruption was 8.4×10^{26} ergs. Write this number in standard form.

Try Problem 43
$$8.4 \times 10^{26} = 840,000,000,000,000,000,000,000,000 \blacktriangleleft$$

EXAMPLE 3   The diameter of a red corpuscle is about 3×10^{-5} inch. Write this number in standard form.

Try Problem 5
$$3 \times 10^{-5} = 3 \times \frac{1}{10^5} = 3 \times \frac{1}{100,000} = 3 \times 0.00001 = 0.00003 \blacktriangleleft$$

When we multiply the number 3 by 10^{-5}, we move the decimal point 5 places to the left. That is,

$$3 \times 10^{-5} = 0.00003.$$

5 places

EXAMPLE 4   The mass of an electron is 9.11×10^{-28} gram. Write this number in standard form.

Try Problem 45
$$9.11 \times 10^{-28} = 0.000000000000000000000000000911 \blacktriangleleft$$

To convert a number from standard form to scientific notation we reverse the process. The table below illustrates the conversion of several different numbers.

Standard Form	Scientific Notation
2,001,000	$= 2.001 \times 10^6$
10,000	$= 1 \quad \times 10^4$ (or 10^4)
5.9	$= 5.9 \quad \times 10^0$ (or 5.9)
0.37	$= 3.7 \quad \times 10^{-1}$
0.004	$= 4 \quad \times 10^{-3}$
0.0000707	$= 7.07 \quad \times 10^{-5}$

It is also true that

$$2{,}001{,}000 = 20.01 \times 10^5,$$

but the number on the right is not in scientific notation, since $a = 20.01$ lies outside the interval $1 \le a < 10$.

To take full advantage of scientific notation, we must be able to perform operations on numbers while they are written in this form.

EXAMPLE 5   Compute $(2000)(34{,}000)$ by first converting to scientific notation. Express your answer in scientific notation.

Write each number in scientific notation.

$$(2000)(34{,}000) = (2 \times 10^3)(3.4 \times 10^4)$$

Group the numbers between 1 and 10 together. Group the powers of 10 together.

$$= (2 \times 3.4)(10^3 \times 10^4)$$

Operate on each group separately.

$$= (6.8)(10^7)$$
$$= 6.8 \times 10^7 \blacktriangleleft$$

Try Problem 47

EXAMPLE 6   Compute $\dfrac{(8{,}000{,}000{,}000)(0.072)}{(0.002)(240{,}000{,}000)}$.

$$\frac{(8{,}000{,}000{,}000)(0.072)}{(0.002)(240{,}000{,}000)}$$

$$= \frac{(8 \times 10^9)(7.2 \times 10^{-2})}{(2 \times 10^{-3})(2.4 \times 10^8)} \qquad \text{Convert to scientific notation}$$

$$= \frac{(8)(7.2)}{(2)(2.4)} \times \frac{10^9 \; 10^{-2}}{10^{-3} \; 10^8} \qquad \text{Group}$$

$$= (4)(3) \times \frac{10^7}{10^5}$$

$$= 12 \times 10^2 \qquad \left.\vphantom{\begin{matrix}a\\a\\a\end{matrix}}\right\} \text{Operate on each group}$$

$$= (1.2 \times 10) \times 10^2 \qquad \text{Write 12 in scientific notation}$$
$$= 1.2 \times 10^3 \qquad \text{Combine powers of 10} \blacktriangleleft$$

Try Problem 61 \\

We write negative numbers in scientific notation by first writing the corresponding positive number in scientific notation and then prefixing a negative sign. For example,

$$-477{,}000 = -4.77 \times 10^5 \quad \text{and} \quad -0.00018 = -1.8 \times 10^{-4}.$$

PROBLEM SET 5.2

Write each number in standard form.

1. 6×10^4
2. 3×10^5
3. 5.5×10^7
4. 8.8×10^6
5. 4×10^{-3}
6. 7×10^{-2}
7. 1.9×10^{-5}
8. 2.4×10^{-6}
9. -8.17×10^8
10. -6.43×10^9
11. -7.95×10^{-4}
12. -1.15×10^{-7}
13. 9.007×10^0
14. 5.003×10^0
15. 6.332×10
16. 4.118×10
17. 1.01×10^{-1}
18. 7.07×10^{-1}

Write each number in scientific notation.

19. 6,000,000
20. 9,000,000
21. 140,000
22. 250,000
23. 0.005
24. 0.003
25. 0.00072
26. 0.00047
27. 809,000,000,000
28. 701,000,000,000
29. 0.00000000003084
30. 0.00000000005068
31. 10,000,000
32. 100,000,000
33. 0.00001
34. 0.000001
35. 5.59
36. 4.41
37. -275.1
38. -834.5
39. 43.07
40. 62.09
41. 0.119
42. 0.331

43. The candlepower of the sun is 3×10^{27}. Write this number in standard form.

44. In 1939, one U.S. dollar bought 3.38 Hungarian pengös. By 1946, however, the inflation in Hungary was so bad (the worst in history) that one U.S. dollar bought 5×10^{20} pengös. Write this last number in standard form.

45. The least stable nuclear particles known have a lifetime of 1.6×10^{-24} second. Write this number in standard form.

46. The probable mass of a hydrogen atom is 1.7×10^{-24} gram. Write this number in standard form.

Compute by first converting to scientific notation. Express your answer in scientific notation.

47. $(3000)(21,000)$

48. $(4000)(22,000)$

49. $(40,000,000,000)(0.000000002)$

50. $(20,000,000,000)(0.000000003)$

51. $(0.000008)(5,000,000)$

52. $(0.000006)(5,000,000)$

53. $\dfrac{320,000}{2000}$

54. $\dfrac{360,000}{3000}$

55. $\dfrac{0.00006}{0.00000002}$

56. $\dfrac{0.00008}{0.00000004}$

57. $\dfrac{0.00000042}{7,000,000,000}$

58. $\dfrac{0.00000035}{5,000,000,000}$

59. $\dfrac{(260,000)(42,000)}{0.000000000004}$

60. $\dfrac{(520,000)(27,000)}{0.000000000006}$

61. $\dfrac{(9,000,000,000)(0.084)}{(0.003)(210,000,000)}$

62. $\dfrac{(6,000,000,000)(0.065)}{(0.002)(130,000,000)}$

63. $\dfrac{(7,500,000,000)(0.00098)}{200,000}$

64. $\dfrac{(8,500,000,000)(0.00078)}{500,000}$

65. $\dfrac{(0.00000004)(65,000,000)}{(50,000,000,000)(0.8)}$ *(handwritten: 4×10^{-8} 6.5×10^{7} 5×10^{10} 8×10^{-1} $\dfrac{4(6.5)}{5(8)} \times 10^{-10}$)*

66. $\dfrac{(0.00000005)(44,000,000)}{(80,000,000,000)(0.5)}$

67. A certain computer can perform 4.5×10^5 calculations in a single second. How many calculations can it perform in 10 hours?

68. A certain computer can perform 8.5×10^6 calculations in a single second. How many calculations can it perform in 10 hours?

69. The mass of the earth is about 6×10^{27} grams. A gram is 1.1×10^{-6} ton. Find the mass of the earth in tons.

70. The mass of the sun is about 1.9×10^{36} grams. A gram is 1.1×10^{-6} ton. Find the mass of the sun in tons.

71. How long does it take light rays from the sun 93,000,000 miles away to reach the earth? (The speed of light is 186,000 mi/sec.)

72. Radio waves travel at the speed of light (186,000 mi/sec). How long would it take a radio message from earth to reach a spaceship 279,000 miles away?

CALCULATOR PROBLEMS

73. Compute to three significant figures by first converting to scientific notation. Express your answer in scientific notation.

a) $\dfrac{(43,720,000,000)(0.6517)}{(0.000000000888)(103,000,000)}$

b) $\dfrac{1}{(84,000,000,000)(70,000,000,000)}$

74. The national debt of the United States in 1986 was $1,960,000,000,000. If the total number of U.S. taxpayers was 80,000,000, what was each taxpayer's share of the debt?

5.3   *Roots and Radicals*

READ
  54 too

If $b^2 = a$, then the number b is called a **square root** of the number a. For example, since

$$3^2 = 9,$$

the number 3 is a square root of 9. But -3 is also a square root of 9 since

$$(-3)^2 = 9.$$

In fact, *every positive number has two square roots.*
    If $b^3 = a$, then b is called the **cube root** of a. For example, since

$$2^3 = 8,$$

the number 2 is the cube root of 8. *Every real number has exactly one cube root.* Based on this discussion, we define the **nth root** of a number.

Definition   Let n be a natural number. The number b is called an **nth root** of the number a if $b^n = a$.

As we have seen, the number 9 has two square roots, 3 and -3. To indicate the positive square root of 9, called the **principal square root** of 9, we use the symbol $\sqrt{}$ and write

$$\sqrt{9} = 3.$$

The symbol $\sqrt{}$ is called a **radical sign,** and $\sqrt{9}$ is often read "radical 9." *
    To indicate the negative square root of 9 we write

$$-\sqrt{9} = -3.$$

To indicate both roots we write

$$\pm\sqrt{9} = \pm 3.$$

That is, ± 3 represents two numbers, namely 3 and -3.
    To denote the principal nth root of the number a we write $\sqrt[n]{a}$. The number

---

* The symbol $\sqrt{}$ is a distortion of the letter r from the Latin word *radix* (meaning root).

n is called the **index,** and a is called the **radicand.**

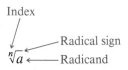

Index

Radical sign

$\sqrt[n]{a}$ ← Radicand

The index 2 is usually omitted and we simply write $\sqrt{a}$ for the principal square root of a.

EXAMPLE 1   Find: **a)** $\sqrt{25}$, **b)** $-\sqrt{25}$, **c)** $\pm\sqrt{25}$, **d)** $\sqrt{-25}$.

a) $\sqrt{25} = 5$

b) $-\sqrt{25} = -5$

c) $\pm\sqrt{25} = \pm 5$

Try Problem 1   **d)** $\sqrt{-25}$ is not a real number ◄

Why isn't $\sqrt{-25}$ a real number? Suppose that there *were* a real number b such that $\sqrt{-25} = b$. Then $b^2 = -25$, which is impossible. The number $\sqrt{-25}$ is actually an *imaginary number.* We shall study imaginary numbers in Section 5.9.

EXAMPLE 2   Find: **a)** $\sqrt[3]{216}$, **b)** $-\sqrt[3]{216}$, **c)** $\sqrt[3]{-216}$, **d)** $-\sqrt[3]{-216}$.

a) $\sqrt[3]{216} = 6$ Since $6^3 = 216$

b) $-\sqrt[3]{216} = -6$

c) $\sqrt[3]{-216} = -6$ Since $(-6)^3 = -216$

Try Problem 13   **d)** $-\sqrt[3]{-216} = -(-6) = 6$ ◄

EXAMPLE 3   Find: **a)** $\sqrt[4]{81}$, **b)** $-\sqrt[4]{81}$, **c)** $\sqrt[4]{-81}$.

a) $\sqrt[4]{81} = 3$ Since $3^4 = 81$

b) $-\sqrt[4]{81} = -3$

Try Problem 19   **c)** $\sqrt[4]{-81}$ is not a real number ◄

Often a principal nth root is not a rational number. For example, $\sqrt{102}$, $\sqrt[3]{49}$, and $\sqrt[4]{17}$ are irrational numbers. We can approximate these roots using a calculator. For example, $\sqrt[3]{49} \approx 3.66$, since $(3.66)^3 \approx 49$.

A certain amount of care must be exercised when finding *even* principal roots of variable expressions. Consider the two statements below.

If $a = 3$, then $\sqrt{a^2} = \sqrt{3^2} = \sqrt{9} = 3 = a.$
If $a = -3$, then $\sqrt{a^2} = \sqrt{(-3)^2} = \sqrt{9} = 3 = -a.$

In general,

$$\sqrt{a^2} = \begin{cases} a & \text{if } a \geq 0 \\ -a & \text{if } a < 0. \end{cases}$$

That is,

$$\sqrt{a^2} = |a|.$$

To avoid writing the absolute-value symbol, we shall assume that all variables in Examples 4 and 5 represent positive real numbers.

EXAMPLE 4 Find: **a)** $\sqrt{x^2}$, **b)** $\sqrt[5]{x^5}$, **c)** $\sqrt{x^8}$, **d)** $\sqrt[4]{x^8}$.

a) $\sqrt{x^2} = x$

b) $\sqrt[5]{x^5} = x$

Try Problem 45 **c)** $\sqrt{x^8} = \sqrt{(x^4)^2} = x^4$

d) $\sqrt[4]{x^8} = \sqrt[4]{(x^2)^4} = x^2$ ◀

We can determine the exponent of each answer to Example 4 simply by dividing the index into the exponent of the radicand.

EXAMPLE 5 Find: **a)** $\sqrt[5]{32x^5}$, **b)** $\sqrt[3]{-27a^6b^{12}}$, **c)** $\sqrt[11]{(s+t)^{77}}$, **d)** $\sqrt{r^2 + 12r + 36}$.

a) $\sqrt[5]{32x^5} = \sqrt[5]{2^5x^5} = \sqrt[5]{(2x)^5} = 2x$

b) $\sqrt[3]{-27a^6b^{12}} = \sqrt[3]{(-3)^3(a^2)^3(b^4)^3} = \sqrt[3]{(-3a^2b^4)^3} = -3a^2b^4$

c) $\sqrt[11]{(s+t)^{77}} = (s+t)^7$

d) $\sqrt{r^2 + 12r + 36} = \sqrt{(r+6)^2}$ Factor the radicand

Try Problem 67 $= r + 6$ ◀

Problem Solving

Estimating an Accident Vehicle's Speed

The speed s_a of a car that was involved in an accident can be estimated as follows: A policeman drives a car (the same car if possible) under similar conditions at some test speed s_t, and then skids to a stop. Then s_a is given by the formula

$$s_a = s_t\sqrt{\frac{l_a}{l_t}},$$

where l_a and l_t are the lengths of the skid marks from the accident and the test, respectively. Determine s_a if $s_t = 36$ mph, $l_a = 125$ ft, and $l_t = 45$ ft.

$$s_a = (36 \text{ mph})\sqrt{\frac{125 \text{ ft}}{45 \text{ ft}}} \qquad \text{Substitute the known values}$$

$$= 36\sqrt{\frac{25}{9}} \text{ mph} \qquad \text{Simplify}$$

$$= 36 \cdot \frac{5}{3} \text{ mph} \qquad \text{Find the principal square root}$$

$$= 60 \text{ mph} \qquad \text{Multiply}$$

The speed of the car was 60 mph.

PROBLEM SET 5.3

Find each root.

1. $\sqrt{36}$
2. $\sqrt{49}$
3. $-\sqrt{36}$
4. $-\sqrt{49}$
5. $\pm\sqrt{36}$
6. $\pm\sqrt{49}$
7. $\sqrt{-36}$
8. $\sqrt{-49}$
9. $\sqrt[3]{125}$
10. $\sqrt[3]{64}$
11. $-\sqrt[3]{125}$
12. $-\sqrt[3]{64}$
13. $\sqrt[3]{-125}$
14. $\sqrt[3]{-64}$
15. $-\sqrt[3]{-125}$
16. $-\sqrt[3]{-64}$
17. $\sqrt[4]{16}$
18. $\sqrt[4]{625}$
19. $-\sqrt[4]{16}$
20. $-\sqrt[4]{625}$
21. $\sqrt[4]{-16}$
22. $\sqrt[4]{-625}$
23. $\sqrt[5]{243}$
24. $\sqrt[5]{32}$
25. $\sqrt[5]{-243}$
26. $\sqrt[5]{-32}$
27. $\sqrt{\frac{4}{25}}$
28. $\sqrt{\frac{9}{49}}$
29. $\pm\sqrt{\frac{1}{81}}$
30. $\pm\sqrt{\frac{1}{36}}$
31. $\sqrt{1}$
32. $-\sqrt{1}$
33. $\sqrt{-1}$
34. $-\sqrt{-1}$
35. $\sqrt{0}$
36. $\sqrt{-0}$
37. $\sqrt{1.21}$
38. $\sqrt{1.44}$
39. $\sqrt{5^2}$
40. $\sqrt{7^2}$
41. $\sqrt{4^2}$
42. $\sqrt{9^2}$
43. $\sqrt{(-4)^2}$
44. $\sqrt{(-9)^2}$

Find each root. Assume all variables represent positive real numbers.

45. $\sqrt[3]{x^3}$
46. $\sqrt[4]{x^4}$
47. $\sqrt[7]{x^7}$
48. $\sqrt[9]{x^9}$
49. $\sqrt{x^6}$
50. $\sqrt{x^{10}}$
51. $\sqrt[3]{x^6}$
52. $\sqrt[5]{x^{10}}$
53. $\sqrt[7]{x^{21}}$
54. $\sqrt[3]{x^{15}}$
55. $\sqrt{49x^2}$
56. $\sqrt{81x^2}$
57. $\sqrt[3]{64x^3}$
58. $\sqrt[3]{125x^3}$
59. $\sqrt[4]{81y^8}$
60. $\sqrt[4]{16y^{12}}$
61. $\sqrt[5]{32p^{25}}$
62. $\sqrt[5]{243p^{30}}$
63. $\sqrt{169b^4c^{16}}$
64. $\sqrt{225b^6c^{12}}$
65. $\sqrt[3]{r^3s^{30}t^{60}}$
66. $\sqrt[3]{r^{15}s^{90}t^3}$
67. $\sqrt[3]{-8a^9b^{18}}$
68. $\sqrt[3]{-64a^{12}b^6}$
69. $-\sqrt[3]{-216a^{33}b^{66}}$
70. $-\sqrt[3]{-512a^{36}b^{99}}$
71. $\sqrt[3]{\frac{27}{64}p^6q^6}$
72. $\sqrt[3]{\frac{8}{125}p^9q^9}$
73. $\sqrt[11]{(s+t)^{99}}$
74. $\sqrt[11]{(s+t)^{55}}$
75. $\sqrt[5]{(v+w)^{10}}$
76. $\sqrt[5]{(v+w)^{15}}$
77. $\sqrt{r^2+4r+4}$
78. $\sqrt{r^2+16r+64}$
79. $\sqrt{p^2+2pq+q^2}$
80. $\sqrt{p^2+6pq+9q^2}$
81. $\sqrt{16r^{6n}s^{4m}}$
82. $\sqrt{36r^{8n}s^{4m}}$

83. For the car in Figure 5.1 to negotiate the loop successfully, its minimum velocity v in feet per second at the top of the loop must be $v = \sqrt{32r}$, where r is the radius of the loop in feet. Determine the minimum velocity required to negotiate a loop of radius 50 feet.

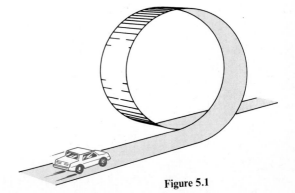

Figure 5.1

84. The time t in seconds that it takes a pendulum l feet long to complete a cycle (see Figure 5.2) is given by the formula

$$t = 2\pi\sqrt{\frac{l}{32}}.$$

How long does it take a pendulum that is 8 feet long to complete a cycle?

Figure 5.2

CALCULATOR PROBLEMS

85. Find each root to the nearest hundredth.

a) $\sqrt{39,601}$ b) $\sqrt[3]{-4913}$ c) $\sqrt[4]{279,841}$ d) $\sqrt[5]{\pi}$

86. The area A of a triangle can be determined from the lengths of its sides a, b, and c using Heron's formula*
$$A = \sqrt{s(s-a)(s-b)(s-c)},$$

where s is one half the perimeter. Find the area of the triangle in Figure 5.3 to the nearest tenth of a square meter.

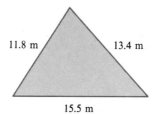

11.8 m   13.4 m

15.5 m

Figure 5.3

5.4 Rational Exponents

We already know that

$$a^2 = a \cdot a, \qquad a^0 = 1, \quad \text{and} \quad a^{-2} = \frac{1}{a^2}.$$

But what does $a^{1/3}$ mean? If we apply the rule $(a^m)^n = a^{mn}$ to cube this expression, we have

$$(a^{1/3})^3 = a^{1/3 \cdot 3} = a^1 = a.$$

---

* Named after Heron of Alexandria (ca. 75 A.D.).

But if $(a^{1/3})^3 = a$, then $a^{1/3}$ must be the cube root of a. That is, $a^{1/3} = \sqrt[3]{a}$. In general, we have the definition below.

Definition

If a is any real number and n is any natural number, then
$$a^{1/n} = \sqrt[n]{a},$$
assuming $\sqrt[n]{a}$ is a real number. (Remember $\sqrt[n]{a}$ is not a real number if n is even and a is negative.)

EXAMPLE 1 Find: **a)** $9^{1/2}$, **b)** $-9^{1/2}$, **c)** $(-9)^{1/2}$.

a) $9^{1/2} = \sqrt{9} = 3$

b) $-9^{1/2} = -(9^{1/2}) = -\sqrt{9} = -3$

Try Problem 1

c) $(-9)^{1/2} = \sqrt{-9}$, which is not a real number. Since $\sqrt{-9}$ is not a real number, the expression $(-9)^{1/2}$ is undefined. ◂

CAUTION The expression $9^{1/2}$ does *not* mean 9 times $\frac{1}{2}$.

EXAMPLE 2 Find: **a)** $1000^{1/3}$, **b)** $625^{1/4}$, **c)** $(-\frac{1}{32})^{1/5}$.

a) $1000^{1/3} = \sqrt[3]{1000} = 10$

b) $625^{1/4} = \sqrt[4]{625} = 5$

Try Problem 7

c) $(-\frac{1}{32})^{1/5} = \sqrt[5]{-\frac{1}{32}} = -\frac{1}{2}$ ◂

The same principles apply when the base involves variables. We shall assume that all variables in the following examples represent positive real numbers.

EXAMPLE 3 Write in radical form: **a)** $(5x)^{1/2}$, **b)** $5x^{1/2}$, **c)** $(x^2 + 1)^{1/2}$.

a) $(5x)^{1/2} = \sqrt{5x}$

b) $5x^{1/2} = 5\sqrt{x}$

Try Problem 23

c) $(x^2 + 1)^{1/2} = \sqrt{x^2 + 1}$ ◂

CAUTION The expression $\sqrt{x^2 + 1}$ cannot be simplified.

Now let's consider the rational power $a^{2/3}$. Applying the rule $a^{m \cdot n} = (a^m)^n$, we have
$$a^{2/3} = a^{(1/3) \cdot 2} = (a^{1/3})^2 = (\sqrt[3]{a})^2.$$

We can also write
$$a^{2/3} = a^{2 \cdot (1/3)} = (a^2)^{1/3} = \sqrt[3]{a^2}.$$

In general, we define **rational-number exponents** as follows:*

Definition	If a is any real number and m and n are natural numbers with no common prime factor, then $$a^{m/n} = (\sqrt[n]{a})^m = \sqrt[n]{a^m},$$ assuming $\sqrt[n]{a}$ is a real number.

EXAMPLE 4   Find: **a)** $64^{2/3}$,   **b)** $(\frac{9}{25})^{3/2}$,   **c)** $(-8)^{4/3}$.

a) $64^{2/3} = (\sqrt[3]{64})^2 = 4^2 = 16$

b) $(\frac{9}{25})^{3/2} = (\sqrt{\frac{9}{25}})^3 = (\frac{3}{5})^3 = \frac{27}{125}$

c) $(-8)^{4/3} = (\sqrt[3]{-8})^4 = (-2)^4 = 16$ ◄

We can also find the powers in Example 4, using the form $\sqrt[n]{a^m}$. For part (a) the computation would be

$$64^{2/3} = \sqrt[3]{64^2} = \sqrt[3]{4096} = 16.$$

Try Problem 29   In this case, however, it was easier to use the form $(\sqrt[n]{a})^m$.

EXAMPLE 5   Write in radical form: **a)** $6y^{3/4}$,   **b)** $(r + 8)^{2/3}$.

a) $6y^{3/4} = 6\sqrt[4]{y^3}$

Try Problem 43   **b)** $(r + 8)^{2/3} = \sqrt[3]{(r + 8)^2} = \sqrt[3]{r^2 + 16r + 64}$ ◄

We extend the definition of negative-integer exponents to negative-rational exponents as follows:

Definition	If $a \neq 0$ then $$a^{-m/n} = \frac{1}{a^{m/n}},$$ assuming $\sqrt[n]{a}$ is a real number.

EXAMPLE 6   Find: **a)** $100^{-1/2}$,   **b)** $8^{-2/3}$,   **c)** $(-27)^{-4/3}$.

a) $100^{-1/2} = \dfrac{1}{100^{1/2}} = \dfrac{1}{\sqrt{100}} = \dfrac{1}{10}$

---

* The English mathematician John Wallis (1616–1703) was the first to fully explain the meaning of rational exponents.

b) $8^{-2/3} = \dfrac{1}{8^{2/3}} = \dfrac{1}{(\sqrt[3]{8})^2} = \dfrac{1}{(2)^2} = \dfrac{1}{4}$

c) $(-27)^{-4/3} = \dfrac{1}{(-27)^{4/3}} = \dfrac{1}{(\sqrt[3]{-27})^4} = \dfrac{1}{(-3)^4} = \dfrac{1}{81}$ ◄

There are several ways to compute a negative rational power. For example, we could also find part (a) of Example 6 as follows:

$$100^{-1/2} = (\sqrt{100})^{-1} = (10)^{-1} = \dfrac{1}{10}.$$

Try Problem 45

Choose the method that is easiest for you.

We can use the rules for exponents stated in Section 2.2 to simplify expressions with rational exponents. By "simplify" we mean "write with positive exponents in simplest form."

EXAMPLE 7   Simplify: **a)** $x^{1/4} \cdot x^{3/4}$, **b)** $\dfrac{y^{11/5}}{y^{1/5}}$, **c)** $(r^{-3/2})^8$.

a) $x^{1/4} \cdot x^{3/4} = x^{1/4+3/4} = x^{4/4} = x$

b) $\dfrac{y^{11/5}}{y^{1/5}} = y^{11/5 - 1/5} = y^{10/5} = y^2$

Try Problem 63

c) $(r^{-3/2})^8 = r^{(-3/2) \cdot 8} = r^{-12} = \dfrac{1}{r^{12}}$ ◄

EXAMPLE 8   Simplify: **a)** $(2r^{1/2}s^{4/3})^6$, **b)** $\left(\dfrac{-8a^3b^{-12}}{c^9}\right)^{1/3}$.

a) $(2r^{1/2}s^{4/3})^6 = 2^6(r^{1/2})^6(s^{4/3})^6$
$$= 64r^{(1/2) \cdot 6} \cdot s^{(4/3) \cdot 6}$$
$$= 64r^3s^8$$

b) $\left(\dfrac{-8a^3b^{-12}}{c^9}\right)^{1/3} = \dfrac{(-8)^{1/3}(a^3)^{1/3}(b^{-12})^{1/3}}{(c^9)^{1/3}}$
$$= \dfrac{-2ab^{-4}}{c^3}$$

Try Problem 73

$$= \dfrac{-2a}{b^4c^3}$$ ◄

EXAMPLE 9   Multiply $w^{1/2}(w^{3/2} + w)$.

$$w^{1/2}(w^{3/2} + w) = w^{1/2} \cdot w^{3/2} + w^{1/2} \cdot w \qquad \text{Distribute } w^{1/2}$$
$$= w^{1/2 + 3/2} + w^{1/2 + 1} \qquad \text{Add exponents}$$

Try Problem 87

$$= w^2 + w^{3/2}$$ ◄

PROBLEM SET 5.4

Find each rational power.

1. $16^{1/2}$ **2.** $36^{1/2}$ **3.** $-16^{1/2}$ **4.** $-36^{1/2}$

5. $(-16)^{1/2}$ **6.** $(-36)^{1/2}$ **7.** $8^{1/3}$ **8.** $64^{1/3}$

9. $(-8)^{1/3}$ **10.** $(-64)^{1/3}$ **11.** $81^{1/4}$ **12.** $16^{1/4}$

13. $1^{1/5}$ **14.** $(-1)^{1/5}$ **15.** $0^{1/6}$ **16.** $0^{1/7}$

17. $(\frac{49}{121})^{1/2}$ **18.** $(\frac{25}{144})^{1/2}$ **19.** $(-\frac{1}{125})^{1/3}$ **20.** $(-\frac{1}{1000})^{1/3}$

Write in radical form. Assume all variables represent positive real numbers.

21. $(3x)^{1/2}$ **22.** $(7x)^{1/2}$ **23.** $3x^{1/2}$ **24.** $7x^{1/2}$

25. $(x+6)^{1/3}$ **26.** $(x+5)^{1/3}$ **27.** $(x^2+4)^{1/2}$ **28.** $(x^2+9)^{1/2}$

Find each rational power.

29. $27^{2/3}$ **30.** $125^{2/3}$ **31.** $(\frac{4}{25})^{3/2}$ **32.** $(\frac{9}{16})^{3/2}$

33. $(-8)^{2/3}$ **34.** $(-27)^{4/3}$ **35.** $(-8)^{5/3}$ **36.** $(-27)^{2/3}$

Write in radical form. Assume all variables represent positive real numbers.

37. $y^{2/3}$ **38.** $y^{2/5}$ **39.** $10y^{3/4}$ **40.** $15y^{3/5}$

41. $(2rs)^{2/3}$ **42.** $(5rs)^{2/3}$ **43.** $(r+3)^{2/5}$ **44.** $(r+4)^{2/3}$

Find each negative rational power.

45. $144^{-1/2}$ **46.** $121^{-1/2}$ **47.** $4^{-3/2}$ **48.** $9^{-3/2}$

49. $625^{-3/4}$ **50.** $256^{-3/4}$ **51.** $(-8)^{-2/3}$ **52.** $(-27)^{-2/3}$

53. $(-125)^{-5/3}$ **54.** $(-1000)^{-5/3}$ **55.** $(\frac{4}{9})^{-1/2}$ **56.** $(\frac{16}{25})^{-1/2}$

Use the rules for exponents to simplify each expression. Assume all variables represent positive real numbers.

57. $x^{1/2} \cdot x^{1/2}$ **58.** $x^{1/3} \cdot x^{2/3}$ **59.** $x^{1/4} \cdot x^{5/4}$ **60.** $x^{1/6} \cdot x^{7/6}$

61. $\dfrac{y^{2/3}}{y^{1/3}}$ **62.** $\dfrac{y^{3/5}}{y^{1/5}}$ **63.** $\dfrac{y^{9/4}}{y^{1/4}}$ **64.** $\dfrac{y^{7/2}}{y^{1/2}}$

65. $\dfrac{p^{-1/2} \cdot p^{3/2}}{p^{1/2}}$ **66.** $\dfrac{p^{-3/2} \cdot p^{5/2}}{p^{1/3}}$ **67.** $\dfrac{p^{-4/3}}{p^{-2/3} \cdot p^{1/3}}$ **68.** $\dfrac{p^{-3/2}}{p^{-3/4} \cdot p^{1/4}}$

69. $(r^{-3/2})^6$ **70.** $(r^{-2/3})^6$ **71.** $(r^2 s^{-4})^{1/8}$ **72.** $(r^2 s^{-5})^{1/10}$

73. $(2r^{3/2}s^{5/3})^6$ **74.** $(3r^{1/2}s^{3/4})^4$ **75.** $(8p^6q^9)^{2/3}$ **76.** $(27p^9q^6)^{2/3}$

77. $\left(\dfrac{m^{10}}{n^{15}}\right)^{1/5}$ **78.** $\left(\dfrac{m^{20}}{n^{25}}\right)^{1/5}$ **79.** $\left(\dfrac{a^{1/2}b^{3/4}}{c^{-3/2}}\right)^4$ **80.** $\left(\dfrac{a^{3/2}b^{5/4}}{c^{-1/2}}\right)^4$

81. $\left(\dfrac{-27a^6b^{-15}}{c^3}\right)^{1/3}$ **82.** $\left(\dfrac{-8a^9b^{-3}}{c^{18}}\right)^{1/3}$ **83.** $\left(\dfrac{16p^{-12}}{q^{16}}\right)^{-1/4}$ **84.** $\left(\dfrac{81p^{-20}}{q^8}\right)^{-1/4}$

85. $(x^{7/2}y^{-3/2})^4(x^{-1/2}y^3)^{-1}$ **86.** $(x^{5/2}y^{-5/3})^6(x^2y^{-1/2})^{-1}$

Multiply. Assume all variables represent positive real numbers.

87. $w^{1/2}(w^{1/2} + w)$ **88.** $w^{3/2}(w^{1/2} - w)$ **89.** $y^{5/2}(y^{1/2} - y)$

90. $y^{1/2}(y^{7/2} - y)$ **91.** $z^{-2/3}(z^{5/3} - z^{2/3})$ **92.** $z^{-3/4}(z^{7/4} + z^{3/4})$

93. A particular Cobb-Douglas production function gives the number of units P produced using x units of labor and y units of capital as

$$P = 48x^{1/3}y^{2/3}.$$

Find P if $x = 10$ and $y = 10$.

94. Find P in Problem 93 if $x = 20$ and $y = 20$.

† 95. Factor $x^{1/2}$ from $x + x^{1/2}$.

† 96. Factor $y^{-1/3}$ from $y^{2/3} + y^{-1/3}$.

CALCULATOR PROBLEMS

97. Compute to the nearest hundredth.
    a) $10^{7/2}$ **b)** $\pi^{0.6185}$ **c)** $7^{-1/5}$

98. Kepler's third law* of planetary motion states that the time T in years that it takes a planet to revolve about the sun is given by $T = d^{3/2}$, where d is the average distance in astronomical units between the planet and the sun. Find T for Jupiter if $d = 5.201$.

5.5   *Simplifying Radical Expressions*

Sometimes we cannot completely eliminate the radical sign from a radical expression, as we did when we wrote $\sqrt{25} = 5$ and $\sqrt[4]{x^8} = x^2$. In these cases the best we can do is simplify the radical expression.

A Radical Expression Is in Simplest Form When

1. The radicand contains no factor raised to a power greater than or equal to the index. ($\sqrt{x^3}$ violates this rule.)

---

* Named after the German mathematician and astronomer Johann Kepler (1571–1630).

2. The radicand contains no fractions. $\left(\sqrt{\dfrac{x}{9}} \text{ violates this rule.} \right)$

3. The expression does not contain a root of a root. ($\sqrt{\sqrt{x}}$ violates this rule.)

4. The index and the exponent of the radicand contain no common factor. ($\sqrt[4]{x^2}$ violates this rule.)

5. No denominator contains a radical. $\left(\dfrac{1}{\sqrt{x}} \text{ violates this rule.} \right)$

To write a radical in simplest form, we use two rules—the product rule for radicals and the quotient rule for radicals. First, note that

$$\sqrt[3]{ab} = (ab)^{1/3} = a^{1/3}b^{1/3} = \sqrt[3]{a} \cdot \sqrt[3]{b}.$$

In general, we have the rule below.

Product Rule for Radicals

If a and b are positive real numbers and n is a natural number, then

$$\sqrt[n]{ab} = \sqrt[n]{a} \cdot \sqrt[n]{b}.$$

CAUTION The expressions $\sqrt[n]{a+b}$ and $\sqrt[n]{a} + \sqrt[n]{b}$ are *not* equal. Neither are the expressions $\sqrt[n]{a-b}$ and $\sqrt[n]{a} - \sqrt[n]{b}$.

EXAMPLE 1 Simplify: **a)** $\sqrt{45}$, **b)** $\sqrt{600}$, **c)** $\sqrt{15}$.

Find the largest perfect-square factor in each radicand. Then apply the product rule.

a) The largest perfect-square factor of 45 is 9. Therefore

$$\sqrt{45} = \underbrace{\sqrt{9 \cdot 5} = \sqrt{9} \cdot \sqrt{5}}_{\text{Product rule}} = 3 \cdot \sqrt{5} = 3\sqrt{5}.$$

b) $\sqrt{600} = \sqrt{100 \cdot 6} = \sqrt{100} \cdot \sqrt{6} = 10 \cdot \sqrt{6} = 10\sqrt{6}$

c) 15 has no perfect-square factor (other than 1), so $\sqrt{15}$ cannot be simplified. ◄

Try Problem 1

It is common to omit the multiplication symbol when writing radical expressions such as $3\sqrt{5}$.

EXAMPLE 2 Simplify: **a)** $\sqrt[3]{32}$, **b)** $\sqrt[4]{162}$.

a) $\sqrt[3]{32} = \sqrt[3]{8 \cdot 4}$ Find the largest perfect-*cube* factor
$\phantom{\sqrt[3]{32}} = \sqrt[3]{8} \cdot \sqrt[3]{4}$ Product rule
$\phantom{\sqrt[3]{32}} = 2\sqrt[3]{4}$ Since $\sqrt[3]{8} = 2$

b) $\sqrt[4]{162} = \sqrt[4]{81 \cdot 2}$ Since 81 is a perfect-*fourth* root

$\qquad\quad = \sqrt[4]{81} \cdot \sqrt[4]{2}$ Product rule

Try Problem 9 $\qquad = 3\sqrt[4]{2}$ Since $\sqrt[4]{81} = 3$ ◄

CAUTION Do not forget to write the index. The expression $3\sqrt{2}$ is an *incorrect* answer to Example 2(b).

These same techniques apply when the radicand contains variables. Again we assume that all variables represent positive real numbers.

EXAMPLE 3   Simplify: a) $\sqrt{50x^2}$, **b)** $4\sqrt{24y^3}$, **c)** $\sqrt[3]{27p^6q^4}$, **d)** $\sqrt[4]{48m^{15}}$.

a) $\sqrt{50x^2} = \sqrt{25x^2 \cdot 2} = \sqrt{25x^2} \cdot \sqrt{2} = 5x\sqrt{2}$

b) $4\sqrt{24y^3} = 4\sqrt{4y^2 \cdot 6y} = 4 \cdot \sqrt{4y^2} \cdot \sqrt{6y} = 4 \cdot 2y \cdot \sqrt{6y} = 8y\sqrt{6y}$

c) $\sqrt[3]{27p^6q^4} = \sqrt[3]{27p^6q^3 \cdot q} = \sqrt[3]{27p^6q^3} \cdot \sqrt[3]{q} = 3p^2q\sqrt[3]{q}$

Try Problem 15 **d)** $\sqrt[4]{48m^{15}} = \sqrt[4]{16m^{12} \cdot 3m^3} = \sqrt[4]{16m^{12}} \cdot \sqrt[4]{3m^3} = 2m^3\sqrt[4]{3m^3}$ ◄

To simplify radicals that involve fractions, we first note that

$$\sqrt[3]{\frac{a}{b}} = \left(\frac{a}{b}\right)^{1/3} = \frac{a^{1/3}}{b^{1/3}} = \frac{\sqrt[3]{a}}{\sqrt[3]{b}}.$$

In general, we have the rule below.

Quotient Rule for Radicals

If a and b are positive real numbers and n is a natural number, then

$$\sqrt[n]{\frac{a}{b}} = \frac{\sqrt[n]{a}}{\sqrt[n]{b}}.$$

EXAMPLE 4   Simplify: a) $\sqrt{\dfrac{18}{49}}$, **b)** $\sqrt{\dfrac{8r^3}{25}}$, **c)** $\sqrt[3]{\dfrac{8t^5}{27}}$.

Apply the quotient rule. Then apply the product rule.

a) $\underbrace{\sqrt{\dfrac{18}{49}} = \dfrac{\sqrt{18}}{\sqrt{49}}}_{\text{Quotient rule}} = \dfrac{\sqrt{9 \cdot 2}}{7} = \dfrac{\sqrt{9} \cdot \sqrt{2}}{7} = \dfrac{3\sqrt{2}}{7}$

b) $\sqrt{\dfrac{8r^3}{25}} = \dfrac{\sqrt{8r^3}}{\sqrt{25}} = \dfrac{\sqrt{4r^2 \cdot 2r}}{5} = \dfrac{\sqrt{4r^2} \cdot \sqrt{2r}}{5} = \dfrac{2r\sqrt{2r}}{5}$

Try Problem 21 **c)** $\sqrt[3]{\dfrac{8t^5}{27}} = \dfrac{\sqrt[3]{8t^5}}{\sqrt[3]{27}} = \dfrac{\sqrt[3]{8t^3 \cdot t^2}}{3} = \dfrac{\sqrt[3]{8t^3} \cdot \sqrt[3]{t^2}}{3} = \dfrac{2t\sqrt[3]{t^2}}{3}$ ◄

When a radical expression involves a root of a root, we convert to rational exponents.

EXAMPLE 5   Simplify: **a)** $\sqrt{\sqrt{5}}$, **b)** $\sqrt{\sqrt[3]{x}}$, **c)** $\sqrt{\sqrt{\sqrt{15}}}$.

a) $\sqrt{\sqrt{5}} = \sqrt{5^{1/2}} = (5^{1/2})^{1/2} = 5^{(1/2)\cdot(1/2)} = 5^{1/4} = \sqrt[4]{5}$

b) $\sqrt{\sqrt[3]{x}} = (x^{1/3})^{1/2} = x^{(1/3)\cdot(1/2)} = x^{1/6} = \sqrt[6]{x}$

Try Problem 33   **c)** $\sqrt{\sqrt{\sqrt{15}}} = [(15^{1/2})^{1/2}]^{1/2} = 15^{(1/2)\cdot(1/2)\cdot(1/2)} = 15^{1/8} = \sqrt[8]{15}$ ◄

We also convert to rational exponents when the index and the exponent of the radicand contain a common factor.

EXAMPLE 6   Simplify: **a)** $\sqrt[10]{z^5}$, **b)** $\sqrt[4]{9}$, **c)** $\sqrt[6]{25x^4y^2}$.

a) $\sqrt[10]{z^5} = z^{5/10} = z^{1/2} = \sqrt{z}$

b) $\sqrt[4]{9} = 9^{1/4} = (3^2)^{1/4} = 3^{2\cdot(1/4)} = 3^{1/2} = \sqrt{3}$

c) $\sqrt[6]{25x^4y^2} = (25x^4y^2)^{1/6} = [(5x^2y)^2]^{1/6} = (5x^2y)^{2\cdot(1/6)}$
$= (5x^2y)^{1/3}$
Try Problem 39 $= \sqrt[3]{5x^2y}$ ◄

Problem Solving

Distance to the Horizon

The distance d in miles to the horizon from an object at an altitude of a feet (see Figure 5.4) is given by the formula

$$d = \sqrt{1.5a}.$$

Find the distance to the horizon from an observer in a hot-air balloon at an altitude of 3000 feet.

$d = \sqrt{1.5(3000)}$ Substitute $a = 3000$

$= \sqrt{4500}$ Multiply

$= \sqrt{900 \cdot 5}$

$= \sqrt{900} \cdot \sqrt{5}$ Simplify the radical

$= 30\sqrt{5}$

The horizon is $30\sqrt{5}$ (approximately 67) miles away.

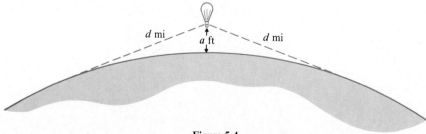

Figure 5.4

PROBLEM SET 5.5

Simplify each radical expression using the product rule. Assume all variables represent positive real numbers.

1. $\sqrt{12}$ 2. $\sqrt{18}$ 3. $\sqrt{50}$ 4. $\sqrt{48}$

5. $\sqrt{500}$ 6. $\sqrt{700}$ 7. $\sqrt{6}$ 8. $\sqrt{10}$

9. $\sqrt[3]{72}$ 10. $\sqrt[3]{200}$ 11. $\sqrt[4]{48}$ 12. $\sqrt[4]{80}$

13. $\sqrt{45x^2}$ 14. $\sqrt{63x^2}$ 15. $2\sqrt{144y^3}$ 16. $3\sqrt{121y^3}$

17. $\sqrt[3]{64p^9q^4}$ 18. $\sqrt[3]{1000p^7q^6}$ 19. $\sqrt[4]{80m^{11}}$ 20. $\sqrt[4]{162m^{19}}$

Simplify each radical expression by first applying the quotient rule. Assume all variables represent positive real numbers.

21. $\sqrt{\dfrac{20}{81}}$ 22. $\sqrt{\dfrac{12}{25}}$ 23. $-\sqrt{\dfrac{200}{169}}$ 24. $-\sqrt{\dfrac{300}{196}}$

25. $\sqrt{\dfrac{49x}{144}}$ 26. $\sqrt{\dfrac{81x}{100}}$ 27. $\sqrt{\dfrac{27r^3}{64}}$ 28. $\sqrt{\dfrac{64r^3}{81}}$

29. $\sqrt[3]{\dfrac{27t^5}{1000}}$ 30. $\sqrt[3]{\dfrac{8t^8}{125}}$ 31. $\sqrt[5]{\dfrac{x}{y^{10}}}$ 32. $\sqrt[5]{\dfrac{x}{y^{15}}}$

Simplify each radical expression by first converting to rational exponents. Assume all variables represent positive real numbers.

33. $\sqrt{\sqrt{3}}$ 34. $\sqrt{\sqrt{7}}$ 35. $\sqrt[3]{\sqrt{x}}$ 36. $\sqrt[4]{\sqrt{x}}$

37. $\sqrt{\sqrt{\sqrt{6}}}$ 38. $\sqrt{\sqrt{\sqrt{10}}}$ 39. $\sqrt[4]{z^2}$ 40. $\sqrt[6]{z^3}$

41. $\sqrt[4]{25}$ 42. $\sqrt[4]{49}$ 43. $\sqrt[6]{36x^2y^4}$ 44. $\sqrt[6]{100x^6y^4}$

Simplify. Assume all variables represent positive real numbers.

45. $\sqrt{99}$ 46. $\sqrt{44}$ 47. $\sqrt{a^4b^5}$ 48. $\sqrt{a^6b^5}$

49. $\sqrt{\dfrac{2}{25}}$ 50. $\sqrt{\dfrac{3}{49}}$ 51. $-\sqrt{60}$ 52. $-\sqrt{24}$

53. $\sqrt[3]{\sqrt[4]{x}}$ 54. $\sqrt[3]{\sqrt[3]{x}}$ 55. $\sqrt{27}$ 56. $\sqrt{125}$

57. $\sqrt[3]{54}$ 58. $\sqrt[3]{81}$ 59. $\sqrt{\dfrac{42y^2}{25}}$ 60. $\sqrt{\dfrac{30y^2}{49}}$

61. $\sqrt[4]{1250}$ 62. $\sqrt[4]{512}$ 63. $5\sqrt{28}$ 64. $2\sqrt{75}$

65. $\sqrt{c^7d^9}$ 66. $\sqrt{c^9d^{11}}$ 67. $\sqrt[4]{9x^2}$ 68. $\sqrt[4]{4x^2}$

69. $\sqrt{\dfrac{x^2y^5}{4z^4}}$ 70. $\sqrt{\dfrac{x^2y^7}{9z^4}}$ 71. $\sqrt{\dfrac{pq}{q^3}}$ 72. $\sqrt{\dfrac{pq}{q^5}}$

73. $\sqrt{242r^3s^6}$ 74. $\sqrt{288r^3s^8}$ 75. $\sqrt[3]{16t^6}$ 76. $\sqrt[3]{54t^6}$

77. $\sqrt[3]{-27k^{14}}$ 78. $\sqrt[3]{-125k^{17}}$ 79. $\sqrt[4]{625w^5}$ 80. $\sqrt[4]{256w^5}$

81. $\sqrt[3]{\dfrac{16}{125}}$ **82.** $\sqrt[3]{\dfrac{81}{64}}$ **83.** $-\sqrt[4]{\dfrac{m^{21}}{256}}$ **84.** $-\sqrt[4]{\dfrac{m^{25}}{1296}}$

85. $\sqrt[8]{16}$ **86.** $\sqrt[6]{27}$ **87.** $\sqrt{25\sqrt{k}}$ **88.** $\sqrt{36\sqrt{k}}$

89. $-\sqrt[3]{-8a^9b^{18}c^{12}}$ **90.** $-\sqrt[3]{-27a^{15}b^{24}c^{21}}$

91. Construct a specific example to show that $\sqrt{a+b}$ and $\sqrt{a}+\sqrt{b}$ are not equal.

92. Construct a specific example to show that $\sqrt{a-b}$ and $\sqrt{a}-\sqrt{b}$ are not equal.

93. Find the distance to the horizon from the top of the Sears Tower, 1400 feet high.

94. Find the distance to the horizon from an airplane at an altitude of 20,000 feet.

CALCULATOR PROBLEMS

95. Verify each statement for $a = 2401$ and $b = 4096$.

    a) $\sqrt{ab} = \sqrt{a} \cdot \sqrt{b}$ **b)** $\sqrt{\dfrac{a}{b}} = \dfrac{\sqrt{a}}{\sqrt{b}}$

96. The minimum velocity v that a rocket must have to escape the earth's gravity is given by $v = \sqrt{2gR}$, where $g = 32$ ft/sec^2 is the earth's gravitational constant, and $R = 3960$ miles is the radius of the earth. Write v in simplest form in ft/sec. [*Hint:* 1 mile = 5280 feet.]

5.6   *Multiplying and Dividing Radicals*

If we turn the product and quotient rules for radicals around, we can use them to multiply and divide radicals.

Assume a and b are positive real numbers and n is a natural number.

Product Rule for Radicals: $\sqrt[n]{a} \cdot \sqrt[n]{b} = \sqrt[n]{ab}$

Quotient Rule for Radicals: $\dfrac{\sqrt[n]{a}}{\sqrt[n]{b}} = \sqrt[n]{\dfrac{a}{b}}$

In other words, two radicals can be multiplied (divided) simply by multiplying (dividing) their radicands, so long as the radicals have the same index and the radicands are positive. We shall assume that all variables in the following examples are positive.

EXAMPLE 1   Multiply and simplify: **a)** $\sqrt{3} \cdot \sqrt{7}$,   **b)** $\sqrt{6} \cdot \sqrt{x}$, **c)** $\sqrt{5} \cdot \sqrt{5}$,   **d)** $(\sqrt{y})^3$.

a) The index for each radical is 2, so multiply the radicands.

$$\underbrace{\sqrt{3} \cdot \sqrt{7} = \sqrt{3 \cdot 7}}_{\text{Product rule}} = \sqrt{21}$$

b) $\sqrt{6} \cdot \sqrt{x} = \sqrt{6x}$

c) $\sqrt{5} \cdot \sqrt{5} = \sqrt{25} = 5$

Try Problem 1

d) $(\sqrt{y})^3 = \sqrt{y} \cdot \sqrt{y} \cdot \sqrt{y} = (\sqrt{y} \cdot \sqrt{y}) \cdot \sqrt{y} = y\sqrt{y}$ ◄

Note that if a is a positive number, then

$$\sqrt{a} \cdot \sqrt{a} = a.$$

Also, it is common practice to omit the multiplication symbol and write $\sqrt{a} \cdot \sqrt{a}$ as $\sqrt{a}\sqrt{a}$.

EXAMPLE 2   Multiply and simplify: **a)** $\sqrt{2x}\sqrt{6x}$,   **b)** $\sqrt[3]{4p^2}\sqrt[3]{16p}$, **c)** $\sqrt[3]{3r^2s^2}\sqrt[3]{4r^5s^2}$.

a) $\sqrt{2x}\sqrt{6x} = \sqrt{12x^2}$ \qquad Multiply the radicands

$\left.\begin{array}{l} = \sqrt{4x^2 \cdot 3} \\ = \sqrt{4x^2} \cdot \sqrt{3} \\ = 2x\sqrt{3} \end{array}\right\}$ Simplify the radical

b) $\sqrt[3]{4p^2}\sqrt[3]{16p} = \sqrt[3]{64p^3} = 4p$

c) $\sqrt[3]{3r^2s^2}\sqrt[3]{4r^5s^2} = \sqrt[3]{12r^7s^4} = \sqrt[3]{r^6s^3 \cdot 12rs}$

$= \sqrt[3]{r^6s^3} \cdot \sqrt[3]{12rs}$

$= r^2s\sqrt[3]{12rs}$ ◄

Try Problem 17

We now illustrate division of radicals using the quotient rule.

EXAMPLE 3   Divide and simplify: **a)** $\dfrac{\sqrt{100x}}{\sqrt{x}}$,   **b)** $\dfrac{\sqrt{2y}}{\sqrt{8y^3}}$.

a) The indexes are the same, so divide the radicands.

$$\underbrace{\frac{\sqrt{100x}}{\sqrt{x}} = \sqrt{\frac{100x}{x}}}_{\text{Quotient rule}} = \sqrt{100} = 10$$

Try Problem 45

b) $\dfrac{\sqrt{2y}}{\sqrt{8y^3}} = \sqrt{\dfrac{2y}{8y^3}} = \sqrt{\dfrac{1}{4y^2}} = \dfrac{\sqrt{1}}{\sqrt{4y^2}} = \dfrac{1}{2y}$ ◄

EXAMPLE 4 Divide and simplify: **a)** $\dfrac{\sqrt{90m^5}}{\sqrt{5m}}$, **b)** $\dfrac{15\sqrt{128c}}{3\sqrt{2}}$, **c)** $\dfrac{\sqrt[4]{72p^3q^3}}{\sqrt[4]{2pq}}$.

a)

$$\frac{\sqrt{90m^5}}{\sqrt{5m}} = \sqrt{\frac{90m^5}{5m}} \qquad \text{Apply the quotient rule}$$

$$= \sqrt{18m^4} \qquad \text{Divide the radicands}$$

$$\left.\begin{array}{l} = \sqrt{9m^4 \cdot 2} \\ = \sqrt{9m^4} \cdot \sqrt{2} \\ = 3m^2\sqrt{2} \end{array}\right\} \text{Simplify}$$

b)

$$\frac{15\sqrt{128c}}{3\sqrt{2}} = \frac{15}{3}\sqrt{\frac{128c}{2}} = 5\sqrt{64c} = 5\sqrt{64} \cdot \sqrt{c}$$

$$= 5 \cdot 8 \cdot \sqrt{c}$$

$$= 40\sqrt{c}$$

c)

$$\frac{\sqrt[4]{72p^3q^3}}{\sqrt[4]{2pq}} = \sqrt[4]{\frac{72p^3q^3}{2pq}} = \sqrt[4]{36p^2q^2} = \sqrt[4]{(6pq)^2}$$

$$= (6pq)^{2/4}$$

$$= (6pq)^{1/2}$$

Try Problem 53

$$= \sqrt{6pq} \blacktriangleleft$$

CAUTION Neither the product rule nor the quotient rule for radicals applies when the indexes are not the same.

To multiply or divide radicals with different indexes, convert to rational exponents.

EXAMPLE 5 Multiply $\sqrt{2} \ \sqrt[3]{2}$.

$$\sqrt{2} \ \sqrt[3]{2} = 2^{1/2} \cdot 2^{1/3} \qquad \text{Convert to rational exponents}$$

$$= 2^{3/6} \cdot 2^{2/6} \qquad \text{Write each exponent with the LCD 6}$$

$$= 2^{5/6} \qquad \text{Add exponents}$$

$$= \sqrt[6]{2^5} \qquad \text{Convert to radical form}$$

Try Problem 69

$$= \sqrt[6]{32} \blacktriangleleft$$

EXAMPLE 6 Divide $\dfrac{\sqrt{k}}{\sqrt[4]{k}}$.

Try Problem 75

$$\frac{\sqrt{k}}{\sqrt[4]{k}} = \frac{k^{1/2}}{k^{1/4}} = \frac{k^{2/4}}{k^{1/4}} = k^{2/4-1/4} = k^{1/4} = \sqrt[4]{k} \blacktriangleleft$$

PROBLEM SET 5.6

Multiply and simplify. Assume all variables represent positive real numbers.

1. $\sqrt{3} \cdot \sqrt{5}$

2. $\sqrt{5} \cdot \sqrt{7}$

3. $\sqrt{10} \cdot \sqrt{x}$

4. $\sqrt{15} \cdot \sqrt{x}$

5. $\sqrt{2} \cdot \sqrt{2}$

6. $\sqrt{3} \cdot \sqrt{3}$

7. $\sqrt{6} \cdot \sqrt{6}$

8. $\sqrt{10} \cdot \sqrt{10}$

9. $\sqrt{5} \cdot \sqrt{5} \cdot \sqrt{5}$

10. $\sqrt{2} \cdot \sqrt{2} \cdot \sqrt{2}$

11. $(\sqrt{y})^4$

12. $(\sqrt{y})^6$

13. $(\sqrt{y})^5$

14. $(\sqrt{y})^7$

15. $\sqrt{2} \cdot \sqrt{7m}$

16. $\sqrt{3} \cdot \sqrt{5m}$

17. $\sqrt{3x} \sqrt{6x}$

18. $\sqrt{2x} \sqrt{10x}$

19. $\sqrt{5y} \sqrt{10y}$

20. $\sqrt{3y} \sqrt{15y}$

21. $\sqrt{3} \sqrt{3z} \sqrt{11}$

22. $\sqrt{2} \sqrt{2z} \sqrt{11}$

23. $(2\sqrt{k})^3$

24. $(3\sqrt{k})^3$

25. $(\sqrt{x+1})^2$

26. $(\sqrt{x+4})^2$

27. $\sqrt{rs^3} \sqrt{rs^4}$

28. $\sqrt{r^2s} \sqrt{rs^5}$

29. $\sqrt{45} \sqrt{27t}$

30. $\sqrt{40} \sqrt{15t}$

31. $\sqrt{2a} \sqrt{3a^3} \sqrt{6a}$

32. $\sqrt{2a^3} \sqrt{5a^3} \sqrt{10a}$

33. $\sqrt[3]{4p^2} \sqrt[3]{2p}$

34. $\sqrt[3]{9p^2} \sqrt[3]{3p}$

35. $\sqrt[3]{6} \sqrt[3]{18q}$

36. $\sqrt[3]{5} \sqrt[3]{40q}$

37. $\sqrt[3]{3r^2s^2} \sqrt[3]{6r^2s^5}$

38. $\sqrt[3]{2r^4s} \sqrt[3]{9r^3s^3}$

39. $\sqrt[4]{2a} \sqrt[4]{8b}$

40. $\sqrt[4]{9a} \sqrt[4]{9b}$

41. $\sqrt{1.5 \times 10^{-2}} \sqrt{6 \times 10^{10}}$

42. $\sqrt{4.5 \times 10^{-3}} \sqrt{8 \times 10^{11}}$

Divide and simplify. Assume all variables represent positive real numbers.

43. $\dfrac{\sqrt{50}}{\sqrt{2}}$

44. $\dfrac{\sqrt{75}}{\sqrt{3}}$

45. $\dfrac{\sqrt{49x}}{\sqrt{x}}$

46. $\dfrac{\sqrt{81x}}{\sqrt{x}}$

47. $\dfrac{\sqrt{42}}{\sqrt{7}}$

48. $\dfrac{\sqrt{30}}{\sqrt{5}}$

49. $\dfrac{\sqrt{363a}}{\sqrt{3}}$

50. $\dfrac{\sqrt{288a}}{\sqrt{2}}$

51. $\dfrac{\sqrt{3y}}{\sqrt{27y^3}}$

52. $\dfrac{\sqrt{4y}}{\sqrt{64y^3}}$

53. $\dfrac{\sqrt{24m^5}}{\sqrt{2m}}$

54. $\dfrac{\sqrt{54m^5}}{\sqrt{3m}}$

55. $\dfrac{15\sqrt{96c}}{5\sqrt{6}}$

56. $\dfrac{12\sqrt{160c}}{4\sqrt{10}}$

57. $\dfrac{\sqrt{75a^3b^5c^7}}{\sqrt{15abc}}$

58. $\dfrac{\sqrt{91a^5b^9c^3}}{\sqrt{13abc}}$

59. $\dfrac{\sqrt{648r^4s}}{\sqrt{8r^2s^5}}$

60. $\dfrac{\sqrt{768r^7s}}{\sqrt{12r^3s^7}}$

61. $\dfrac{\sqrt[3]{81t^4}}{\sqrt[3]{3t}}$

62. $\dfrac{\sqrt[3]{32t^4}}{\sqrt[3]{4t}}$

63. $\dfrac{\sqrt[4]{98p^3q^3}}{\sqrt[4]{2pq}}$

64. $\dfrac{\sqrt[4]{108p^5q^5}}{\sqrt[4]{3p^3q^3}}$

65. $\dfrac{\sqrt{2r^5}\sqrt{2rs}}{\sqrt{25s^3}}$

66. $\dfrac{\sqrt{3r^3}\sqrt{3r^3s}}{\sqrt{49s^3}}$

67. $\dfrac{\sqrt{6.4 \times 10^9}}{\sqrt{1.6 \times 10}}$

68. $\dfrac{\sqrt{9.9 \times 10^7}}{\sqrt{1.1 \times 10}}$

Multiply or divide as indicated and simplify. Assume all variables represent positive real numbers.

69. $\sqrt{3} \sqrt[3]{3}$

70. $\sqrt{5} \sqrt[3]{5}$

71. $\sqrt[4]{x} \sqrt{x}$

72. $\sqrt[6]{x} \sqrt{x}$

73. $\dfrac{\sqrt{6}}{\sqrt[4]{6}}$ **74.** $\dfrac{\sqrt{10}}{\sqrt[4]{10}}$ **75.** $\dfrac{\sqrt{k}}{\sqrt[6]{k}}$ **76.** $\dfrac{\sqrt{k}}{\sqrt[8]{k}}$

77. $\sqrt{a}\,\sqrt[4]{b}$ **78.** $\sqrt{a}\,\sqrt[6]{b}$

† **79.** Construct specific examples to show that a rational number raised to a rational power may result in a rational number or an irrational number.

† **80.** Construct specific examples to show that an irrational number raised to a rational power may result in a rational number or an irrational number.

CALCULATOR PROBLEMS

81. Verify that $\sqrt[3]{a}\,\sqrt[3]{b} = \sqrt[3]{ab}$ for $a = 3375$ and $b = 2744$.

82. Verify that $\dfrac{\sqrt[4]{a}}{\sqrt[4]{b}} = \sqrt[4]{\dfrac{a}{b}}$ for $a = 20{,}736$ and $b = 1296$.

5.7   *Adding and Subtracting Radicals*

Radical expressions are added and subtracted in much the same way that like terms are added and subtracted.

EXAMPLE 1   Add: **a)** $3x + 5x$,   **b)** $3\sqrt{2} + 5\sqrt{2}$.

We use the distributive law in both cases.

Try Problem 1

   a) $3x + 5x = (3 + 5)x = 8x$ Like terms

   b) $3\sqrt{2} + 5\sqrt{2} = (3 + 5)\sqrt{2} = 8\sqrt{2}$ Radical expressions ◄

We can combine like terms only if they have the same *variable* factor. We can combine radical expressions only if they have the same *radical* factor.

CAUTION For radical factors to be the same, they must have the same index *and* the same radicand.

EXAMPLE 2   Add: **a)** $3x + 5y$,   **b)** $3\sqrt{2} + 5\sqrt{7}$.

a) $3x + 5y$ cannot be simplified because the variables x and y are different.

b) $3\sqrt{2} + 5\sqrt{7}$ cannot be simplified because the radicands 2 and 7 are different. ◄

Try Problem 5

EXAMPLE 3   Add: **a)** $x^2 + x^3$,   **b)** $\sqrt{6} + \sqrt[3]{6}$.

a) $x^2 + x^3$ cannot be simplified because the exponents 2 and 3 are different.

b) $\sqrt{6} + \sqrt[3]{6}$ cannot be simplified because the indexes 2 and 3 are different. ◄

Try Problem 7

Assume all variables in the following examples represent positive real numbers.

EXAMPLE 4   Add or subtract as indicated: **a)** $2\sqrt{x} + \sqrt{x}$,   **b)** $6\sqrt{3y} - 2\sqrt{3y}$,
c) $3\sqrt{10} + 4\sqrt{10} - 9\sqrt{10}$.

a) $2\sqrt{x} + \sqrt{x} = 2\sqrt{x} + 1\sqrt{x} = (2+1)\sqrt{x} = 3\sqrt{x}$

b) $6\sqrt{3y} - 2\sqrt{3y} = (6-2)\sqrt{3y} = 4\sqrt{3y}$

Try Problem 9   **c)** $3\sqrt{10} + 4\sqrt{10} - 9\sqrt{10} = (3+4-9)\sqrt{10} = -2\sqrt{10}$ ◄

Sometimes the radicals must be simplified before they can be combined.

EXAMPLE 5   Combine: **a)** $\sqrt{18} + \sqrt{2}$,   **b)** $5\sqrt{28} - \sqrt{63} - \sqrt{112}$,
c) $\sqrt[3]{16p} - \sqrt[3]{2p} + \sqrt[3]{54p}$.

a) $\left.\begin{aligned} \sqrt{18} + \sqrt{2} &= \sqrt{9 \cdot 2} + \sqrt{2} \\ &= \sqrt{9} \cdot \sqrt{2} + \sqrt{2} \\ &= 3\sqrt{2} + \sqrt{2} \end{aligned}\right\}$ Simplify the first radical

$\qquad\qquad = 4\sqrt{2}$ Combine

b) $\begin{aligned} 5\sqrt{28} - \sqrt{63} - \sqrt{112} &= 5\sqrt{4 \cdot 7} - \sqrt{9 \cdot 7} - \sqrt{16 \cdot 7} \\ &= 5\sqrt{4} \cdot \sqrt{7} - \sqrt{9} \cdot \sqrt{7} - \sqrt{16} \cdot \sqrt{7} \\ &= 5 \cdot 2\sqrt{7} - 3\sqrt{7} - 4\sqrt{7} \\ &= (10 - 3 - 4)\sqrt{7} \\ &= 3\sqrt{7} \end{aligned}$

c) $\begin{aligned} \sqrt[3]{16p} - \sqrt[3]{2p} + \sqrt[3]{54p} &= \sqrt[3]{8 \cdot 2p} - \sqrt[3]{2p} + \sqrt[3]{27 \cdot 2p} \\ &= \sqrt[3]{8} \cdot \sqrt[3]{2p} - \sqrt[3]{2p} + \sqrt[3]{27} \cdot \sqrt[3]{2p} \\ &= 2\sqrt[3]{2p} - \sqrt[3]{2p} + 3\sqrt[3]{2p} \end{aligned}$

Try Problem 21 $\qquad\qquad\qquad\qquad = 4\sqrt[3]{2p}$ ◄

Now that we know how to add and subtract radical expressions, we can tackle multiplication of radical expressions that involve more than one term.

EXAMPLE 6   Multiply and simplify $\sqrt{2}(5\sqrt{6} + \sqrt{2})$.

$\begin{aligned} \sqrt{2}(5\sqrt{6} + \sqrt{2}) &= \sqrt{2} \cdot 5\sqrt{6} + \sqrt{2} \cdot \sqrt{2} \qquad &\text{Distribute } \sqrt{2} \\ &= 5\sqrt{12} + 2 &\text{Multiply radicals} \\ &= 5\sqrt{4 \cdot 3} + 2 &\left.\vphantom{\begin{aligned}a\\b\\c\end{aligned}}\right\} \\ &= 5\sqrt{4} \cdot \sqrt{3} + 2 & \\ &= 5 \cdot 2 \cdot \sqrt{3} + 2 &\text{Simplify} \end{aligned}$

Try Problem 55 $\qquad\qquad\qquad\qquad = 10\sqrt{3} + 2$ ◄

EXAMPLE 7 Multiply and simplify $(\sqrt{6} + 1)(\sqrt{5} + 3)$.

We use the FOIL method.

$$
\begin{array}{cccc}
\text{First} & \text{Outer} & \text{Inner} & \text{Last}
\end{array}
$$

$$(\sqrt{6} + 1)(\sqrt{5} + 3) = \sqrt{6} \cdot \sqrt{5} + \sqrt{6} \cdot 3 + 1 \cdot \sqrt{5} + 1 \cdot 3$$
$$= \sqrt{30} + 3\sqrt{6} + \sqrt{5} + 3$$

Try Problem 63

This answer cannot be simplified further. ◄

EXAMPLE 8 Multiply and simplify.

a) $(\sqrt{7} + \sqrt{3})(\sqrt{7} - \sqrt{3}) = \sqrt{7}\sqrt{7} - \sqrt{7}\sqrt{3} + \sqrt{3}\sqrt{7} - \sqrt{3}\sqrt{3}$
$$= 7 - \sqrt{21} + \sqrt{21} - 3$$
$$= 4$$

b) $(\sqrt{5} + 6\sqrt{2})(\sqrt{5} + \sqrt{3}) = \sqrt{5}\sqrt{5} + \sqrt{5}\sqrt{3} + 6\sqrt{2}\sqrt{5} + 6\sqrt{2}\sqrt{3}$
$$= 5 + \sqrt{15} + 6\sqrt{10} + 6\sqrt{6}$$

c) $(\sqrt{x} - \sqrt{y})^2 = (\sqrt{x} - \sqrt{y})(\sqrt{x} - \sqrt{y})$
$$= \sqrt{x}\sqrt{x} - \sqrt{x}\sqrt{y} - \sqrt{y}\sqrt{x} + \sqrt{y}\sqrt{y}$$
$$= x - \sqrt{xy} - \sqrt{xy} + y$$
$$= x - 2\sqrt{xy} + y \blacktriangleleft$$

Try Problem 65

PROBLEM SET 5.7

Add or subtract as indicated. All variables represent positive real numbers.

1. $3\sqrt{7} + 5\sqrt{7}$ 2. $2\sqrt{5} + 4\sqrt{5}$ 3. $17\sqrt{6} - 4\sqrt{6}$

4. $19\sqrt{10} - 7\sqrt{10}$ 5. $2\sqrt{5} + 5\sqrt{2}$ 6. $3\sqrt{7} + 7\sqrt{3}$

7. $\sqrt{10} + \sqrt[4]{10}$ 8. $\sqrt[3]{6} + \sqrt[5]{6}$ 9. $4\sqrt{x} + \sqrt{x}$

10. $\sqrt{x} + 5\sqrt{x}$ 11. $9\sqrt{2y} - 5\sqrt{2y}$ 12. $7\sqrt{5y} - 3\sqrt{5y}$

13. $6\sqrt{x} + 4\sqrt{y}$ 14. $8\sqrt{x} + 2\sqrt{y}$ 15. $5\sqrt{11} + 2\sqrt{11} - 10\sqrt{11}$

16. $3\sqrt{13} + 2\sqrt{13} - 8\sqrt{13}$ 17. $7\sqrt{m} - \sqrt{m} + 5m$ 18. $9\sqrt{m} - \sqrt{m} + 3m$

19. $6a\sqrt{5} + 3a\sqrt{5}$ 20. $5a\sqrt{3} + 2a\sqrt{3}$

Simplify each radical expression and then combine. All variables represent positive real numbers.

21. $\sqrt{12} + \sqrt{3}$

22. $\sqrt{20} + \sqrt{5}$

23. $\sqrt{20} + \sqrt{45}$

24. $\sqrt{12} + \sqrt{48}$

25. $5\sqrt{8} - \sqrt{2}$

26. $3\sqrt{32} - \sqrt{2}$

27. $2\sqrt{7} - 3\sqrt{28}$

28. $3\sqrt{7} - 2\sqrt{63}$

29. $4\sqrt{18} - \sqrt{72} - \sqrt{50}$

30. $2\sqrt{175} - \sqrt{28} - \sqrt{63}$

31. $2\sqrt{40} + 3\sqrt{90} + 4\sqrt{160}$

32. $5\sqrt{250} + 6\sqrt{360} + 7\sqrt{490}$

33. $\sqrt{6x} + \sqrt{24x}$

34. $\sqrt{3x} + \sqrt{75x}$

35. $\sqrt{27a} + 6\sqrt{3a} - 2\sqrt{75a}$

36. $\sqrt{8a} + 3\sqrt{2a} - 5\sqrt{18a}$

37. $5\sqrt{11r^3} + \sqrt{44r^3}$

38. $2\sqrt{11r^3} + \sqrt{99r^3}$

39. $\sqrt[3]{81} + \sqrt[3]{3}$

40. $\sqrt[3]{16} + \sqrt[3]{2}$

41. $\sqrt[3]{64y} - \sqrt[3]{8y}$

42. $\sqrt[3]{125y} - \sqrt[3]{27y}$

43. $\sqrt[3]{24p} - \sqrt[3]{3p} + \sqrt[3]{192p}$

44. $\sqrt[3]{40p} - \sqrt[3]{5p} + \sqrt[3]{625p}$

45. $5\sqrt[3]{54} - 3\sqrt[3]{128} - 5\sqrt[3]{16}$

46. $2\sqrt[3]{48} - 5\sqrt[3]{162} - 2\sqrt[3]{384}$

47. $\sqrt[4]{16t} + 3\sqrt[4]{t}$

48. $\sqrt[4]{81t} + 2\sqrt[4]{t}$

Multiply and simplify. All variables represent positive real numbers.

49. $5(\sqrt{3} + 4)$

50. $7(\sqrt{2} + 4)$

51. $2(6\sqrt{10} - \sqrt{5})$

52. $2(4\sqrt{14} - \sqrt{7})$

53. $\sqrt{3}(\sqrt{2} + \sqrt{5})$

54. $\sqrt{5}(\sqrt{3} + \sqrt{7})$

55. $\sqrt{5}(3\sqrt{10} + \sqrt{5})$

56. $\sqrt{3}(2\sqrt{6} + \sqrt{3})$

57. $\sqrt{x}(\sqrt{x} - \sqrt{xy})$

58. $\sqrt{x}(\sqrt{x} + \sqrt{xy})$

59. $2\sqrt{3}(5\sqrt{6} + 4\sqrt{3})$

60. $3\sqrt{2}(4\sqrt{6} + 5\sqrt{2})$

61. $\sqrt[3]{2}(\sqrt[3]{4} - \sqrt[3]{3})$

62. $\sqrt[3]{3}(\sqrt[3]{9} - \sqrt[3]{4})$

Multiply and simplify. All variables represent positive real numbers.

63. $(\sqrt{3} + 1)(\sqrt{2} + 5)$

64. $(\sqrt{5} + 1)(\sqrt{2} + 7)$

65. $(\sqrt{5} + \sqrt{2})(\sqrt{5} - \sqrt{2})$

66. $(\sqrt{7} + \sqrt{2})(\sqrt{7} - \sqrt{2})$

67. $(\sqrt{7} - \sqrt{3})(\sqrt{7} - \sqrt{2})$

68. $(\sqrt{5} - \sqrt{3})(\sqrt{5} - \sqrt{2})$

69. $(\sqrt{11} + \sqrt{5})(\sqrt{2} + \sqrt{3})$

70. $(\sqrt{13} + \sqrt{5})(\sqrt{2} + \sqrt{3})$

71. $(\sqrt{3} + 4\sqrt{5})(\sqrt{3} + \sqrt{2})$

72. $(\sqrt{2} + 10\sqrt{7})(\sqrt{2} + \sqrt{3})$

73. $(t + \sqrt{5})(t - \sqrt{5})$

74. $(t + \sqrt{7})(t - \sqrt{7})$

75. $(\sqrt{s} - 2)(\sqrt{s} + 2)$

76. $(\sqrt{s} - 3)(\sqrt{s} + 3)$

77. $(4\sqrt{10} - 9)(3\sqrt{2} + 1)$

78. $(2\sqrt{6} - 10)(4\sqrt{3} + 1)$

79. $(\sqrt{5} + 2)^2$

80. $(\sqrt{7} + 3)^2$

81. $(\sqrt{3} + \sqrt{7})^2$

82. $(\sqrt{2} + \sqrt{5})^2$

83. $(\sqrt{x} + \sqrt{y})^2$

84. $(\sqrt{x} + 2\sqrt{y})^2$

85. $(\sqrt{x + 4} - 1)^2$

86. $(\sqrt{x + 9} - 1)^2$

87. $(\sqrt{2x + 1} + \sqrt{x})^2$

88. $(\sqrt{2x + 3} + \sqrt{x})^2$

89. $(\sqrt{2} + 1)^3$

90. $(\sqrt{3} + 1)^3$

CALCULATOR PROBLEMS

Compute to the nearest tenth. Then perform the indicated operation, compute the value of the new expression, and compare.

91. $\sqrt{63} + 5\sqrt{28}$

92. $(\sqrt{3} + \sqrt{5})^2$

5.8   *Rationalizing Denominators*

In Section 5.5 we stated that a radical expression in simplest form contains no radicals in the denominator. The process of rewriting a radical expression as an equivalent expression without radicals in the denominator is called **rationalizing the denominator.**

EXAMPLE 1   Rationalize the denominator and simplify: **a)** $\dfrac{1}{\sqrt{2}}$,   **b)** $\dfrac{2}{\sqrt{14}}$,   **c)** $\dfrac{10}{\sqrt[3]{2}}$.

a)   Multiply numerator and denominator by $\sqrt{2}$, since $\sqrt{2} \cdot \sqrt{2} = 2$.

$$\frac{1}{\sqrt{2}} = \frac{1 \cdot \sqrt{2}}{\sqrt{2} \cdot \sqrt{2}} = \frac{\sqrt{2}}{2}$$

b) $\dfrac{2}{\sqrt{14}} = \dfrac{2 \cdot \sqrt{14}}{\sqrt{14} \cdot \sqrt{14}} = \dfrac{2\sqrt{14}}{14} = \dfrac{\sqrt{14}}{7}$

c)   To make the radicand in the denominator a perfect *cube*, we multiply by $\sqrt[3]{4}$.

Try Problem 1

$$\frac{10 \cdot \sqrt[3]{4}}{\sqrt[3]{2} \cdot \sqrt[3]{4}} = \frac{10\sqrt[3]{4}}{\sqrt[3]{8}} = \frac{10\sqrt[3]{4}}{2} = 5\sqrt[3]{4} \blacktriangleleft$$

Assume that all variables in the following examples represent positive real numbers.

EXAMPLE 2   Rationalize the denominator and simplify.

a) $\dfrac{\sqrt{20x^3}}{\sqrt{15xy}} = \sqrt{\dfrac{20x^3}{15xy}}$ Quotient rule for radicals

$\qquad = \sqrt{\dfrac{4x^2}{3y}}$ Simplify the radicand

$\qquad = \dfrac{\sqrt{4x^2}}{\sqrt{3y}}$ Quotient rule for radicals

$\qquad = \dfrac{2x}{\sqrt{3y}}$

$\qquad = \dfrac{2x \cdot \sqrt{3y}}{\sqrt{3y} \cdot \sqrt{3y}}$ Rationalize the denominator

$\qquad = \dfrac{2x\sqrt{3y}}{3y}$

b) $\dfrac{\sqrt[3]{10}}{\sqrt[3]{18p}} = \sqrt[3]{\dfrac{10}{18p}}$

$= \sqrt[3]{\dfrac{5}{9p}}$

$= \dfrac{\sqrt[3]{5}}{\sqrt[3]{9p}}$

$= \dfrac{\sqrt[3]{5} \cdot \sqrt[3]{3p^2}}{\sqrt[3]{9p} \cdot \sqrt[3]{3p^2}}$ Rationalize the denominator

$= \dfrac{\sqrt[3]{15p^2}}{\sqrt[3]{27p^3}}$

Try Problem 25 $= \dfrac{\sqrt[3]{15p^2}}{3p}$ ◄

To rationalize a denominator that contains two terms, we multiply numerator and denominator by the conjugate of the denominator. The expressions $\sqrt{5} + \sqrt{2}$ and $\sqrt{5} - \sqrt{2}$ are called **conjugates** of each other. Note that their product does not contain a radical.

$$(\sqrt{5} + \sqrt{2}) \cdot (\sqrt{5} - \sqrt{2}) = \sqrt{5}\sqrt{5} - \sqrt{5}\sqrt{2} + \sqrt{2}\sqrt{5} - \sqrt{2}\sqrt{2}$$
$$= 5 - \sqrt{10} + \sqrt{10} - 2$$
$$= 3$$

Hence this procedure eliminates any radicals in the denominator.

EXAMPLE 3   Rationalize the denominator and simplify: **a)** $\dfrac{3}{\sqrt{7} - 2}$,

b) $\dfrac{4 + \sqrt{6}}{\sqrt{2} + \sqrt{3}}$.

a)   We multiply numerator and denominator by the conjugate of the denominator.

$$\dfrac{3}{\sqrt{7} - 2} = \dfrac{3 \cdot (\sqrt{7} + 2)}{(\sqrt{7} - 2) \cdot (\sqrt{7} + 2)}$$ The conjugate of $\sqrt{7} - 2$ is $\sqrt{7} + 2$

$$= \dfrac{3(\sqrt{7} + 2)}{\sqrt{7}\sqrt{7} + \sqrt{7} \cdot 2 - 2\sqrt{7} - 4}$$

$$= \dfrac{3(\sqrt{7} + 2)}{7 + 2\sqrt{7} - 2\sqrt{7} - 4}$$

$$= \dfrac{3(\sqrt{7} + 2)}{3}$$

$$= \sqrt{7} + 2$$

b) $\dfrac{4 + \sqrt{6}}{\sqrt{2} + \sqrt{3}} = \dfrac{(4 + \sqrt{6}) \cdot (\sqrt{2} - \sqrt{3})}{(\sqrt{2} + \sqrt{3}) \cdot (\sqrt{2} - \sqrt{3})}$

$= \dfrac{4\sqrt{2} - 4\sqrt{3} + \sqrt{12} - \sqrt{18}}{2 - \sqrt{6} + \sqrt{6} - 3}$

$= \dfrac{4\sqrt{2} - 4\sqrt{3} + 2\sqrt{3} - 3\sqrt{2}}{-1}$ Simplify $\sqrt{12}$ and $\sqrt{18}$

$= \dfrac{\sqrt{2} - 2\sqrt{3}}{-1}$ Combine radicals

Try Problem 33 $= -\sqrt{2} + 2\sqrt{3}$ Divide each term by -1 ◂

It is easier to add and subtract fractions involving radicals if we first rationalize their denominators.

EXAMPLE 4 Subtract $\dfrac{\sqrt{7}}{3} - \dfrac{1}{\sqrt{7}}$.

$\dfrac{\sqrt{7}}{3} - \dfrac{1}{\sqrt{7}} = \dfrac{\sqrt{7}}{3} - \dfrac{1 \cdot \sqrt{7}}{\sqrt{7} \cdot \sqrt{7}}$ Rationalize the denominator

$= \dfrac{\sqrt{7}}{3} - \dfrac{\sqrt{7}}{7}$

$= \dfrac{7 \cdot \sqrt{7}}{7 \cdot 3} - \dfrac{3 \cdot \sqrt{7}}{3 \cdot 7}$ Write with LCD 21

$= \dfrac{7\sqrt{7} - 3\sqrt{7}}{21}$ Subtract numerators

Try Problem 51 $= \dfrac{4\sqrt{7}}{21}$ ◂

EXAMPLE 5 Add $\sqrt{40m} + \sqrt{\dfrac{2m}{5}}$.

$\sqrt{40m} + \sqrt{\dfrac{2m}{5}} = 2\sqrt{10m} + \dfrac{\sqrt{2m}}{\sqrt{5}}$ Simplify

$= 2\sqrt{10m} + \dfrac{\sqrt{2m} \cdot \sqrt{5}}{\sqrt{5} \cdot \sqrt{5}}$ Rationalize the denominator

$= \dfrac{2\sqrt{10m}}{1} + \dfrac{\sqrt{10m}}{5}$

$= \dfrac{5 \cdot 2\sqrt{10m}}{5 \cdot 1} + \dfrac{\sqrt{10m}}{5}$ Write with LCD 5

$= \dfrac{10\sqrt{10m} + \sqrt{10m}}{5}$ Add numerators

Try Problem 59 $= \dfrac{11\sqrt{10m}}{5}$ ◂

PROBLEM SET 5.8

Rationalize the denominator and simplify. All variables represent positive real numbers.

1. $\dfrac{1}{\sqrt{5}}$ $\dfrac{\sqrt{5}}{5}$

2. $\dfrac{1}{\sqrt{3}}$

3. $\dfrac{\sqrt{5}}{\sqrt{2}}$ $\dfrac{\sqrt{10}}{2}$

4. $\dfrac{\sqrt{3}}{\sqrt{7}}$

5. $\dfrac{6}{\sqrt{2}}$ $3\sqrt{2}$

6. $\dfrac{21}{\sqrt{7}}$

7. $\dfrac{2}{\sqrt{6}}$

8. $\dfrac{3}{\sqrt{6}}$

9. $\dfrac{3}{\sqrt{8}}$

10. $\dfrac{5}{\sqrt{8}}$

11. $\dfrac{2\sqrt{3}}{\sqrt{x}}$

12. $\dfrac{2\sqrt{5}}{\sqrt{x}}$

13. $\dfrac{4\sqrt{2}}{\sqrt{10}}$

14. $\dfrac{6\sqrt{5}}{\sqrt{10}}$

15. $\dfrac{15}{\sqrt[3]{3}}$

16. $\dfrac{10}{\sqrt[3]{5}}$

17. $\dfrac{8}{\sqrt[3]{4}}$

18. $\dfrac{6}{\sqrt[3]{9}}$

19. $\sqrt{\dfrac{2}{5}}$

20. $\sqrt{\dfrac{7}{3}}$

21. $\sqrt{\dfrac{3}{r}}$

22. $\sqrt{\dfrac{2}{r}}$

23. $\sqrt{\dfrac{7}{10t}}$

24. $\sqrt{\dfrac{3}{10t}}$

25. $\dfrac{\sqrt{45x^3}}{\sqrt{10xy}}$

26. $\dfrac{\sqrt{27x^3}}{\sqrt{15xy}}$

27. $\dfrac{\sqrt[3]{14}}{\sqrt[3]{63p}}$

28. $\dfrac{\sqrt[3]{15}}{\sqrt[3]{20p}}$

Rationalize the denominator and simplify. All variables represent positive real numbers.

29. $\dfrac{1}{\sqrt{2}+1}$

30. $\dfrac{1}{\sqrt{2}-1}$

31. $\dfrac{1}{2-\sqrt{3}}$

32. $\dfrac{1}{2+\sqrt{3}}$

33. $\dfrac{2}{\sqrt{6}-2}$

34. $\dfrac{2}{\sqrt{11}-3}$

35. $\dfrac{10}{\sqrt{5}-1}$

36. $\dfrac{14}{\sqrt{5}+1}$

37. $\dfrac{1}{3\sqrt{5}+1}$

38. $\dfrac{1}{2\sqrt{7}+1}$

39. $\dfrac{8}{\sqrt{5}+\sqrt{3}}$

40. $\dfrac{12}{\sqrt{7}-\sqrt{5}}$

41. $\dfrac{\sqrt{2}}{\sqrt{13}+\sqrt{2}}$

42. $\dfrac{\sqrt{2}}{\sqrt{17}+\sqrt{2}}$

43. $\dfrac{\sqrt{7}+\sqrt{3}}{\sqrt{7}-\sqrt{3}}$

44. $\dfrac{\sqrt{11}+\sqrt{5}}{\sqrt{11}-\sqrt{5}}$

45. $\dfrac{4+\sqrt{6}}{\sqrt{2}-\sqrt{3}}$

46. $\dfrac{2+\sqrt{30}}{\sqrt{5}-\sqrt{6}}$

47. $\dfrac{4\sqrt{x}}{4\sqrt{x}+3\sqrt{y}}$

48. $\dfrac{3\sqrt{y}}{3\sqrt{x}+5\sqrt{y}}$

49. $\dfrac{m-4}{\sqrt{m}+2}$

50. $\dfrac{m-9}{\sqrt{m}+3}$

Add or subtract as indicated and simplify. All variables represent positive real numbers.

51. $\dfrac{\sqrt{5}}{3}-\dfrac{1}{\sqrt{5}}$

52. $\dfrac{\sqrt{3}}{2}-\dfrac{1}{\sqrt{3}}$

53. $\dfrac{1}{\sqrt{2}}+\sqrt{2}$

54. $\dfrac{1}{\sqrt{7}} + \sqrt{7}$ 　　　　　　**55.** $\dfrac{\sqrt{8}}{\sqrt{6}} + \dfrac{\sqrt{6}}{\sqrt{8}}$ 　　　　　　**56.** $\dfrac{\sqrt{8}}{\sqrt{10}} + \dfrac{\sqrt{10}}{\sqrt{8}}$

57. $\sqrt{50} + \sqrt{\dfrac{1}{2}}$ 　　　　　　**58.** $\sqrt{48} + \sqrt{\dfrac{1}{3}}$ 　　　　　　**59.** $\sqrt{90m} + \sqrt{\dfrac{2m}{5}}$

60. $\sqrt{60m} + \sqrt{\dfrac{3m}{5}}$ 　　　　　　**61.** $\dfrac{1}{\sqrt{x}} + \dfrac{1}{\sqrt{y}}$ 　　　　　　**62.** $\dfrac{1}{\sqrt{x}} - \dfrac{1}{\sqrt{y}}$

CALCULATOR PROBLEMS

Compute to the nearest hundredth. Then rationalize the denominator, compute the value of the new expression, and compare.

63. $\dfrac{1}{\sqrt{11}}$ 　　　　　　　　　　　　**64.** $\dfrac{\sqrt{5} + \sqrt{2}}{\sqrt{5} - \sqrt{2}}$

5.9　　　　　*Complex Numbers*

Since there is no real number whose square is -1, the expression $\sqrt{-1}$ is not a real number. Instead, $\sqrt{-1}$ is called an **imaginary number.***

Definition	The **imaginary unit** is denoted by i, where $$i = \sqrt{-1} \quad \text{and} \quad i^2 = -1.$$

If we extend the product rule for radicals so that it also applies when one of the radicands is negative, we can write square roots of negative numbers in terms of i.

EXAMPLE 1　Write in terms of i: **a)** $\sqrt{-4}$, **b)** $\sqrt{-\frac{1}{9}}$, **c)** $\sqrt{-5}$.

---

* René Descartes (1596–1650) introduced the term imaginary number, thus giving imaginary numbers a permanent stigma. Of course, these numbers really do exist and, in fact, have a wide variety of useful applications in physics and engineering. The first application of imaginary numbers was introduced by Charles Steinmetz (1865–1923), who used them to explain the behavior of electric circuits.

a) $\sqrt{-4} = \underbrace{\sqrt{4 \cdot (-1)}}_{\text{Product rule}} = \sqrt{4} \cdot \sqrt{-1} = 2 \cdot i = 2i$

Product rule

b) $\sqrt{-\frac{1}{9}} = \sqrt{\frac{1}{9} \cdot (-1)} = \sqrt{\frac{1}{9}} \cdot \sqrt{-1} = \frac{1}{3} \cdot i = \frac{1}{3}i$

Try Problem 1

c) $\sqrt{-5} = \sqrt{5 \cdot (-1)} = \sqrt{5} \cdot \sqrt{-1} = \sqrt{5} \cdot i = \sqrt{5}i$ ◄

CAUTION The number $\sqrt{5}i$ means $(\sqrt{5})i$ and *not* $\sqrt{5i}$.

When we combine the set of imaginary numbers with the set of real numbers we obtain the set of **complex numbers.**

Definition

A **complex number** is a number of the form

$$a + bi,$$

where a and b are both real numbers.

If $b = 0$, the complex number $a + bi$ is simply the real number a. If $a = 0$, the complex number $a + bi$ is simply bi, which is called a **pure imaginary number.** Figure 5.5 illustrates the relationships between the set of complex numbers and its subsets.

Complex numbers $a + bi$

Real numbers $(b = 0)$
$8, 0, -\frac{2}{3}, 0.\overline{27}, \sqrt{5}, \pi$

Imaginary numbers $(b \neq 0)$
$7 + 4i$ $\frac{1}{2} - \frac{\sqrt{3}}{2}i$

Pure imaginary numbers $(a = 0, b \neq 0)$
$2i, \frac{1}{3}i, \sqrt{5}i$

Figure 5.5

Now let's examine how we add, subtract, multiply, and divide the two complex numbers $a + bi$ and $c + di$.

Operations on Complex Numbers

Addition: $(a + bi) + (c + di) = (a + c) + (b + d)i$
Subtraction: $(a + bi) - (c + di) = (a - c) + (b - d)i$
Multiplication: $(a + bi) \cdot (c + di) = (ac - bd) + (ad + bc)i$
Division: $\dfrac{a + bi}{c + di} = \dfrac{ac + bd}{c^2 + d^2} + \dfrac{bc - ad}{c^2 + d^2}i$

Memorizing the four definitions above would be a formidable task. Fortunately, that will not be necessary. You can get the same result simply by treating each complex number as a binomial in i.

EXAMPLE 2   Add $6 + 2i$ and $4 + 5i$.

$$(6 + 2i) + (4 + 5i) = 6 + 2i + 4 + 5i \qquad \text{Remove parentheses}$$
$$= 10 + 7i \qquad \text{Combine like terms} \blacktriangleleft$$

Try Problem 19

EXAMPLE 3   Add $-7 - i$ and $3i$.

Try Problem 27

$$(-7 - i) + 3i = -7 - i + 3i = -7 + 2i \blacktriangleleft$$

EXAMPLE 4   Subtract $2 + 8i$ from $5 + 9i$.

$$(5 + 9i) - (2 + 8i) = 5 + 9i - 2 - 8i \qquad \text{Remove parentheses}$$
$$= 3 + i \qquad \text{Combine like terms}$$

We can check this answer by addition as follows:

$$(3 + i) + (2 + 8i) = 3 + i + 2 + 8i$$
$$= 5 + 9i \blacktriangleleft$$

Try Problem 33

EXAMPLE 5   Subtract $2 + 4i$ from 6.

Try Problem 37

$$6 - (2 + 4i) = 6 - 2 - 4i = 4 - 4i \blacktriangleleft$$

Before we illustrate multiplication of complex numbers, recall that $i^2 = -1$. Therefore whenever i^2 appears, we replace it by -1.

EXAMPLE 6   Multiply $3 + 4i$ and $2 + 7i$.

$$(3 + 4i) \cdot (2 + 7i) = 3 \cdot 2 + 3 \cdot 7i + 4i \cdot 2 + 4i \cdot 7i \qquad \text{FOIL method}$$
$$= 6 + 21i + 8i + 28i^2$$
$$= 6 + 21i + 8i + 28(-1) \qquad \text{Replace } i^2 \text{ by } -1$$
$$= -22 + 29i \qquad \text{Combine like terms} \blacktriangleleft$$

Try Problem 45

EXAMPLE 7   Multiply $2i$ and $5 - 9i$.

$$2i \cdot (5 - 9i) = 2i \cdot 5 - 2i \cdot 9i \qquad \text{Distribute } 2i$$
$$= 10i - 18i^2$$
$$= 10i - 18(-1) \qquad \text{Replace } i^2 \text{ by } -1$$
$$= 10i + 18$$
$$= 18 + 10i \blacktriangleleft$$

Try Problem 57

We can find the quotient of two complex numbers by multiplying numerator and denominator by the conjugate of the denominator. The complex numbers $a + bi$ and $a - bi$ are called **conjugates** of each other. This procedure eliminates the i in the denominator, since

$$(a + bi) \cdot (a - bi) = a^2 - a \cdot bi + bi \cdot a - b^2 i^2$$
$$= a^2 - abi + abi - b^2(-1)$$
$$= a^2 + b^2.$$

EXAMPLE 8   Divide $5 + 10i$ by $2 + i$.

Multiply numerator and denominator by the conjugate of the denominator.

$$\frac{5 + 10i}{2 + i} = \frac{(5 + 10i)(2 - i)}{(2 + i)(2 - i)} \qquad \text{The conjugate of } 2 + i \text{ is } 2 - i$$

$$= \frac{10 - 5i + 20i - 10i^2}{4 - 2i + 2i - i^2}$$

$$= \frac{10 - 5i + 20i - 10(-1)}{4 - 2i + 2i - (-1)} \qquad \text{Replace } i^2 \text{ by } -1$$

$$= \frac{20 + 15i}{5} \qquad \text{Combine like terms}$$

$$= \frac{20}{5} + \frac{15i}{5} \qquad \text{Divide each term by 5}$$

$$= 4 + 3i$$

We can check this answer by multiplication as follows:

$$(2 + i) \quad (4 + 3i) = 8 + 6i + 4i + 3i^2$$
$$= 8 + 6i + 4i + 3(-1)$$
$$= 5 + 10i \blacktriangleleft$$

Try Problem 73 ＼

EXAMPLE 9   Divide $7 + 6i$ by $3i$.

$$\frac{7 + 6i}{3i} = \frac{(7 + 6i) \cdot (-3i)}{3i \cdot (-3i)} \qquad \text{The conjugate of } 3i \text{ is } -3i$$

$$= \frac{-21i - 18i^2}{-9i^2}$$

$$= \frac{-21i - 18(-1)}{-9(-1)} \qquad \text{Replace } i^2 \text{ by } -1$$

$$= \frac{18 - 21i}{9}$$

$$= \frac{18}{9} - \frac{21i}{9} \qquad \text{Divide each term by 9}$$

$$= 2 - \frac{7}{3}i \blacktriangleleft$$

Try Problem 79 ＼

PROBLEM SET 5.9

Write each number in terms of the imaginary unit i.

1. $\sqrt{-9}$ 2. $\sqrt{-25}$ 3. $\sqrt{-36}$ 4. $\sqrt{-64}$

5. $\sqrt{-\frac{1}{4}}$ 6. $\sqrt{-\frac{1}{16}}$ 7. $\sqrt{-2}$ 8. $\sqrt{-3}$

9. $\sqrt{-12}$ 10. $\sqrt{-18}$ 11. $5\sqrt{-49}$ 12. $6\sqrt{-81}$

13. $-\sqrt{-100}$ 14. $-\sqrt{-121}$ 15. $\sqrt{-\frac{3}{4}}$ 16. $\sqrt{-\frac{5}{9}}$

17. True or false.
   a) Every real number is also a complex number.
   b) No real number is also an imaginary number.
   c) Every pure imaginary number is also an imaginary number.

18. True or false.
   a) Every imaginary number is also a complex number.
   b) No imaginary number is also a real number.
   c) Some imaginary numbers are not pure imaginary numbers.

Add or subtract as indicated.

19. $(6 + 2i) + (4 + 7i)$ 20. $(8 + 3i) + (5 + 9i)$

21. $(-9 + 6i) + (4 - 3i)$ 22. $(-7 + 9i) + (3 - 4i)$

23. $(5 + 4i) + (-3 - 4i)$ 24. $(6 + 3i) + (-2 - 3i)$

25. $(-7 - 12i) + (7 - 8i)$ 26. $(-8 - 14i) + (8 - 7i)$

27. $(-8 - i) + 4i$ 28. $(-5 - i) + 6i$

29. $(16 + i) + 9$ 30. $(17 + i) + 11$

31. $6i + 2i$ 32. $8i + 2i$

33. $(10 + 6i) - (3 + 5i)$ 34. $(7 + 5i) - (2 + 3i)$

35. $(-3 - 7i) - (-3 - 7i)$ 36. $(-8 - 4i) - (-8 - 4i)$

37. $11 - (4 + 5i)$ 38. $13 - (2 + 7i)$

39. $(-6 - 9i) - (-17)$ 40. $(-5 - 6i) - (-15)$

41. $i - 13i$ 42. $i - 19i$

43. $(-5 - i) - [(7 + 6i) - (8 - 2i)]$ 44. $(-4 - 3i) - [(6 + 9i) - (7 - i)]$

Multiply.

45. $(5 + 4i)(2 + 3i)$ 46. $(2 + 6i)(4 + 5i)$

47. $(2 + 3i)(6 - 5i)$ 48. $(4 + 2i)(8 - 3i)$

49. $(-8 + i)(2 - 4i)$ 50. $(-7 + i)(2 - 5i)$

51. $(-1 - 9i)(-6 - 3i)$ 52. $(-8 - 6i)(-1 - 4i)$

53. $(4 - 5i)(4 + 5i)$ 54. $(6 - 3i)(6 + 3i)$

55. $(-4i)(-3i)$

56. $(-5i)(-2i)$

57. $2i(5-3i)$

58. $4i(3-2i)$

59. $5(-9+6i)$

60. $7(-3+2i)$

61. $(3-2i)^2$

62. $(5-3i)^2$

63. $(1+\sqrt{2}i)^2$

64. $(1+\sqrt{3}i)^2$

65. $(1+i)^3$

66. $(1-i)^3$

Divide. Check some of your answers by multiplication.

67. $\dfrac{8}{1-i}$

68. $\dfrac{6}{1+i}$

69. $\dfrac{5i}{1+2i}$

70. $\dfrac{10i}{1+3i}$

71. $\dfrac{7-3i}{1+i}$

72. $\dfrac{5+3i}{1-i}$

73. $\dfrac{2+11i}{2+i}$

74. $\dfrac{10+10i}{3+i}$

75. $\dfrac{6+4i}{3-2i}$

76. $\dfrac{2+5i}{4-3i}$

77. $\dfrac{2i}{-5+3i}$

78. $\dfrac{5i}{-3+4i}$

79. $\dfrac{4+10i}{5i}$

80. $\dfrac{1+4i}{2i}$

81. $\dfrac{15i}{-3i}$

82. $\dfrac{18i}{-3i}$

83. $\dfrac{-1}{-8i}$

84. $\dfrac{-1}{-7i}$

85. $\dfrac{6-8i}{2}$

86. $\dfrac{8-12i}{4}$

87. Show that $2i$ and $-2i$ are both square roots of -4 by calculating $(2i)^2$ and $(-2i)^2$.

88. Show that $5i$ and $-5i$ are both square roots of -25 by calculating $(5i)^2$ and $(-5i)^2$.

† 89. Show that the sum of any complex number $a+bi$ and its conjugate $a-bi$ is a real number.

† 90. Show that the product of any complex number $a+bi$ and its conjugate $a-bi$ is a real number.

CALCULATOR PROBLEMS

Multiply or divide as indicated.

91. $(4.82-6.25i)(3.15+9.88i)$

92. $\dfrac{1.1+9.9i}{3.5-0.5i}$

CHAPTER 5 REVIEW

Assume all variables in this chapter review represent positive real numbers.

[5.1] *Simplify.*

1. 3^{-2}

2. $(-3)^{-3}$

3. $\dfrac{1}{2r^{-2}}$

4. $x^{-8}\cdot x^{13}$

5. $\dfrac{y^{-2}}{y^{-6}}$

6. $x + x^{-2}$

7. $(a^{-2}b^3)^{-2}$

8. $\dfrac{r-s}{r^{-1}-s^{-1}}$

[5.2] *Write each number in standard form.*

9. 6.1×10^5

10. 3.02×10^{-4}

Write each number in scientific notation.

11. 34,000,000

12. 0.00000253

Compute by first converting to scientific notation. Express your answer in scientific notation.

13. $\dfrac{(6,000,000,000)(32,000)}{500,000}$

14. $\dfrac{(0.000005)(0.0000066)}{20,000}$

[5.3] *Find each root.*

15. $\sqrt[6]{64}$

16. $\sqrt[3]{-27}$

17. $\sqrt[3]{-8x^6y^9}$

18. $\sqrt[8]{(a+b)^{16}}$

19. $\sqrt[4]{81a^{12}}$

20. $\pm\sqrt{121}$

21. $\sqrt{-81}$

22. $\sqrt{25x^{2n}y^{6m}}$

[5.4] *Find each rational power.*

23. $243^{3/5}$

24. $\left(\dfrac{25}{49}\right)^{1/2}$

25. $36^{-3/2}$

Write in radical form.

26. $x^{4/3}$

27. $4y^{2/3}$

28. $(y+2)^{1/2}$

Simplify each expression.

29. $x^{5/4} \cdot x^{-1/4}$

30. $(16p^8q^{16})^{3/4}$

31. $(x^{-3/4}y^{3/2})^{-4}(x^{3/2}y^{-4})^{-1}$

32. $y^{7/2}(y^{1/2} - y)$

[5.5] *Simplify each radical expression.*

33. $\sqrt{20}$

34. $\sqrt[3]{54}$

35. $\sqrt{\dfrac{8a^5}{121}}$

36. $\sqrt[3]{\dfrac{27x^4}{125}}$

37. $\sqrt{\sqrt{64x^4}}$

38. $\sqrt[3]{a^6b^4}$

39. $\sqrt[4]{\sqrt[3]{x}}$

40. $\sqrt{16\sqrt{a}}$

41. $\sqrt{\dfrac{x^3y^5}{z^6}}$

[5.6] *Multiply or divide as indicated. Simplify your answer.*

42. $\sqrt{5} \cdot \sqrt{10}$

43. $(\sqrt{y})^8$

44. $\sqrt{7x}\,\sqrt{14x}$

45. $\sqrt[3]{x^2y^4}\,\sqrt[3]{xy^2}$

46. $\sqrt{3a^2}\,\sqrt{5a}\,\sqrt{15a^4}$

47. $\sqrt[5]{2a}\,\sqrt[5]{16b}$

48. $\dfrac{\sqrt{98}}{\sqrt{2}}$

49. $\dfrac{\sqrt{40x}}{\sqrt{2x}}$

50. $\dfrac{\sqrt{3a}}{\sqrt{147a^3}}$

51. $\dfrac{\sqrt[4]{32x^6}}{\sqrt[4]{2x^2}}$

52. $\dfrac{\sqrt[3]{35a^8b^{13}c^4}}{\sqrt[3]{5a^2bc}}$

53. $\dfrac{\sqrt{15p^4q^4}\,\sqrt{5pq^2}}{\sqrt{3p^2q^3}}$

54. $\sqrt[3]{2} \cdot \sqrt[6]{2}$

55. $\dfrac{\sqrt{a}}{\sqrt[3]{a}}$

[5.7] *Add or subtract as indicated.*

56. $5\sqrt{3} + 2\sqrt{3}$

57. $\sqrt{4x} + \sqrt{9x}$

58. $\sqrt[3]{54y} - \sqrt[3]{16y}$

59. $3\sqrt{20} + 4\sqrt{45} - 5\sqrt{80}$

Multiply and simplify.

60. $\sqrt{2}(\sqrt{14} - \sqrt{2})$

61. $(\sqrt{x} + \sqrt{y})(\sqrt{x} - \sqrt{y})$

62. $\sqrt[3]{2}(\sqrt[3]{4} - \sqrt[3]{2})$

63. $(t + \sqrt{3})^2$

64. $(\sqrt{2} + \sqrt{3})^2$

65. $(2\sqrt{15} - 5)(3\sqrt{3} + 2)$

[5.8] *Rationalize the denominator and simplify.*

66. $\dfrac{2}{\sqrt{3}}$

67. $\sqrt{\dfrac{3}{5}}$

68. $\dfrac{8}{\sqrt[3]{2}}$

69. $\dfrac{1}{3 - \sqrt{5}}$

70. $\dfrac{\sqrt{7} - \sqrt{5}}{\sqrt{7} + \sqrt{5}}$

71. $\dfrac{1}{\sqrt{2}} + \dfrac{1}{\sqrt{3}}$

[5.9] *Perform the indicated operation.*

72. $(3 + 2i) + (7 + 2i)$

73. $(3 - 4i) - (3 - 2i)$

74. $11 - (3 - 2i)$

75. $(3 + 2i)(4 - i)$

76. $3i(4 + i)$

77. $\dfrac{5}{2 + 3i}$

78. $\dfrac{8i}{-2i}$

79. $\dfrac{3 - 2i}{5 + i}$

CHAPTER 5 TEST

Assume all variables represent positive real numbers.

1. Write each number in scientific notation.
   a) 160,000 **b)** 0.000769

Simplify.

2. 3^{-4}

3. $(-5)^{-3}$

4. $(2a^2b^{-1})^{-2}$

5. $\dfrac{1}{x^{-1} + y^{-1}}$

Find each root.

6. $\sqrt[3]{-64}$

7. $\sqrt{\dfrac{9x^2y^3}{25y}}$

Simplify each expression.

8. $(8a^3b^{-6})^{1/3}$

9. $\sqrt{\dfrac{12a^3}{b^4}}$

10. $\dfrac{\sqrt{242}}{\sqrt{2}}$

11. $27^{-2/3}$

12. $(x^{1/3}y^{-1/6})^{-12}$

13. $\sqrt[3]{m^5n^3}$

Perform the indicated operations and simplify.

14. $\sqrt{6} \cdot \sqrt{15} \cdot \sqrt{10}$

15. $\dfrac{\sqrt{3}}{\sqrt{75}}$

16. $\dfrac{\sqrt[3]{x^2}}{\sqrt[6]{x}}$

17. $(\sqrt{3} + \sqrt{5})(\sqrt{3} - \sqrt{5})$

18. $4\sqrt{3} + 2\sqrt{27}$

19. $\sqrt{2}(\sqrt{18} + \sqrt{3})$

Rationalize the denominator and simplify.

20. $\dfrac{10}{\sqrt[3]{5}}$

21. $\dfrac{1}{\sqrt{5} + 3}$

Perform the indicated operations.

22. $6 + (9 + 8i)$

23. $(7 - 2i) - (3 - 5i)$

24. $(5 - 2i)(4 + i)$

25. $\dfrac{4 + i}{3 + 2i}$

In Chapter 3 you learned how to solve first-degree equations and first-degree inequalities. In this chapter you will learn how to solve second-degree equations and second-degree inequalities. Second-degree equations, also called quadratic equations, can be used to measure the speed of a current, determine when an object that has been projected upward will strike the ground, and solve right-triangle problems involving the Pythagorean theorem.

There are several ways to solve a quadratic equation—by factoring, by extraction of roots, by completing the square, and by the quadratic formula. You will learn which method works best in a given situation. You will also learn how to solve certain types of nonquadratic equations and equations involving radicals.

6

Second-Degree Equations and Inequalities

6.1 *Factoring and Extraction of Roots*

In Section 3.1 we learned that first-degree (or linear) equations in the variable x take the form

$$ax + b = 0,$$

where $a \neq 0$. Now let's define equations of the second degree.

Definition

A **second-degree equation,** or **quadratic equation,** in the variable x is an equation that can be expressed in the form

$$ax^2 + bx + c = 0,$$

where $a \neq 0$. This form is called the **standard form** of a quadratic equation.

Note that if $a = 0$, then $ax^2 + bx + c = 0$ becomes $bx + c = 0$, which is not a quadratic equation.*

The equations

$$(x - 3)(x + 5) = 0 \quad \text{and} \quad 3y^2 + 11y = 4$$

are examples of quadratic equations.

FACTORING METHOD

We can solve many quadratic equations by writing the equation in standard form, factoring the left side, and applying the zero factor property.

Zero Factor Property

$$p \cdot q = 0 \quad \text{if and only if} \quad p = 0 \ \text{or} \ q = 0.$$

In other words, a product of two numbers is 0 if and only if one or both of the numbers is 0.

EXAMPLE 1 Solve $(x - 3)(x + 5) = 0$.

Apply the zero factor property. That is,

$$\underbrace{(x - 3)}_{p} \cdot \underbrace{(x + 5)}_{q} = 0$$
$$ = 0$$

---

* A procedure for solving quadratic equations was known by the Babylonians, perhaps as early as 2000 B.C.

if and only if
$$x - 3 = 0 \quad \text{or} \quad x + 5 = 0.$$
$$\underbrace{}_{p} = 0 \quad \text{or} \quad \underbrace{}_{q} = 0$$

Therefore the original equation is true if and only if
$$x = 3 \quad \text{or} \quad x = -5.$$

Try Problem 1   The solution is 3 or -5. Check each solution in the original equation. ◄

CAUTION The zero factor property applies only when a product is equal to *zero*. The fact that a product equals 6, for example, does not mean that either factor is 6.

EXAMPLE 2   Solve $3y^2 + 11y = 4$.

Write the equation in standard form so you can apply the zero factor property.
$$3y^2 + 11y = 4$$
$$3y^2 + 11y - 4 = 0 \qquad \text{Subtract 4 from each side}$$

Then factor the left side.
$$(3y - 1)(y + 4) = 0$$

Apply the zero factor property.
$$3y - 1 = 0 \quad \text{or} \quad y + 4 = 0$$

Solve each linear equation.
$$3y = 1 \quad \text{or} \quad y = -4$$
$$y = \tfrac{1}{3}$$

Try Problem 15   The solution is $\frac{1}{3}$ or -4. Check in the original equation. ◄

Now let's look at a quadratic equation that has only one solution.

EXAMPLE 3   Solve $z^2 = 2z - 1$.
$$z^2 - 2z + 1 = 0 \qquad \text{Write in standard form}$$
$$(z - 1)(z - 1) = 0 \qquad \text{Factor the left side}$$
$$z - 1 = 0 \quad \text{or} \quad z - 1 = 0 \qquad \text{Set each factor equal to 0}$$
$$z = 1 \quad \text{or} \quad z = 1 \qquad \text{Solve each linear equation}$$

Try Problem 21   The number 1 checks, so the solution is 1. ◄

A quadratic equation may have *no* real solutions. For example, the equation $x^2 = -4$ has no real solutions, since no real number squared is -4. A quadratic equation can have zero, one, or two real solutions.

If $c = 0$, the equation $ax^2 + bx + c = 0$ becomes $ax^2 + bx = 0$. Quadratic equations of this type always have two real solutions, one of which is always 0.

EXAMPLE 4   Solve $5x^2 = 10x$.

$$5x^2 - 10x = 0 \qquad \text{Write in standard form}$$
$$5x(x - 2) = 0 \qquad \text{Factor the left side}$$
$$5x = 0 \quad \text{or} \quad x - 2 = 0 \qquad \text{Set each factor equal to 0}$$
$$x = 0 \quad \text{or} \qquad x = 2 \qquad \text{Solve each linear equation}$$

Try Problem 27   Both numbers check, so the solution is 0 or 2. ◄

CAUTION Do not divide both sides of the equation $5x^2 = 10x$ by $5x$ and get $x = 2$. In doing so, you lose the solution $x = 0$. The danger in dividing by a variable expression such as $5x$ is that you may be dividing by zero.

To Solve a Quadratic Equation by Factoring

1. Write the equation in standard form.
2. Factor the left side.
3. Set each factor equal to zero.
4. Solve each linear equation resulting from step 3.
5. Check each solution in the original equation.

EXTRACTION-OF-ROOTS METHOD

If $b = 0$, the equation $ax^2 + bx + c = 0$ becomes $ax^2 + c = 0$. We can solve equations of this type using either factoring or **extraction of roots**. Both methods are illustrated in Example 5.

EXAMPLE 5   Solve $x^2 - 36 = 0$.

Factoring Method

$$x^2 - 36 = 0 \qquad \text{Equation is already in standard form}$$
$$(x - 6)(x + 6) = 0 \qquad \text{Factor the left side}$$
$$x - 6 = 0 \quad \text{or} \quad x + 6 = 0 \qquad \text{Set each factor equal to 0}$$
$$x = 6 \quad \text{or} \qquad x = -6 \qquad \text{Solve each linear equation}$$

The solution is 6 or -6.

Extraction-of-Roots Method

$$x^2 - 36 = 0$$
$$x^2 = 36 \qquad \text{Write in the form } x^2 = k$$
$$x = 6 \quad \text{or} \quad x = -6 \qquad \text{Extract both square roots of 36}$$

Try Problem 33   The solution is 6 or -6. ◄

CAUTION Be sure to extract both the positive and the negative square root when using the extraction-of-roots method.

EXAMPLE 6   Solve $(3p + 4)(p + 2) = 5(2p + 1) + 9$.

First write the equation in standard form.

$$(3p + 4)(p + 2) = 5(2p + 1) + 9$$
$$3p^2 + 10p + 8 = 10p + 5 + 9 \qquad \text{Remove parentheses}$$
$$3p^2 + 10p + 8 = 10p + 14 \qquad \text{Combine } 5 + 9$$
$$3p^2 - 6 = 0 \qquad \text{Subtract } 10p + 14 \text{ from each side}$$

Then solve by extraction of roots.

$$\left.\begin{array}{l} 3p^2 = 6 \\ p^2 = 2 \end{array}\right\} \text{Write in the form } x^2 = k$$
$$p = \pm\sqrt{2} \qquad \text{Extract both square roots of 2}$$

Try Problem 57   The solution is $\sqrt{2}$ or $-\sqrt{2}$. ◄

To Solve a Quadratic Equation by Extraction of Roots

1. You must be able to write the equation in the form $x^2 = k$.
2. Extract both square roots of k and get $x = \pm\sqrt{k}$.
3. Simplify $\pm\sqrt{k}$ if possible.

We can use the methods of factoring and extraction of roots to solve an equation in several variables for one of its variables.

EXAMPLE 7   Solve $x^2 - 9a^2 = 0$ for x.

Use extraction of roots.

$$x^2 = 9a^2 \qquad \text{Add } 9a^2 \text{ to each side}$$

Try Problem 59 $$x = \pm 3a \qquad \text{Extract both square roots of } 9a^2 \text{ ◄}$$

EXAMPLE 8   Solve $x^2 - 3ax = 4a^2$ for x.

Solve by factoring.

$$x^2 - 3ax - 4a^2 = 0 \qquad \text{Write in standard form}$$
$$(x - 4a)(x + a) = 0 \qquad \text{Factor the left side}$$
$$x - 4a = 0 \quad \text{or} \quad x + a = 0 \qquad \text{Set each factor equal to 0}$$

Try Problem 65 $$x = 4a \quad \text{or} \qquad x = -a \qquad \text{Solve each equation for } x \text{ ◄}$$

Problem Solving

Measuring the Speed of a Current

The speed of the current in a stream can be determined using an open-ended, L-shaped tube as shown in Figure 6.1. The current speed s in feet per second is given by Torricelli's law as $s^2 = 64h$, where h is the height in feet of the water in the tube. Determine the current speed if $h = 6$ inches.

Since h must be in feet, write

$$h = 6 \text{ in.} = \tfrac{6}{12} \text{ ft} = \tfrac{1}{2} \text{ ft.}$$

Substitute $h = \tfrac{1}{2}$ into the formula $s^2 = 64h$.

$$s^2 = 64(\tfrac{1}{2}) = 32$$

Extract both roots and simplify the radical.

$$s = \pm\sqrt{32} = \pm\sqrt{16 \cdot 2} = \pm\sqrt{16}\sqrt{2} = \pm 4\sqrt{2}$$

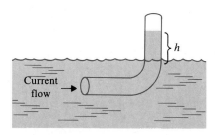

Figure 6.1

We reject the negative answer, since the speed cannot be a negative number. Therefore the speed of the current is $4\sqrt{2}$ (about 5.7) feet per second.

PROBLEM SET 6.1

Use the zero factor property to solve each equation.

1. $(x - 2)(x + 6) = 0$

2. $(x - 4)(x + 5) = 0$

3. $(2y + 1)(y + 4) = 0$

4. $(3y + 1)(y + 2) = 0$

5. $(z + 7)(z - 7) = 0$

6. $(z + 3)(z - 3) = 0$

7. $(r - 8)(r - 8) = 0$

8. $(r - 9)(r - 9) = 0$

9. $4m(m + 3) = 0$

10. $6m(m + 7) = 0$

Solve each equation.

11. $x^2 - 5x + 4 = 0$

12. $x^2 - 7x + 6 = 0$

13. $x^2 + 4x = 21$

14. $x^2 + 2x = 35$

15. $3y^2 + 10y = 8$

16. $3y^2 + 16y = 12$

17. $y^2 = 16 - 6y$

18. $y^2 = 12 - 4y$

19. $z^2 + 10z + 25 = 0$

20. $z^2 + 8z + 16 = 0$

21. $z^2 = 4z - 4$

22. $z^2 = 6z - 9$

23. $2m^2 + 5m = 3$

24. $2m^2 + 9m = 5$

25. $r^2 - 17r = 0$

26. $r^2 - 19r = 0$

27. $4x^2 = 20x$

28. $6x^2 = 18x$

29. $6t^2 + 15 = 19t$

30. $6t^2 + 10 = 23t$

31. $p^2 = 4$

32. $p^2 = 9$

33. $x^2 - 16 = 0$

34. $x^2 - 25 = 0$

35. $2y^2 - 72 = 0$

36. $2y^2 - 98 = 0$

37. $16s^2 + 9 = 24s$

38. $25s^2 + 9 = 30s$

39. $9z^2 - 4 = 0$

40. $4z^2 - 9 = 0$

41. $6m^2 - 3m = 0$

42. $8m^2 - 2m = 0$

43. $r^2 = r$

44. $r^2 = -r$

45. $5x^2 = 0$

46. $7x^2 = 0$

47. $y^2 - 3 = 0$

48. $y^2 - 2 = 0$

49. $4z^2 - 48 = 0$

50. $5z^2 - 90 = 0$

51. $(r - 9)(r + 11) = 21$

52. $(r - 8)(r + 11) = 42$

53. $m(9m + 7) = m - 1$

54. $m(4m + 5) = m - 1$

55. $(2x + 1)^2 + 3x^2 = 6x + 1$

56. $(3x + 1)^2 - x^2 = 9x + 1$

57. $(3p + 2)(p + 4) = 7(2p + 1) + 16$

58. $(3p + 5)(p + 1) = 4(2p + 3) + 14$

Solve each equation for x.

59. $x^2 - 4a^2 = 0$

60. $x^2 - 25a^2 = 0$

61. $5x^2 - 10ax = 0$

62. $7x^2 - 21ax = 0$

63. $x^2 - 6ax + 9a^2 = 0$

64. $x^2 - 8ax + 16a^2 = 0$

65. $x^2 - 4ax = 5a^2$

66. $x^2 - 5ax = 6a^2$

67. $6x^2 + 7ax - 5a^2 = 0$

68. $12x^2 + 7ax - 10a^2 = 0$

69. Solve $c^2 = a^2 + b^2$ for a.

70. Solve $c^2 = a^2 + b^2$ for b.

71. Determine the current speed in Figure 6.1 if $h = 9$ inches.

72. Determine the current speed in Figure 6.1 if $h = 15$ inches.

73. An air gun with a muzzle velocity of 96 feet per second fires a pellet upward from the top of a building that is 112 feet tall. If we ignore air resistance, the height h in feet of the pellet after t seconds is given by the formula $h = -16t^2 + 96t + 112$. Determine when the pellet strikes the ground. [*Hint:* Set $h = 0$.]

74. The revenue equation for a certain business is $R = 125x - 4x^2$, and the cost equation is $C = 25x + 600$. Determine the break-even points. [*Hint:* Set $R = C$.]

75. The square of a number x is equal to 6 more than 5 times x. Find x.

76. The square of a number x is equal to 8 more than 7 times x. Find x.

By reversing the steps for solving a quadratic equation, write a quadratic equation in standard form with integer coefficients having the given solutions.

† **77.** 3 and 5   † **78.** 4 and -4 † **79.** 7 and 0   † **80.** $\frac{2}{3}$

CALCULATOR PROBLEMS

81. Solve $9.090x^2 - 721.035x = 0$ to the nearest hundredth.

82. Find x to the nearest hundredth if the area of the square in Figure 6.2 is 1171.

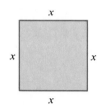

Figure 6.2

6.2   *Completing the Square*

Factoring and extraction of roots are the easiest ways to solve many quadratic equations. Unfortunately, they don't always work. For example, try to solve the equation below using one of these methods.

$$x^2 - 6x + 2 = 0$$

Every quadratic equation, however, can be solved using a method known as **completing the square.** We illustrate this method in Example 1.

EXAMPLE 1   Solve $x^2 - 8x + 7 = 0$ by completing the square.

Subtract 7 from each side.

$$x^2 - 8x = -7$$

Make the left side a perfect-square trinomial by adding to it *the square of one half the coefficient of x.* That is, add $(\frac{-8}{2})^2 = 16$ to the left side. To maintain equality, you must add 16 to the right side as well.

$$x^2 - 8x + 16 = -7 + 16$$

Factor the left side and simplify the right side.

$$(x - 4)^2 = 9$$

Extract both square roots of 9.

$$x - 4 = \pm 3$$

Add 4 to each side.

$$x = 4 \pm 3$$

Separate into two equations.

$$x = 4 + 3 \quad \text{or} \quad x = 4 - 3$$
$$x = 7 \qquad \text{or} \quad x = 1$$

Try Problem 25

Both numbers check, so the solution is 7 or 1. ◄

The fact that the solutions to Example 1 were rational numbers means that we could have solved the equation by factoring. This is not the case in Example 2.

EXAMPLE 2   Solve $x^2 - 6x + 2 = 0$.

$x^2 - 6x \quad\;\; = -2$	Subtract 2 from each side
$x^2 - 6x + 9 = -2 + 9$	Add $(\frac{-6}{2})^2 = 9$ to each side
$(x - 3)^2 = 7$	Factor the left side
$x - 3 = \pm\sqrt{7}$	Extract both square roots of 7
$x = 3 \pm\sqrt{7}$	Add 3 to each side

Try Problem 33

The solution is $3 + \sqrt{7}$ or $3 - \sqrt{7}$. ◄

CAUTION Before you complete the square on $ax^2 + bx$, the coefficient a should equal 1. If $a \neq 1$, divide each side of the equation by a.

EXAMPLE 3   Solve $2x^2 + 2x - 1 = 0$.

$$2x^2 + 2x = 1 \qquad\qquad \text{Add 1 to each side}$$

$$\frac{2x^2 + 2x}{2} = \frac{1}{2}$$

$$\left.\frac{2x^2}{2} + \frac{2x}{2} = \frac{1}{2}\right\} \quad \text{Divide each side by 2}$$

$$x^2 + x = \frac{1}{2}$$

$$x^2 + x + \frac{1}{4} = \frac{1}{2} + \frac{1}{4} \qquad \text{Add } (\tfrac{1}{2})^2 = \tfrac{1}{4} \text{ to each side}$$

$$\left(x + \frac{1}{2}\right)^2 = \frac{3}{4} \qquad\qquad \text{Factor the left side}$$

$$x + \frac{1}{2} = \pm\frac{\sqrt{3}}{2} \qquad\qquad \text{Extract both square roots of } \tfrac{3}{4}$$

$$x = -\frac{1}{2} \pm \frac{\sqrt{3}}{2} \qquad\qquad \text{Subtract } \tfrac{1}{2} \text{ from each side}$$

The solution is $-\frac{1}{2} + \frac{\sqrt{3}}{2}$ or $-\frac{1}{2} - \frac{\sqrt{3}}{2}$. These solutions can also be written as

Try Problem 45

$\dfrac{-1 + \sqrt{3}}{2}$ and $\dfrac{-1 - \sqrt{3}}{2}$. ◄

To Solve a Quadratic Equation by Completing the Square

1. Write the equation in the form $ax^2 + bx = c$.
2. If $a \neq 1$, divide each side by a.
3. Add the square of one half the coefficient of x to each side.
4. Factor the left side and simplify the right side.
5. Solve by extraction of roots.

PROBLEM SET 6.2

Determine the number that must be added to each expression to make it a perfect-square trinomial.

1. $x^2 + 8x$	**2.** $x^2 + 10x$	**3.** $y^2 - 6y$	**4.** $y^2 - 4y$
5. $p^2 + 2p$	**6.** $p^2 + 12p$	**7.** $r^2 - 3r$	**8.** $r^2 - 5r$
9. $m^2 + m$	**10.** $m^2 - m$	**11.** $t^2 - \frac{2}{5}t$	**12.** $t^2 + \frac{2}{3}t$

Solve by extraction of roots.

13. $(x - 3)^2 = 4$	**14.** $(x - 4)^2 = 9$	**15.** $(x + 2)^2 = 36$	**16.** $(x + 5)^2 = 49$
17. $(y - 1)^2 = 2$	**18.** $(y - 2)^2 = 3$	**19.** $(p + 4)^2 = 27$	**20.** $(p + 1)^2 = 8$
21. $(r - \frac{1}{2})^2 = \frac{25}{4}$	**22.** $(r - \frac{1}{2})^2 = \frac{9}{4}$	**23.** $(m + \frac{3}{2})^2 = \frac{5}{4}$	**24.** $(m + \frac{3}{2})^2 = \frac{7}{4}$

Solve each equation a) by completing the square, b) by factoring.

25. $x^2 - 10x + 9 = 0$	**26.** $x^2 - 10x + 21 = 0$
27. $x^2 - 4x - 5 = 0$	**28.** $x^2 - 4x - 12 = 0$
29. $x^2 + 2x + 1 = 0$	**30.** $x^2 + 6x + 9 = 0$
31. $x^2 + 6x = 0$	**32.** $x^2 + 2x = 0$

Solve each equation by completing the square.

33. $x^2 - 6x + 3 = 0$	**34.** $x^2 - 6x + 4 = 0$
35. $y^2 - 4y - 3 = 0$	**36.** $y^2 - 4y - 2 = 0$
37. $z^2 + 10z + 13 = 0$	**38.** $z^2 + 10z + 17 = 0$

39. $p^2 - 8p = 2$

40. $p^2 - 8p = 11$

41. $r^2 + r = 1$

42. $r^2 + r = 3$

43. $2m^2 + 24m + 8 = 0$

44. $2m^2 + 28m + 10 = 0$

45. $2x^2 + 2x - 3 = 0$

46. $2x^2 - 2x - 1 = 0$

47. $4y^2 - 4y - 3 = 0$

48. $4y^2 + 4y - 3 = 0$

49. $2t^2 + t - 2 = 0$

50. $2t^2 - t - 2 = 0$

51. $3r^2 - 6r = 1$

52. $3r^2 + 6r = 1$

53. $3p^2 - 2p = 1$

54. $3p^2 - 4p = 4$

55. If twice a positive number x is subtracted from its square, the result is 4. Find x.

56. If twice a positive number x is added to its square, the result is 2. Find x.

57. Solve $x^2 - 2ax + a^2 - 1 = 0$ for x. [*Hint:* Add 1 to each side. Then factor the left side.]

58. Solve $x^2 - 4ax + 4a^2 - 9 = 0$ for x. [*Hint:* Add 9 to each side. Then factor the left side.]

CALCULATOR PROBLEMS

59. Solve $3x^2 - 16{,}134x + 14{,}951{,}160 = 0$ by completing the square.

60. Check the solutions to Example 3 by first approximating each solution to the nearest thousandth.

6.3   *The Quadratic Formula*

Factoring and extraction of roots don't work on some quadratic equations, and completing the square is sometimes long and involved. Therefore we shall now use the method of completing the square to develop a formula that can be used to solve any quadratic equation.

$$ax^2 + bx + c = 0 \qquad \text{Write the general quadratic equation}$$

$$ax^2 + bx = -c \qquad \text{Subtract } c \text{ from each side}$$

$$x^2 + \frac{b}{a}x = -\frac{c}{a} \qquad \text{Divide each side by } a$$

Add the square of one half the coefficient of x to each side. That is, add $\left[\frac{1}{2}\left(\frac{b}{a}\right)\right]^2 = \left[\frac{b}{2a}\right]^2 = \frac{b^2}{4a^2}$ to each side.

$$x^2 + \frac{b}{a}x + \frac{b^2}{4a^2} = -\frac{c}{a} + \frac{b^2}{4a^2}$$

Factor the left side, and combine the two fractions on the right side.

$$\left(x+\frac{b}{2a}\right)^2 = -\frac{c \cdot 4a}{a \cdot 4a} + \frac{b^2}{4a^2}$$

$$\left(x+\frac{b}{2a}\right)^2 = \frac{b^2 - 4ac}{4a^2}$$

Extract roots.

$$x+\frac{b}{2a} = \pm\sqrt{\frac{b^2 - 4ac}{4a^2}}$$

Simplify the radical.

$$x+\frac{b}{2a} = \pm\frac{\sqrt{b^2 - 4ac}}{\sqrt{4a^2}} \qquad \text{Quotient rule for radicals}$$

$$x+\frac{b}{2a} = \pm\frac{\sqrt{b^2 - 4ac}}{2a}$$

Subtract $\dfrac{b}{2a}$ from each side.

$$x = -\frac{b}{2a} \pm \frac{\sqrt{b^2 - 4ac}}{2a}$$

Write as one fraction.

$$x = \frac{-b \pm \sqrt{b^2 - 4ac}}{2a}$$

The Quadratic Formula*

The solution to the quadratic equation $ax^2 + bx + c = 0$ is

$$x = \frac{-b + \sqrt{b^2 - 4ac}}{2a} \quad \text{or} \quad x = \frac{-b - \sqrt{b^2 - 4ac}}{2a}.$$

EXAMPLE 1   Solve $5x^2 + 3x = 2$ using the quadratic formula.

Write the equation in standard form.

$$5x^2 + 3x - 2 = 0$$

Identify a, b, and c.

$$a = 5, \qquad b = 3, \qquad c = -2$$

---

* This formula has been known since about the year 1000.

Substitute the values of a, b, and c into the quadratic formula.

$$x = \frac{-b \pm \sqrt{b^2 - 4ac}}{2a}$$

$$x = \frac{-3 \pm \sqrt{3^2 - 4(5)(-2)}}{2(5)}$$

Simplify the right side.

$$x = \frac{-3 \pm \sqrt{9 + 40}}{10}$$

$$x = \frac{-3 \pm \sqrt{49}}{10}$$

$$x = \frac{-3 \pm 7}{10}$$

Separate the two equations.

$$x = \frac{-3 + 7}{10} \quad \text{or} \quad x = \frac{-3 - 7}{10}$$

$$x = \frac{2}{5} \quad \quad \text{or} \quad x = -1$$

Try Problem 3   Both numbers check, so the solution is $\frac{2}{5}$ or -1. ◄

That fact that the solutions to Example 1 were rational numbers means that we could have solved the equation by factoring. This is not the case in Example 2.

EXAMPLE 2   Solve $2x^2 + 20x + 28 = 0$.

We could use $a = 2$, $b = 20$, and $c = 28$. However, we can simplify our calculations by first dividing each side by 2.

$$\frac{2x^2 + 20x + 28}{2} = \frac{0}{2}$$

$$\frac{2x^2}{2} + \frac{20x}{2} + \frac{28}{2} = \frac{0}{2} \qquad \text{Divide each term by 2}$$

$$x^2 + 10x + 14 = 0$$

Substitute $a = 1$, $b = 10$, and $c = 14$ into the quadratic formula.

$$x = \frac{-b \pm \sqrt{b^2 - 4ac}}{2a}$$

$$x = \frac{-10 \pm \sqrt{10^2 - 4(1)(14)}}{2(1)}$$

Simplify the right side.

$$x = \frac{-10 \pm \sqrt{100 - 56}}{2}$$

$$x = \frac{-10 \pm \sqrt{44}}{2}$$

But $\sqrt{44} = \sqrt{4 \cdot 11} = \sqrt{4} \cdot \sqrt{11} = 2\sqrt{11}$. Therefore

$$x = \frac{-10 \pm 2\sqrt{11}}{2}$$

$$x = \frac{-10}{2} \pm \frac{2\sqrt{11}}{2}$$

$$x = -5 \pm \sqrt{11}$$

Try Problem 19 The solution is $-5 + \sqrt{11}$ or $-5 - \sqrt{11}$. ◄

CAUTION The expression $\dfrac{-10 \pm 2\sqrt{11}}{2}$ is *not* equal to $-5 \pm 2\sqrt{11}$. You must divide 2 into *both* terms in the numerator.

EXAMPLE 3 Solve $\dfrac{y^2}{4} - \dfrac{y}{3} - \dfrac{5}{12} = 0$.

We could use $a = \frac{1}{4}$, $b = -\frac{1}{3}$, and $c = -\frac{5}{12}$. However, we can simplify our calculations by first multiplying each side by the LCD 12.

$$12 \cdot \left(\frac{y^2}{4} - \frac{y}{3} - \frac{5}{12} \right) = 12 \cdot 0$$

$$12 \cdot \frac{y^2}{4} - 12 \cdot \frac{y}{3} - 12 \cdot \frac{5}{12} = 12 \cdot 0 \qquad \text{Distribute 12 over every term}$$

$$3y^2 - 4y - 5 = 0$$

Substitute $a = 3$, $b = -4$, and $c = -5$ into the quadratic formula.

$$y = \frac{-b \pm \sqrt{b^2 - 4ac}}{2a}$$

$$y = \frac{-(-4) \pm \sqrt{(-4)^2 - 4(3)(-5)}}{2(3)}$$

Simplify the right side.

$$y = \frac{4 \pm \sqrt{16 + 60}}{6}$$

$$y = \frac{4 \pm \sqrt{76}}{6}$$

But $\sqrt{76} = \sqrt{4 \cdot 19} = \sqrt{4} \cdot \sqrt{19} = 2\sqrt{19}$. Therefore

$$y = \frac{4 \pm 2\sqrt{19}}{6}$$

$$y = \frac{2(2 \pm \sqrt{19})}{2 \cdot 3} \qquad \text{Factor numerator and denominator}$$

$$y = \frac{2 \pm \sqrt{19}}{3}. \qquad \text{Divide out 2}$$

Try Problem 33 The solution is $\dfrac{2 + \sqrt{19}}{3}$ or $\dfrac{2 - \sqrt{19}}{3}$. ◂

If you are trying to solve a quadratic equation, first try factoring or extraction of roots. If you are having no luck, go straight to the quadratic formula. It always works.

To Solve a Quadratic Equation Using the Quadratic Formula

1. Write the equation in standard form. Be sure to clear fractions and divide out common *constant* factors.
2. Identify *a, b,* and *c.*
3. Substitute the values of *a, b,* and *c* into the quadratic formula,

$$x = \frac{-b \pm \sqrt{b^2 - 4ac}}{2a}.$$

4. Simplify the right side.

Recall from Section 2.3 that when an object is projected upward and air resistance is ignored, the height *h* of the object in feet after *t* seconds is given by the formula $h = -16t^2 + rt + s$, where *r* is the initial velocity in feet per second, and *s* is the initial height in feet.

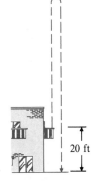

20 ft

Figure 6.3

EXAMPLE 4 A slingshot propels a stone upward with an initial velocity of 80 feet per second from a balcony that is 20 feet high (see Figure 6.3). If air resistance is ignored, when will the stone be at a height of 108 feet?

Substitute $h = 108$, $r = 80$, and $s = 20$ into the formula $h = -16t^2 + rt + s$.

$$108 = -16t^2 + 80t + 20$$

Then solve this equation for *t.*

$$16t^2 - 80t + 88 = 0 \qquad \text{Write in standard form}$$

$$2t^2 - 10t + 11 = 0 \qquad \text{Divide each side by 8}$$

Use the quadratic formula with $a = 2$, $b = -10$, and $c = 11$.

$$t = \frac{-(-10) \pm \sqrt{(-10)^2 - 4(2)(11)}}{2(2)}$$

$$t = \frac{10 \pm \sqrt{100 - 88}}{4}$$

$$t = \frac{10 \pm \sqrt{12}}{4}$$

But $\sqrt{12} = \sqrt{4 \cdot 3} = \sqrt{4} \cdot \sqrt{3} = 2\sqrt{3}$. Therefore

$$t = \frac{10 \pm 2\sqrt{3}}{4} = \frac{2(5 \pm \sqrt{3})}{2 \cdot 2} = \frac{5 \pm \sqrt{3}}{2}.$$

The stone will be at a height of 108 feet at $t = \dfrac{5 - \sqrt{3}}{2}$ (approximately 1.6)

seconds on the way up, and at $t = \dfrac{5 + \sqrt{3}}{2}$ (approximately 3.4) seconds on the

Try Problem 43 \ way down. ◄

PROBLEM SET 6.3

Solve each equation **a)** *using the quadratic formula,* **b)** *by factoring.*

1. $x^2 + x - 2 = 0$
2. $x^2 + x - 6 = 0$
3. $3x^2 + 2x = 1$
4. $5x^2 + 4x = 1$
5. $x^2 - 4x + 4 = 0$
6. $x^2 - 10x + 25 = 0$
7. $x^2 - 2x = 0$
8. $x^2 - 3x = 0$
9. $x^2 - 1 = 0$
10. $4x^2 - 1 = 0$

Use the quadratic formula to solve each equation.

11. $x^2 + x - 3 = 0$
12. $x^2 + x - 4 = 0$
13. $y^2 - 3y - 6 = 0$
14. $y^2 - 5y - 2 = 0$
15. $p^2 + 7p + 1 = 0$
16. $p^2 + 3p - 9 = 0$
17. $r^2 = 4r - 1$
18. $r^2 = 4r + 7$
19. $2x^2 + 24x + 50 = 0$
20. $2x^2 + 20x + 36 = 0$
21. $2m^2 - m - 2 = 0$
22. $2m^2 - m - 4 = 0$
23. $2z^2 - 2z = 1$
24. $2z^2 - 4z = 3$
25. $3t^2 - 4t - 5 = 0$
26. $3t^2 - 2t - 4 = 0$

27. $4p(p - 2) = -1$

28. $(2p + 3)(2p + 1) = 2$

29. $(5r + 7)(r + 1) = 1$

30. $5r(r + 2) = -4$

31. $90m^2 = 60m + 20$

32. $90m^2 = 20 - 60m$

33. $\dfrac{y^2}{4} - \dfrac{y}{3} - \dfrac{1}{12} = 0$

34. $\dfrac{y^2}{4} + \dfrac{y}{3} - \dfrac{5}{12} = 0$

35. $\frac{2}{3}x^2 - \frac{4}{9}x - \frac{1}{3} = 0$

36. $\frac{2}{3}x^2 + \frac{4}{9}x - \frac{1}{3} = 0$

37. $0.2z^2 + 1.2z - 0.5 = 0$

38. $0.3z^2 - 1.4z - 0.4 = 0$

39. $3x^2 + \sqrt{3}x - 2 = 0$

40. $2x^2 - \sqrt{3}x = \frac{3}{4}$

41. The square of a positive number x is 4 more than twice x. Find x.

42. The square of a positive number x is 2 less than four times x. Find x.

43. When will the stone of Example 4 be at a height of 92 feet?

44. When will the stone of Example 4 be at a height of 76 feet?

45. Given the revenue equation $R = 125x - \frac{1}{4}x^2$ and the cost equation $C = 15x$, determine the break-even points. [*Hint:* Set $R = C$.]

46. Given the revenue equation $R = 175x - \frac{1}{4}x^2$ and the cost equation $C = 20x$, determine the break-even points. [*Hint:* Set $R = C$.]

CALCULATOR PROBLEMS

Solve each equation to the nearest hundredth.

47. $2173p^2 + 740p - 8633 = 0$

48. $1.47x^2 - 7.04x + 6.32 = 0$

6.4   *Word Problems*

We solve the word problems in this section using the same five steps we used in Section 3.3. The only difference in this section is that the equation we write in Step 3 is a quadratic equation.

EXAMPLE 1   Two numbers have a sum of 12 and a product of 35. Find the numbers.

Step 1   Write down the quantities to be determined.

first number
second number

Step 2   Represent one quantity by x, and then represent the other quantity in terms of x. The sum of the two numbers is 12. Therefore if the first number is 8, the second number must be $12 - 8$. Similarly, if the first number is x, the second number must be $12 - x$.

$$x = \text{first number}$$
$$12 - x = \text{second number}$$

Step 3   Write an equation involving x.

$$\text{The product of the two numbers} = 35$$
$$x \cdot (12 - x) = 35$$

Step 4   Solve the equation.

Collect terms on the right side and solve by factoring.

$$12x - x^2 = 35$$
$$0 = x^2 - 12x + 35$$
$$0 = (x - 5)(x - 7)$$
$$x = 5 \quad \text{or} \quad x = 7$$

If $x = 5$, the second number is $12 - 5 = 7$. If $x = 7$, the second number is $12 - 7 = 5$. In either case, the two numbers are 5 and 7.

Step 5   Check in the original problem.

Try Problem 1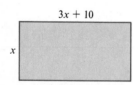

The numbers 5 and 7 have a sum of 12 and a product of 35. ◄

EXAMPLE 2   The length of a rectangle is 10 meters more than 3 times the width. The area of the rectangle is 168 square meters. Find the width and the length of the rectangle (see Figure 6.4).

$$x = \text{width of rectangle}$$
$$3x + 10 = \text{length of rectangle}$$
$$\text{Area of rectangle} = 168$$
$$x(3x + 10) = 168 \qquad \text{Since area} = \text{width} \cdot \text{length}$$
$$3x^2 + 10x = 168$$
$$3x^2 + 10x - 168 = 0$$

$3x + 10$

x

Figure 6.4

Instead of wasting time trying to factor the left side, use the quadratic formula with $a = 3$, $b = 10$, and $c = -168$.

$$x = \frac{-b \pm \sqrt{b^2 - 4ac}}{2a}$$
$$x = \frac{-10 \pm \sqrt{10^2 - 4(3)(-168)}}{2(3)}$$
$$x = \frac{-10 \pm \sqrt{100 + 2016}}{6}$$
$$x = \frac{-10 \pm \sqrt{2116}}{6}$$

Using a calculator, we find that $\sqrt{2116} = 46$. Therefore

$$x = \frac{-10 + 46}{6} \quad \text{or} \quad x = \frac{-10 - 46}{6}$$

$$x = 6 \quad \text{or} \quad x = -\tfrac{28}{3}.$$

We reject the negative answer, since the width of a rectangle cannot be a negative number. Therefore the width of the rectangle is 6 meters, and the length is $3 \cdot 6 + 10 = 28$ meters. ◄

Try Problem 11

EXAMPLE 3   Mary has a flower bed that is 4 feet by 8 feet (see Figure 6.5). She wants to surround the bed with a border of mulch of uniform width. She has enough mulch to cover 28 square feet. How wide should the border of mulch be?

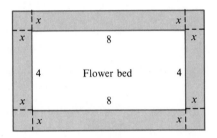

Figure 6.5

$$x = \text{width of the border}$$

$$\text{Area of mulch} = 28$$

$$\underset{\substack{\downarrow}}{\underset{\text{4 corners}}{\text{Area of}}} + \underset{\substack{\downarrow}}{\underset{\text{2 side strips}}{\text{area of}}} + \underset{\substack{\downarrow}}{\underset{\text{bottom strips}}{\text{area of top and}}} = 28$$

$$4x^2 \quad + \quad 2(4x) \quad + \quad 2(8x) \quad = 28$$

$$4x^2 + 8x + 16x = 28$$

$$4x^2 + 24x - 28 = 0$$

$$x^2 + 6x - 7 = 0 \qquad \text{Divide each side by 4}$$

$$(x - 1)(x + 7) = 0$$

$$x = 1 \quad \text{or} \quad x = -7$$

Try Problem 13

The border cannot be -7 feet wide, so it must be 1 foot wide. ◄

One of the most important formulas in plane geometry is the **Pythagorean theorem.***

---

* The Pythagorean theorem is named after the Greek mathematician Pythagoras (ca. 572–495 B.C.)

The Pythagorean Theorem

In any right triangle, the square of the hypotenuse equals the sum of the squares of the two legs. That is,

$$c^2 = a^2 + b^2.$$

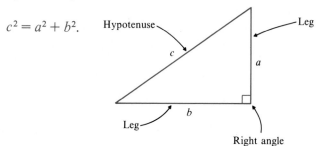

We use the Pythagorean theorem to write the equations in the next two examples.

EXAMPLE 4   Find the lengths of the sides of a right triangle if one leg is 1 inch longer than the other leg, and the hypotenuse is 2 inches longer than the shorter leg.

$x =$ length of shorter leg
$x + 1 =$ length of longer leg
$x + 2 =$ length of hypotenuse

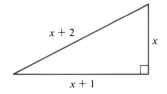

Figure 6.6

Since the triangle is a right triangle (see Figure 6.6), use the Pythagorean theorem to write the equation.

$$c^2 = a^2 + b^2$$
$$(x + 2)^2 = x^2 + (x + 1)^2$$

Solve this equation.

$$x^2 + 4x + 4 = x^2 + x^2 + 2x + 1$$
$$0 = x^2 - 2x - 3 \qquad \text{Collect terms on the right side}$$
$$0 = (x - 3)(x + 1)$$
$$x = 3 \quad \text{or} \quad x = -1$$

The shorter leg cannot be -1 inch, so it must be 3 inches. Therefore the longer leg is $3 + 1 = 4$ inches, and the hypotenuse is $3 + 2 = 5$ inches. ◄

Try Problem 17

EXAMPLE 5   A jogger and a cyclist leave the same point simultaneously. The jogger travels north and the cyclist travels east. After 1 hour, they are 26 miles apart. If the cyclist travels 14 mph faster than the jogger, find the rate of each.

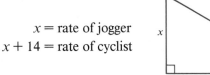
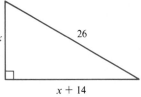

$$x = \text{rate of jogger}$$
$$x + 14 = \text{rate of cyclist}$$

Figure 6.7

By the Pythagorean theorem (see Figure 6.7), we have

$$(\text{Distance of jogger})^2 + (\text{distance of cyclist})^2 = 26^2$$

Since distance = rate · time, the jogger's distance is $x \cdot 1 = x$, and the cyclist's distance is $(x + 14) \cdot 1 = x + 14$. Therefore

$$x^2 + (x + 14)^2 = 26^2$$
$$x^2 + x^2 + 28x + 196 = 676$$
$$2x^2 + 28x - 480 = 0$$
$$x^2 + 14x - 240 = 0 \qquad \text{Divide each side by 2}$$
$$(x + 24)(x - 10) = 0$$
$$x = -24 \quad \text{or} \quad x = 10.$$

The jogger's rate cannot be -24 mph, so it must be 10 mph. The cyclist's rate is $10 + 14 = 24$ mph. ◄

Try Problem 21

PROBLEM SET 6.4

Solve the following number problems.

1. Two numbers have a sum of 13 and a product of 42. Find the numbers.

2. Two numbers have sum of 14 and a product of 45. Find the numbers.

3. One positive number is three less than twice another. Find both numbers if their product is 5.

4. One positive number is five less than twice another. Find both numbers if their product is 7.

5. The sum of the squares of two consecutive positive integers is 61. Find the integers.

6. The sum of the squares of two consecutive positive integers is 113. Find the integers.

7. The square of the sum of two consecutive even integers is 36. Find the integers.

8. The square of the sum of two consecutive odd integers is 64. Find the integers.

9. Find three consecutive odd integers such that the square of the first added to the product of the other two is 268.

10. Find three consecutive even integers such that the square of the first added to the product of the other two is 224.

Solve the following area and volume problems.

11. The length of a rectangle is 4 meters more than 3 times the width. The area of the rectangle is 175 square meters. Find the width and the length of the rectangle.

12. The length of a rectangle is 2 meters more than 3 times the width. The area of the rectangle is 208 square meters. Find the width and the length of the rectangle.

13. Sue has a flower bed that is 4 feet by 8 feet. She wants to surround the bed with a border of mulch of uniform width. She has enough mulch to cover 64 square feet. How wide should the border of mulch be?

14. Dave has a swimming pool that is 20 feet by 30 feet. He wants to surround the pool with a concrete border of uniform width. He has enough concrete to cover 336 square feet. How wide should he make the border?

15. An open pan is to be made from a square piece of sheet metal by cutting a 2-inch square from each corner and then turning up the sides (see Figure 6.8). What size piece of sheet metal is needed if the volume of the pan is to be 242 cubic inches?

16. A box without a top is to be constructed from a square piece of cardboard by cutting a 5-inch square from each corner and then turning up the sides (see Figure 6.9). What size piece of cardboard should be used if the volume of the box is to be 720 cubic inches?

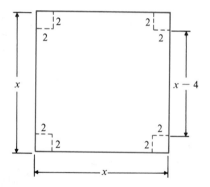

Figure 6.8

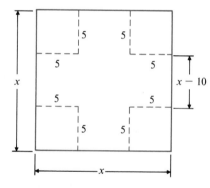

Figure 6.9

Solve the following Pythagorean theorem problems.

17. Find the lengths of the sides of a right triangle if one leg is 2 inches longer than the other leg, and the hypotenuse is 2 inches shorter than twice the shorter leg.

18. Find the lengths of the sides of a right triangle if one leg is 2 inches longer than twice the other leg, and the hypotenuse is 2 inches shorter than three times the shorter leg.

19. Determine the length of one side of a square whose diagonal is 4 feet long.

20. Determine the length of one side of a square whose diagonal is 6 feet long.

21. Two ships leave the same port simultaneously. Ship A travels north and ship B travels east. After 1 hour, the ships are 25 miles apart. If ship B travels 5 mph faster than ship A, find the rate of each.

22. Two traffic helicopters leave the top of a building simultaneously. Helicopter A travels south and helicopter B travels west. After 1 hour, the helicopters are 50 miles apart. If helicopter B travels 10 mph faster than helicopter A, find the rate of each.

† **23.** Find x in Figure 6.10.

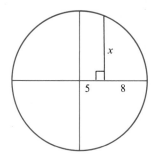

Figure 6.10

† **24.** Determine the height of the pile of logs in Figure 6.11 if the diameter of each log is 6 inches.

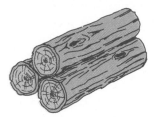

Figure 6.11

CALCULATOR PROBLEMS

25. Sections of railroad track are laid with a small space between them to allow for expansion when the metal becomes hot. Suppose two 1-mile rails are joined and the outer ends are fixed. If each expands 1 inch, and no space is allowed for this expansion, by how much x would the junction be displaced if both remained straight? (See Figure 6.12.)

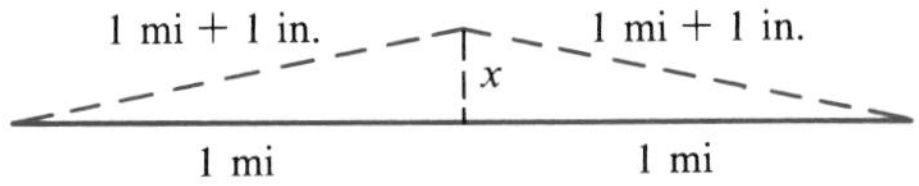

$$1 \text{ mi} + 1 \text{ in.} \qquad 1 \text{ mi} + 1 \text{ in.}$$
$$x$$
$$1 \text{ mi} \qquad\qquad 1 \text{ mi}$$

Figure 6.12

26. Show that the triangle with sides of lengths 577, 997, and 1151 could not be a right triangle.

6.5	*Radical Equations*

To solve an equation with a radical, first remove the radical. This is done using the **power rule for equality.**

Power Rule for Equality

Assume a and b are real numbers and n is a positive integer.

$$\text{If} \quad a = b, \quad \text{then} \quad a^n = b^n.$$

You will never lose solutions when you apply the power rule, but you may gain solutions. For example, the equation $x = 5$ has only the one solution 5. But if you square each side, you get the equation $x^2 = 25$. This equation has two solutions, namely 5 and -5.

CAUTION Whenever you raise each side of an equation to an *even* power, you must check all solutions in the original equation. Those solutions that do not check are called **extraneous solutions** and must be discarded. Raising each side of an equation to an *odd* power does not introduce extraneous solutions.

EXAMPLE 1 Solve $\sqrt{x-3} = 2$.

Remove the radical by using the power rule to square each side.

$$(\sqrt{x-3})^2 = 2^2$$
$$x - 3 = 4$$

This is a linear equation, which we solve by isolating x.

$$x = 7$$

CHECK:
$$\sqrt{7-3} \overset{?}{=} 2$$
$$\sqrt{4} \overset{?}{=} 2$$
$$2 \overset{\checkmark}{=} 2$$

Try Problem 3

The number 7 checks, so the solution is 7. ◄

EXAMPLE 2 Solve $\sqrt{4z+1} + 3 = 0$.

Isolate the radical by subtracting 3 from each side.

$$\sqrt{4z+1} = -3$$

Square each side.

$$(\sqrt{4z+1})^2 = (-3)^2$$
$$4z + 1 = 9$$

Solve for z.

$$4z = 8$$
$$z = 2$$

CHECK:
$$\sqrt{4 \cdot 2 + 1} + 3 \overset{?}{=} 0$$
$$\sqrt{9} + 3 \overset{?}{=} 0$$
$$6 \neq 0$$

Try Problem 11

Since 2 does not check, it is an extraneous solution and must be discarded. Therefore there is no solution to the original equation. ◄

CAUTION Be sure to square each *side* of the equation. Do *not* square each *term*. It is *wrong* to write the equation of Example 2 as $(\sqrt{4z+1})^2 + 3^2 = 0^2$.

To solve Example 3, we must remember that the square of a binomial is the square of the first term plus twice the product of the two terms plus the square of the last term.

EXAMPLE 3 Solve $\sqrt{m + 16} = m - 4$.

Square each side. Note that you are squaring a binomial on the right side.

$$(\sqrt{m + 16})^2 = (m - 4)^2$$
$$m + 16 = m^2 - 8m + 16$$

$\qquad$ Square the last term

$\qquad$ Twice the product of m and -4

$\qquad$ Square the first term

This is a quadratic equation. We collect terms on the right side and solve by factoring.

$$0 = m^2 - 9m$$
$$0 = m(m - 9)$$
$$m = 0 \quad \text{or} \quad m = 9$$

The number 9 checks, but 0 does not. Therefore 0 is an extraneous solution. Hence the only solution to the original equation is 9. ◄

Try Problem 43

EXAMPLE 4 Solve $\sqrt{4p - 3} - \sqrt{2p - 5} = 2$.

Isolate one of the radicals.

$$\sqrt{4p - 3} = 2 + \sqrt{2p - 5}$$

Square each side, treating the right side as a binomial.

$$(\sqrt{4p - 3})^2 = (2 + \sqrt{2p - 5})^2$$
$$4p - 3 = 4 + 4\sqrt{2p - 5} + (2p - 5)$$

$\qquad$ Square $\sqrt{2p - 5}$

$\qquad$ Twice the product of 2 and $\sqrt{2p - 5}$

$\qquad$ Square 2

Isolate the remaining radical.

$$4p - 3 = 4\sqrt{2p - 5} + 2p - 1 \qquad \text{Combine terms}$$
$$2p - 2 = 4\sqrt{2p - 5} \qquad \text{Subtract } 2p \text{ and add 1 to each side}$$
$$p - 1 = 2\sqrt{2p - 5} \qquad \text{Divide each side by 2}$$

Square each side again.

$$(p - 1)^2 = (2\sqrt{2p - 5})^2$$
$$p^2 - 2p + 1 = 4(2p - 5)$$

Solve this quadratic equation by factoring.

$$p^2 - 2p + 1 = 8p - 20$$
$$p^2 - 10p + 21 = 0$$
$$(p - 3)(p - 7) = 0$$
$$p = 3 \quad \text{or} \quad p = 7$$

Try Problem 51 \   Both numbers check, so the solution is 3 or 7. ◄

If the index on the radical is 3, we must *cube* each side to remove the radical.

EXAMPLE 5   Solve $\sqrt[3]{6t-3} = \sqrt[3]{8t+5}$.

$$(\sqrt[3]{6t-3})^3 = (\sqrt[3]{8t+5})^3 \left.\right\} \text{ Cube each side}$$
$$6t - 3 = 8t + 5$$
$$-2t = 8 \left.\right\}$$
$$t = -4 \quad\left.\right\} \text{ Solve for } t$$

Try Problem 61 \   The solution is -4. Check in the original equation. ◄

To Solve a Radical Equation

1.   Isolate one of the radicals on one side of the equation.
2.   Raise each *side* (not each term) of the equation to a power equal to the index of the radical.
3.   If the equation still has a radical, repeat steps 1 and 2.
4.   Solve the resulting equation.
5.   Check all solutions in the original equation and discard any extraneous solutions.

We can use these same techniques to solve a formula with radicals for one of its variables.

EXAMPLE 6   Solve $r = \sqrt{\dfrac{3V}{\pi h}}$ for V.

$$r^2 = \left(\sqrt{\frac{3V}{\pi h}}\right)^2 \left.\right\}$$
$$r^2 = \frac{3V}{\pi h} \qquad\left.\right\} \text{ Square each side}$$
$$\pi h r^2 = 3V \qquad\qquad \text{Multiply each side by } \pi h$$

Try Problem 69 \
$$\frac{\pi h r^2}{3} = V \qquad\qquad \text{Divide each side by 3 ◄}$$

Problem Solving   **Period of a Pendulum**

The time t in seconds that it takes a pendulum l feet long to complete a cycle is called the period of the pendulum and is given by the formula

$$t = 2\pi\sqrt{\frac{l}{32}}.$$

What length should a pendulum have if it is to take 2 seconds to complete a cycle?

$$2 = 2\pi\sqrt{\frac{l}{32}}$$ Replace t by 2

$$\frac{2}{2\pi} = \sqrt{\frac{l}{32}}$$ Divide each side by 2π

$$\frac{1}{\pi} = \sqrt{\frac{l}{32}}$$ Simplify the left side

$$\left.\begin{array}{c} \left(\dfrac{1}{\pi}\right)^2 = \left(\sqrt{\dfrac{l}{32}}\right)^2 \\[2mm] \dfrac{1}{\pi^2} = \dfrac{l}{32} \end{array}\right\}$$ Square each side

$$\frac{32}{\pi^2} = l$$ Multiply each side by 32

The pendulum should be $\dfrac{32}{\pi^2}$ (approximately 3.24) feet long. ◄

PROBLEM SET 6.5

Solve each equation.

1. $\sqrt{x} = 4$

2. $\sqrt{x} = 9$

3. $\sqrt{x-1} = 2$

4. $\sqrt{x-2} = 3$

5. $\sqrt{2x-3} = 9$

6. $\sqrt{2x-5} = 11$

7. $\sqrt{x-2} = 5$

8. $\sqrt{x-3} = 5$

9. $\sqrt{3y+7} - 8 = 0$

10. $\sqrt{3y+1} - 10 = 0$

11. $\sqrt{4z-3} + 5 = 0$

12. $\sqrt{4z-7} + 1 = 0$

13. $\sqrt{5p+4} + 1 = 3$

14. $\sqrt{5p+9} + 2 = 5$

15. $\sqrt{v-6} = \sqrt{6-v}$

16. $\sqrt{v-7} = \sqrt{7-v}$

17. $\sqrt{10r} - 2\sqrt{r+3} = 0$

18. $\sqrt{6r} - 2\sqrt{r+5} = 0$

19. $\sqrt{s+1} = \sqrt{s+2}$

20. $\sqrt{s+3} = \sqrt{s+4}$

21. $\sqrt{9+4\sqrt{x}} = 7$

22. $\sqrt{4+9\sqrt{x}} = 7$

23. $\sqrt{m^2+24} = m-2$

24. $\sqrt{m^2+45} = m-3$

25. $\sqrt{r^2-6r+12} = r$

26. $\sqrt{r^2-7r+21} = r$

27. $\sqrt{x-3} = \sqrt{x}-1$

28. $\sqrt{x-5} = \sqrt{x}-1$

29. $\sqrt{y+2} + 1 = \sqrt{y+4}$

30. $\sqrt{y+3} + 1 = \sqrt{y+5}$

Solve each equation.

31. $\sqrt{x^2 + 3} = 2$

32. $\sqrt{x^2 + 5} = 3$

33. $\sqrt{x^2 - 1} = 1$

34. $\sqrt{x^2 - 1} = 2$

35. $\sqrt{x + 2} = x$

36. $\sqrt{x + 6} = x$

37. $\sqrt{5y - 1} - 2y = 0$

38. $\sqrt{10y - 1} - 3y = 0$

39. $\sqrt{p} = p - 6$

40. $\sqrt{p} = p - 2$

41. $\sqrt{r^2 + 12r} = 3\sqrt{5}$

42. $\sqrt{r^2 + 12r} = 2\sqrt{7}$

43. $\sqrt{m + 4} = m - 2$

44. $\sqrt{m + 9} = m - 3$

45. $\sqrt{t - 1} + t - 3 = 0$

46. $\sqrt{t - 2} + t - 4 = 0$

47. $\sqrt{2x^2 - 14x + 21} = 4 - x$

48. $\sqrt{2x^2 - 11x + 13} = 3 - x$

49. $\sqrt{3y - 2} - \sqrt{y} = 2$

50. $\sqrt{3y + 4} - \sqrt{y} = 2$

51. $\sqrt{3p + 4} - \sqrt{2p - 4} = 2$

52. $\sqrt{5p + 6} - \sqrt{3p + 4} = 2$

53. $\sqrt{5r - 1} + \sqrt{r + 3} = 4$

54. $\sqrt{5r + 6} + \sqrt{r + 3} = 3$

Solve each equation.

55. $\sqrt[3]{x} = 4$

56. $\sqrt[3]{x} = 5$

57. $\sqrt[4]{x - 6} = 1$

58. $\sqrt[4]{x - 1} = 2$

59. $\sqrt[3]{x + 2} + 3 = 0$

60. $\sqrt[3]{x + 5} + 1 = 0$

61. $\sqrt[3]{2t + 5} = \sqrt[3]{4t + 11}$

62. $\sqrt[3]{3t - 2} = \sqrt[3]{5t + 2}$

63. $\sqrt[3]{y^2 - 17} = 2$

64. $\sqrt[3]{y^2 + 23} = 3$

65. $\sqrt[4]{\sqrt{r} + 1} = 3$

66. $\sqrt[4]{\sqrt{r} - 44} = 2$

Solve each equation for the specified variable. Assume all variables represent positive real numbers.

67. $s = \sqrt{A}$ for A

68. $v = \sqrt{2gR}$ for R

69. $r = \sqrt{\dfrac{2V}{\pi h}}$ for V

70. $r = \sqrt{\dfrac{S}{4\pi}}$ for S

71. $e = \sqrt[3]{V}$ for V

72. $r = \sqrt[3]{\dfrac{3V}{4\pi}}$ for V

73. $c = \sqrt{a^2 + b^2}$ for a

74. $c = \sqrt{a^2 + b^2}$ for b

75. The principal square root of the sum of a number x and 2 is 4 less than x. Find x.

76. The principal square root of the sum of a number x and 7 is 5 less than x. Find x.

77. What length should a pendulum have if it is to take 4 seconds to complete a cycle?

78. What length should a pendulum have if it is to take 6 seconds to complete a cycle?

CALCULATOR PROBLEMS

79. Solve $2.43\sqrt{8.77x - 6.19} - 4.88 = 15.75$ to the nearest hundredth.

80. Because of the curvature of the earth's surface, the distance d in miles between an object at an altitude of a feet and the horizon is given by $d = \sqrt{1.5a}$. At what altitude in miles is an airplane whose horizon is 178 miles away? [*Hint:* 1 mile $=$ 5280 feet.]

6.6   *Equations That Lead to Quadratic Equations*

Sometimes equations involving rational expressions lead to quadratic equations.

EXAMPLE 1   Solve $1 + \dfrac{2}{r} - \dfrac{8}{r^2} = 0$.

Multiply each side by the LCD r^2 to clear the equation of fractions.

$$r^2 \cdot \left(1 + \frac{2}{r} - \frac{8}{r^2}\right) = r^2 \cdot 0$$

$$r^2 \cdot 1 + r^2 \cdot \frac{2}{r} - r^2 \cdot \frac{8}{r^2} = r^2 \cdot 0 \qquad \text{Distribute } r^2 \text{ over every term}$$

$$r^2 + 2r - 8 = 0 \qquad \text{Find each product}$$

$$(r + 4)(r - 2) = 0 \qquad \text{Factor}$$

$$r = -4 \quad \text{or} \quad r = 2$$

Try Problem 3   Both numbers check, so the solution is -4 or 2. ◂

EXAMPLE 2   Solve $\dfrac{m}{m + 2} - \dfrac{2}{m - 1} = 2$.

Distribute the LCD $(m + 2)(m - 1)$ over every term.

$$(m + 2)(m - 1) \cdot \frac{m}{m + 2} - (m + 2)(m - 1) \cdot \frac{2}{m - 1} = (m + 2)(m - 1) \cdot 2$$

$$(m - 1) \cdot m - (m + 2) \cdot 2 = (m^2 + m - 2) \cdot 2$$

$$m^2 - m - 2m - 4 = 2m^2 + 2m - 4$$

$$0 = m^2 + 5m$$

$$0 = m(m + 5)$$

$$m = 0 \quad \text{or} \quad m = -5$$

Try Problem 13   Both numbers check, so the solution is 0 or -5. ◂

    Certain types of nonquadratic equations become quadratic equations when we make an appropriate substitution. Such equations are said to be of **quadratic form**.

EXAMPLE 3   Solve $x^4 - 10x^2 + 9 = 0$.

Let $u = x^2$. Then $u^2 = (x^2)^2 = x^4$. Substituting u^2 for x^4 and u for x^2, the original equation becomes

$$u^2 - 10u + 9 = 0.$$

This equation is quadratic in u. Solve by factoring.

$$(u - 1)(u - 9) = 0$$
$$u = 1 \quad \text{or} \quad u = 9$$

Since $u = x^2$, the last two equations become

$$x^2 = 1 \quad \text{or} \quad x^2 = 9.$$

Extract roots and get

$$x = \pm 1 \quad \text{or} \quad x = \pm 3.$$

Try Problem 21   The solution is $1, -1, 3,$ or -3. Check in the original equation. ◄

You can solve the equation of Example 3 without making a substitution.

$$x^4 - 10x^2 + 9 = 0$$
$$(x^2 - 1)(x^2 - 9) = 0 \qquad \text{Factor}$$
$$x^2 - 1 = 0 \quad \text{or} \quad x^2 - 9 = 0 \qquad \text{Set each factor equal to 0}$$
$$x^2 = 1 \quad \text{or} \quad x^2 = 9 \qquad \left.\right\}\text{Solve each equation by extracting roots}$$
$$x = \pm 1 \quad \text{or} \quad x = \pm 3$$

Choose the method that works best for you.

EXAMPLE 4   Solve $p^{2/3} - 3p^{1/3} - 4 = 0$.

Let $u = p^{1/3}$. Then $u^2 = (p^{1/3})^2 = p^{2/3}$. Hence the original equation becomes

$$u^2 - 3u - 4 = 0.$$

Solve by factoring.

$$(u - 4)(u + 1) = 0$$
$$u = 4 \quad \text{or} \quad u = -1$$

Since $u = p^{1/3}$, write these equations as

$$p^{1/3} = 4 \quad \text{or} \quad p^{1/3} = -1.$$

Cube each side to find p.

$$(p^{1/3})^3 = 4^3 \quad \text{or} \quad (p^{1/3})^3 = (-1)^3$$
$$p = 64 \quad \text{or} \quad p = -1$$

Try Problem 29   Both numbers check, so the solution is 64 or -1. ◄

CAUTION To solve $p^{1/3} = 4$ you must *cube* each side. Do not take the cube *root* of each side.

To Solve an Equation of Quadratic Form ($ax^{2n} + bx^n + c = 0$)

1. Let $u = x^n$. Then $u^2 = (x^n)^2 = x^{2n}$.
2. Rewrite the equation as $au^2 + bu + c = 0$.
3. Solve the quadratic equation in step 2.
4. Substitute x^n for u and solve the resulting equation.
5. Check each solution in the original equation.

Certain types of word problems result in equations that lead to quadratic equations. Recall the following rule from Section 4.6.

Rule for Solving Work Problems

If it takes t hours to do a job, then the portion of the job completed in 1 hour is $\dfrac{1}{t}$.

EXAMPLE 5   An old computer takes 5 hours longer to process the company payroll than does a new computer. If it takes the two computers 6 hours to process the payroll working together, how long would it take each computer working alone?

$$t = \text{time it takes new computer}$$
$$t + 5 = \text{time it takes old computer}$$

$$\begin{pmatrix} \text{Portion of job done} \\ \text{by new computer} \\ \text{in 1 hour} \end{pmatrix} + \begin{pmatrix} \text{portion of job done} \\ \text{by old computer} \\ \text{in 1 hour} \end{pmatrix} = \begin{pmatrix} \text{portion of job done} \\ \text{by both computers} \\ \text{in 1 hour} \end{pmatrix}$$

$$\frac{1}{t} \qquad + \qquad \frac{1}{t+5} \qquad = \qquad \frac{1}{6}$$

Multiply each term by the LCD $6t(t + 5)$.

$$6t(t+5) \cdot \frac{1}{t} + 6t(t+5) \cdot \frac{1}{t+5} = 6t(t+5) \cdot \frac{1}{6}$$
$$6(t+5) + 6t = t(t+5)$$
$$6t + 30 + 6t = t^2 + 5t$$
$$0 = t^2 - 7t - 30$$
$$0 = (t - 10)(t + 3)$$
$$t = 10 \quad \text{or} \quad t = -3$$

Discard the negative solution. The new computer takes 10 hours to process the payroll, and the old computer takes $10 + 5 = 15$ hours. ◄

Try Problem 33

To solve motion problems recall the formula $d = rt$, which can also be written as $r = \dfrac{d}{t}$ or as $t = \dfrac{d}{r}$.

EXAMPLE 6 On the third Sunday of each month, Violet drives 70 miles to visit her granddaughter in the country. If she drove 5 mph faster than her normal speed, she would arrive 15 minutes sooner. Find Violet's normal speed.

$$x = \text{Violet's normal speed}$$

	d	r	t
Normal trip	70	x	$\dfrac{70}{x}$
Fast trip	70	$x + 5$	$\dfrac{70}{x + 5}$

Since $t = \dfrac{d}{r}$

Convert 15 minutes to a fraction of an hour as follows:

$$15 \text{ min} = \frac{15}{60} \text{ hr} = \frac{1}{4} \text{ hr.}$$

$$(\text{Time for normal trip}) - (\text{time for fast trip}) = (\tfrac{1}{4} \text{ hour})$$

$$\frac{70}{x} - \frac{70}{x + 5} = \frac{1}{4}$$

Multiply each term by the LCD $4x(x + 5)$.

$$4x(x + 5) \cdot \frac{70}{x} - 4x(x + 5) \cdot \frac{70}{x + 5} = 4x(x + 5) \cdot \frac{1}{4}$$

$$280(x + 5) - 280x = x(x + 5)$$
$$280x + 1400 - 280x = x^2 + 5x$$
$$0 = x^2 + 5x - 1400$$

Instead of trying to factor the right side, use the quadratic formula with $a = 1$, $b = 5$, and $c = -1400$.

$$x = \frac{-5 \pm \sqrt{5^2 - 4(1)(-1400)}}{2(1)}$$

$$x = \frac{-5 \pm \sqrt{5625}}{2}$$

From a calculator, $\sqrt{5625} = 75$. Therefore

$$x = \frac{-5 + 75}{2} \quad \text{or} \quad x = \frac{-5 - 75}{2}$$

$$x = 35 \quad \text{or} \quad x = -40.$$

Try Problem 35

Violet's normal speed cannot be -40 mph, so it must be 35 mph. ◄

PROBLEM SET 6.6

Solve each equation.

1. $x - \dfrac{1}{x} = 0$

2. $x - \dfrac{4}{x} = 0$

3. $1 + \dfrac{6}{r} + \dfrac{8}{r^2} = 0$

4. $1 + \dfrac{8}{r} + \dfrac{12}{r^2} = 0$

5. $4y + 21 = \dfrac{18}{y}$

6. $6y + 13 = \dfrac{15}{y}$

7. $1 - \dfrac{3}{2p} = \dfrac{1}{p^2}$

8. $1 + \dfrac{5}{6p} = \dfrac{1}{p^2}$

9. $\dfrac{t}{t-4} + \dfrac{2}{t-4} + t = 0$

10. $\dfrac{t}{t-5} + \dfrac{3}{t-5} + t = 0$

11. $\dfrac{1}{x} + \dfrac{1}{x+1} = \dfrac{5}{6}$

12. $\dfrac{1}{x} + \dfrac{1}{x-1} = \dfrac{7}{12}$

13. $\dfrac{m}{m+1} - \dfrac{3}{m-3} = 2$

14. $\dfrac{m}{m+2} + \dfrac{4}{m-1} = 2$

15. $1 = \dfrac{1}{y+2} + \dfrac{12}{(y+2)^2}$

16. $1 = \dfrac{1}{y+3} + \dfrac{20}{(y+3)^2}$

17. $\dfrac{2}{r^2+r} + \dfrac{2}{r+1} = 1$

18. $\dfrac{3}{r^2-3r} - \dfrac{1}{r-3} = 1$

Solve each equation.

19. $x^4 - 2x^2 + 1 = 0$

20. $x^4 - 8x^2 + 16 = 0$

21. $x^4 - 5x^2 + 4 = 0$

22. $x^4 - 13x^2 + 36 = 0$

23. $t^4 - 11t^2 + 18 = 0$

24. $t^4 - 7t^2 + 12 = 0$

25. $4y^4 - 13y^2 + 9 = 0$

26. $4y^4 - 29y^2 + 25 = 0$

27. $x^{1/2} - 5x^{1/4} + 6 = 0$

28. $x^{1/2} - 3x^{1/4} + 2 = 0$

29. $p^{2/3} - 2p^{1/3} - 3 = 0$

30. $p^{2/3} - p^{1/3} - 6 = 0$

31. $6r^{-2} + 29r^{-1} - 5 = 0$

32. $7r^{-2} + 41r^{-1} - 6 = 0$

33. An old computer takes 6 hours longer to process the company payroll than does a new computer. If it takes the two computers 4 hours to process the payroll working together, how long does it take each computer working alone?

34. An old printing press takes 8 hours longer to put out a daily paper than a new press. If the two presses work together, the job takes 3 hours. How long does it take each press to put out the paper on its own?

35. A family moves to a new home 150 miles away. The parents drive a rented truck while their children drive the family car. Find the speed of each vehicle if the kids drive 10 mph faster than their parents and arrive a half hour sooner.

36. A trucker drives to a city 60 miles away, unloads, and then returns. The trucker drives 20 mph faster on the return trip, which takes a half hour less time. How fast did the trucker drive each way?

37. A kayaker who paddles 5 mph in still water takes 1 hour longer to paddle 12 miles upstream to a camping ground than she does to make the return trip. Find the speed of the current.

38. A pontoon boat that travels 10 mph in still water takes 1 hour longer to travel 24 miles against the current than it does to paddle the same distance with the current. Find the speed of the current.

39. Solve $F = \dfrac{Gm_1m_2}{r^2}$ for r. All variables are positive.

40. Solve $I = \dfrac{W}{4\pi r^2}$ for r. All variables are positive.

CALCULATOR PROBLEMS

41. Solve $3.17x^4 - 5.24x^2 + 1.66 = 0$ to the nearest hundredth.

42. A small plane flies 220 miles into a headwind and then returns in a total time of 8 hours and 15 minutes. How fast does the plane travel in still air if the speed of the wind is 20 mph?

6.7   *Complex Solutions and the Discriminant*

COMPLEX SOLUTIONS

The solutions to a quadratic equation are not always real numbers.

EXAMPLE 1   Solve $x^2 + 25 = 0$.

Solve by extraction of roots.

$$x^2 = -25 \qquad \text{Write in the form } x^2 = k$$
$$x = \pm\sqrt{-25} \qquad \text{Extract both roots of } -25$$

But $\pm\sqrt{-25} = \pm\sqrt{25 \cdot (-1)} = \pm\sqrt{25} \cdot \sqrt{-1} = \pm 5i$. Therefore

$$x = \pm 5i.$$

Try Problem 1   The solution is the complex number $5i$ or $-5i$. ◄

EXAMPLE 2   Solve $p^2 + 18 = 0$.

$$p^2 = -18 \qquad \text{Write in the form } x^2 = k$$
$$p = \pm\sqrt{-18} \qquad \text{Extract both roots of } -18$$

But $\pm\sqrt{-18} = \pm\sqrt{9\cdot2\cdot(-1)} = \pm\sqrt{9}\cdot\sqrt{2}\cdot\sqrt{-1} = \pm3\sqrt{2}i$. Therefore

$$p = \pm3\sqrt{2}i.$$

Try Problem 11 \\ The solution is $3\sqrt{2}i$ or $-3\sqrt{2}i$. ◄

If you are having no luck solving a quadratic equation by factoring or extraction of roots, use the quadratic formula.

EXAMPLE 3 Solve $r^2 - 6r + 25 = 0$.

Apply the quadratic formula with $a = 1$, $b = -6$, and $c = 25$.

$$r = \frac{-b \pm \sqrt{b^2 - 4ac}}{2a}$$

$$r = \frac{-(-6) \pm \sqrt{(-6)^2 - 4(1)(25)}}{2(1)}$$

$$r = \frac{6 \pm \sqrt{36 - 100}}{2}$$

$$r = \frac{6 \pm \sqrt{-64}}{2}$$

$$r = \frac{6 \pm 8i}{2} \qquad \text{Since } \sqrt{-64} = \sqrt{64}\sqrt{-1} = 8i$$

$$r = 3 \pm 4i$$

Try Problem 21 \\ The solution is $3 + 4i$ or $3 - 4i$. ◄

The next example illustrates an equation of quadratic form that has complex solutions.

EXAMPLE 4 Solve $x^4 + 5x^2 - 36 = 0$.

$$\begin{array}{lll}
(x^2 - 4)(x^2 + 9) = 0 & & \text{Factor} \\
x^2 - 4 = 0 \quad \text{or} \quad x^2 + 9 = 0 & & \text{Set each factor equal to 0} \\
x^2 = 4 \quad \text{or} \quad x^2 = -9 & \\
x = \pm2 \quad \text{or} \quad x = \pm\sqrt{-9} & & \left.\begin{array}{c} \\ \\ \end{array}\right\} \text{Solve by extraction of roots} \\
\qquad\qquad\qquad x = \pm3i &
\end{array}$$

Try Problem 31 \\ The solution is 2, -2, $3i$, or $-3i$. ◄

THE DISCRIMINANT

We can determine whether the solutions to the quadratic equation $ax^2 + bx + c = 0$ are rational numbers, irrational numbers, or complex numbers without actually solving the equation. We do this by calculating the value of $b^2 - 4ac$. If we determine that the solutions are rational numbers, this means that the trinomial $ax^2 + bx + c$ is factorable. Therefore this calculation also serves as a test for factorability.

The Discriminant

The **discriminant** of the quadratic equation $ax^2 + bx + c = 0$ is $b^2 - 4ac$. If a, b, and c are integers, then the nature of the solutions to this equation is determined as follows:

Discriminant	*Solutions*	
Square of an integer	Two rational solutions	⎫
Zero	One rational solution	⎬ $ax^2 + bx + c$ is factorable
Positive, but not the square of an integer	Two irrational solutions	
Negative	Two complex solutions	

CAUTION The discriminant is $b^2 - 4ac$, *not* $\sqrt{b^2 - 4ac}$.

EXAMPLE 5   Use the discriminant to determine the nature of the solutions to each equation: **a)** $6x^2 - 5x - 4 = 0$,   **b)** $2m^2 + 5 = 3m$.

a)   In this case, $a = 6$, $b = -5$, and $c = -4$. Therefore the discriminant is
$$b^2 - 4ac = (-5)^2 - 4(6)(-4) = 25 + 96 = 121.$$

The discriminant is the square of an integer (since $121 = 11^2$), so the equation has two rational solutions. This means that $6x^2 - 5x - 4$ is factorable. Note that
$$6x^2 - 5x - 4 = (3x - 4)(2x + 1).$$

b)   First write the equation in standard form.
$$2m^2 - 3m + 5 = 0$$
Since $a = 2$, $b = -3$, and $c = 5$, the discriminant is
$$b^2 - 4ac = (-3)^2 - 4(2)(5) = 9 - 40 = -31.$$

Try Problem 39

The discriminant is negative, so the equation has two complex solutions. ◄

PROBLEM SET 6.7

Solve each equation.

1. $x^2 + 4 = 0$ **2.** $x^2 + 9 = 0$ **3.** $x^2 + 1 = 0$ **4.** $x^2 + 16 = 0$

5. $x^2 + 64 = 0$ **6.** $x^2 + 36 = 0$ **7.** $x^2 + 2 = 0$ **8.** $x^2 + 3 = 0$

9. $y^2 + 5 = 0$

10. $y^2 + 7 = 0$

11. $p^2 + 12 = 0$

12. $p^2 + 20 = 0$

13. $r^2 + 45 = 0$

14. $r^2 + 63 = 0$

15. $4t^2 + 1 = 0$

16. $9t^2 + 1 = 0$

17. $25m^2 + 8 = 0$

18. $49m^2 + 27 = 0$

Solve each equation.

19. $x^2 + 2x + 2 = 0$

20. $x^2 + 4x + 5 = 0$

21. $r^2 - 6r + 13 = 0$

22. $r^2 - 6r + 10 = 0$

23. $y^2 - 4y + 13 = 0$

24. $y^2 - 6y + 34 = 0$

25. $2m^2 + 2m + 1 = 0$

26. $2m^2 - 2m + 1 = 0$

27. $z^2 - z + 1 = 0$

28. $z^2 + z + 1 = 0$

29. $t^2 - 2t + 3 = 0$

30. $t^2 - 4t + 10 = 0$

Solve each equation.

31. $x^4 + 35x^2 - 36 = 0$

32. $x^4 + 21x^2 - 100 = 0$

33. $y^4 - y^2 - 2 = 0$

34. $y^4 - 2y^2 - 3 = 0$

35. $z^4 + 29z^2 + 100 = 0$

36. $z^4 + 25z^2 + 144 = 0$

Use the discriminant to determine the nature of the solutions to each equation. Do not solve the equation.

37. $x^2 + 5x + 6 = 0$

38. $x^2 + 5x + 4 = 0$

39. $4x^2 - 3x - 10 = 0$

40. $4x^2 + 5x - 6 = 0$

41. $x^2 + x - 7 = 0$

42. $x^2 + x - 9 = 0$

43. $m^2 = 4m - 4$

44. $m^2 = 2m - 1$

45. $2m^2 + 4 = 5m$

46. $2m^2 + 3 = 4m$

47. $9m^2 = 6m - 1$

48. $9m^2 = 12m - 4$

By setting the discriminant equal to zero, determine the value(s) of k that cause the given equation to have exactly one solution.

49. $x^2 + 6x + k = 0$

50. $x^2 + 8x + k = 0$

51. $2x^2 - kx + 8 = 0$

52. $2x^2 - kx + 2 = 0$

53. The sum of a number x and twice its reciprocal is 2. Find x.

54. The sum of a number x and twice its reciprocal is -2. Find x.

55. An arrow is shot upward with an initial velocity of 64 feet per second. If we ignore air resistance, the height h in feet of the arrow after t seconds is given by the formula $h = -16t^2 + 64t$. When will the arrow be at a height of 80 feet?

56. When will the arrow of Problem 55 be at a height of 128 feet?

CALCULATOR PROBLEMS

Solve each equation.

57. $x^2 + 6241 = 0$ **58.** $25x^2 - 2x + 74 = 0$

6.8   *Quadratic Inequalities*

Now that we have learned to solve quadratic equations, we turn our attention to solving quadratic inequalities.

Definition

A **quadratic inequality** in the variable x is an inequality that can be expressed in the form

$$ax^2 + bx + c < 0,$$

where $a \neq 0$. This form is called the **standard form** of a quadratic inequality.

We can replace the symbol $<$ in the definition above with any of the symbols $\leq$, $>$, or $\geq$.

We now illustrate one method for solving quadratic inequalities.

EXAMPLE 1   Solve $(x - 2)(x + 4) > 0$.

Replace the symbol $>$ with the symbol $=$ and solve the resulting equation.

$$(x - 2)(x + 4) = 0$$
$$x = 2 \quad \text{or} \quad x = -4$$

Graph the solutions 2 and -4 on a number line. These numbers divide the number line into the three regions A, B, and C shown in Figure 6.13.

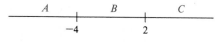

Figure 6.13

Choose any value in region A, say -6, and substitute this value for x in the expression $(x - 2)(x + 4)$.

$$\text{Test } x = -6: \quad (-6 - 2)(-6 + 4) = (-8)(-2) = 16$$

It can be shown that since $(x - 2)(x + 4)$ is positive (namely positive 16) when $x = -6$, then $(x - 2)(x + 4)$ is positive for *every* value of x in region A. Therefore label region A with the symbol $+$ (see Figure 6.14). Now choose test values from regions B and C, say $x = 0$ from B and $x = 3$ from C.

$$\text{Test } x = 0: \quad (0 - 2)(0 + 4) = (-2)(4) = -8$$
$$\text{Test } x = 3: \quad (3 - 2)(3 + 4) = (1)(7) = 7$$

We conclude that $(x - 2)(x + 4)$ is negative throughout region B and positive throughout region C. Therefore label region B with the symbol $+$ and region C with the symbol $-$ (see Figure 6.14).

Figure 6.14

Since we are solving the inequality $(x - 2)(x + 4) > 0$, we are looking for those x-values that make $(x - 2)(x + 4)$ positive. Figure 6.14 shows that $(x - 2)(x + 4)$ is positive when $x < -4$ or when $x > 2$. Therefore the solution is $x < -4$ or $x > 2$. The solution is graphed in Figure 6.15.

Try Problem 1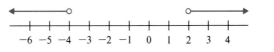

Figure 6.15

CAUTION In Example 1, do not choose either 2 or -4 as one of your test points. These numbers make $(x - 2)(x + 4)$ equal to zero.

EXAMPLE 2   Solve $(x - 2)(x + 4) < 0$.

In this case we are looking for all those x-values that make $(x - 2)(x + 4)$ negative. Figure 6.14 shows that $(x - 2)(x + 4)$ is negative when $-4 < x < 2$. The solution is graphed in Figure 6.16.

Try Problem 3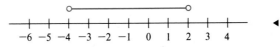

Figure 6.16

EXAMPLE 3   Solve $(x - 2)(x + 4) \leq 0$.

This solution is the same as the solution to Example 2, with one change. In this case we also include those x-values that make $(x - 2)(x + 4)$ equal to zero, namely $x = 2$ and $x = -4$. Therefore the solution is $-4 \leq x \leq 2$ (see Figure 6.17).

Try Problem 5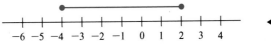

Figure 6.17

EXAMPLE 4   Solve $y(y - 5) < y - 5$.

Write the inequality in standard form.

$$y(y - 5) < y - 5$$
$$y^2 - 5y < y - 5$$
$$y^2 - 6y + 5 < 0$$

Replace $<$ by $=$ and solve the resulting equation.

$$y^2 - 6y + 5 = 0$$
$$(y - 1)(y - 5) = 0$$
$$y = 1 \quad \text{or} \quad y = 5$$

Graph 1 and 5 on a number line as shown in Figure 6.18.

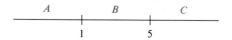

Figure 6.18

Test a value from each region in $y^2 - 6y + 5$, or in the factored form $(y - 1)(y - 5)$.

$$\text{Test } y = 0: \quad (0 - 1)(0 - 5) = (-1)(-5) = 5$$
$$\text{Test } y = 3: \quad (3 - 1)(3 - 5) = (2)(-2) = -4$$
$$\text{Test } y = 6: \quad (6 - 1)(6 - 5) = (5)(1) = 5$$

These three tests tell us that $(y - 1)(y - 5)$ is positive in regions A and C, and negative in region B. Label the number line as shown in Figure 6.19.

Figure 6.19

Try Problem 35

Since we are looking for those y-values that make $(y - 1)(y - 5)$ negative, the solution is $1 < y < 5$. ◄

EXAMPLE 5   Solve $x^2 - 2x - 1 \geq 0$.

Replace $\geq$ with $=$.

$$x^2 - 2x - 1 = 0$$

Solve this equation using the quadratic formula with $a = 1$, $b = -2$, and $c = -1$.

$$x = \frac{-b \pm \sqrt{b^2 - 4ac}}{2a}$$
$$x = \frac{-(-2) \pm \sqrt{(-2)^2 - 4(1)(-1)}}{2(1)}$$
$$x = \frac{2 \pm \sqrt{8}}{2}$$
$$x = \frac{2 \pm 2\sqrt{2}}{2}$$
$$x = 1 \pm \sqrt{2}$$

Graph $1 + \sqrt{2}$ (approximately 2.4) and $1 - \sqrt{2}$ (approximately -0.4) on a number line (see Figure 6.20). Test a value from each region in $x^2 - 2x - 1$.

$$\text{Test } x = -1: \qquad (-1)^2 - 2(-1) - 1 = 1 + 2 - 1 = 2$$
$$\text{Test } x = 0: \qquad 0^2 - 2(0) - 1 = 0 - 0 - 1 = -1$$
$$\text{Test } x = 3: \qquad 3^2 - 2(3) - 1 = 9 - 6 - 1 = 2$$

The results of these tests are shown in Figure 6.20.

$$1 - \sqrt{2} \approx -0.4 \qquad 1 + \sqrt{2} \approx 2.4$$

Figure 6.20

Try Problem 39

Comparing Figure 6.20 with the inequality $x^2 - 2x - 1 \geq 0$, we determine that the solution is $x \leq 1 - \sqrt{2}$ or $x \geq 1 + \sqrt{2}$. ◄

EXAMPLE 6   Solve $x^2 - 6x + 9 > 0$.

Replace $>$ with $=$ and solve the resulting equation.

$$x^2 - 6x + 9 = 0$$
$$(x - 3)(x - 3) = 0$$
$$x = 3$$

Graph $x = 3$, dividing the number line into just *two* regions (see Figure 6.21). Test a value from each region.

$$\text{Test } x = 2: \qquad (2 - 3)(2 - 3) = (-1)(-1) = 1$$
$$\text{Test } x = 4: \qquad (4 - 3)(4 - 3) = (1)(1) = 1$$

The results of these tests are shown in Figure 6.21.

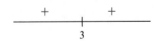

$$3$$

Figure 6.21

Comparing Figure 6.21 with the inequality $x^2 - 6x + 9 > 0$, we determine that the solution is $x < 3$ or $x > 3$. The graph of this solution is shown in Figure 6.22.

Try Problem 43

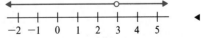

Figure 6.22

You can also solve $x^2 - 6x + 9 > 0$ by writing it as $(x - 3)^2 > 0$. Since $(x - 3)^2$ is positive for every value of x except 3, the solution is $x < 3$ or $x > 3$.

EXAMPLE 7   Solve $x^2 - 6x + 9 < 0$.

Figure 6.21 shows that no value of x makes $x^2 - 6x + 9$ negative. Therefore there is no solution to this inequality (and hence no graph). ◄

Try Problem 47

You can also solve $x^2 - 6x + 9 < 0$ by writing it as $(x - 3)^2 < 0$. Since no value of x makes $(x - 3)^2$ negative, there is no solution.

EXAMPLE 8   Solve $x^2 + x + 1 > 0$.

Replace $>$ with $=$ and solve the resulting equation.

$$x^2 + x + 1 = 0$$

Use the quadratic formula with $a = 1$, $b = 1$, and $c = 1$.

$$x = \frac{-1 \pm \sqrt{1^2 - 4(1)(1)}}{2(1)}$$

$$x = \frac{-1 \pm \sqrt{-3}}{2}$$

Neither solution is a real number, because $\sqrt{-3}$ is not a real number. Therefore neither can be graphed on the real number line. This means that there is only one region (namely, the entire number line) from which to choose a test value.

$$\text{Test } x = 0: \qquad 0^2 + 0 + 1 = 1$$

This test leads to Figure 6.23.

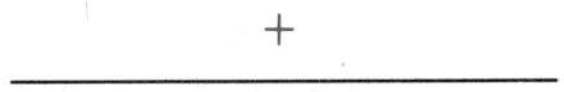

$$+$$
_____

Figure 6.23

Try Problem 51

Figure 6.23 indicates that $x^2 + x + 1$ is positive for every real value of x. Therefore the solution consists of all real numbers. ◄

To Solve a Quadratic Inequality

1. Write the inequality in one of the standard forms $ax^2 + bx + c < 0$, $ax^2 + bx + c \leq 0$, $ax^2 + bx + c > 0$, or $ax^2 + bx + c \geq 0$.
2. Replace the inequality symbol with the symbol $=$ and solve the resulting equation. You may have to use the quadratic formula.
3. Graph all real solutions found in step 2 on a number line. This will divide the number line into one, two, or three regions.
4. Test a value from each region in the expression $ax^2 + bx + c$. (It may be easier to use the factored form of $ax^2 + bx + c$.) Use the test results to label each region with the symbol $-$ or the symbol $+$.
5. If the original inequality symbol is $<$, the solution consists of all numbers in the region labeled $-$. If the symbol is $\leq$, include in your solution the numbers found in step 2.
6. If the original inequality symbol is $>$, the solution consists of all numbers in the regions labeled $+$. If the symbol is $\geq$, include in your solution the numbers found in step 2.

PROBLEM SET 6.8

Solve each inequality. Graph your solution.

1. $(x - 2)(x + 3) > 0$

2. $(x - 4)(x + 5) > 0$

3. $(x - 2)(x + 3) < 0$

4. $(x - 4)(x + 5) < 0$

5. $(x - 2)(x + 3) \leq 0$

6. $(x - 4)(x + 5) \leq 0$

7. $(x - 2)(x + 3) \geq 0$

8. $(x - 4)(x + 5) \geq 0$

Solve each inequality. Graph your solution.

9. $y^2 - 6y + 8 \geq 0$

10. $y^2 - 5y + 4 \geq 0$

11. $z^2 + 7z + 6 > 0$

12. $z^2 + 6z + 5 > 0$

13. $x^2 - 4 < 0$

14. $x^2 - 9 < 0$

15. $x^2 - 4x < 0$

16. $x^2 - 9x < 0$

17. $2p^2 - 11p + 5 < 0$

18. $2p^2 - 13p + 6 < 0$

19. $4y^2 + 5y - 6 > 0$

20. $4y^2 + 9y - 9 > 0$

21. $m^2 + 4m \geq 0$

22. $m^2 + 5m \geq 0$

23. $t^2 - 25 > 0$

24. $t^2 - 16 > 0$

25. $r^2 - r \leq 0$

26. $r^2 + r \leq 0$

27. $9x^2 - 1 \geq 0$

28. $4x^2 - 1 \geq 0$

29. $y^2 - 2 \leq 0$

30. $y^2 - 3 \leq 0$

Solve each inequality.

31. $m^2 \geq 36$

32. $m^2 \geq 49$

33. $p^2 \leq 9p$

34. $p^2 \leq 4p$

35. $y(y - 3) < y - 3$

36. $y(y - 4) < y - 4$

37. $6y(y - 2) + y > 5 + 2y$

38. $6y(y - 3) + y > 7 + 2y$

39. $x^2 - 2x - 2 \geq 0$

40. $x^2 - 2x - 4 \geq 0$

41. $2r^2 + 4r - 1 \leq 0$

42. $2r^2 + 4r + 1 \leq 0$

Solve each inequality.

43. $x^2 - 4x + 4 > 0$

44. $x^2 - 10x + 25 > 0$

45. $x^2 - 4x + 4 \geq 0$

46. $x^2 - 10x + 25 \geq 0$

47. $x^2 - 4x + 4 < 0$

48. $x^2 - 10x + 25 < 0$

49. $x^2 - 4x + 4 \leq 0$

50. $x^2 - 10x + 25 \leq 0$

Solve each inequality.

51. $x^2 + x + 2 > 0$

52. $x^2 + x + 3 > 0$

53. $x^2 + x + 2 < 0$

54. $x^2 + x + 3 < 0$

55. $y^2 + 4 \geq 0$

56. $y^2 + 9 \geq 0$

57. $y^2 + 4 \leq 0$

58. $y^2 + 9 \leq 0$

59. $(r + 1)^2 > r^2 + 2r$

60. $(r + 2)^2 > r^2 + 4r$

61. $(r + 1)^2 < r^2 + 2r$

62. $(r + 2)^2 < r^2 + 4r$

Solve each inequality.

63. $(x - 2)(x - 4)(x - 6) \geq 0$

64. $(y + 1)(y - 3)^2 > 0$

65. $p^3 + p^2 - 6p < 0$

66. $m^4 - 10m^2 + 9 \leq 0$

67. A company's daily profit P in thousands of dollars is given by $P = -4x^2 + 20x$, where x is the number of units produced in hundreds. Find those values of x for which P is at least 16.

68. If we ignore air resistance, the height h in feet of an object projected upward with an initial velocity of 176 feet per second is given by $h = -16t^2 + 176t$. Over what time interval is the object's height at least 288 feet?

CALCULATOR PROBLEMS

Solve each inequality.

69. $x^2 - 34x - 2847 < 0$

70. $55y^2 - 288y + 377 > 0$

CHAPTER 6 REVIEW

[6.1] *Solve each equation.*

1. $(x + 4)(x - 3) = 0$ **2.** $3y(y - 5) = 0$ **3.** $x^2 + 7x + 10 = 0$ **4.** $x^2 + x = 6$

5. $6y^2 - 17y - 14 = 0$ **6.** $25z^2 - 16 = 0$ **7.** $3r^2 = 4r$ **8.** $(t + 2)(t - 2) = 5$

Solve each equation for x.

9. $5x^2 - 15bx = 0$

10. $x^2 - ax = 6a^2$

[6.2] *Determine the number that must be added to each expression to make it a perfect-square trinomial.*

11. $x^2 - 8x$

12. $y^2 + y$

Solve each equation by extraction of roots.

13. $(y - 3)^2 = 25$

14. $(p + \frac{1}{3})^2 = \frac{49}{9}$

Solve each equation by completing the square.

15. $x^2 + 8x + 15 = 0$

16. $w^2 - 2w - 2 = 0$

17. $y^2 + y - 1 = 0$

18. $3m^2 - 18m - 1 = 0$

[6.3] *Use the quadratic formula to solve each equation.*

19. $x^2 + 2x - 24 = 0$

20. $p^2 - 6p + 9 = 0$

21. $2y^2 + y - 2 = 0$

22. $8m^2 + 2m = 3$

23. $\dfrac{x^2}{3} + \dfrac{x}{6} - 1 = 0$

24. $4t^2 - 4t - 11 = 0$

25. $x^2 + 2x = 5x + 12$

26. $4x(1 - x) = 1$

[6.4]

27. The sum of the squares of two consecutive odd integers is 74. Find the integers.

28. A rectangle is 4 meters longer than twice its width. Its area is 70 square meters. Find the length and width of the rectangle.

29. Kerry has a garden that is 6 feet by 4 feet. She wants to surround the garden with a brick walk of uniform width. She has enough bricks to cover 24 square feet. How wide should the brick walk be?

30. Find the lengths of the sides of a right triangle if one leg is 1 inch less than twice the shorter leg, and the hypotenuse is 1 inch more than twice the shorter leg.

[6.5] *Solve each equation.*

31. $\sqrt{x - 1} = 5$

32. $\sqrt{2y - 1} + 1 = 0$

33. $\sqrt{m + 8} = \sqrt{m} + 2$

34. $t - \sqrt{t + 3} = 3$

35. $\sqrt{w + 2} + \sqrt{2w + 5} = 1$

36. $\sqrt[3]{4y^2 - 36} - 4 = 0$

Solve each equation for the specified variable. Assume all variables represent positive real numbers.

37. $v = \sqrt{2gR}$ for g

38. $t = 2\pi\sqrt{\dfrac{l}{32}}$ for l

[6.6] *Solve each equation.*

39. $3 + \dfrac{2}{x} - \dfrac{1}{x^2} = 0$

40. $\dfrac{2t}{t + 2} - \dfrac{3}{t} = 1$

41. $y^4 - 29y^2 + 100 = 0$

42. $m^4 - 4m^2 + 3 = 0$

43. $x^{2/3} + 2x^{1/3} - 3 = 0$

44. Jim and Bill can do a piece of work in 12 days if they work together. Working alone Bill takes 10 days longer than Jim to do the same piece of work. How long would it take each to do the work alone?

45. When Maria rides her moped to her cabin 30 miles away, it takes one-half hour less time than when she rides her bicycle. If her moped travels 10 mph faster than her bike, find the rate of each.

[6.7] *Solve each equation.*

46. $x^2 + 25 = 0$

47. $4m^2 + 9 = 0$

48. $2w^2 - 3w + 2 = 0$

49. $z^2 - 12z + 37 = 0$

50. $m^4 + 4m^2 - 5 = 0$

Use the discriminant to determine the nature of the solutions to each equation. Do not solve the equation.

51. $3t^2 - 2t + 4 = 0$

52. $2x^2 - 7x + 3 = 0$

53. By setting the discriminant equal to zero, determine the value(s) of k that cause $x^2 + kx + 9 = 0$ to have exactly one solution.

[6.8] *Solve each inequality. Graph your solution.*

54. $t^2 - 9t \geq 0$

55. $25 - 9x^2 \leq 0$

56. $w^2 - w + 12 < 0$

57. $y(y - 5) > y - 5$

58. $x^2 - 8x + 16 \geq 0$

59. $x^2 - 6x + 3 < 0$

CHAPTER 6 TEST

Solve each equation.

1. $(x + 6)(x - 3) = 0$

2. $4t^2 - 81 = 0$

3. $5r^2 = 6r$

4. $y^2 - 20y + 100 = 0$

5. $x^2 + x = 4$

6. $x^2 + 8x = 12$

7. $z^2 + 6z + 3 = 0$

8. $p^2 + 1 = 0$

9. $4m^2 + 8m + 5 = 0$

10. Determine the number that must be added to $x^2 - 5x$ to make it a perfect-square trinomial.

Solve each inequality. Graph your solution.

11. $x^2 + 2x - 3 > 0$

12. $x^2 - 2x - 8 \leq 0$

Solve each equation.

13. $\sqrt{p - 3} = 2$

14. $t^4 - 14t^2 + 45 = 0$

15. $5 + \dfrac{7}{x} - \dfrac{6}{x^2} = 0$

16. $\dfrac{m}{m + 4} - \dfrac{2}{m} = 0$

17. $\sqrt{x + 20} = x$

18. $\sqrt{2y - 1} = 1 + \sqrt{y + 3}$

19. One leg of a right triangle is two inches longer than twice the other leg. The hypotenuse is three inches longer than twice the shorter leg. Find the length of each side.

20. Fred can jog the 3-mile distance to his boat dock in one fourth of an hour less time than he can walk it. If he jogs 2 mph faster than he walks, how fast does he walk?

CUMULATIVE REVIEW FOR CHAPTERS 4–6

[4] *Simplify each rational expression.*

1. $\dfrac{x^2 - 9y^2}{2x - 6y}$

2. $\dfrac{3x^2 + 11x + 10}{3x^2 - 4x - 15}$

Perform the indicated operation. Write your answer in simplest form.

3. $n - \dfrac{1}{n + 5}$

4. $\dfrac{2}{t^2 - 4} + \dfrac{t + 2}{t - 2}$

5. $\dfrac{x^2 - 9}{x - 9} \div (x + 3)$

6. $\dfrac{p^2 + p - 6}{p^2 - p - 12} \cdot \dfrac{p^2 - 3p - 4}{p^2 - p - 2}$

Simplify each complex fraction.

7. $\dfrac{1 + \dfrac{1}{p}}{1 - \dfrac{1}{p^2}}$

8. $\dfrac{1 - \dfrac{y^3}{x^3}}{\dfrac{y^2}{x^3} + \dfrac{y}{x^2} + \dfrac{1}{x}}$

Solve each equation.

9. $\dfrac{x}{2} + \dfrac{1}{4} = \dfrac{1}{8} - \dfrac{x}{4}$

10. $\dfrac{2}{r + 3} = \dfrac{3}{r + 2}$

11. $\dfrac{t + 2}{t + 1} = \dfrac{1}{t + 1} + 2$

12. $\dfrac{x + 2}{x - 3} = \dfrac{2x - 3}{x^2 - 9} + 1$

13. A formula used for enlargements in photography is $\dfrac{1}{x} + \dfrac{1}{n} = \dfrac{1}{f}$. Solve for f.

14. Divide $\dfrac{12a^2b^3 + 36a^2b^2 - 6ab}{6ab}$.

Use long division of polynomials to find the quotient and remainder.

15. $(x^5 + x^3 + 2x^2 - 6x + 4) \div (x^2 + 4)$

16. $(6x^4 - x^3 + 8x^2 + x - 3) \div (2x + 1)$

Use synthetic division to divide.

17. $(x^6 - 4x^4 + 3x^3 + 2x + 1) \div (x + 2)$

18. One inlet pipe can fill an empty tank in 20 minutes, and another takes 25 minutes. With both pipes open how long will it take to fill the tank?

19. An express train travels 135 miles in the same time that a freight train travels 75 miles. If the express travels 20 mph faster than the freight, find the speed of each.

[5] *Simplify.*

20. $(-2)^{-4}$

21. $4^{-3/2}$

22. $a^3 \cdot a^{-2}$

23. $\dfrac{p^{-4}}{p^{-6}}$

24. $\dfrac{ax^{-2}}{by^{-3}}$

25. $(2x^{-3})^{-3}$

26. $(3a^{-9}b^4)^2$

27. $\dfrac{1}{a^{-2} + b^{-2}}$

Write each number in scientific notation.

28. 0.000000963

29. $-640{,}000$

Find each root. Assume all variables represent positive real numbers.

30. $\sqrt[5]{-243}$

31. $\sqrt[3]{\dfrac{125}{8}}$

32. $\sqrt[3]{64x^9}$

33. $\sqrt{r^2 + 10r + 25}$

Simplify.

34. $\left(\dfrac{a^{20}}{b^{15}}\right)^{1/5}$

35. $\left(\dfrac{8x^{-3}b^6}{c^{18}}\right)^{1/3}$

36. $w^{-1/6}(w^{2/3} + w^{1/6})$

37. $\dfrac{\sqrt{162}}{\sqrt{2}}$

38. $\sqrt{\sqrt{a}}$

39. $\sqrt[4]{625x^6}$

Perform the indicated operations and simplify.

40. $\sqrt[3]{6q}\sqrt[3]{12q^2}$

41. $\dfrac{\sqrt{2m}}{\sqrt{50m^3}}$

42. $\sqrt{3} - \sqrt{12}$

43. $\sqrt{24t} + \sqrt{6t}$

44. $\sqrt{2}(3\sqrt{6} - \sqrt{2})$

45. $(\sqrt{2} + \sqrt{7})(\sqrt{2} - \sqrt{7})$

Rationalize the denominator and simplify.

46. $\dfrac{3}{\sqrt[3]{3}}$

47. $\dfrac{6}{\sqrt{5} - \sqrt{3}}$

Perform the indicated operations.

48. $(6 - 2i) - (2 + 3i)$

49. $(3 + i)(2 - i)$

50. $\dfrac{3i}{-2 + 3i}$

51. $\dfrac{3}{2i}$

[6] *Solve each equation.*

52. $2r^2 + 5r - 12 = 0$

53. $6t^2 - 18t = 0$

Solve by extraction of roots.

54. $x^2 - 81 = 0$

55. $s^2 + 9 = 0$

Solve each equation by completing the square.

56. $x^2 + 4x - 6 = 0$

57. $4t^2 + 12t + 7 = 0$

Use the quadratic formula to solve each equation.

58. $3m^2 - 4m + 1 = 0$

59. $w^2 + 8w - 1 = 0$

60. Determine the number that must be added to $u^2 - \frac{1}{2}u$ to make it a perfect-square trinomial.

61. Solve $4x^2 - 20dx = 0$ for x.

Solve each equation.

62. $\dfrac{2}{r^2} - \dfrac{1}{r} - 6 = 0$

63. $\dfrac{x}{x-1} + \dfrac{x}{x+1} = \dfrac{2x}{x^2-1}$

64. $\sqrt{x^2 - 6x} - 4 = 0$

65. $y^4 - 6y^2 - 27 = 0$

66. $\sqrt{2m+5} - \sqrt{m+2} = 1$

67. Use the discriminant to determine the nature of the solutions to $3x^2 + x - 2 = 0$.

Solve each inequality. Graph your solution.

68. $x^2 - 3x < 0$

69. $x^2 - x - 2 \geq 0$

70. One leg of a right triangle is 1 inch longer than the other. The hypotenuse is 1 inch shorter than twice the shorter leg. Find the length of each side.

71. A wading pool that is 6 feet by 8 feet is surrounded by a concrete border of uniform width. If the area of the concrete border is 72 square feet, what is the width?

You have probably heard the expression, "A picture is worth a thousand words." In mathematics, we often draw pictures called graphs to clarify the relationship between two variables. In this chapter you will learn to graph linear equations in two variables, such as $y = x + 2$. They are called linear equations because their graphs are straight lines. You will also learn to graph linear inequalities in two variables, such as $x + y < 3$. The graph of a linear inequality is a portion of a plane.

The method of graphing that is introduced in this chapter paved the way for, among other things, the development of calculus during the latter part of the seventeenth century. In short, it marked the birth of modern mathematics.

7

Graphing Linear Equations and Inequalities

7.1 *The Rectangular Coordinate System*

To graph the solution to an equation involving two variables we need two number lines, one for each variable. Experience has shown that the clearest graph is obtained when the two number lines are placed perpendicular to each other as shown in Figure 7.1.

Notice that the number lines, now called **axes,** divide the plane into four disjoint regions, called **quadrants.** These quadrants are always numbered as shown in Figure 7.1. The axes themselves are not considered to be part of any quadrant. The point of intersection of the two axes is called the **origin.**

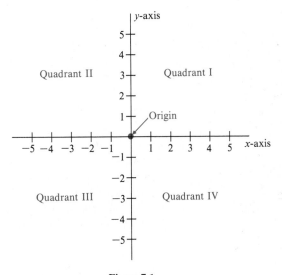

Figure 7.1

Each axis is labeled with the variable it represents. Variables are usually assigned alphabetically: first to the horizontal axis, then to the vertical axis. Therefore if the equation is $y = x + 2$, we label the horizontal axis as the x-axis and the vertical axis as the y-axis.

To locate a point in the plane we need two numbers. The *first* number tells us how far and in which direction to move *horizontally.* The *second* number tells us how far and in which direction to move *vertically.* These two numbers are separated by a comma and enclosed in parentheses. They are referred to as an **ordered pair** of numbers.

EXAMPLE 1   Graph the ordered pairs (4, 2) and (2, 4).

Consider Figure 7.2. The ordered pair (4, 2) gives the location of the point P that is reached by starting at the origin, traveling 4 units to the right, and then traveling 2 units up. The ordered pair (2, 4) directs us to the point Q that is 2 units to the right and 4 units up. ◄

Try Problem 1

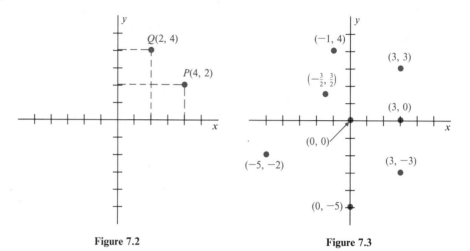

Figure 7.2   Figure 7.3

The numbers 4 and 2 are called the **coordinates** of point P; similarly, 2 and 4 are the coordinates of point Q. The points P and Q are called the **graphs** of the given ordered pairs. As you can see, the order of the coordinates in an ordered pair is significant. That is, (4, 2) and (2, 4) represent entirely different points.

The coordinates of several other points are shown in Figure 7.3. Note that the point (3, 3) lies in quadrant I, the point (3, −3) lies in quadrant IV, and the point (3, 0) lies on the positive part of the x-axis between quadrants I and IV.

This system of assigning ordered pairs of real numbers to points in a plane is called a **rectangular coordinate system.***

Problem Solving

Probabilities in Dice

We can use a rectangular coordinate system to display all the possible outcomes that can occur when two fair dice (one red and one green) are rolled. We agree that the first coordinate denotes the outcome of the red die, and the second coordinate denotes the outcome of the green die. Then the ordered pair (4, 1) represents a roll that produced a 4 on the red die and a 1 on the green die. This is a different outcome from (1, 4), which represents a 1 on the red die and a 4 on the green die. The 36 possible outcomes are graphed in Figure 7.4. The proba-

---

* It is also called a Cartesian coordinate system in honor of its inventor, René Descartes (1596–1650).

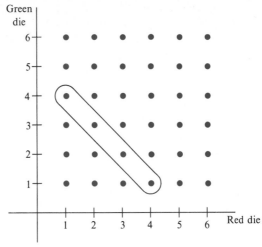

Figure 7.4

bility of rolling a total of n is denoted $P(n)$, and is given by the formula

$$P(n) = \frac{\text{number of ways to roll a total of } n}{36}.$$

Use this formula and Figure 7.4 to determine the probability of rolling a total of 5.

There are four ways to roll a total of 5 (see Figure 7.4). These four ways are represented by the ordered pairs (1, 4), (2, 3), (3, 2), and (4, 1). Therefore the probability of rolling a 5 is

$$P(5) = \tfrac{4}{36} = \tfrac{1}{9}.$$

PROBLEM SET 7.1

Graph each point in a rectangular coordinate system.

1. (4, 3)
2. (5, 1)
3. (−2, 5)
4. (−3, 4)
5. (1, −1)
6. (2, −2)
7. (−2, −3)
8. (−1, −4)
9. (6, 0)
10. (4, 0)
11. (−4, 0)
12. (−6, 0)
13. (0, $\tfrac{5}{2}$)
14. (0, $\tfrac{7}{2}$)
15. (0, −5)
16. (0, −1)

Graph each point. Let each hash mark represent 5 units.

17. (15, 25)
18. (20, 30)
19. (−10, 20)
20. (−25, −15)
21. (−5, −5)
22. (5, −10)
23. (0, 35)
24. (0, −35)

Graph each point. Let each hash mark on the horizontal axis represent 1 unit and each hash mark on the vertical axis represent 10 units.

25. $(2, 40)$ **26.** $(3, 30)$ **27.** $(-3, -30)$

28. $(-4, 50)$ **29.** $(1, -10)$ **30.** $(1, -20)$

31. $(0, -45)$ **32.** $(0, 55)$

Given that $m > 0$ and $n < 0$, determine the quadrant in which each point lies.

33. (m, n) **34.** (n, m) **35.** $(-m, -n)$

36. $(-n, -m)$ **37.** $(mn, n - m)$ **38.** $\left(m - n, \dfrac{m}{n}\right)$

39. $(m, 0)$ **40.** $(0, n)$

Two fair dice are rolled. Determine the probability of each of the following events.

41. A total of 6   **42.** A total of 8

43. A total of 10   **44.** A total of 4

45. A total greater than 9   **46.** A total less than 5

47. A total between 1 and 13   **48.** A total greater than 0

49. A total less than 2   **50.** A total of 16

51. What is the easiest number to roll with two dice?   **52.** What is the hardest number to roll with two dice?

53. a) Graph three points whose first coordinate is 5.
   b) Graph all points whose first coordinate is 5.

54. a) Graph three points whose second coordinate is 2.
   b) Graph all points whose second coordinate is 2.

55. a) Graph three points whose first and second coordinates are equal.
   b) Graph all points whose first and second coordinates are equal.

56. a) Graph three points whose first and second coordinates are opposites.
   b) Graph all points whose first and second coordinates are opposites.

CALCULATOR PROBLEMS

Two fair dice are rolled. Write the probability of each event as a percent.

57. Any total except 2   **58.** A total of 7 or 11

7.2 *Graphing Linear Equations*

Equations like $2x + 5y = 10$, where both variables appear to the first power, are called **linear equations** in two variables.

Definition

A **linear equation** in the two variables x and y is an equation that can be expressed in the form

$$ax + by = c,$$

where neither a nor b is zero. This form is called the **standard form** of a linear equation.

EXAMPLE 1   Which of the following equations are linear: **a)** $y = x + 2$, **b)** $y = x^2 + 5$, **c)** $y = x$, **d)** $y = \frac{1}{x}$.

a) $y = x + 2$ is linear, since it can be written as $-x + y = 2$.
b) $y = x^2 + 5$ is not linear, because x has an exponent of 2.
c) $y = \sqrt{x}$ is not linear, because $\sqrt{x} = x^{1/2}$.

Try Problem 1

d) $y = \frac{1}{x}$ is not linear, because $\frac{1}{x} = x^{-1}$. ◄

A **solution** to a linear equation in two variables is an ordered pair that satisfies the equation. To **graph** a linear equation means to graph all of its solutions.

EXAMPLE 2   Graph $y = x + 2$.

We find solutions to this equation by substituting any value we choose for x, and then solving for y. For example, if we let $x = 0$, then

$$y = 0 + 2 = 2. \qquad \text{Means (0, 2) is a solution}$$

If we let $x = 1$, then

$$y = 1 + 2 = 3. \qquad \text{Means (1, 3) is a solution}$$

If we let $x = 2$, then

$$y = 2 + 2 = 4. \qquad \text{Means (2, 4) is a solution}$$

We can also substitute any value for y and solve for x. For example, if $y = 0$, then

$$0 = x + 2$$
$$-2 = x. \qquad \text{Means } (-2, 0) \text{ is a solution}$$

The most convenient way to collect ordered-pair solutions is in a **table of values** like the one on the following page.

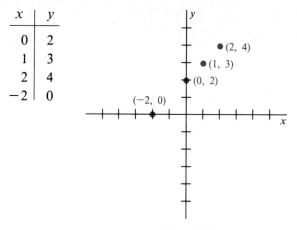

x	y
0	2
1	3
2	4
−2	0

Figure 7.5

Plotting these four solutions in a rectangular coordinate system produces the graph in Figure 7.5. It is more than a coincidence that all four points lie on the same straight line. In fact the equation $y = x + 2$ has an *infinite* number of solutions, yet the graph of every solution lies on this same straight line. Likewise, every point on this line represents a solution to $y = x + 2$. Therefore to complete the graph of this equation we draw a solid straight line through the four points (see Figure 7.6).

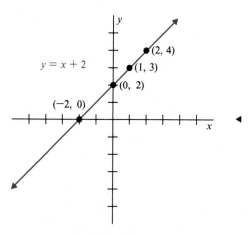

Try Problem 9

Figure 7.6

The graph of every linear equation in two variables is a straight line. Since a straight line is determined when any two points on the line are known, we need to find only two solutions to the equation to graph the entire line. It is a good idea, however, to find a third solution as a check. If the three points do not line up, we have made a mistake.

EXAMPLE 3   Graph $2x + 5y = 10$.

If $x = 0$, then

$$2 \cdot 0 + 5y = 10$$
$$5y = 10$$
$$y = 2.$$

x	y
0	2
5	0
3	$\frac{4}{5}$

If $y = 0$, then

$$2x + 5 \cdot 0 = 10$$
$$2x = 10$$
$$x = 5.$$

If $x = 3$, then

$$2 \cdot 3 + 5y = 10$$
$$5y = 4$$
$$y = \tfrac{4}{5}.$$

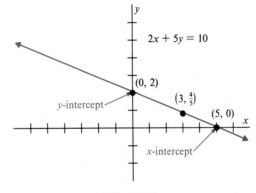

Figure 7.7

Graph the three points and draw a straight line through them to complete the graph (see Figure 7.7). ◄

Since the line in Figure 7.7 crosses the x-axis at (5, 0), we call 5 the **x-intercept** of the line. Since the line crosses the y-axis at (0, 2), we call 2 the **y-intercept.** Whenever possible, we should draw enough of the line to display both intercepts.

Try Problem 17

Intercepts of a Line

To find the x-intercept, set $y = 0$.
To find the y-intercept, set $x = 0$.

If a linear equation in *one* variable is graphed in the xy-plane, the result is either a horizontal line or a vertical line.

EXAMPLE 4   Graph $y = 3$ in the xy-plane.

This equation can be written as

$$0 \cdot x + y = 3.$$

Now if any value is substituted for x, the corresponding y-value will be 3. For example, if $x = 4$, then

$$0 \cdot 4 + y = 3$$
$$y = 3. \qquad \text{Means (4, 3) is a solution}$$

Other solutions to this equation are $(-3, 3)$, $(0, 3)$, and $(3, 3)$. Graphing these points and joining them with a straight line produces the graph in Figure 7.8.

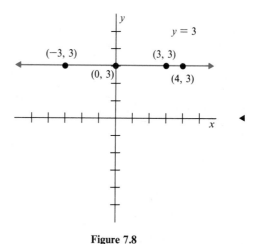

Try Problem 33

Figure 7.8

We see from Example 4 that the graph of $y = 3$ is a horizontal line through the point $(0, 3)$. This illustrates the following rule.

Horizontal Line

Suppose k is a constant. The graph of $y = k$ in the xy-plane is a horizontal line through the point $(0, k)$.

This rule allows us to graph equations of the form $y = k$ without constructing a table of values. A similar rule holds for equations of the form $x = k$.

Vertical Line

Suppose k is a constant. The graph of $x = k$ in the xy-plane is a vertical line through the point $(k, 0)$.

EXAMPLE 5   Graph $2x + 5 = 0$ in the xy-plane.

$$2x + 5 = 0$$
$$2x = -5 \left.\begin{array}{r} \\ \\ \end{array}\right\} \text{Solve for } x$$
$$x = -\tfrac{5}{2}$$

The graph is the horizontal line through $(-\tfrac{5}{2}, 0)$ shown in Figure 7.9.

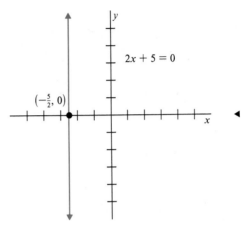

Figure 7.9

Try Problem 45

PROBLEM SET 7.2

Which of the following equations are linear?

1. $y = 2x + 3$ 2. $y = 3x + 4$ 3. $y = x^2$ 4. $y = x^2 - 1$

5. $y = \sqrt{x} + 4$ 6. $y = \sqrt[3]{x}$ 7. $y = \dfrac{6}{x}$ 8. $y = \dfrac{8}{x}$

Complete each table of values using the given equation. Then graph the points and draw a straight line through them.

9. $y = x + 3$

x	y
0	
1	
2	
	0

10. $y = x + 4$

x	y
0	
1	
2	
	0

11. $y = 2x - 2$

x	y
-1	
0	
	0
	3

12. $y = 3x - 3$

x	y
-1	
0	
	0
	4

Graph each equation. Label the x- and the y-intercept.

13. $x + y = 4$

14. $x + y = 5$

15. $x + 2y = 10$

16. $2x + y = 8$

17. $2x + 3y = 6$

18. $3x + 4y = 12$

19. $x - y = 1$

20. $y - x = 1$

21. $4y - x = 8$

22. $x - 5y = 10$

23. $5x - 3y = 10$

24. $5x - 2y = 15$

25. $y = x$

26. $y = -x$

27. $y = -2x$

28. $y = 2x$

29. $x = 2y$

30. $x = 3y$

31. $y = -2 - \frac{1}{3}x$

32. $y = -3 - \frac{1}{2}x$

Complete each table of values using the given equation. Then graph the points and draw a straight line through them.

33. $y = 4$

x	y
-3	
0	
1	
4	

34. $y = 5$

x	y
-2	
0	
3	
5	

35. $x = 3$

x	y
	-2
	0
	3
	4

36. $x = 1$

x	y
	-3
	0
	1
	4

Graph each equation in the xy-plane.

37. $y = -3$

38. $y = -2$

39. $x = -5$

40. $x = -4$

41. $y - 1 = 0$

42. $y - 6 = 0$

43. $x - 6 = 0$

44. $x - 4 = 0$

45. $2x + 3 = 0$

46. $2x + 7 = 0$

47. $2y - 5 = 0$

48. $2y - 3 = 0$

49. $x = 0$

50. $y = 0$

51. $4y = 0$

52. $\frac{1}{4}x = 0$

53. Scientists have shown that the Fahrenheit temperature T and the number of chirps per minute of a house cricket n are related by the equation $T = 40 + 0.25n$. Graph this equation for all values of n in the interval $0 \le n \le 160$.

54. The quantity q of ringside fight tickets demanded is related to the price p of the ticket by the equation $q = 125 - 2.5p$. Graph this equation for all values of p in the interval $0 \le p \le 50$.

CALCULATOR PROBLEMS

55. Determine whether the given point lies on the graph of $12.7x + 18.1y = 178.8$.

  a) $(5.1, 6.3)$ **b)** $(2.4, 8.2)$

56. Graph $12.7x + 18.1y = 178.8$.

7.3   *Distance and Slope*

THE DISTANCE FORMULA

Suppose P and Q are two different points in the xy-plane. Suppose further that the coordinates of P are (x_1, y_1) and the coordinates of Q are (x_2, y_2) as shown in Figure 7.10. (Remember that the subscripts 1 and 2 do not indicate an operation.) Then the coordinates of point R in Figure 7.10 must be (x_2, y_1).

Now the distance between P and R is $x_2 - x_1$. The distance between Q and R is $y_2 - y_1$. If we denote the distance between P and Q as d, then by the Pythagorean theorem we have

$$d^2 = (x_2 - x_1)^2 + (y_2 - y_1)^2.$$

Since d represents a positive number, we extract only the positive square root of the right side and get

$$d = \sqrt{(x_2 - x_1)^2 + (y_2 - y_1)^2}.$$

Although we chose points P and Q in the first quadrant, this **distance formula** holds for any two points in the plane.

The Distance Formula

The distance d between the two points $P\ (x_1, y_1)$ and $Q\ (x_2, y_2)$ is given by

$$d = \sqrt{(x_2 - x_1)^2 + (y_2 - y_1)^2}.$$

The distance d can also be denoted by PQ.

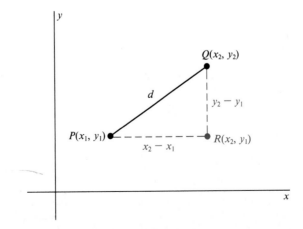

Figure 7.10

Since $7 - 3$ and $3 - 7$ are opposites, their squares are equal. That is, $(7 - 3)^2 = (3 - 7)^2$. By the same token, $(x_2 - x_1)^2 = (x_1 - x_2)^2$ and $(y_2 - y_1)^2 = (y_1 - y_2)^2$. Therefore it does not matter which point we call (x_1, y_1) and which we call (x_2, y_2).

EXAMPLE 1   Find the distance between: **a)** $(2, 3)$ and $(5, 7)$,   **b)** $(6, -3)$ and $(4, 5)$.

a)   Choose $(x_1, y_1) = (2, 3)$ and $(x_2, y_2) = (5, 7)$.
$$d = \sqrt{(x_2 - x_1)^2 + (y_2 - y_1)^2}$$
$$d = \sqrt{(5 - 2)^2 + (7 - 3)^2}$$
$$d = \sqrt{3^2 + 4^2} = \sqrt{9 + 16} = \sqrt{25} = 5$$

b)   Choose $(x_1, y_1) = (6, -3)$ and $(x_2, y_2) = (4, 5)$.
$$d = \sqrt{(4 - 6)^2 + [5 - (-3)]^2}$$
$$d = \sqrt{(-2)^2 + 8^2} = \sqrt{4 + 64} = \sqrt{68} = \sqrt{4 \cdot 17} = 2\sqrt{17} \blacktriangleleft$$

Try Problem 1

CAUTION: Remember that $\sqrt{9 + 16}$ does *not* equal $\sqrt{9} + \sqrt{16}$.

EXAMPLE 2   Find x if the distance between $(3, 4)$ and $(x, 5)$ is $\sqrt{2}$.

Apply the distance formula with $d = \sqrt{2}$, $(x_1, y_1) = (3, 4)$, and $(x_2, y_2) = (x, 5)$.
$$\sqrt{2} = \sqrt{(x - 3)^2 + (5 - 4)^2}$$

Square each side to remove the radicals.
$$2 = (x - 3)^2 + 1^2$$

Solve the resulting quadratic equation.
$$2 = x^2 - 6x + 9 + 1$$
$$0 = x^2 - 6x + 8$$
$$0 = (x - 2)(x - 4)$$
$$x = 2 \quad \text{or} \quad x = 4$$

Try Problem 55

Both values check, so $x = 2$ or $x = 4$. $\blacktriangleleft$

THE SLOPE FORMULA

If a roof rises 5 inches for every horizontal run of 12 inches, we say that the **slope** of the roof is $\frac{5}{12}$ (see Figure 7.11). When the run is 24 inches, the roof rises 10 inches.

Now consider the line passing through the two points (x_1, y_1) and (x_2, y_2), as shown in Figure 7.12. If we call the slope of this line m,* then

---

* From the French word *monter*, meaning "to climb."

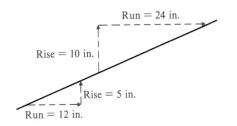

$$\text{Slope} = \frac{\text{rise}}{\text{run}} = \frac{5 \text{ in.}}{12 \text{ in.}} = \frac{5}{12}$$

Figure 7.11

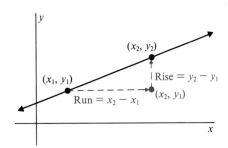

Figure 7.12

$$m = \frac{\text{rise}}{\text{run}} = \frac{y_2 - y_1}{x_2 - x_1}.$$

The Slope Formula

The slope m of the line passing through the two points (x_1, y_1) and (x_2, y_2) is given by

$$m = \frac{y_2 - y_1}{x_2 - x_1}.$$

It can be shown using similar triangles that the slope of a line is the same no matter which two points on the line are used to calculate the slope.

EXAMPLE 3 Find the slope of the line passing through $(3, 6)$ and $(5, 10)$.

Try Problem 25

$$m = \frac{y_2 - y_1}{x_2 - x_1} = \frac{10 - 6}{5 - 3} = \frac{4}{2} = 2 \blacktriangleleft$$

You can also compute the slope in Example 3 by letting $(x_1, y_1) = (5, 10)$ and $(x_2, y_2) = (3, 6)$.

$$m = \frac{6 - 10}{3 - 5} = \frac{-4}{-2} = 2$$

EXAMPLE 4 Find the slope of the line passing through each pair of points; then graph the lines on the same coordinate axes: **a)** $(0, 0)$, $(4, 1)$;   **b)** $(0, 0)$, $(4, 4)$;   **c)** $(0, 0)$, $(1, 4)$;   **d)** $(0, 0)$, $(-1, 4)$;   **e)** $(0, 0)$, $(-4, 4)$;   **f)** $(0, 0)$, $(-4, 1)$.

The solutions are as follows:

a) $m = \dfrac{1-0}{4-0} = \dfrac{1}{4}$

b) $m = \dfrac{4-0}{4-0} = 1$

c) $m = \dfrac{4-0}{1-0} = 4$

d) $m = \dfrac{4-0}{-1-0} = -4$

e) $m = \dfrac{4-0}{-4-0} = -1$

f) $m = \dfrac{1-0}{-4-0} = -\dfrac{1}{4}$

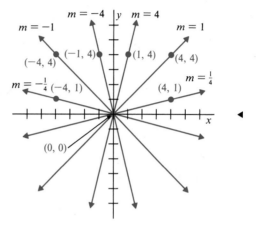

Figure 7.13

Try Problem 41

As you can see in Figure 7.13, a *positive* slope means the line is *increasing,* and a *negative* slope means the line is *decreasing.* Also, the *steeper* the line, the *greater* its slope in absolute value.

EXAMPLE 5 Find the slope of the line passing through each pair of points; then graph the lines on the same coordinate axes: **a)** $(-4, 4)$, $(4, 4)$; **b)** $(-4, -4)$, $(4, -4)$; **c)** $(4, 4)$, $(4, -4)$; **d)** $(-4, 4)$, $(-4, -4)$.

The solutions are as follows:

a) $m = \dfrac{4-4}{4-(-4)} = \dfrac{0}{8} = 0$

b) $m = \dfrac{-4-(-4)}{4-(-4)} = \dfrac{0}{8} = 0$

c) $m = \dfrac{-4-4}{4-4} = \dfrac{-8}{0}$ undefined

d) $m = \dfrac{-4-4}{-4-(-4)} = \dfrac{-8}{0}$ undefined

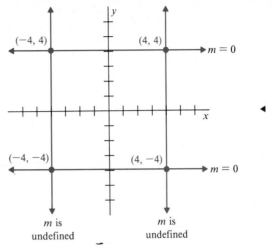

Try Problem 43

Figure 7.14

From Figure 7.14 you can see that *the slope of a horizontal line is zero,* and *the slope of a vertical line is undefined.*

Figure 7.15 summarizes some of our findings about slope.

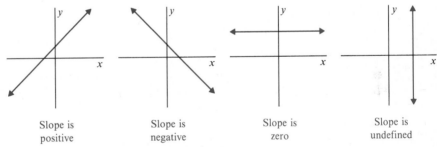

Figure 7.15

Using methods from geometry, it can be shown that parallel lines have the same slope. It can also be shown that perpendicular lines have slopes that are negative reciprocals of each other.

Parallel and Perpendicular Lines

Suppose m_1 is the slope of line L_1, and m_2 is the slope of line L_2.

1. L_1 is **parallel** to L_2 if and only if $m_1 = m_2$.

2. L_1 is **perpendicular** to L_2 if and only if $m_1 = -\dfrac{1}{m_2}$.

EXAMPLE 6 Suppose L_1 is the line through $(2, -7)$ and $(8, -11)$. Find the slope of line L_2 given that: **a)** L_2 is parallel to L_1, **b)** L_2 is perpendicular to L_1.

First find the slope of line L_1.

$$m_1 = \frac{-11 - (-7)}{8 - 2} = \frac{-4}{6} = -\frac{2}{3}$$

a) If L_2 is parallel to L_1, then $m_2 = m_1$. That is,

$$m_2 = -\frac{2}{3}.$$

b) If L_2 is perpendicular to L_1, then $m_2 = -\dfrac{1}{m_1}$. That is,

Try Problem 57

$$m_2 = -\frac{1}{-\dfrac{2}{3}} = \frac{3}{2}. \blacktriangleleft$$

PROBLEM SET 7.3

Find the distance between each pair of points.

1. $(2, 1)$ and $(6, 4)$
2. $(1, 2)$ and $(9, 8)$
3. $(-5, 10)$ and $(7, -6)$
4. $(-3, 7)$ and $(2, -5)$
5. $(-2, 3)$ and $(1, 1)$
6. $(2, -2)$ and $(1, 1)$
7. $(0, 0)$ and $(-5, 1)$
8. $(0, 0)$ and $(-4, 1)$
9. $(3, 5)$ and $(6, 2)$
10. $(4, 7)$ and $(8, 11)$
11. $(7, 6)$ and $(9, -2)$
12. $(5, 4)$ and $(9, -2)$
13. $(-3, 9)$ and $(3, -1)$
14. $(-1, 3)$ and $(2, -6)$
15. $(7.1, -8.3)$ and $(6.6, -9.5)$
16. $(5.9, -4.3)$ and $(6.7, -2.8)$
17. $(0, -\frac{1}{2})$ and $(\frac{1}{3}, -\frac{1}{4})$
18. $(-\frac{1}{4}, \frac{1}{3})$ and $(-\frac{1}{2}, 0)$
19. (a^2, ab) and $(b^2, -ab)$
20. $(-ab, a^2)$ and (ab, b^2)

Use the distance formula to find the distance between each pair of points. Then graph the points and determine the distance using the graph. Why isn't it necessary to use the distance formula for these problems?

21. $(5, 2)$ and $(8, 2)$
22. $(-3, 1)$ and $(5, 1)$
23. $(-4, -1)$ and $(-4, 6)$
24. $(-5, -9)$ and $(-5, -2)$

Find the slope of the line passing through each pair of points.

25. $(2, 4)$ and $(4, 8)$
26. $(2, 6)$ and $(3, 9)$
27. $(-4, 2)$ and $(-1, -16)$
28. $(-3, 2)$ and $(-1, -8)$
29. $(-1, -3)$ and $(7, 3)$
30. $(-4, -3)$ and $(8, 7)$
31. $(5, 9)$ and $(6, 9)$
32. $(4, -11)$ and $(5, -11)$
33. $(-7, 1)$ and $(-7, -6)$

34. $(1, -2)$ and $(1, 3)$ **35.** $(-3.8, 4.6)$ and $(2.9, 5.7)$ **36.** $(-4.2, 6.5)$ and $(1.9, 7.8)$

37. $\left(-\frac{1}{2}, \frac{11}{24}\right)$ and $\left(-\frac{1}{3}, 0\right)$ **38.** $\left(-\frac{1}{4}, 0\right)$ and $\left(-\frac{1}{2}, \frac{11}{12}\right)$ **39.** $(a + b, 1)$ and $(a - b, 3)$

40. $(6, a + b)$ and $(2, a - b)$

Find the slope of the line passing through each pair of points. Then graph the lines on the same coordinate axes.

41. a) $(0, 0), (1, 3)$ **b)** $(0, 0), (3, 3)$ **42. a)** $(0, 0), (1, 2)$ **b)** $(0, 0), (2, 2)$
     c) $(0, 0), (3, 1)$ **d)** $(0, 0), (-1, 3)$ **c)** $(0, 0), (2, 1)$ **d)** $(0, 0), (-1, 2)$
     e) $(0, 0), (-3, 3)$ **f)** $(0, 0), (-3, 1)$ **e)** $(0, 0), (-2, 2)$ **f)** $(0, 0), (-2, 1)$

43. a) $(-2, 2), (2, 2)$ **b)** $(-2, -2), (2, -2)$ **44. a)** $(-3, 3), (3, 3)$ **b)** $(-3, -3), (3, -3)$
     c) $(2, 2), (2, -2)$ **d)** $(-2, 2), (-2, -2)$ **c)** $(3, 3), (3, -3)$ **d)** $(-3, 3), (-3, -3)$

State the slope of each line graphed in Problems 45–52.

45.

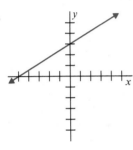

46.

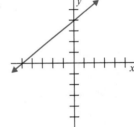

47.

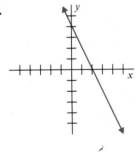

48.

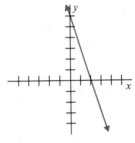

49.

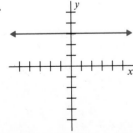

50.

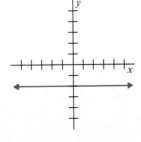

51.

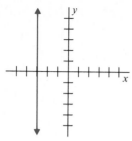

52.

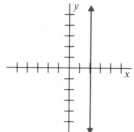

53. An isosceles triangle is a triangle with two equal sides. Show that the triangle with vertices at $A(-1, 3)$, $B(2, 6)$, and $C(3, 2)$ is an isosceles triangle.

54. An equilateral triangle is a triangle with three equal sides. Show that the triangle with vertices at $A(0, \sqrt{3})$, $B(1, 0)$, and $C(-1, 0)$ is an equilateral triangle.

55. Find x if the distance between $(3, 2)$ and $(x, 1)$ is $\sqrt{5}$.

56. Find y if the distance between $(1, 4)$ and $(5, y)$ is 5.

57. Suppose L_1 is the line through $(3, -9)$ and $(11, -15)$. Find the slope of line L_2 given that
a) L_2 is parallel to L_1.
b) L_2 is perpendicular to L_1.

58. Suppose L_1 is the line through $(2, -5)$ and $(12, -11)$. Find the slope of line L_2 given that
a) L_2 is parallel to L_1.
b) L_2 is perpendicular to L_1.

59. Suppose the slope of line L_1 is zero. What is the slope of line L_2 given that
a) L_2 is parallel to L_1?
b) L_2 is perpendicular to L_1?

60. Suppose the slope of line L_1 is undefined. What is the slope of line L_2 given that
a) L_2 is parallel to L_1?
b) L_2 is perpendicular to L_1?

61. A parallelogram is a four-sided plane figure with opposite sides parallel. Show that the four-sided figure with vertices at $A(1, 5)$, $B(3, 8)$, $C(4, 1)$, and $D(6, 4)$ is a parallelogram. Is it also a rectangle?

62. Show that the four-sided figure with vertices at $A(3, 2)$, $B(5, 5)$, $C(3, 8)$, and $D(1, 5)$ is a parallelogram (see Problem 61). Is it also a square?

63. Find t if the line through $(7, t^2)$ and $(4, 2t)$ is parallel to the line with slope 5.

64. Find t if the line through $(9, t^2)$ and $(2, 3t)$ is perpendicular to the line with slope $-\frac{1}{4}$.

† 65. Determine the total area of the roof in Figure 7.16, given that the slope of the roof is $\frac{5}{12}$.

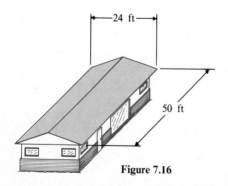

Figure 7.16

† 66. The grade of a road is the ratio of the change in elevation of the road to the corresponding horizontal change (see Figure 7.17). If you drive on a road that has a constant grade of 2%, by how much will you increase your elevation if your horizontal change is 6000 feet?

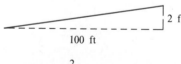

$$\text{Grade} = \frac{2}{100} = 0.02 = 2\%$$

Figure 7.17

CALCULATOR PROBLEMS

67. Determine the distance between (1.4, 10.6) and (−8.2, 5.9) to the nearest tenth of a unit.

68. Find the slope of the line passing through (20.288, 10.144) and (28.213, 4.755).

7.4   *Forms of a Linear Equation*

In Section 7.2 we stated that the standard form of a linear equation in two variables is

$$ax + by = c.$$

In this section we develop two other forms of a linear equation in two variables.

Consider the line with slope m that contains the point (x_1, y_1), as shown in Figure 7.18. If (x, y) is any other point on the line, then the slope of the line is

$$\frac{y - y_1}{x - x_1}.$$

Since the slope of the line is also m, we can write the equation

$$\frac{y - y_1}{x - x_1} = m.$$

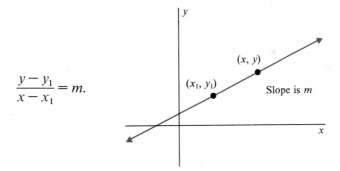

Figure 7.18

Multiplying each side of this last equation by $x - x_1$ gives the **point-slope form** of a linear equation.

Point-Slope Form

A line has slope m and passes through the point (x_1, y_1) if and only if its equation can be written as

$$y - y_1 = m(x - x_1).$$

EXAMPLE 1 Find an equation for the line that passes through $(-3, 4)$ and has slope 2. Then write the equation in standard form.

Use the point-slope form with $(x_1, y_1) = (-3, 4)$ and $m = 2$.

$$y - y_1 = m(x - x_1)$$
$$y - 4 = 2[x - (-3)]$$

Then write the equation in standard form.

$$y - 4 = 2(x + 3)$$
$$y - 4 = 2x + 6 \qquad \text{Distribute 2}$$
$$-2x + y = 10$$

Try Problem 3

$$2x - y = -10 \qquad \text{Multiply each side by } -1 \blacktriangleleft$$

EXAMPLE 2 Find an equation for the line through $(-5, -6)$ and $(1, -8)$. Then write the equation in standard form.

First find the slope of the line.

$$m = \frac{-8 - (-6)}{1 - (-5)} = \frac{-8 + 6}{1 + 5} = \frac{-2}{6} = -\frac{1}{3}$$

Then use the point-slope form with $(x_1, y_1) = (-5, -6)$ and $m = -\frac{1}{3}$.

$$y - (-6) = -\frac{1}{3}[x - (-5)]$$
$$y + 6 = -\frac{1}{3}(x + 5)$$
$$3y + 18 = -(x + 5) \qquad \text{Multiply each side by 3}$$
$$3y + 18 = -x - 5$$

Try Problem 19

$$x + 3y = -23 \blacktriangleleft$$

You should verify that choosing the other point $(1, -8)$ as (x_1, y_1) produces the same answer for Example 2.

We can use the point-slope form to develop another useful form of a linear equation. Consider the line with slope m and y-intercept b, as shown in Figure 7.19. (This is a different b from the one in $ax + by = c$.) Since the y-intercept is b, the point $(0, b)$ lies on the line. Using the point-slope form with $(x_1, y_1) = (0, b)$ and slope $= m$, we have

$$y - b = m(x - 0)$$
$$y - b = mx$$
$$y = mx + b.$$

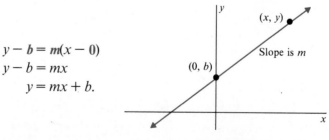

Figure 7.19

This equation is known as the **slope-intercept form** of a linear equation.

Slope-Intercept Form

A line has slope m and y-intercept b if and only if its equation can be written as

$$y = mx + b.$$

When an equation of a line is written in slope-intercept form, the coefficient of x is the slope of the line, and the constant term is the y-intercept. This means that we can graph the line without constructing a table of values.

EXAMPLE 3 Graph $y = 2x + 4$.

The slope is 2 and the y-intercept is 4.

$$y = 2\,x + 4$$
$$\overset{\nearrow}{\text{slope}} \qquad \overset{\nwarrow}{y\text{-intercept}}$$

To graph the line, plot the point $(0, 4)$, and then use the slope to locate a second point (see Figure 7.20).

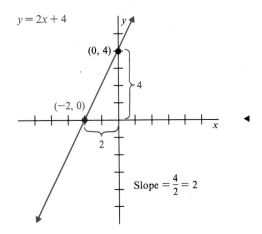

Figure 7.20

Try Problem 29

EXAMPLE 4 Determine the slope and the y-intercept of the line $2x - 3y = 3$. Then graph the line.

Solve for y to put the equation in slope-intercept form.

$$2x - 3y = 3$$

$$-3y = -2x + 3 \qquad \text{Subtract } 2x \text{ from each side}$$

$$y = \frac{-2x}{-3} + \frac{3}{-3} \qquad \text{Divide each side by } -3$$

$$y = \frac{2}{3}x - 1 \qquad \text{Simplify}$$

Therefore the slope is $\frac{2}{3}$ and the y-intercept is -1. To graph the line, plot $(0, -1)$ and then use the slope to locate a second point (see Figure 7.21).

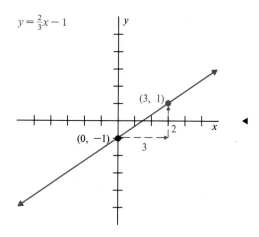

Try Problem 33

Figure 7.21

EXAMPLE 5 Find an equation for the line with y-intercept $\frac{4}{5}$ that is: **a)** parallel to $2x + 3y = 6$, **b)** perpendicular to $2x + 3y = 6$. Then write the equation in standard form.

First find the slope of $2x + 3y = 6$ by solving for y.

$$3y = -2x + 6 \qquad \text{Subtract } 2x$$

$$y = -\tfrac{2}{3}x + 2 \qquad \text{Divide by } 3$$

Therefore the slope is $-\frac{2}{3}$.

a) Use the slope-intercept form with $m = -\frac{2}{3}$ and $b = \frac{4}{5}$.

$$y = mx + b$$

$$y = -\tfrac{2}{3}x + \tfrac{4}{5}$$

Write in standard form.

$$15y = -10x + 12 \qquad \text{Multiply by } 15$$

$$10x + 15y = 12$$

b) Use the slope-intercept form with $m = \frac{3}{2}$ and $b = \frac{4}{5}$.

$$y = \tfrac{3}{2}x + \tfrac{4}{5}$$
$$10y = 15x + 8 \qquad \text{Multiply by 10}$$
$$-15x + 10y = 8$$
$$15x - 10y = -8 \qquad \text{Multiply by } -1 \blacktriangleleft$$

Try Problem 57

The various forms of a linear equation are summarized below.

Forms of a Linear Equation

Standard form	$ax + by = c$
Horizontal line	$y = k$
Vertical line	$x = k$
Point-slope form	$y - y_1 = m(x - x_1)$
Slope-intercept form	$y = mx + b$

PROBLEM SET 7.4

Find an equation for the line that passes through the given point and has the given slope. Then write the equation in standard form.

1. $(2, 5)$, $m = -3$

2. $(1, 4)$, $m = -3$

3. $(-6, 7)$, $m = 2$

4. $(-5, 8)$, $m = 2$

5. $(3, -1)$, $m = \frac{1}{4}$

6. $(6, -2)$, $m = \frac{1}{4}$

7. $(-4, -8)$, $m = 1$

8. $(-3, -7)$, $m = 1$

9. $(0, 10)$, $m = -\frac{5}{6}$

10. $(0, 11)$, $m = -\frac{5}{6}$

11. $(4, 2)$, $m = 0$

12. $(5, 3)$, $m = 0$

13. $(1, 3)$, m is undefined

14. $(4, 1)$, m is undefined

Find an equation for the line through each pair of points. Then write the equation in standard form.

15. $(7, 5)$, $(8, 8)$

16. $(4, 3)$, $(5, 5)$

17. $(0, 0)$, $(6, -6)$

18. $(0, 0)$, $(7, -7)$

19. $(-4, -7)$, $(2, -9)$

20. $(-2, -9)$, $(4, -12)$

21. $(3, 7)$, $(6, 7)$

22. $(2, 9)$, $(5, 9)$

23. $(-8, 5)$, $(-8, -2)$

24. $(-6, 4)$, $(-6, -1)$

25. $(1, \sqrt{3})$, $(5, 3\sqrt{3})$

26. $(2, -\sqrt{5})$, $(6, \sqrt{5})$

Determine the slope and the y-intercept of each line. Then graph the line.

27. $y = x + 1$

28. $y = x + 2$

29. $y = 2x + 6$

30. $y = 3x + 6$

31. $4x + y = 2$

32. $5x + y = 3$

33. $2x - 3y = 6$

34. $3x - 4y = 8$

35. $x + 2y = 3$ **36.** $x + 2y = 5$ **37.** $4y - 3x = 0$ **38.** $5y - 4x = 0$

39. $y - 3 = 0$ **40.** $y - 4 = 0$

Write an equation for each line graphed in Problems 41–44.

41.

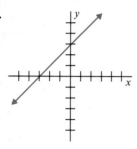

42.

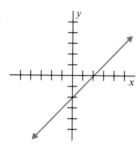

43.

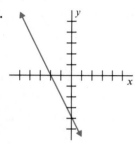

44.

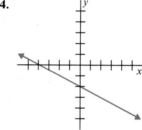

45. What are the slope and the y-intercept of the line $x = 5$?

46. What are the slope and the y-intercept of the line $x = -4$?

47. Using the same coordinate axes, graph $y = mx$ for
  a) $m = \frac{1}{2}$. **b)** $m = 1$. **c)** $m = 2$.

48. Using the same coordinate axes, graph $y = mx$ for
  a) $m = -\frac{1}{2}$. **b)** $m = -1$. **c)** $m = -2$.

49. Using the same coordinate axes, graph $y = -x + b$ for
  a) $b = -2$. **b)** $b = 0$. **c)** $b = 2$.

50. Using the same coordinate axes, graph $y = x + b$ for
  a) $b = -2$. **b)** $b = 0$. **c)** $b = 2$.

Determine whether the lines with the given equations are parallel, perpendicular, or neither.

51. $y - 3x = 0$
  $6x - 2y = 5$

52. $y - x = 2$
  $3x - 3y = 7$

53. $y = 4x + 1$
  $3x + 12y = 8$

54. $y = 5x - 1$
  $2x + 10y = 0$

55. $4x + 5y = 15$
  $4x - 5y = -10$

56. $5x + 6y = 18$
  $5x - 6y = -12$

57. Find an equation for the line with y-intercept $\frac{2}{5}$ that is
  a) parallel to $2x + 3y = 12$.
  b) perpendicular to $2x + 3y = 12$.
  Then write the equation in standard form.

58. Find an equation for the line with y-intercept $\frac{3}{7}$ that is
  a) parallel to $3x + 4y = 20$.
  b) perpendicular to $3x + 4y = 20$.
  Then write the equation in standard form.

† **59.** Find an equation for the line with x-intercept 7 that is
    a) parallel to $2x + 5y = 9$.
    b) perpendicular to $2x + 5y = 9$.

† **60.** The relationship between the Fahrenheit temperature F and the Celsius temperature C is linear. Given that water freezes at $0°C$ or $32°F$ and boils at $100°C$ or $212°F$, find an equation that describes that relationship.

CALCULATOR PROBLEMS

61. Find an equation for the line through (4.71, 6.98) and (2.91, 1.796).

62. Determine whether the lines $1.4579x + 3.05y = 26.047$ and $2.99375x + 6.25y = 26.047$ are parallel, perpendicular, or neither.

7.5   *Graphing Linear Inequalities*

If the equals sign in a linear equation is replaced by an inequality sign, the result is a **linear inequality.** Therefore

$$x > 2, \qquad y \le 4, \quad \text{and} \quad x + y < 3$$

are linear inequalities.

EXAMPLE 1   Graph $x > 2$ in the xy-plane.

The solution to this inequality consists of all those ordered pairs (x, y) whose x-value is greater than 2. We graph this solution by shading the region to the right of the vertical line $x = 2$ (see Figure 7.22). The boundary line $x = 2$ is

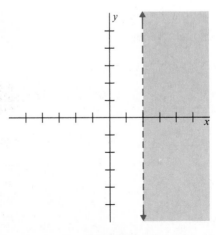

Figure 7.22

Try Problem 1

shown as a dashed line to signify that points on this line are not part of the solution (since the inequality sign is $>$ and not $\geq$). ◄

Thus just as a cut across a piece of paper divides the paper into two pieces, so each line in a plane divides the plane into two pieces, called **half planes.** A half plane that includes its boundary line is called a **closed half plane.** A half plane that does not include its boundary line is an **open half plane.** The graph in Figure 7.22 is an open half plane.

EXAMPLE 2   Graph $y \leq 4$ in the xy-plane.

The graph is the closed half plane that lies below and on the horizontal line $y = 4$ (see Figure 7.23). The boundary line $y = 4$ is solid, since the inequality sign is $\leq$ and not $<$.

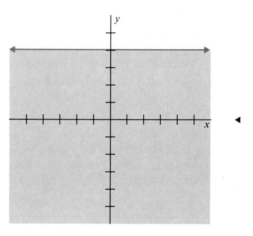

Try Problem 9

Figure 7.23

The graph of a linear inequality in *two* variables is also a half plane.

EXAMPLE 3   Graph $x + y < 3$.

First we graph the boundary line $x + y = 3$ as shown in Figure 7.24. This line divides the plane into two half planes, one above the line and one below. We must now determine which half plane corresponds to the original inequality $x + y < 3$. We do this by choosing a test point that does *not* lie on the boundary line itself. If the test point satisfies the original inequality, then the half plane that contains the test point represents the solution. If the test point does not

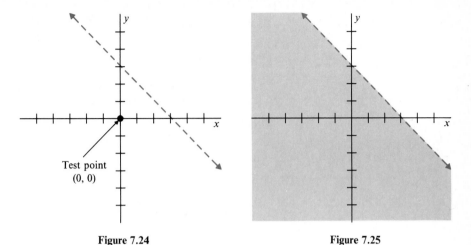

Figure 7.24   Figure 7.25

satisfy the original inequality, then the half plane that does not contain the test point represents the solution.

We choose $(0, 0)$ as our test point to make the arithmetic easy, but any point not on the boundary line may be used. We then substitute $(0, 0)$ for (x, y) in the original inequality.

$$x + y < 3$$
$$0 + 0 < 3$$
$$0 < 3 \qquad \text{True}$$

Try Problem 23

Since $(0, 0)$ satisfies $x + y < 3$, we shade the open half plane that contains $(0, 0)$ as shown in Figure 7.25. ◄

If the boundary line passes through the point $(0, 0)$, we cannot use $(0, 0)$ as our test point.

EXAMPLE 4   Graph $y \geq 2x$.

First graph the boundary line $y = 2x$ as shown in Figure 7.26. Then substitute a test point into the original inequality. We cannot use $(0, 0)$, since it lies on the boundary line, so we use $(4, 0)$.

$$y \geq 2x$$
$$0 \geq 2(4)$$
$$0 \geq 8 \qquad \text{False}$$

Since $(4, 0)$ does *not* satisfy $y \geq 2x$, shade the closed half plane that does *not* contain $(4, 0)$ as shown in Figure 7.27.

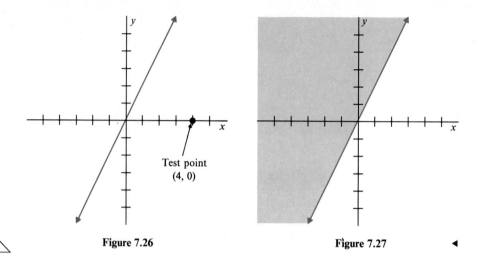

Try Problem 29   **Figure 7.26**   **Figure 7.27**   ◄

We can summarize our strategy for graphing a linear inequality as follows:

To Graph a Linear Inequality

1. Graph the boundary line obtained by replacing the inequality sign with an equals sign. Use a solid line if the inequality sign is ≤ or ≥. Use a dashed line if the inequality sign is < or >.
2. Choose a test point that is *not* on the boundary line and substitute it into the original inequality.
3. If the result of step 2 is a *true* statement, shade the half plane that *contains* the test point. If the result of step 2 is a *false* statement, shade the half plane that does *not* contain the test point.

The next example illustrates that the graph of an absolute-value inequality may be a portion of a plane.

EXAMPLE 5   Graph $|x| < 3$ in the xy-plane.

From Section 3.6 we know that $|x| < 3$ is equivalent to $-3 < x < 3$. Therefore the graph consists of all those points between the two vertical boundary lines $x = -3$ and $x = 3$ (see Figure 7.28).

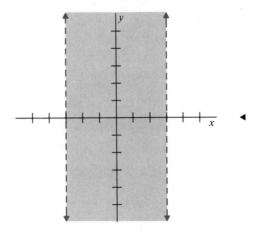

Try Problem 39

Figure 7.28

Problem Solving

Diet Preparation

Suppose that an egg contains 3 grams of fat, and a strip of bacon contains 6 grams of fat. For breakfast, a hospital patient consumes x eggs and y strips of bacon. Then the total number of grams of fat consumed is given by the expression below.

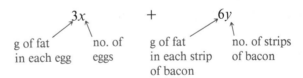

Suppose the patient is to consume no more than 18 grams of fat at breakfast. Then

$$3x + 6y \leq 18.$$

The graph of this inequality is the half plane shown in Figure 7.29. But neither x nor y can be negative. Therefore only that portion of the half plane shown in Figure 7.30 applies in this situation. Any ordered pair taken from the shaded region in Figure 7.30 represents a breakfast possibility. For example, $(4, \frac{1}{2})$ represents 4 eggs and $\frac{1}{2}$ strip of bacon, for a total of

$$3(4) + 6(\tfrac{1}{2}) = 12 + 3 = 15 \text{ grams of fat.}$$

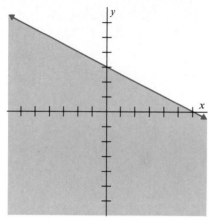

Figure 7.29

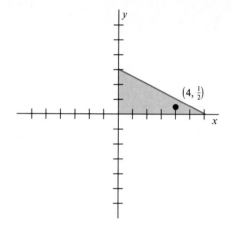

Figure 7.30

PROBLEM SET 7.5

Graph each inequality in the xy-plane.

1. $x > 3$
2. $x > 1$
3. $x \geq 3$
4. $x \geq 1$

5. $x < 3$
6. $x < 1$
7. $x \leq 3$
8. $x \leq 1$

9. $y \leq 2$
10. $y \leq 3$
11. $y + 1 > 0$
12. $y + 4 > 0$

13. $2x + 3 \leq 0$
14. $2x + 5 \leq 0$
15. $x \geq 0$
16. $x \leq 0$

17. $y < 0$
18. $y > 0$

Graph each inequality.

19. $y < x + 1$
20. $y < x + 3$
21. $y < 2x - 3$
22. $y < 2x - 1$

23. $x + y < 4$
24. $x + y < 5$
25. $3x - 4y \geq 12$
26. $4x - 5y \geq 20$

27. $2x - 5y \leq 10$
28. $2x - 3y \leq 6$
29. $y \geq 3x$
30. $y \geq 4x$

31. $x - y < 0$
32. $x - y > 0$
33. $x + 2y \leq 0$
34. $x + 3y \geq 0$

Graph each inequality in the xy-plane.

35. $-2 < x \leq 5$
36. $-3 < x \leq 4$
37. $-1 \leq y < 3$
38. $-2 \leq y < 1$

39. $|x| < 1$
40. $|x| < 2$
41. $|x| > 1$
42. $|x| > 2$

43. $|y| \leq 3$
44. $|y| \leq 4$
45. $|y| \geq 3$
46. $|y| \geq 4$

47. $|x - 2| \leq 2$
48. $|x - 3| \leq 3$
49. $|2y + 1| > 5$
50. $|2y + 1| > 3$

51. Suppose a slice of wheat bread contains 4 grams of protein, and a meatball contains 8 grams of protein. For dinner, a hospital patient consumes x slices of bread and y meatballs. The patient is to consume no more than 24 grams of protein at dinner. Write an inequality that describes this situation. Then graph the inequality.

52. Repeat Problem 51 if the patient is to consume at least 24 grams of protein at dinner.

Graph all ordered pairs (x, y) that satisfy the given conditions.

53. $xy \geq 0$

54. $|x| \leq 2$ and $|y| \leq 1$

CALCULATOR PROBLEMS

Graph each inequality in the xy-plane.

55. $18.45x - 6.15y \leq 51.66$

56. $|3.42x - 10.26| < 5.13$

CHAPTER 7 REVIEW

[7.1] *Graph each point in a rectangular coordinate system and state the quadrant in which it lies.*

1. $(3, -4)$

2. $(0, 7)$

3. $(-6, -3)$

4. $(-4, 0)$

5. Graph all points whose first coordinate is -2.

6. Graph all points whose second coordinate is 0.

[7.2] *Graph each equation. Label the x- and the y-intercept if they exist.*

7. $x + y = 2$

8. $3y - x = 9$

9. $y = 3x$

10. $3x - 5y = 15$

11. $y = -6$

12. $3x - 9 = 0$

13. $y = -x$

14. $y = \frac{1}{2}x - 2$

[7.3] *Find the distance between each pair of points.*

15. $(2, 8)$ and $(7, -4)$

16. $(1, -1)$ and $(-4, 3)$

Find the slope of the line passing through each pair of points.

17. $(3, 8)$ and $(1, 2)$

18. $(3, -2)$ and $(-6, 4)$

19. $(-4, 7)$ and $(-4, 4)$

20. $(0, \frac{1}{2})$ and $(-\frac{1}{3}, 0)$

State the slope of each line graphed below.

21.

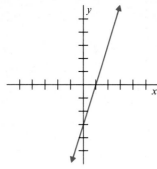

22.

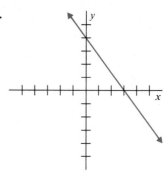

23. Determine whether the line passing through $(-4, 2)$ and $(2, -4)$ is parallel, perpendicular, or neither to the line that passes through $(-2, 3)$ and $(1, 6)$.

24. Show that the four-sided figure with vertices at $A(4, 3)$, $B(-8, 5)$, $C(-6, -2)$, and $D(6, -4)$ is a parallelogram.

[7.4] *Find an equation for the line that passes through the given point and has the given slope. Then write the equation in standard form.*

25. $(2, -3)$, $m = -1$

26. $(-2, 5)$, $m = \frac{3}{4}$

27. $(2, 4)$, $m = 0$

28. $(4, 2)$, m is undefined

Find an equation for the line through each pair of points. Then write the equation in standard form.

29. $(-4, 2)$, $(2, 8)$

30. $(2, 3)$, $(-6, -6)$

31. $(0, 0)$, $(-4, 4)$

Determine the slope and the y-intercept of each line. Then graph the line.

32. $y = 2x - 3$

33. $y = -3x + 4$

34. $3x + 5y = 15$

35. $y + 2 = 0$

36. $7x - 3y = 0$

37. $-y = 2x + 3$

Write an equation for each line graphed in Problems 38–41.

38.

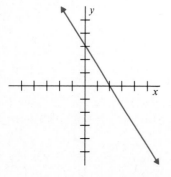

39.

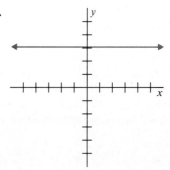

40.

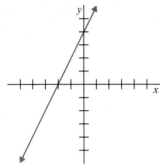

41.

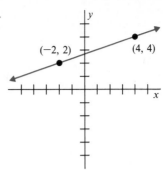

(−2, 2)   (4, 4)

Determine whether the lines with the given equations are parallel, perpendicular, or neither.

42. $y - 2x = 0$
    $x + 2y = 4$

43. $x + y = 2$
    $3x + 3y = 5$

[7.5]   *Graph each inequality in the xy-plane.*

44. $x \geq -2$ **45.** $y < 3$ **46.** $3x - 5y \leq 15$

47. $y < 2x + 1$ **48.** $y \geq x + 2$ **49.** $x + y \leq 0$

50. $|x| \leq 2$ **51.** $|y| > 1$ **52.** $|2x + 1| < 5$

CHAPTER 7 TEST

1. Graph each point in the same rectangular coordinate system.
    a) $(3, 4)$ **b)** $(-3, 0)$ **c)** $(0, 5)$ **d)** $(-2, -3)$

Graph each equation. Label the x- and the y-intercept.

2. $y - x = 1$ **3.** $y = x$ **4.** $2x - 3y = 12$ **5.** $y = -\frac{1}{2}x$

Graph each equation in the xy-plane.

6. $x = -5$ **7.** $y - 7 = 0$

8. Find the distance between $(-4, 3)$ and $(-7, -2)$.   **9.** Find the slope of the line passing through $(3, -4)$ and $(9, -1)$.

10. Find an equation for the line that passes through $(-2, 2)$ and has slope 3. Then write the equation in standard form.

11. Find an equation for the line with y-intercept -3 that is parallel to $3x - 2y = 4$. Then write the equation in standard form.

12. Find an equation for the line through $(3, 4)$ and $(-1, -2)$. Then write the equation in standard form.

Graph each inequality in the xy-plane.

13. $x - y \geq 2$ **14.** $y < x$ **15.** $x > -1$ **16.** $|y - 1| < 3$

17. Determine the slope and the y-intercept of the line $3x + 2y = 12$.

Write an equation for each line graphed in Problems 18 and 19.

18.

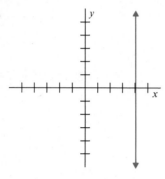

19.

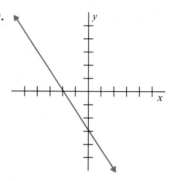

20. Determine whether the lines with the given equations are parallel, perpendicular, or neither.

$$y = 3x - 4$$
$$x + 3y = 6$$

Suppose that a shirt and tie together cost $24, and the shirt costs $9 more
than the tie. If x is the cost of the shirt and y is the cost of the tie, then
we can write the two equations below.

$$x + y = 24$$
$$x - y = 9$$

When two or more equations are true simultaneously, they form a *system
of equations.* When the equations are linear, as these are, they form a
linear system. In this chapter we discuss three methods for solving a
linear system — by graphing, by substitution, and by addition. You will
see how systems of equations can be used to solve problems ranging from
diet preparation to finding the equilibrium point in economics. We
conclude the chapter by introducing a mathematical tool known as a
determinant, and we show how determinants can be used to solve a linear
system.

8

Systems of Linear Equations

8.1   The Graphing Method

A **solution** to a system of two equations in the variables x and y is an ordered pair of values (x, y) that satisfies both equations in the system.

EXAMPLE 1   Determine whether $(5, 7)$ is a solution of the following system.

$$y = 2x - 3$$
$$y = x + 2$$

Substitute $x = 5$ and $y = 7$ in each equation.

$$\begin{array}{ll} y = 2x - 3 & y = x + 2 \\ 7 \stackrel{?}{=} 2 \cdot 5 - 3 & 7 \stackrel{?}{=} 5 + 2 \\ 7 \le 7 & 7 \le 7 \end{array}$$

Try Problem 1 \\   Since $(5, 7)$ satisfies both equations, it is a solution of the system. ◄

EXAMPLE 2   Determine whether $(6, -1)$ is a solution of the following system.

$$3x - 4y = 22$$
$$x + 2y = 8$$

Substitute $(x, y) = (6, -1)$ into both equations.

$$\begin{array}{ll} 3x - 4y = 22 & x + 2y = 8 \\ 3(6) - 4(-1) \stackrel{?}{=} 22 & 6 + 2(-1) \stackrel{?}{=} 8 \\ 18 + 4 \le 22 & 6 - 2 \ne 8 \end{array}$$

Try Problem 5 \\   Since $(6, -1)$ does not satisfy *both* equations, it is *not* a solution of the system. ◄

One way to solve a system of equations is to graph both equations using the same coordinate axes and then identify any point of intersection.

EXAMPLE 3   Solve the following system by graphing.

$$x + y = 4$$
$$x - y = 6$$

Each equation has an infinite number of solutions. Each solution is represented by a point on one of the lines in Figure 8.1. But since the intersection point $P(5, -1)$ is the only point that lies on both lines, $(5, -1)$ is the only ordered pair that satisfies both equations. Checking in both equations verifies that $(5, -1)$ is the solution of the system.

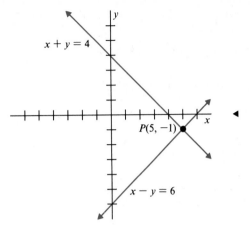

Figure 8.1

Try Problem 7

EXAMPLE 4   Solve the following system by graphing.

$$2x - y = -1$$
$$4x - 2y = 6$$

The graphs of the two equations are parallel lines, as shown in Figure 8.2. If we write each equation in slope-intercept form, we see that the lines have the same slope but different y-intercepts.

$$2x - y = -1 \qquad\qquad 4x - 2y = 6$$
$$-y = -2x - 1 \qquad\qquad -2y = -4x + 6$$
$$y = 2x + 1 \qquad\qquad y = 2x - 3$$

Since there is no intersection point, there is no solution to the system.

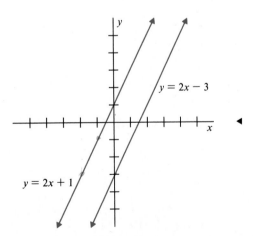

Try Problem 15

Figure 8.2

EXAMPLE 5   Solve the following system by graphing.

$$6x + 6y = 12$$
$$5x + 5y = 10$$

The graphs of the two equations coincide, as shown in Figure 8.3. Writing each equation in slope-intercept form reveals that the equations are equivalent.

$$6x + 6y = 12 \qquad\qquad 5x + 5y = 10$$
$$6y = -6x + 12 \qquad\qquad 5y = -5x + 10$$
$$y = -x + 2 \qquad\qquad y = -x + 2$$

Any ordered pair that satisfies one equation also satisfies the other. Therefore there are an infinite number of solutions, namely all those ordered pairs that satisfy $y = -x + 2$.

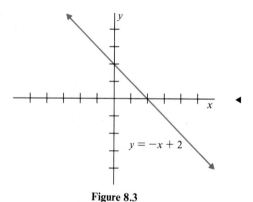

$y = -x + 2$

Try Problem 17

Figure 8.3

Examples 3, 4, and 5 illustrate that there are three possibilities for the solution to a linear system of two equations. These possibilities are categorized in Figure 8.4.

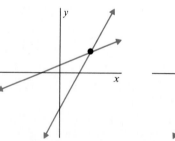

a)   Independent system
One solution

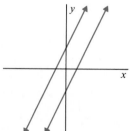

b)   Inconsistent system
No solution

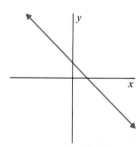

c)   Dependent system
Infinite number
of solutions

Figure 8.4

We can summarize the graphing method as follows:

To Solve a Linear System of Two Equations by Graphing

1. Graph both equations using the same coordinate axes. You may want to write each equation in slope-intercept form to determine whether the lines are intersecting, parallel, or identical.
2. If the lines intersect (different slopes), the system is **independent** and has a unique solution. The coordinates of the intersection point give the solution.
3. If the lines are parallel (same slope but different y-intercepts), the system is **inconsistent**. There is no solution to the system.
4. If the lines coincide (same slope and same y-intercept), the system is **dependent**. Any point on the line is a solution.

PROBLEM SET 8.1

Determine whether the given ordered pair is a solution of the given system.

1. $(4, 7)$ $y = 2x - 1$
   $y = x + 3$

2. $(3, 5)$ $y = 3x - 4$
   $y = x + 2$

3. $(-2, -3)$ $4x - y = -11$
   $2x - 5y = 19$

4. $(-1, -4)$ $5x - y = -9$
   $3x - 6y = 27$

5. $(5, -1)$ $3x - 7y = 22$
   $x + 2y = 7$

6. $(6, -2)$ $2x - 7y = 26$
   $x + 4y = 14$

Solve each system by graphing. Classify each system as independent, inconsistent, or dependent.

7. $x + y = 2$
   $x - y = 4$

8. $x + y = 3$
   $x - y = 5$

9. $y = -x$
   $x = -3$

10. $y = x$
    $x = 3$

11. $y = \frac{1}{2}x$
    $y = 2$

12. $y = -2x$
    $y = -4$

13. $x + y = 2$
    $y = 3$

14. $x + y = 3$
    $y = 4$

15. $2x - y = -2$
    $6x - 3y = 12$

16. $x - 2y = -2$
    $5x - 10y = 20$

17. $4x - 4y = 8$
    $7x - 7y = 14$

18. $3x + 3y = 12$
    $2x + 2y = 8$

19. $x + y = 4$
    $7x - 5y = 10$

20. $x - y = 5$
    $9x + 7y = 21$

21. Two numbers, x and y, have a sum of 4 and a difference of 6. Write a system of equations involving x and y, and solve by graphing.

22. Two numbers, x and y, have a sum of 5 and a difference of 7. Write a system of equations involving x and y, and solve by graphing.

CALCULATOR PROBLEMS

23. Determine whether $(5.13, -9.67)$ is a solution of the following system.

$$13x + 8y = -10.67$$
$$7x - 3y = 64.92$$

24. Classify the following system as independent, inconsistent, or dependent.

$$12.03x + 1.5y = 6.6$$
$$17.644x + 2.2y = 9.9$$

8.2   *The Substitution Method and the Addition Method*

One major disadvantage of the graphing method for solving a system of equations is that it may be difficult to determine the exact coordinates of the intersection point. This is particularly true when the coordinates are not integers. Therefore in this section we present two algebraic methods for solving a system —the substitution method and the addition method.

THE SUBSTITUTION METHOD

In the **substitution method** we solve either equation for either of its variables. Then we substitute the result into the other equation. This produces an equation in one variable.

EXAMPLE 1   Solve the following system using the substitution method.

$$x + 5y = 17$$
$$2x - 3y = -18$$

We can solve either equation for either of its variables. To avoid fractions, we solve the first equation for x.

$$x = 17 - 5y$$

Then we substitute $17 - 5y$ for x in the second equation.

$$2(17 - 5y) - 3y = -18$$

This is an equation in one variable and is easily solved.

$$34 - 10y - 3y = -18$$
$$34 - 13y = -18$$
$$-13y = -52$$
$$y = 4$$

To find x, we substitute 4 for y in the equation $x = 17 - 5y$. (The substitution can also be made in either of the original equations.)

$$x = 17 - 5 \cdot 4$$
$$x = -3$$

Try Problem 9

The ordered pair $(-3, 4)$ checks in both equations, so the solution is $(-3, 4)$. ◄

EXAMPLE 2   Solve the following system by substitution.

$$6x - 3y = 0$$
$$y = 2x$$

The second equation is already solved for y, so simply replace y in the first equation by $2x$.

$$6x - 3(2x) = 0$$
$$6x - 6x = 0$$
$$0 = 0$$

Try Problem 15

The fact that we arrived at an identity means the original system is dependent. Therefore any ordered pair satisfying $y = 2x$ is a solution. ◄

EXAMPLE 3   Solve the following system by substitution.

$$12x - 4y = 11$$
$$-6x + 2y = 7$$

In this case we cannot avoid fractions when solving either equation for either variable. We decide to solve the second equation for y.

$$2y = 7 + 6x$$
$$y = \tfrac{7}{2} + 3x$$

Then we substitute $\tfrac{7}{2} + 3x$ for y in the first equation.

$$12x - 4(\tfrac{7}{2} + 3x) = 11$$
$$12x - 14 - 12x = 11$$
$$-14 = 11$$

Try Problem 17

The fact that we arrived at a contradiction means the original system is inconsistent. Therefore there is no solution. ◄

We can summarize the substitution method as follows:

To Solve a Linear System of Two Equations by Substitution

1. Solve either equation for either variable.
2. Substitute the result of step 1 into the other equation.

3. Solve the equation in one variable that results from step 2. If a contradiction results, the system is inconsistent. If an identity results, the system is dependent.
4. Substitute the result from step 3 into the equation resulting from step 1 to find the value of the other variable.
5. Check your solution in both of the original equations.

THE ADDITION METHOD

We can avoid fractions when solving a system by using the **addition method.** In this method we add corresponding sides of the two equations with the intent of eliminating one of the variables.

EXAMPLE 4 Solve the following system using the addition method.

$$2x + y = 13$$
$$x - y = 5$$

If we add corresponding sides of the two equations, the variable y is eliminated.

$$\begin{array}{r} 2x + y = 13 \\ \underline{x - y = 5} \\ 3x = 18 \end{array}$$

Dividing each side of the equation $3x = 18$ by 3 gives

$$x = 6.$$

To find y, replace x with 6 in either of the original equations. If we use the second equation, we have

$$6 - y = 5$$
$$-y = -1$$
$$y = 1.$$

Try Problem 21

The solution is (6, 1). Check in both of the original equations. ◄

EXAMPLE 5 Solve the following system by addition.

$$-x + 4y = 5$$
$$2x - 3y = -10$$

Adding the equations will not eliminate x or y. However, if we multiply the first equation by 2, then x will be eliminated when the equations are added.

$$\begin{array}{r} -2x + 8y = 10 \\ \underline{2x - 3y = -10} \\ 5y = 0 \\ y = 0 \qquad \text{Divide each side by 5} \end{array}$$

To find x, substitute $y = 0$ into either of the original equations. Using the first equation, we have

$$-x + 4 \cdot 0 = 5$$
$$-x = 5$$
$$x = -5.$$

Try Problem 25

The solution is $(-5, 0)$. Check in both of the original equations. ◄

EXAMPLE 6   Solve the following system by addition.

$$\frac{x}{4} - \frac{y}{6} = -\frac{1}{12}$$
$$4x - 5y + 6 = 0$$

Multiply the first equation by 12 to clear the fractions.

$$3x - 2y = -1$$

Subtract 6 from each side of the second equation to put the equation in standard form.

$$4x - 5y = -6$$

In this case, both equations must be multiplied by a suitable number so that either x or y is eliminated when the equations are added. We decide to multiply $3x - 2y = -1$ by 4 and $4x - 5y = -6$ by -3.

$$\begin{array}{r} 12x - 8y = -4 \\ -12x + 15y = 18 \\ \hline 7y = 14 \\ y = 2 \end{array} \quad \text{Add the two equations}$$

Substitute $y = 2$ into the second of the original equations.

$$4x - 5(2) + 6 = 0$$
$$4x - 4 = 0$$
$$4x = 4$$
$$x = 1$$

Try Problem 35

The solution is $(1, 2)$. Check in both of the original equations. ◄

To Solve a Linear System of Two Equations by Addition

1.   Write each equation in standard form with integer coefficients.
2.   If necessary, multiply one or both equations by a suitable number so that one variable is eliminated when the equations are added.
3.   Add the two equations and solve the resulting equation in one variable. If a contradiction results, the system is inconsistent. If an identity results, the system is dependent.

4. Substitute the result from step 3 into either of the original equations to find the value of the other variable.

5. Check your solution in both of the original equations.

PROBLEM SET 8.2

Solve each system using the substitution method.

1. $5x + y = 6$
   $y = x$

2. $7x + y = 8$
   $y = x$

3. $\quad x = y + 1$
   $x + 3y = 9$

4. $\quad x = y + 2$
   $x + 2y = 8$

5. $2x + 5y = 18$
   $y = 2$

6. $3x + 4y = 19$
   $y = 1$

7. $\quad x = -1$
   $6x + 5y = -6$

8. $\quad x = -2$
   $2x + 5y = -4$

9. $\quad x + 4y = 16$
   $3x - 2y = -22$

10. $\quad x + 3y = 16$
    $4x - 5y = -38$

11. $2x - y = 5$
    $4x + y = 7$

12. $2x - y = 7$
    $6x + y = 17$

13. $2x + 6y = 5$
    $4x = 3y$

14. $3x + 10y = 5$
    $6x = 5y$

15. $8x - 2y = 0$
    $y = 4x$

16. $12x - 4y = 0$
    $y = 3x$

17. $24x - 6y = 13$
    $-8x + 2y = 9$

18. $24x - 8y = 17$
    $-6x + 2y = 5$

19. $5x + 4y = 2$
    $2x + 3y = 3$

20. $5x + 4y = 1$
    $6x + 3y = 2$

Solve each system using the addition method.

21. $3x + y = 7$
    $x - y = 1$

22. $\quad x + y = 9$
    $4x - y = 6$

23. $3x - 6y = 1$
    $5x + 6y = -9$

24. $2x - 8y = 2$
    $9x + 8y = -13$

25. $-x + 4y = 4$
    $2x - 5y = -8$

26. $-x + 4y = -3$
    $2x - 3y = 6$

27. $2x + 3y = 11$
    $4x + 5y = 19$

28. $3x + 4y = 19$
    $6x + 2y = 32$

29. $7x - 5y = 2$
    $6x + 2y = -3$

30. $3x - 5y = 6$
    $7x + 3y = -8$

31. $-x + 2y = 3$
    $3x - 6y = -9$

32. $\quad x - 3y = 2$
    $-4x + 12y = -8$

33. $-9x + 15y = 3$
    $6x - 10y = -1$

34. $\quad 8x - 10y = 2$
    $-12x + 15y = -1$

35. $\dfrac{x}{4} - \dfrac{y}{6} = \dfrac{1}{4}$
    $4x - 5y + 10 = 0$

36. $\dfrac{x}{3} - \dfrac{y}{4} = \dfrac{1}{3}$
    $5x - 2y - 19 = 0$

Set up a system of equations to solve each word problem.

37. Two numbers, x and y, have a sum of 23 and a difference of 9. Find the numbers.

38. Two numbers, x and y, have a sum of 21 and a difference of 5. Find the numbers.

39. A number x added to twice a number y is 18. Find the numbers if x is 3 more than y.

40. A number x added to three times a number y is 22. Find the numbers if x is 2 less than y.

Solve each system by first making the substitutions $u = \dfrac{1}{x}$ and $v = \dfrac{1}{y}$.

† **41.** $\dfrac{1}{x} + \dfrac{1}{y} = 3$

$\dfrac{2}{x} - \dfrac{1}{y} = 3$

† **42.** $\dfrac{3}{x} + \dfrac{4}{y} = \dfrac{5}{2}$

$\dfrac{5}{x} - \dfrac{3}{y} = \dfrac{7}{4}$

Solve each system for the variables x and y in terms of the constants a, b, and c.

† **43.** $5ax - y = 6$

$y = 2ax$

† **44.** $ax + by = c$

$x - y = 0$

CALCULATOR PROBLEMS

Solve each system. Give the value of each variable to the nearest hundredth.

45. $4.31x + 5.87y = 2.55$

$y = 1.39x - 8.91$

46. $29.173x + 37.986y = 88.736$

$42.654x - 18.112y = 14.504$

8.3   *Word Problems*

Many word problems with two unknown quantities are easier to solve by using two variables rather than one variable. Of course, when two variables are used, a system of two equations must be written to determine the solution to the problem.

To Solve a Word Problem Using a System of Equations

1. Write down the quantities to be determined, and represent each by a different variable.
2. Write a system of equations involving the variables.
3. Solve the system.
4. Check your solution in the original problem.

We illustrate these steps in Example 1.

EXAMPLE 1   Suppose each slice of cheese contains 5 grams of protein and 3 grams of fat, and each slice of wheat bread contains 6 grams of protein and 2 grams of fat. How many slices of each food should be used to prepare a lunch containing 32 grams of protein and 16 grams of fat?

Step 1   Write down the quantities to be determined, and represent each by a different variable.

$$x = \text{number of slices of cheese}$$
$$y = \text{number of slices of bread}$$

Step 2   Write a system of equations involving the variables. For the first equation we write a statement about protein.

$$
\begin{array}{c}
\text{No. of g of} \\
\text{protein in } x \\
\text{slices of cheese}
\end{array}
\;+\;
\begin{array}{c}
\text{No. of g of} \\
\text{protein in } y \\
\text{slices of bread}
\end{array}
\;=\; 32
$$

$$
\downarrow \qquad\qquad \downarrow
$$

$$5x \quad + \quad 6y \quad = 32$$

For the second equation we write a statement about fat.

$$
\begin{array}{c}
\text{No. of g of} \\
\text{fat in } x \\
\text{slices of cheese}
\end{array}
\;+\;
\begin{array}{c}
\text{No. of g of} \\
\text{fat in } y \\
\text{slices of bread}
\end{array}
\;=\; 16
$$

$$
\downarrow \qquad\qquad \downarrow
$$

$$3x \quad + \quad 2y \quad = 16$$

Therefore the following system describes the original problem.

$$5x + 6y = 32$$
$$3x + 2y = 16$$

Step 3   Solve the system. Multiply the second equation by -3, and add the result to the first equation.

$$
\begin{array}{rl}
5x + 6y = & 32 \\
-9x - 6y = & -48 \\
\hline
-4x \qquad = & -16 \\
x = & 4
\end{array}
$$

Substitute $x = 4$ into the first equation to obtain the value of y.

$$5 \cdot 4 + 6y = 32$$
$$6y = 12$$
$$y = 2$$

The lunch should consist of 4 slices of cheese and 2 slices of bread.

Step 4   Check your solution in the original problem. Four slices of cheese contain $4 \cdot 5 = 20$ grams of protein and $4 \cdot 3 = 12$ grams of fat. Two slices of bread contain $2 \cdot 6 = 12$ grams of protein and $2 \cdot 2 = 4$ grams of fat. This makes a total of $20 + 12 = 32$ grams of protein and $12 + 4 = 16$ grams of fat. ◄

Try Problem 3

EXAMPLE 2   A square and an equilateral triangle (three equal sides) have the same perimeter. The sum of one side of the square and one side of the triangle is 21 centimeters. Find the length of one side of each figure.

$$x = \text{length of one side of the square}$$
$$y = \text{length of one side of the triangle}$$

Figure 8.5

Since the perimeters are the same (see Figure 8.5), we can write the equation $4x = 3y$. Since the sum of one side of the square and one side of the triangle is 21, we can write the equation $x + y = 21$. Therefore we have the following system.

$$4x = 3y$$
$$x + y = 21$$

Solve the first equation for x.

$$x = \frac{3y}{4}$$

Replace x by $\frac{3y}{4}$ in the second equation.

$$\frac{3y}{4} + y = 21$$
$$3y + 4y = 84 \qquad \text{Multiply by 4}$$
$$7y = 84$$
$$y = 12$$

Substitute $y = 12$ into $x + y = 21$ to find x.

$$x + 12 = 21$$
$$x = 9$$

The square has sides of length 9 centimeters, and the triangle has sides of length 12 centimeters. Check in the original problem. ◄

Try Problem 7

EXAMPLE 3   A plane travels 320 miles against the wind in 4 hours, then turns and travels 220 miles with the wind in half that time. Find the speed of the plane and the speed of the wind.

$$p = \text{speed of the plane in mph}$$
$$w = \text{speed of the wind in mph}$$

Sometimes a table like the one below is helpful in solving problems involving distance, rate, and time.

	d	r	t
Against the wind	320	$p - w$	4
With the wind	220	$p + w$	2

Since rate $\cdot$ time = distance, we can write the following system.

$$(p - w)4 = 320$$
$$(p + w)2 = 220$$

Divide the first equation by 4 and the second equation by 2. This gives the simpler system below.

$$p - w = 80$$
$$p + w = 110$$

Add the two equations and solve for p.

$$2p = 190$$
$$p = 95$$

Substitute $p = 95$ into the equation $p + w = 110$.

$$95 + w = 110$$
$$w = 15$$

Try Problem 9

The speed of the plane is 95 mph, and the speed of the wind is 15 mph. Check in the original problem. ◄

To solve the next example, we use the following principle concerning place value. A two-digit number, say 37, can be written as $3 \cdot 10 + 7$. Similarly, the two-digit number tu can be written as $t \cdot 10 + u = 10t + u$. If we reverse the digits, the resulting two-digit number ut can be written as $10u + t$.

EXAMPLE 4   The sum of the digits of a two-digit number is 10. If the digits are reversed, the new number is 18 less than the original number. Find the original number.

$$t = \text{tens digit of original number}$$
$$u = \text{units digit of original number}$$

The first equation is $t + u = 10$. The second equation is

new number = original number $- 18$

$10u + t = (10t + u) - 18$	
$9u - 9t = -18$	Subtract $10t + u$ from each side.
$u - t = -2$	Divide by 9
$t - u = 2.$	Multiply by -1

Hence we have the following system.
$$t + u = 10$$
$$t - u = 2$$
Add the two equations and solve for t.
$$2t = 12$$
$$t = 6$$
Substitute $t = 6$ into the equation $t + u = 10$.
$$6 + u = 10$$
$$u = 4$$

Try Problem 11   The original number is 64. Check in the original problem. ◄

Problem Solving   **Finding the Equilibrium Point**

A demand equation for a product gives the quantity D demanded at price p. A supply equation gives the quantity S that firms are willing to supply at price p. The point at which supply equals demand is called the equilibrium point, and the price at that point is called the equilibrium price. Determine the equilibrium point for the system of supply and demand equations below (assume p is in dollars).

$$D = 253 - 6p$$
$$S = 16p$$

The equilibrium point occurs when supply equals demand. Therefore write
$$S = D.$$

Substitute for S and D
from the original system.

$$16p = 253 - 6p$$

Solve for p.

$$22p = 253$$
$$p = 11.5$$

Replace p with 11.5 in the
supply equation.

$$S = 16(\mathbf{11.5}) = 184$$

Therefore the equilibrium
price is $11.50. Supply (as
well as demand) at that
price is 184 units, as shown
in Figure 8.6.

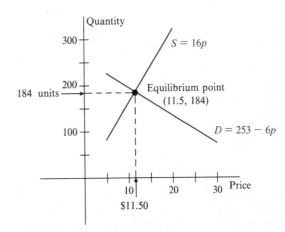

Figure 8.6

PROBLEM SET 8.3

Solve each word problem by setting up a system of equations.

1. One marigold plant and four geranium plants cost $19. Seven marigolds and two geraniums cost $29. Find the cost of each marigold and each geranium.

2. One ballpoint pen and three fountain pens cost $29. Five ballpoints and two fountain pens cost $28. Find the cost of each ballpoint and each fountain pen.

3. Suppose each ounce of macaroni contains 4 grams of protein and 3 grams of fat, and each ounce of tuna contains 6 grams of protein and 2 grams of fat. How many ounces of each food should be used to prepare a lunch containing 46 grams of protein and 27 grams of fat?

4. Suppose each ounce of potatoes contains 2 grams of protein and 4 grams of fat, and each ounce of meatloaf contains 5 grams of protein and 3 grams of fat. How many ounces of each food should be used to prepare a dinner containing 31 grams of protein and 27 grams of fat?

5. To manufacture one teddy bear requires 2 units of capital and 3 units of labor. One doll requires 4 units of capital and 5 units of labor. How many teddy bears and how many dolls can be produced with 80 units of capital and 105 units of labor?

6. To manufacture one baseball requires 3 units of capital and 4 units of labor. One football requires 5 units of capital and 7 units of labor. How many baseballs and how many footballs can be produced with 85 units of capital and 115 units of labor?

7. A square and an equilateral triangle (three equal sides) have the same perimeter. The sum of one side of the square and one side of the triangle is 28 centimeters. Find the length of one side of each figure.

8. Do Problem 7 if the sum of one side of the square and one side of the triangle is 35 centimeters.

9. A motorboat travels 21 miles upstream in 3 hours, then turns and travels 22 miles downstream in 2 hours. Find the speed of the boat and the speed of the current.

10. A plane travels 300 miles against the wind in 5 hours, then turns and travels 330 miles with the wind in 3 hours. Find the speed of the plane and the speed of the wind.

11. The sum of the digits of a two-digit number is 11. If the digits are reversed, the new number is 45 less than the original number. Find the original number.

12. The sum of the digits of a two-digit number is 12. If the digits are reversed, the new number is 54 less than the original number. Find the original number.

13. How old are Mickey and Minnie if 4 years ago Mickey was 6 times as old as Minnie, but 4 years from now he will be only twice as old?

14. How old are Donald and Daisy if 7 years ago Donald was 6 times as old as Daisy, but 1 year from now he will be only twice as old?

15. Glenn and Jon have just started collecting baseball cards. If Glenn gives Jon one card, Jon will have twice as many cards as Glenn. If Jon gives Glenn one card, each will have the same number of cards. How many cards does each have?

16. Chefs Paul and Julia each have some potatoes to peel. If Paul gives Julia two potatoes, Julia will have three times as many potatoes as Paul. If Julia gives Paul two potatoes, each will have the same number. How many potatoes does each chef have?

Determine the equilibrium point for each system of supply and demand equations. Assume p is in dollars.

17. $D = 225 - 4p$
$S = 14p$

18. $D = 216 - 6p$
$S = 10p$

19. $D = 1500 - 25p$
$S = 150 + 5p$

20. $D = 1750 - 35p$
$S = 100 + 15p$

CALCULATOR PROBLEMS

Set up a system of equations to solve each problem.

21. A sum of $115,600 is invested, part in a low-risk fund paying 9.42% and part in a higher-risk fund paying 13.65%. How much was invested in each fund if the total yearly interest from the two investments was $12,924.15?

22. Fourteen-karat gold is $\dfrac{14 \text{ k}}{24 \text{ k}} = \dfrac{7}{12}$ pure gold. One alloy contains 41% gold, and another contains 67% gold. How many ounces of each should be melted together to produce 101 ounces of an alloy that is 14 k gold?

8.4 *Linear Systems of Three Equations*

To solve an applied problem with three unknown quantities, we often use three variables. In this section we discuss how to solve a linear system of three equations in three variables.

First, consider the following equation in three variables.

$$2x + 3y + z = 6$$

One **solution** to this equation is $x = 1$, $y = -2$, and $z = 10$. This solution is usually written as the **ordered triple** $(x, y, z) = (1, -2, 10)$. We can check this solution as follows:

$$2x + 3y + z = 6$$
$$2(1) + 3(-2) + 10 \stackrel{?}{=} 6$$
$$2 - 6 + 10 \stackrel{?}{=} 6$$
$$6 \stackrel{?}{\leq} 6.$$

You may verify that the ordered triples $(3, 0, 0)$, $(0, 2, 0)$, and $(0, 0, 6)$ also satisfy the equation.

If every ordered triple that satisfies the equation $2x + 3y + z = 6$ is graphed in a **three-dimensional coordinate system,** the result is a *plane.* It is difficult to think about graphing in three dimensions. It may be helpful to think of the **origin** as a corner of a room, where the **x-axis** is the intersection of one wall with the floor, the **y-axis** is the intersection of the adjacent wall with the floor, and the **z-axis** is the intersection of the two walls. The graph of $2x + 3y + z = 6$ in a three-dimensional coordinate system would appear as the plane shown in Figure 8.7.

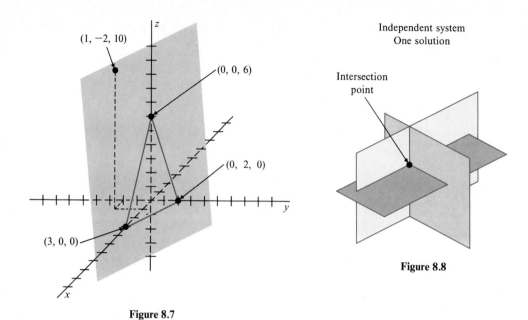

Figure 8.7

Figure 8.8

To solve a linear system of three equations by graphing would mean graphing three planes and then estimating the coordinates of their intersection point (see Figure 8.8). Clearly this is impractical. Instead, we extend the algebraic methods discussed in Section 8.2 to solve linear systems of three equations in three variables.

EXAMPLE 1 Solve the following system.

$$\begin{aligned} x + y + z &= 4 \\ x - y + 3z &= 6 \\ 5x + 4y + 2z &= 9 \end{aligned}$$

Adding the first two equations eliminates the variable y.

$$\begin{array}{r} x + y + z = 4 \\ \underline{x - y + 3z = 6} \\ 2x \phantom{{}-y} + 4z = 10 \\ x \phantom{{}-y} + 2z = 5 \quad \text{Divide by 2} \end{array}$$

To get another equation in x and z, we multiply the second equation by 4 and add the result to the third equation.

$$\begin{array}{r} 4x - 4y + 12z = 24 \\ \underline{5x + 4y + 2z = 9} \\ 9x \phantom{{}-4y} + 14z = 33 \end{array}$$

Combine this last equation with the other equation in x and z to get the following system of two equations in two variables.

$$x + 2z = 5$$
$$9x + 14z = 33$$

We can solve this system using any of the methods discussed in Section 8.2. For example, multiplying the first equation by -7 and adding the result to the second equation gives

$$-7x - 14z = -35$$
$$\underline{9x + 14z = \quad 33}$$
$$2x \qquad\quad = -2$$
$$x = -1.$$

To get z, substitute $x = -1$ into the equation $x + 2z = 5$ (you could use $9x + 14z = 33$ instead).

$$-1 + 2z = 5$$
$$2z = 6$$
$$z = 3$$

Finally, substitute $x = -1$ and $z = 3$ in the equation $x + y + z = 4$ (you could use $x - y + 3z = 6$ or $5x + 4y + 2z = 9$ instead) to get y.

$$-1 + y + 3 = 4$$
$$y = 2$$

To check our solution we substitute $(x, y, z) = (-1, 2, 3)$ into all three original equations.

$$x + y + z = 4 \qquad\qquad x - y + 3z = 6 \qquad\qquad 5x + 4y + 2z = 9$$
$$-1 + 2 + 3 \overset{?}{=} 4 \qquad -1 - 2 + 3 \cdot 3 \overset{?}{=} 6 \qquad 5(-1) + 4 \cdot 2 + 2 \cdot 3 \overset{?}{=} 9$$
$$4 \overset{\checkmark}{=} 4 \qquad\qquad\qquad 6 \overset{\checkmark}{=} 6 \qquad\qquad\qquad\qquad 9 \overset{\checkmark}{=} 9$$

Try Problem 11 \\ Since $(-1, 2, 3)$ satisfies all three equations, it is the solution to the system. ◄

EXAMPLE 2   Solve the following system.

$$2x + y - 3z = 4$$
$$5y - 4z = 12$$
$$3x - y + z = -7$$

In this system the second equation contains only the variables y and z. To produce another equation in y and z, multiply the first equation by 3 and the third equation by -2, and add the resulting equations.

$$6x + 3y - \quad 9z = 12$$
$$\underline{-6x + 2y - \quad 2z = 14}$$
$$5y - 11z = 26$$

Combine this last equation with the second equation in the original system to produce the following system of two equations.

$$5y - 4z = 12$$
$$5y - 11z = 26$$

Multiply the second equation by -1 and add.

$$
\begin{array}{r}
5y - 4z = 12 \\
-5y + 11z = -26 \\
\hline
7z = -14 \\
z = -2
\end{array}
$$

Substitute $z = -2$ in $5y - 4z = 12$.

$$5y - 4(-2) = 12$$
$$5y + 8 = 12$$
$$5y = 4$$
$$y = \tfrac{4}{5}$$

Substitute $y = \tfrac{4}{5}$ and $z = -2$ in $2x + y - 3z = 4$.

$$2x + \tfrac{4}{5} - 3(-2) = 4$$
$$2x + \tfrac{4}{5} + 6 = 4$$
$$2x + \tfrac{4}{5} = -2$$
$$10x + 4 = -10 \qquad \text{Multiply by 5}$$
$$10x = -14$$
$$x = -\tfrac{7}{5}$$

Try Problem 19

The solution is $(-\tfrac{7}{5}, \tfrac{4}{5}, -2)$. Check in the original system. ◄

A linear system of three equations is **inconsistent** when there is no point that is common to all three planes. This can occur in a variety of ways as illustrated in Figure 8.9.

Inconsistent systems
No solution

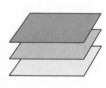

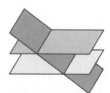

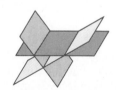

a) 3 parallel planes   b) 2 parallel planes   c) No parallel planes   d) 2 planes coincide, with third plane parallel

Figure 8.9

EXAMPLE 3   Solve the following system.

$$x - 2y + 3z = 1$$
$$-2x + 4y - 6z = 3$$
$$5x + y + z = 8$$

Eliminate x by multiplying the first equation by 2 and adding the result to the second equation.

$$2x - 4y + 6z = 2$$
$$\underline{-2x + 4y - 6z = 3}$$
$$0 = 5$$

Try Problem 25

The fact that we arrived at a contradiction means that the original system is inconsistent and there is no solution. ◄

Figure 8.9(b) illustrates what happened in Example 3. The first two equations of the original system graph into parallel planes. The third plane intersects each of the other two planes along a line.

A linear system of three equations is **dependent** when there are an infinite number of points that are common to all three planes. This can occur in a variety of ways, as illustrated in Figure 8.10.

EXAMPLE 4   Solve the following system.

$$x + 4y - z = 0$$
$$-x + 2y - z = 0$$
$$2x - y + z = 0$$

Eliminate z by adding the first equation to the third.

$$x + 4y - z = 0$$
$$\underline{2x - y + z = 0}$$
$$3x + 3y \quad\;\; = 0$$
$$x + \;\; y \quad\;\; = 0 \qquad \text{Divide by 3}$$

Dependent systems
Infinite number of solutions

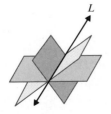

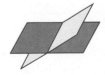

a) 3 planes intersect
    along line L

b) 3 planes coincide

c) 2 planes coincide, with
    third plane intersecting

Figure 8.10

Eliminate z again by adding the second equation to the third.

$$
\begin{array}{rcl}
-x + 2y - z &=& 0 \\
\underline{2x - \ y + z} &=& \underline{0} \\
x + \ y \quad\ &=& 0
\end{array}
$$

This gives the following system of two equations.

$$
\begin{array}{rcl}
x + y &=& 0 \\
x + y &=& 0
\end{array}
$$

If we multiply the second equation by -1 and add the result to the first equation, we obtain the identity $0 = 0$. This system is dependent, and there are an infinite number of solutions. ◄

Try Problem 29

Figure 8.10(a) illustrates what happened in Example 4. The solutions to the system are represented by the points on line L.

We summarize our work with linear systems of three equations as follows:

To Solve a Linear System of Three Equations

1. Write each equation in standard form with integer coefficients.
2. Use any two equations to get an equation in two variables.
3. Use a different pair of equations to get another equation in the same two variables as the equation resulting from step 2.
4. Solve the system of two equations that results from steps 2 and 3.
5. Substitute the values found in step 4 into one of the original equations to find the value of the third variable.
6. Check your solution in all three original equations.
7. If at any time a contradiction is obtained, the system is inconsistent. If at any time an identity is obtained, the system is either inconsistent in the manner of Figure 8.9(d), or dependent. If a system is inconsistent or dependent, we shall simply state that there is no unique solution.

PROBLEM SET 8.4

Determine whether the given ordered triple is a solution of the given system.

1. $(2, -1, 4)$

$$
\begin{array}{rcl}
x + \ y - \ z &=& -3 \\
2x - 3y + 4z &=& 23 \\
5x - 2y - 6z &=& -16
\end{array}
$$

2. $(3, 1, -5)$

$$
\begin{array}{rcl}
x - \ y + \ z &=& -3 \\
3x + 5y - 2z &=& 24 \\
4x - 7y - 3z &=& -10
\end{array}
$$

3. $(-5, \frac{2}{3}, 0)$

$$x - 3y + 8z = -7$$
$$2x + 15y - z = 0$$
$$-3x \qquad + 5z = 15$$

4. $(\frac{3}{4}, 0, -2)$

$$-4x + 9y + z = -5$$
$$16x - y + 6z = 0$$
$$8y - 5z = 10$$

Solve each system.

5. $2x - 3y - z = 2$
$x + y + z = 3$
$x + 2y = 7$

6. $3x - y + z = 1$
$x + 2y - z = 3$
$x - y = 6$

7. $-x + 4y + z = 3$
$2x - 3y - z = 0$
$x + 2y + z = 5$

8. $-x + 3y + z = 3$
$2x - 2y - z = 0$
$x + y + z = 5$

9. $3x + y - z = 2$
$3x - y + 2z = 7$
$-3x + 2y - z = -2$

10. $2x - y + z = 5$
$2x + 2y - z = 9$
$-2x - y + 2z = -6$

11. $x + y + z = 6$
$x - y + 3z = 4$
$5x + 4y + 2z = 13$

12. $x + y + z = 8$
$x - y + 3z = 2$
$5x + 4y + 2z = 21$

13. $2x - 3y + 4z = -2$
$3x + y - 4z = -8$
$x + 4y - 2z = -3$

14. $2x - y + 6z = -2$
$3x - 2y - 6z = -8$
$x + 3y - 3z = -3$

15. $x + 5y + 6z = 17$
$2x - 4y - 3z = 13$
$3x - 7y - 5z = 22$

16. $x - 8y + 4z = 20$
$2x + 7y - 6z = 14$
$3x - 5y + 2z = 26$

17. $3x - 4y = 28$
$5x + y - 2z = 47$
$2x + 3y + 2z = 5$

18. $2x - 5y = 30$
$6x + y - 3z = 5$
$5x + 4y + 3z = 30$

19. $2x + y - 3z = 3$
$5y - 2z = 7$
$3x - y + z = -8$

20. $3x + y - 2z = 2$
$7y - 4z = 17$
$4x - y + z = -10$

21. $4x - 3y = -13$
$2y + 5z = -12$
$6x - 7z = -10$

22. $3x - 5y = -5$
$4y + 2z = -10$
$7x - 9z = -26$

23. $3x - 4y - 6z = 12$
$-\frac{1}{4}x + \frac{1}{3}y + \frac{1}{2}z = -1$
$\frac{1}{2}x - \frac{2}{3}y - z = 2$

24. $x + 5y - 10z = 20$
$-\frac{1}{5}x - \frac{1}{4}y + \frac{1}{2}z = -1$
$\frac{2}{5}x + \frac{1}{2}y - z = 2$

25. $x - 3y + 2z = 1$
$-2x + 6y - 4z = 5$
$5x + y + z = 9$

26. $x + 2y - 3z = 1$
$-4x - 8y + 12z = 5$
$5x + y + z = 8$

27. $2x + 2y - 3z = 0$
$x - y + z = 0$
$3x + 5y - 6z = 0$

28. $x + 3y - 2z = 0$
$3x - y + z = 0$
$2x + 4y - 3z = 0$

29. $x + 5y - z = 0$
$-2x + 2y - z = 0$
$3x - y + z = 0$

30. $x + 6y - z = 0$
$-3x + 2y - z = 0$
$4x - y + z = 0$

31. $z = 1 - x + 3y$
$x + 2z = 8y - 1$
$2x - 11y + 3z = 2$

32. $z = 2 - x + 2y$
$x + 2z = 11y - 2$
$2x - 13y + 3z = 4$

33. $x + 7y + 3z = 35$
$-x - 5y - 3z = -17$
$2x + 4y + 5z = -7$

34. $x + 8y + 4z = 57$
$-x - 5y - 4z = -27$
$3x + 4y + 6z = 25$

35. $x - y = 3$
$x + z = 7$
$y = 3z$

36. $y + z = 5$
$x + z = 2$
$y = 4x$

Select a variable to represent each unknown quantity. Then set up a system of three equations involving these variables. Finally, solve the system.

37. The sum of three numbers is 4. The first plus the second minus the third is -10. Twice the first plus the sum of the other two is 0. Find the numbers.

38. The sum of three numbers is 6. The first plus the second minus the third is -10. Twice the first plus the sum of the other two is 0. Find the numbers.

39. One muffin, two pies, and three cakes cost $23. One muffin, three pies, and two cakes cost $21. One muffin, four pies, and five cakes cost $39. Find the cost of each.

40. Do Problem 39 if the respective costs are $26, $20, and $42.

41. The perimeter of a rectangle is twice the perimeter of an equilateral triangle. The rectangle's width plus its length plus a side of the triangle is 100 centimeters. Four times the rectangle's width equals its length minus one side of the triangle. Find each dimension.

42. Do Problem 41 if four times the rectangle's width equals its length plus one side of the triangle.

43. The average of Jennifer's three test scores is 76. She scored 4 points better on the second test than the first, and 7 points better on the third test than the second. What was her score on each test?

44. Do Problem 43 if Jennifer's average on the three tests is 73.

45. Larry, Curly, and Moe each have some bananas. If Larry gives Curly a banana, Larry and Curly will have the same number. If Larry gives Moe a banana, Moe will have twice as many as Curly. If Larry eats four of Moe's bananas, Moe will still have three more bananas than Larry. How many bananas does each have?

46. Do Problem 45 assuming that if Larry eats five of Moe's bananas then Moe will still have three more bananas than Larry.

47. The sum of the digits of a three-digit number is 16. The tens digit is 3 more than the units digit. If the order of the digits is reversed and the new number is subtracted from the original number, the result is 99. Find the original number.

48. Do Problem 47 if the sum of the digits is 19.

CALCULATOR PROBLEMS

Solve each system.

49. $23x + 17y - 21z = 808$
    $19x - 15y + 25z = 336$
    $11x - 13y - 18z = 688$

50. $0.95x - 3.76z = 20.7$
    $7.85x - 4.78y = 1.36$
    $-5.55y + 2.53z = -29.3$

8.5   *Determinants*

Another method for solving a system of equations utilizes a mathematical tool known as a **determinant.** We shall demonstrate this method in Section 8.6, but first we need to define a determinant and learn how to calculate its value.

Definition	A **determinant** is a square array of numbers enclosed between two vertical lines.

In the determinant

$$\begin{vmatrix} 7 & 2 \\ 5 & 3 \end{vmatrix}$$

the numbers 7, 2, 5, and 3 are called the **elements** of the determinant. Since this determinant has two rows and two columns, it is called a **2 × 2** (read "two by two") **determinant.**

Definition	The **value** of a 2 × 2 determinant is calculated as follows: $$\begin{vmatrix} a_1 & b_1 \\ a_2 & b_2 \end{vmatrix} = a_1 b_2 - a_2 b_1.$$

EXAMPLE 1   Evaluate each determinant:

$$\textbf{a)}\ \begin{vmatrix} 7 & 2 \\ 5 & 3 \end{vmatrix}, \qquad \textbf{b)}\ \begin{vmatrix} 6 & -1 \\ -4 & -2 \end{vmatrix}.$$

Multiply and subtract as shown below.

a) $\begin{vmatrix} 7 & 2 \\ 5 & 3 \end{vmatrix} = 7\cdot 3 - 5\cdot 2 = 21 - 10 = 11$

Try Problem 1   **b)** $\begin{vmatrix} 6 & -1 \\ -4 & -2 \end{vmatrix} = 6(-2) - (-4)(-1) = -12 - 4 = -16$ ◂

To calculate the value of a 3 × 3 determinant, we use what are called **minors.**

Definition	The **minor** of an element in a 3 × 3 determinant is the 2 × 2 determinant that results when both the row and the column that contain that element are deleted.

Applying this definition, we obtain the following minors.

$$\text{minor of } a_1 = \begin{vmatrix} a_1 & b_1 & c_1 \\ a_2 & b_2 & c_2 \\ a_3 & b_3 & c_3 \end{vmatrix} = \begin{vmatrix} b_2 & c_2 \\ b_3 & c_3 \end{vmatrix}$$

$$\text{minor of } a_2 = \begin{vmatrix} a_1 & b_1 & c_1 \\ a_2 & b_2 & c_2 \\ a_3 & b_3 & c_3 \end{vmatrix} = \begin{vmatrix} b_1 & c_1 \\ b_3 & c_3 \end{vmatrix}$$

$$\text{minor of } b_2 = \begin{vmatrix} a_1 & b_1 & c_1 \\ a_2 & b_2 & c_2 \\ a_3 & b_3 & c_3 \end{vmatrix} = \begin{vmatrix} a_1 & c_1 \\ a_3 & c_3 \end{vmatrix}$$

EXAMPLE 2 Determine the minor of 5 in the 3×3 determinant

$$\begin{vmatrix} 1 & 2 & 3 \\ 4 & 5 & 6 \\ 7 & 8 & 9 \end{vmatrix}.$$

We have

Try Problem 21

$$\text{minor of } 5 = \begin{vmatrix} 1 & 2 & 3 \\ 4 & 5 & 6 \\ 7 & 8 & 9 \end{vmatrix} = \begin{vmatrix} 1 & 3 \\ 7 & 9 \end{vmatrix}. \blacktriangleleft$$

Definition

The **value** of a 3×3 determinant can be calculated as follows:

$$\begin{vmatrix} a_1 & b_1 & c_1 \\ a_2 & b_2 & c_2 \\ a_3 & b_3 & c_3 \end{vmatrix} = a_1(\text{minor of } a_1) - a_2(\text{minor of } a_2) + a_3(\text{minor of } a_3)$$

$$= a_1 \begin{vmatrix} b_2 & c_2 \\ b_3 & c_3 \end{vmatrix} - a_2 \begin{vmatrix} b_1 & c_1 \\ b_3 & c_3 \end{vmatrix} + a_3 \begin{vmatrix} b_1 & c_1 \\ b_2 & c_2 \end{vmatrix}.$$

This method is called **expansion by minors** about the first column.

EXAMPLE 3 Evaluate the given determinant.

$$\begin{vmatrix} 3 & -1 & -2 \\ 5 & 0 & -4 \\ 2 & -3 & 6 \end{vmatrix} = 3 \begin{vmatrix} 0 & -4 \\ -3 & 6 \end{vmatrix} - 5 \begin{vmatrix} -1 & -2 \\ -3 & 6 \end{vmatrix} + 2 \begin{vmatrix} -1 & -2 \\ 0 & -4 \end{vmatrix}$$

$$= 3[0 \cdot 6 - (-3)(-4)] - 5[(-1)6 - (-3)(-2)]$$

$$+ 2[(-1)(-4) - 0(-2)]$$

$$= 3[-12] - 5[-12] + 2[4]$$

$$= -36 + 60 + 8$$

Try Problem 29

$$= 32 \blacktriangleleft$$

Actually we can calculate the value of a 3×3 determinant using expansion by minors about *any* column or *any* row. When expanding about a particular row or column, prefix the terms of the expansion with the signs from the corresponding row or column in the **sign array** below.

$$\begin{vmatrix} + & - & + \\ - & + & - \\ + & - & + \end{vmatrix}$$

This sign array is easy to remember. Simply start with the first-row, first-column position (upper left-hand corner) and alternate signs as you travel along any row or column.

EXAMPLE 4   Evaluate the determinant of Example 3 using expansion by minors about the second row.

Since we are expanding about the second row, use the signs given in the second row of the sign array.

$$\begin{vmatrix} 3 & -1 & -2 \\ 5 & 0 & -4 \\ 2 & -3 & 6 \end{vmatrix} = -5\begin{vmatrix} -1 & -2 \\ -3 & 6 \end{vmatrix} + 0\begin{vmatrix} 3 & -2 \\ 2 & 6 \end{vmatrix} - (-4)\begin{vmatrix} 3 & -1 \\ 2 & -3 \end{vmatrix}$$

Signs from sign array

We do not need to evaluate the middle 2×2 determinant since its multiplier is zero.

$$= -5(-6-6) + 0 + 4(-9+2)$$
$$= 60 + 0 - 28$$
$$= 32$$

Try Problem 33

This is the same value we obtained in Example 3. ◀

In summary, we evaluate a 3×3 determinant as follows:

To Evaluate a 3×3 Determinant Using Expansion by Minors

1.   Choose a row or column to expand about. Choose the row or column with the most zeros to simplify your computations.
2.   Multiply each element in the row or column chosen in step 1 by its minor.
3.   Prefix the terms in step 2 with the signs from the corresponding row or column in the sign array.

The method of expansion by minors can be extended to 4×4 determinants, as well as to determinants of even higher order.

PROBLEM SET 8.5

Evaluate each 2 × 2 determinant.

1. $\begin{vmatrix} 5 & 4 \\ 2 & 3 \end{vmatrix}$

2. $\begin{vmatrix} 8 & 2 \\ 5 & 3 \end{vmatrix}$

3. $\begin{vmatrix} 2 & 9 \\ 3 & 6 \end{vmatrix}$

4. $\begin{vmatrix} 4 & 6 \\ 3 & 2 \end{vmatrix}$

5. $\begin{vmatrix} 9 & 6 \\ 3 & 2 \end{vmatrix}$

6. $\begin{vmatrix} 6 & 3 \\ 4 & 2 \end{vmatrix}$

7. $\begin{vmatrix} -10 & \frac{1}{2} \\ 8 & -2 \end{vmatrix}$

8. $\begin{vmatrix} -3 & \frac{1}{3} \\ 9 & -10 \end{vmatrix}$

9. $\begin{vmatrix} 7 & -4 \\ -1 & -3 \end{vmatrix}$

10. $\begin{vmatrix} 8 & -1 \\ -4 & -2 \end{vmatrix}$

11. $\begin{vmatrix} -2 & -2 \\ 5 & 6 \end{vmatrix}$

12. $\begin{vmatrix} -4 & -4 \\ 3 & 5 \end{vmatrix}$

13. $\begin{vmatrix} \frac{1}{5} & \frac{2}{5} \\ -\frac{1}{5} & \frac{3}{5} \end{vmatrix}$

14. $\begin{vmatrix} \frac{4}{5} & \frac{2}{5} \\ -\frac{4}{5} & \frac{3}{5} \end{vmatrix}$

15. $\begin{vmatrix} 1 & 0 \\ 0 & 1 \end{vmatrix}$

16. $\begin{vmatrix} 0 & 1 \\ 1 & 0 \end{vmatrix}$

17. $\begin{vmatrix} 0 & 1 \\ 0 & 1 \end{vmatrix}$

18. $\begin{vmatrix} 0 & 0 \\ 1 & 1 \end{vmatrix}$

19. $\begin{vmatrix} x & y \\ y & x \end{vmatrix}$

20. $\begin{vmatrix} y & x \\ x & y \end{vmatrix}$

Using the determinant below, state the minor of each element given in Problems 21–28.

$$\begin{vmatrix} 1 & 2 & 3 \\ 4 & 5 & 6 \\ 7 & 8 & 9 \end{vmatrix}$$

21. 1

22. 2

23. 3

24. 6

25. 4

26. 7

27. 8

28. 9

Evaluate each 3 × 3 determinant using expansion by minors about the first column.

29. $\begin{vmatrix} 4 & -6 & -1 \\ 3 & 0 & -2 \\ 1 & -3 & 2 \end{vmatrix}$

30. $\begin{vmatrix} 3 & -1 & -2 \\ 5 & 0 & -4 \\ 2 & -3 & 6 \end{vmatrix}$

31. $\begin{vmatrix} 2 & 4 & 1 \\ -3 & 5 & 2 \\ -1 & 3 & -2 \end{vmatrix}$

32. $\begin{vmatrix} 1 & 5 & 1 \\ -2 & 2 & 4 \\ -4 & 3 & -1 \end{vmatrix}$

33. Evaluate the determinant of Problem 29 using expansion by minors about the second row.

34. Evaluate the determinant of Problem 30 using expansion by minors about the second row.

Evaluate each 3 × 3 determinant. Expand about the row or column of your choice.

35. $\begin{vmatrix} 0 & 0 & 4 \\ 4 & 7 & 8 \\ 2 & 6 & -5 \end{vmatrix}$

36. $\begin{vmatrix} 0 & 0 & 5 \\ 5 & 9 & 4 \\ 4 & 8 & -6 \end{vmatrix}$

37. $\begin{vmatrix} -3 & 2 & 1 \\ -1 & 4 & 0 \\ 5 & -2 & 1 \end{vmatrix}$

38. $\begin{vmatrix} -3 & 2 & 1 \\ -2 & 8 & 0 \\ 4 & -1 & 1 \end{vmatrix}$

39. $\begin{vmatrix} 2 & -1 & 1 \\ -2 & 2 & -1 \\ 1 & 1 & 3 \end{vmatrix}$

40. $\begin{vmatrix} 4 & -1 & 2 \\ -4 & 1 & -3 \\ 1 & 1 & 2 \end{vmatrix}$

41. $\begin{vmatrix} -1 & 2 & -1 \\ -3 & 1 & -3 \\ -6 & -2 & 5 \end{vmatrix}$

42. $\begin{vmatrix} -5 & 1 & 2 \\ -3 & 1 & -1 \\ 6 & -2 & -2 \end{vmatrix}$

43. $\begin{vmatrix} 3 & 1 & -1 \\ 2 & -1 & -3 \\ 6 & 2 & -2 \end{vmatrix}$

44. $\begin{vmatrix} 2 & 1 & -1 \\ 3 & -1 & -4 \\ 6 & 3 & -3 \end{vmatrix}$

45. $\begin{vmatrix} a & 1 & 1 \\ b & 2 & 1 \\ c & 3 & 0 \end{vmatrix}$

46. $\begin{vmatrix} a & b & c \\ 1 & 2 & 3 \\ 1 & 1 & 0 \end{vmatrix}$

47. $\begin{vmatrix} 0 & 0 & 1 \\ 0 & 1 & 0 \\ 1 & 0 & 0 \end{vmatrix}$

48. $\begin{vmatrix} 1 & 0 & 0 \\ 0 & 1 & 0 \\ 0 & 0 & 1 \end{vmatrix}$

Solve each determinant equation for x.

49. $\begin{vmatrix} x & 1 \\ 2x & 3 \end{vmatrix} = 6$

50. $\begin{vmatrix} x & 1 \\ 3x & 4 \end{vmatrix} = 8$

51. $\begin{vmatrix} x+1 & -4 \\ x-2 & -5 \end{vmatrix} = x - 3$

52. $\begin{vmatrix} x-1 & -7 \\ x+2 & -4 \end{vmatrix} = x + 4$

53. $\begin{vmatrix} x & 2 & -7 \\ 1 & 0 & 2 \\ 3 & x & -1 \end{vmatrix} = 10$

54. $\begin{vmatrix} x & 3 & -5 \\ 1 & 0 & 2 \\ 4 & x & -1 \end{vmatrix} = 24$

55. Evaluate each determinant. Can you draw a general conclusion?

a) $\begin{vmatrix} 0 & 0 & 0 \\ a & b & c \\ x & y & z \end{vmatrix}$

b) $\begin{vmatrix} a & b & c \\ 0 & 0 & 0 \\ x & y & z \end{vmatrix}$

c) $\begin{vmatrix} a & b & c \\ x & y & z \\ 0 & 0 & 0 \end{vmatrix}$

56. Evaluate each determinant. Can you draw a general conclusion?

a) $\begin{vmatrix} 0 & a & x \\ 0 & b & y \\ 0 & c & z \end{vmatrix}$

b) $\begin{vmatrix} a & 0 & x \\ b & 0 & y \\ c & 0 & z \end{vmatrix}$

c) $\begin{vmatrix} a & x & 0 \\ b & y & 0 \\ c & z & 0 \end{vmatrix}$

CALCULATOR PROBLEMS

Evaluate each determinant.

57. $\begin{vmatrix} 13 & 17 & 19 \\ 23 & 29 & 31 \\ 37 & 41 & 43 \end{vmatrix}$

58. $\begin{vmatrix} 1.06 & -2.68 & 0 \\ 4.93 & 3.47 & -5.26 \\ 6.55 & -7.94 & -8.13 \end{vmatrix}$

8.6   *Cramer's Rule*

We will now demonstrate how determinants can be used to solve a linear system of equations. To begin, consider the general system

$$a_1 x + b_1 y = c_1$$
$$a_2 x + b_2 y = c_2,$$

where a_1, b_1, c_1 and a_2, b_2, c_2 are constants. We use the addition method to solve this system for x and y. To eliminate y, multiply the first equation by b_2 and the second equation by $-b_1$. Then add the resulting equations and solve for x.

$$a_1b_2x + b_1b_2y = c_1b_2$$
$$-a_2b_1x - b_1b_2y = -c_2b_1$$
$$(a_1b_2 - a_2b_1)x = c_1b_2 - c_2b_1$$
$$x = \frac{c_1b_2 - c_2b_1}{a_1b_2 - a_2b_1}$$

To eliminate x, multiply the first equation by $-a_2$ and the second equation by a_1. Then add the resulting equations and solve for y.

$$-a_1a_2x - a_2b_1y = -a_2c_1$$
$$a_1a_2x + a_1b_2y = a_1c_2$$
$$(a_1b_2 - a_2b_1)y = a_1c_2 - a_2c_1$$
$$y = \frac{a_1c_2 - a_2c_1}{a_1b_2 - a_2b_1}$$

Note that the values of x and y are fractions with the same denominator. We can express this common denominator and the two numerators by determinants as follows:

$$x = \frac{\begin{vmatrix} c_1 & b_1 \\ c_2 & b_2 \end{vmatrix}}{\begin{vmatrix} a_1 & b_1 \\ a_2 & b_2 \end{vmatrix}}, \qquad y = \frac{\begin{vmatrix} a_1 & c_1 \\ a_2 & c_2 \end{vmatrix}}{\begin{vmatrix} a_1 & b_1 \\ a_2 & b_2 \end{vmatrix}}.$$

For convenience, we denote the three determinants as shown below.

$$\begin{vmatrix} a_1 & b_1 \\ a_2 & b_2 \end{vmatrix} = D \qquad \begin{vmatrix} c_1 & b_1 \\ c_2 & b_2 \end{vmatrix} = D_x \qquad \begin{vmatrix} a_1 & c_1 \\ a_2 & c_2 \end{vmatrix} = D_y$$

Note that the elements of D consist of the coefficients of x and y in the original system. To form D_x from D, replace the coefficients of x (a_1 and a_2) with the constant terms c_1 and c_2. To form D_y from D, replace the coefficients of y with the constant terms.

This method for solving a system is called **Cramer's rule** and is summarized as follows:

Cramer's Rule* for a Linear System of Two Equations

The solution to the system

$$a_1x + b_1y = c_1,$$
$$a_2x + b_2y = c_2$$

---

* Named after the Swiss mathematician Gabriel Cramer (1704–1752).

is given by

$$x = \frac{D_x}{D} = \frac{\begin{vmatrix} c_1 & b_1 \\ c_2 & b_2 \end{vmatrix}}{\begin{vmatrix} a_1 & b_1 \\ a_2 & b_2 \end{vmatrix}} \quad \text{and} \quad y = \frac{D_y}{D} = \frac{\begin{vmatrix} a_1 & c_1 \\ a_2 & c_2 \end{vmatrix}}{\begin{vmatrix} a_1 & b_1 \\ a_2 & b_2 \end{vmatrix}},$$

so long as $D \neq 0$.

EXAMPLE 1   Use Cramer's rule to solve the following system.
$$6x - 2y = 7$$
$$3x + 4y = -9$$

We begin by finding D, D_x, and D_y.

$$D = \begin{vmatrix} 6 & -2 \\ 3 & 4 \end{vmatrix} = 6 \cdot 4 - 3(-2) = 30$$

$$D_x = \begin{vmatrix} 7 & -2 \\ -9 & 4 \end{vmatrix} = 7 \cdot 4 - (-9)(-2) = 10$$

$$D_y = \begin{vmatrix} 6 & 7 \\ 3 & -9 \end{vmatrix} = 6(-9) - 3 \cdot 7 = -75$$

Then we apply Cramer's rule and get

$$x = \frac{D_x}{D} = \frac{10}{30} = \frac{1}{3} \quad \text{and} \quad y = \frac{D_y}{D} = \frac{-75}{30} = -\frac{5}{2}.$$

Try Problem 5   Therefore the solution is $(\frac{1}{3}, -\frac{5}{2})$. Check in the original system. ◄

CAUTION You should always calculate D before calculating D_x or D_y. If $D = 0$, then Cramer's rule does not apply and the system is inconsistent or dependent. In either case, there is no unique solution.

EXAMPLE 2   Use Cramer's rule to solve the following system.
$$x = \tfrac{1}{3}y - \tfrac{2}{3}$$
$$y = 3x + \tfrac{5}{2}$$

Multiply the first equation by 3 and the second equation by 2 to clear fractions.
$$3x = y - 2$$
$$2y = 6x + 5$$

Then write the system in standard form.
$$3x - y = -2$$
$$-6x + 2y = 5$$

Calculate D before calculating D_x or D_y.

$$D = \begin{vmatrix} 3 & -1 \\ -6 & 2 \end{vmatrix} = 3 \cdot 2 - (-6)(-1) = 0$$

Since $D = 0$, Cramer's rule does not apply. There is no unique solution to the system. ◄

Try Problem 9

Cramer's rule can be extended to a linear system of three equations as follows:

Cramer's Rule for a Linear System of Three Equations

The solution to the system

$$a_1 x + b_1 y + c_1 z = d_1$$
$$a_2 x + b_2 y + c_2 z = d_2$$
$$a_3 x + b_3 y + c_3 z = d_3,$$

is given by

$$x = \frac{D_x}{D}, \qquad y = \frac{D_y}{D}, \qquad z = \frac{D_z}{D},$$

where

$$D = \begin{vmatrix} a_1 & b_1 & c_1 \\ a_2 & b_2 & c_2 \\ a_3 & b_3 & c_3 \end{vmatrix}, \qquad D_x = \begin{vmatrix} d_1 & b_1 & c_1 \\ d_2 & b_2 & c_2 \\ d_3 & b_3 & c_3 \end{vmatrix},$$

$$D_y = \begin{vmatrix} a_1 & d_1 & c_1 \\ a_2 & d_2 & c_2 \\ a_3 & d_3 & c_3 \end{vmatrix}, \qquad D_z = \begin{vmatrix} a_1 & b_1 & d_1 \\ a_2 & b_2 & d_2 \\ a_3 & b_3 & d_3 \end{vmatrix},$$

so long as $D \neq 0$.

As before, if $D = 0$, then Cramer's rule does not apply and the system does not have a unique solution.

EXAMPLE 3   Use Cramer's rule to solve the following system.

$$x + y + z = 2$$
$$x + y - z = 0$$
$$2x - 3y - 5z = 7$$

Calculate D, D_x, D_y, and D_z and obtain the following results.

$$D = \begin{vmatrix} 1 & 1 & 1 \\ 1 & 1 & -1 \\ 2 & -3 & -5 \end{vmatrix} = -10, \qquad D_x = \begin{vmatrix} 2 & 1 & 1 \\ 0 & 1 & -1 \\ 7 & -3 & -5 \end{vmatrix} = -30,$$

$$D_y = \begin{vmatrix} 1 & 2 & 1 \\ 1 & 0 & -1 \\ 2 & 7 & -5 \end{vmatrix} = 20, \qquad D_z = \begin{vmatrix} 1 & 1 & 2 \\ 1 & 1 & 0 \\ 2 & -3 & 7 \end{vmatrix} = -10.$$

Apply Cramer's rule.

$$x = \frac{D_x}{D} = \frac{-30}{-10} = 3$$

$$y = \frac{D_y}{D} = \frac{20}{-10} = -2$$

$$z = \frac{D_z}{D} = \frac{-10}{-10} = 1$$

Try Problem 13   The solution is $(3, -2, 1)$. Check in the original system. ◄

PROBLEM SET 8.6

Use Cramer's rule to solve each system.

1. $2x + 3y = 8$
$\quad 5x + 4y = 13$

2. $4x + 3y = 11$
$\quad 2x + 5y = 9$

3. $x + y = 1$
$\quad 3x + 4y = -2$

4. $x + y = 1$
$\quad 5x + 6y = -2$

5. $6x - 2y = 9$
$\quad 3x + 4y = -13$

6. $8x - 3y = 7$
$\quad 4x + 6y = -9$

7. $7x - y = 21$
$\quad -2x + 5y = -6$

8. $7x - y = 28$
$\quad -3x + 4y = -12$

9. $x = \frac{1}{2}y - 3$
$\quad y = 2x + \frac{4}{3}$

10. $x = \frac{1}{4}y - 2$
$\quad y = 4x + \frac{5}{2}$

11. $4x + 7y = 16$
$\quad x \quad\quad = -3$

12. $\quad\quad y = -2$
$\quad 8x + 5y = 14$

Use Cramer's rule to solve each system.

13. $x + y + z = 4$
$\quad x + y - z = 0$
$\quad 3x - 4y - 5z = 3$

14. $x + y + z = 6$
$\quad x + y - z = 0$
$\quad 2x - 4y - 3z = 3$

15. $x - 3y + z = 2$
$\quad 3x - y + 2z = 2$
$\quad 2x + y + z = 0$

16. $x - 2y + z = 4$
$\quad 4x - y + 2z = 4$
$\quad 3x + y + z = 0$

17. $x + y \quad\quad = -1$
$\quad 2y + 8z = -6$
$\quad 2x \quad - 4z = 7$

18. $x + y \quad\quad = -1$
$\quad 2y + 5z = -8$
$\quad 2x \quad - 10z = 4$

19. $5x - 2z + 2 = 0$
$\quad x - 2y + 3z = 3$
$\quad 3y + z - 1 = 0$

20. $4x - 3z + 3 = 0$
$\quad x - 2y + 2z = 2$
$\quad 5y + z - 1 = 0$

21. $x - 2y + z = 1$
$\quad 2x - 4y + 3z = 2$
$\quad -x + 2y - 3z = 3$

22. $x - 3y + z = 1$
$\quad 2x - 6y + 2z = 3$
$\quad -x + 3y - 2z = 4$

23. $-x + y - 2z = 0$
$\quad 2x - 3y + z = 0$
$\quad 4x - y - 3z = 0$

24. $-x + 2y - z = 0$
$\quad 3x - y + 4z = 0$
$\quad 5x - 3y - z = 0$

CALCULATOR PROBLEMS

Use Cramer's rule to solve each system.

25. $4.71x + 6.25y = 62.68$
$9.34x - 5.88y = 51.2$

26. $161x - 30y + 17z = 527$
$43x + 79y - 22z = -340$
$101x - 88y - 99z = 59$

CHAPTER 8 REVIEW

[8.1] *Solve each system by graphing. Classify each system as independent, inconsistent, or dependent.*

1. $x + y = 5$
$x - y = 3$

2. $y - x = 0$
$2y - 2x = 5$

3. $2x - 3y = 6$
$4x - 6y = 12$

[8.2] *Solve each system using the substitution method.*

4. $x + y = 6$
$y = x$

5. $3x + 4y = 1$
$y = x + 2$

6. $x - 3y = 3$
$3x - 4y = -1$

7. $4x + 5y = 2$
$3x + 2y = 5$

Solve each system using the addition method.

8. $4x + y = 10$
$x - y = 0$

9. $x + 3y = 8$
$x - y = 4$

10. $3x + 2y = 1$
$2x + y = -1$

11. $7x - 3y = 0$
$x + 6y = 0$

12. $2x - 3y = 5$
$-4x + 6y = 3$

13. $2x + 3y = 2$
$\dfrac{x}{3} + \dfrac{y}{2} = \dfrac{1}{3}$

[8.3]

14. A number x added to five times a number y is 17. Find the numbers if x is 1 less than four times y.

15. The sum of the digits of a two-digit number is 14. If the digits are reversed, the new number is 18 less than the original number. Find the original number.

16. One ballpoint pen and two fountain pens cost $17. Four ballpoint pens and one fountain pen cost $19. Find the cost of each ballpoint and each fountain pen.

17. To manufacture one toy car requires 3 units of capital and 5 units of labor. One toy truck requires 5 units of capital and 6 units of labor. How many toy cars and how many toy trucks can be produced with 156 units of capital and 211 units of labor?

18. A mother is six times as old as her daughter. In six years she will be only three times as old as her daughter. Find their present ages.

[8.4] *Solve each system.*

19. $\begin{aligned} x - y + 2z &= 3 \\ x + 2y - 2z &= 12 \\ 2x - 3y + z &= 3 \end{aligned}$

20. $\begin{aligned} x + y + z &= 6 \\ x + 2y - 3z &= 1 \\ x + 3y - 4z &= 2 \end{aligned}$

21. $\begin{aligned} 2x + y + 3z &= 2 \\ 3x + 2y + 4z &= 0 \\ 5x + 2y &= 0 \end{aligned}$

22. $\begin{aligned} x + 2y + z &= 7 \\ x + y - z &= 7 \\ 2x + 4y + 2z &= 5 \end{aligned}$

23. $\begin{aligned} 3x - y + z &= 8 \\ x + y - z &= 4 \\ x + 2y - z &= 6 \end{aligned}$

24. $\begin{aligned} x + 2y - z &= 1 \\ x - 2y + z &= 1 \\ 2x - 4y + 2z &= 2 \end{aligned}$

25. The sum of the digits of a three-digit number is 14. The sum of the units and hundreds digits is equal to the tens digit, and four times the units digit is three more than three times the tens digit. Find the original number.

[8.5] *Evaluate each 2 × 2 determinant.*

26. $\begin{vmatrix} 5 & 4 \\ 3 & 2 \end{vmatrix}$

27. $\begin{vmatrix} 3 & -5 \\ 0 & 2 \end{vmatrix}$

28. $\begin{vmatrix} \frac{1}{2} & -\frac{1}{4} \\ \frac{1}{4} & \frac{3}{4} \end{vmatrix}$

Evaluate each 3 × 3 determinant. Expand about the row or column of your choice.

29. $\begin{vmatrix} 1 & 2 & -4 \\ -1 & 0 & 1 \\ 3 & 0 & 2 \end{vmatrix}$

30. $\begin{vmatrix} 3 & 4 & 5 \\ 2 & 1 & 0 \\ 0 & 4 & -1 \end{vmatrix}$

31. $\begin{vmatrix} 1 & 2 & 1 \\ 1 & 1 & -1 \\ 2 & 4 & 2 \end{vmatrix}$

Solve each determinant equation for x.

32. $\begin{vmatrix} x & 2 \\ 4x & 3 \end{vmatrix} = 5$

33. $\begin{vmatrix} x & \frac{1}{2} \\ 3x & 2 \end{vmatrix} = 3$

[8.6] *Use Cramer's rule to solve each system.*

34. $\begin{aligned} 3x + 4y &= 6 \\ 5x + 3y &= -1 \end{aligned}$

35. $\begin{aligned} 2x - 3y &= 0 \\ 3x + 2y &= 13 \end{aligned}$

36. $\begin{aligned} 2x + 5y &= -3 \\ -3x + 10y &= 8 \end{aligned}$

37. $\begin{aligned} x + 2y + z &= 3 \\ 2x - y - z &= 0 \\ 3x - 2y - z &= 5 \end{aligned}$

38. $\begin{aligned} 2x - y + z &= 5 \\ 4x - 3y &= 5 \\ 6x + 2y + 2z &= 7 \end{aligned}$

39. $\begin{aligned} 3x + 3y - z &= 3 \\ x + 6y + 5z &= 0 \\ x - 3y - 3z &= 1 \end{aligned}$

CHAPTER 8 TEST

Classify each system as independent, inconsistent, or dependent.

1. $\begin{aligned} x + y &= 1 \\ x + y &= 4 \end{aligned}$

2. $\begin{aligned} y &= 1 - 2x \\ 4x + 2y &= 2 \end{aligned}$

3. Determine whether $(2, -3)$ is a solution of the following system.

$$3x + y = 3$$
$$x - 2y = 8$$

Evaluate each determinant.

4. $\begin{vmatrix} 6 & 2 \\ 5 & 3 \end{vmatrix}$

5. $\begin{vmatrix} 5 & 5 \\ 2 & 2 \end{vmatrix}$

Solve each system by graphing.

6. $x + y = 0$
  $x - y = 0$

7. $x + y = 6$
  $x - 2y = 0$

Solve each system using the substitution method.

8. $2x - y = 5$
  $5x + 6y = 4$

9. $y = 2x + 3$
  $4x - 2y = 5$

Solve each system using the addition method.

10. $3x - 2y = 3$
  $x - y = -1$

11. $5x + y = 2$
  $3x - y = 2$

Evaluate each determinant.

12. $\begin{vmatrix} 3 & -2 & 2 \\ 1 & 0 & -1 \\ 1 & -1 & 3 \end{vmatrix}$

13. $\begin{vmatrix} 2 & -1 & 3 \\ 1 & 2 & -1 \\ 3 & -2 & 3 \end{vmatrix}$

14. Solve the determinant equation $\begin{vmatrix} x & x \\ 3 & 5 \end{vmatrix} = 6$.

Solve each system of equations.

15. $x + y + z = 3$
   $y + 2z = 3$
  $2x - z = 2$

16. $2x + 3y + z = 4$
  $x - z = -5$
  $x + 6y = 4$

Use Cramer's rule to solve each system.

17. $3x - y = 9$
  $2x + y = 1$

18. $x - y + z = -4$
  $2x + 3y + z = 5$
  $4x + y - z = 9$

19. A motorboat travels 33 miles upstream in 3 hours, then turns and travels 28 miles downstream in 2 hours. Find the speed of the boat and the speed of the current.

20. Two magazines and four books cost $14. Three magazines and one book cost $6. Find the cost of each magazine and each book.

In Chapter 7 we saw that the graph of a first-degree equation in two variables is a straight line. In this chapter we graph second-degree equations, which are equations with one or more second-degree terms. The graph of a second-degree equation in two variables may be a *parabola*, a *circle*, an *ellipse*, or a *hyperbola*. Collectively, these four curves are known as the *conic sections.** They are called conic sections because each can be formed by intersecting a plane with a cone as shown in Figure 9.1 (see following page). As you will see, the conic sections have many interesting and useful applications in a variety of fields, from mathematics, science, and engineering to astronomy, business, and architecture. Later, in Section 9.4, you will solve systems of second-degree equations. Finally, in Section 9.5, you will graph second-degree inequalities.

---

* The conic sections were first written about in detail by Appollonius of Alexandria in the third century B.C.

9

The Conic Sections

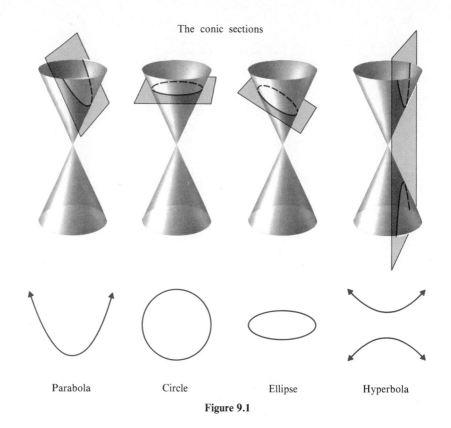

The conic sections

Parabola   Circle   Ellipse   Hyperbola

Figure 9.1

The Parabola

When an object is projected upward (but not vertically), the trajectory of the object is essentially a parabola. Solar furnaces use parabolas as an integral part of their design. The reflectors in spotlights and satellite dishes have parabolic cross sections, and the 200-inch mirror in the telescope at Mount Palomar Observatory uses its parabolic shape to reflect light to its focus 55 feet away (see Figure 9.2).

Any equation of the form

$$y = ax^2 + bx + c \qquad (a \neq 0)$$

represents a parabola. For example if $a = 1$, $b = 0$, and $c = -4$, we have the parabola whose equation is $y = x^2 - 4$.

Ignoring air resistance, the path of this cannon-ball is a portion of a parabola.

When a light is placed at point F (the *focus*), the parabolic reflector will direct the rays into parallel beams.

Television signals from a satellite are collected by the parabolic dish and bounced to the "feedhorn."

Figure 9.2

EXAMPLE 1   Graph $y = x^2 - 4$.

Construct a table of values by substituting values for x and computing the corresponding values of y.

When $x = 3$, $y = 3^2 - 4 = 9 - 4 = 5$.
When $x = 2$, $y = 2^2 - 4 = 4 - 4 = 0$.
When $x = 1$, $y = 1^2 - 4 = 1 - 4 = -3$.
When $x = \frac{1}{2}$, $y = (\frac{1}{2})^2 - 4 = \frac{1}{4} - 4 = -\frac{15}{4}$.

If we continue in this fashion, we obtain the following table of values.

x	3	2	1	$\frac{1}{2}$	0	$-\frac{1}{2}$	-1	-2	-3
y	5	0	-3	$-\frac{15}{4}$	-4	$-\frac{15}{4}$	-3	0	5

If we graph these nine points, they fall into the U-shaped pattern shown in Figure 9.3. In fact, the graph of *every* solution to $y = x^2 - 4$ lies in this pattern.

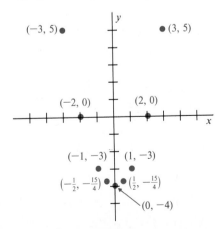

Figure 9.3

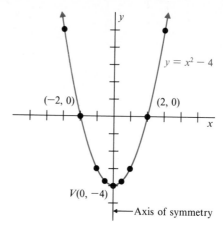

Figure 9.4

Likewise, every point in the pattern represents a solution to $y = x^2 - 4$. Therefore to complete the graph, we join the nine points with a smooth curve to get

Try Problem 1

the parabola shown in Figure 9.4. ◄

If we "fold" Figure 9.4 along the y-axis, the two halves of the parabola will coincide. Therefore the y-axis is called the **axis of symmetry** for this parabola. The point at which the axis of symmetry intersects the parabola is called the **vertex** of the parabola. In this case, the vertex V is $(0, -4)$, and it is the *lowest* point of the parabola.

The next example illustrates that when a is a *negative* number, the parabola $y = ax^2 + bx + c$ opens *downward*.

EXAMPLE 2 Graph $y = -x^2 + 1$.

Substitute values for x and compute the corresponding values of y.

$$\text{When } x = 3, \ y = -3^2 + 1 = -9 + 1 = -8.$$
$$\text{When } x = 2, \ y = -2^2 + 1 = -4 + 1 = -3.$$

Continue in this fashion and obtain the following table of values.

x	y
3	-8
2	-3
1	0
0	1
-1	0
-2	-3
-3	-8

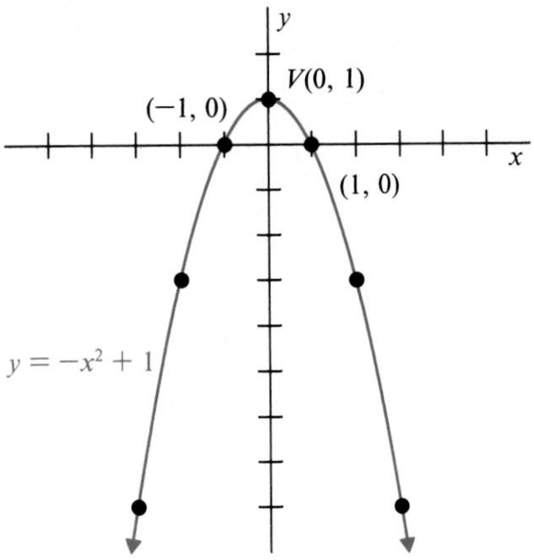

$$V(0, 1)$$

$$(-1, 0)$$

$$(1, 0)$$

$$y = -x^2 + 1$$

Figure 9.5

Graph these seven points and join them with a smooth curve to get the parabola shown in Figure 9.5. Note that in this case the vertex (0, 1) is the *highest* point of the parabola. ◄

Try Problem 5

The most important points of the parabola $y = ax^2 + bx + c$ are the y-intercept, the x-intercepts, and the vertex. To find the y-intercept, set $x = 0$.

$$y = a(0)^2 + b(0) + c$$
$$y = c$$

Therefore there is one y-intercept, and it is c. To find the x-intercepts, set $y = 0$.

$$0 = ax^2 + bx + c$$

Since this is a quadratic equation, it will have two, one, or no real solutions. Therefore the parabola $y = ax^2 + bx + c$ will have two, one, or no x-intercepts (see Figure 9.6). To find the x-value of the vertex, use the formula

$$x = \frac{-b}{2a}.$$

This formula can be proved by rewriting the right side of $y = ax^2 + bx + c$, but we leave that proof to a more advanced course.

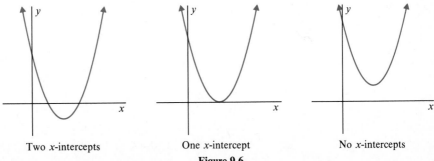

Two x-intercepts   One x-intercept   No x-intercepts

Figure 9.6

EXAMPLE 3   Graph $y = x^2 - 6x + 8$. Label the y-intercept, the x-intercepts (if any), and the vertex.

Since $c = 8$, the y-intercept is 8. Set $y = 0$ to find the x-intercepts.

$$0 = x^2 - 6x + 8$$
$$0 = (x - 2)(x - 4) \qquad \text{Factor}$$
$$x = 2 \quad \text{or} \quad x = 4 \qquad \text{Two } x\text{-intercepts}$$

The x-value of the vertex is

$$x = \frac{-b}{2a} = \frac{-(-6)}{2(1)} = \frac{6}{2} = 3.$$

To find the y-value of the vertex, substitute $x = 3$ into the original equation.

$$y = 3^2 - 6(3) + 8 = 9 - 18 + 8 = -1$$

Since $a = 1$ (a positive number), the parabola opens upward as shown in Figure 9.7. The axis of symmetry for this parabola is the vertical line $x = 3$. Using symmetry, we conclude that the point $(6, 8)$ must be on the parabola, since the point $(0, 8)$ is on the parabola. ◄

Try Problem 15

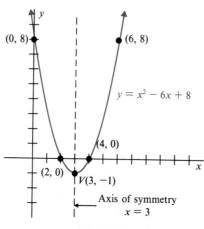

Figure 9.7

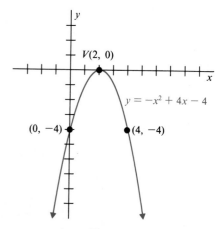

Figure 9.8

EXAMPLE 4   Graph $y = -x^2 + 4x - 4$. Label the intercepts and the vertex.

The y-intercept is -4. Set $y = 0$ to find the x-intercepts.

$$0 = -x^2 + 4x - 4$$
$$0 = x^2 - 4x + 4 \qquad \text{Multiply each side by } -1$$
$$0 = (x - 2)^2 \qquad \text{Factor}$$
$$x = 2 \qquad \text{Only one } x\text{-intercept}$$

The x-value of the vertex is

$$x = \frac{-b}{2a} = \frac{-4}{2(-1)} = \frac{-4}{-2} = 2.$$

The y-value of the vertex is

$$y = -2^2 + 4 \cdot 2 - 4 = -4 + 8 - 4 = 0.$$

Since $a = -1$ (a negative number), the parabola opens downward as shown in Figure 9.8. Since $(0, -4)$ is on the parabola, by symmetry $(4, -4)$ must be on the parabola. ◄

Try Problem 21

EXAMPLE 5   Graph $y = \frac{1}{3}x^2 + x + 1$. Label the intercepts and the vertex.

The y-intercept is 1. Set $y = 0$ to find the x-intercepts.

$$0 = \frac{1}{3}x^2 + x + 1$$

$$0 = x^2 + 3x + 3 \qquad \text{Multiply each side by 3}$$

Solve this equation using the quadratic formula.

$$x = \frac{-3 \pm \sqrt{3^2 - 4(1)(3)}}{2(1)} = \frac{-3 \pm \sqrt{-3}}{2}$$

These solutions are not real numbers; therefore the parabola has no x-intercepts. The x-value of the vertex is

$$x = \frac{-b}{2a} = \frac{-1}{2\left(\frac{1}{3}\right)} = \frac{-1}{\frac{2}{3}} = -\frac{3}{2}.$$

The y-value of the vertex is

$$y = \frac{1}{3}\left(-\frac{3}{2}\right)^2 + \left(-\frac{3}{2}\right) + 1 = \frac{3}{4} - \frac{6}{4} + \frac{4}{4} = \frac{1}{4}.$$

Since $a = \frac{1}{3}$, the parabola opens upward as shown in Figure 9.9. You can improve the accuracy of this parabola by graphing additional points.

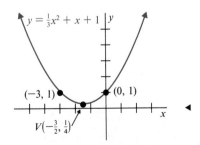

Try Problem 29

Figure 9.9

We can summarize the procedure for graphing a parabola as follows:

To Graph the Parabola $y = ax^2 + bx + c$

1. Note that the y-intercept is c.
2. Set $y = 0$ to find the x-intercepts. You may have to use the quadratic formula. There will be two, one, or no x-intercepts.
3. Use the formula $x = \dfrac{-b}{2a}$ to find the x-value of the vertex. Substitute this x-value into the original equation to find the y-value of the vertex.
4. Remember that when $a > 0$, the parabola opens upward; when $a < 0$, the parabola opens downward.
5. Draw a smooth curve through the intercepts and the vertex. Graph additional points if necessary.

We will now illustrate how a parabola can be used to obtain maximum area when constructing a rectangular pasture.

Problem Solving

Maximizing the Area of a Pasture

A farmer plans to use 12 miles of fencing to construct a rectangular pasture. One side of the farmer's property lies at the foot of a straight cliff and needs no fence. What should the dimensions of the pasture be to maximize its area? What is the maximum area?

Let x be the width of the pasture. Since there are 12 miles of fencing, the length of the pasture must be $12 - 2x$ (see Figure 9.10). Therefore the area A of the pasture is given by

$A = \text{width} \cdot \text{length}$

$A = x(12 - 2x)$

$A = 12x - 2x^2$

$A = -2x^2 + 12x.$

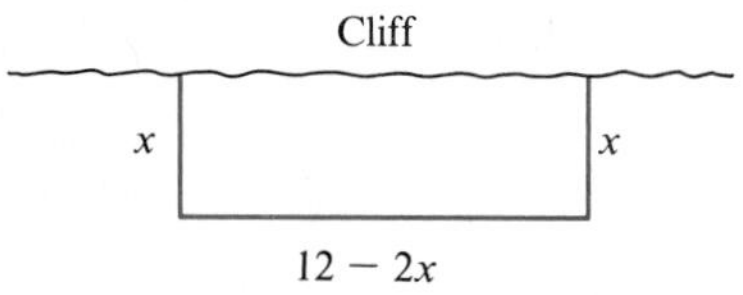

Figure 9.10

The graph of this equation is a parabola with A-intercept 0. Set $A = 0$ to find the x-intercepts.

$$0 = -2x^2 + 12x$$
$$0 = -2x(x - 6) \qquad \text{Factor}$$
$$-2x = 0 \quad \text{or} \quad x - 6 = 0$$
$$x = 0 \quad \text{or} \qquad x = 6$$

The x-value of the vertex is

$$x = \frac{-12}{2(-2)} = \frac{-12}{-4} = 3.$$

The A-value of the vertex is

$$A = -2(3)^2 + 12(3) = -18 + 36 = 18.$$

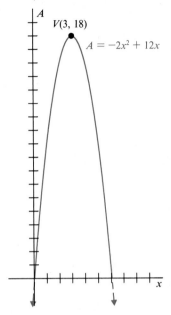

$V(3, 18)$

$A = -2x^2 + 12x$

Figure 9.11

Since $a = -2$, the parabola opens downward as shown in Figure 9.11. There-fore the vertex is the highest point of the parabola, and it represents the maxi-mum value of the area A. The maximum area of 18 square miles occurs when the width is $x = 3$ miles, and the length is $12 - 2x = 12 - 2(3) = 6$ miles.

PROBLEM SET 9.1

Graph each parabola. Label the intercepts and the vertex.

1. $y = x^2 - 1$
2. $y = x^2 - 9$
3. $y = 2x^2 - 8$
4. $y = 2x^2 - 2$
5. $y = -x^2 + 9$
6. $y = -x^2 + 4$
7. $y = 2 - x^2$
8. $y = 3 - x^2$
9. $y = x^2 + 1$
10. $y = x^2 + 2$
11. $y = x^2 - 4x$
12. $y = x^2 + 4x$
13. $y = -6x - 3x^2$
14. $y = 6x - 3x^2$
15. $y = x^2 - 6x + 5$
16. $y = x^2 - 4x + 3$
17. $y = x^2 - 2x - 3$
18. $y = x^2 - 2x - 8$
19. $y = -x^2 - 4x + 5$
20. $y = -x^2 - 8x - 7$
21. $y = -x^2 + 2x - 1$
22. $y = -x^2 + 6x - 9$
23. $y = 3x^2 - 2x - 1$
24. $y = 3x^2 + 2x - 1$
25. $y = x^2 + 3x + 1$
26. $y = x^2 - 3x + 1$
27. $y = \frac{1}{3}x^2 - 2x + 3$
28. $y = \frac{1}{4}x^2 - 2x + 4$
29. $y = \frac{1}{2}x^2 + x + 1$
30. $y = \frac{1}{2}x^2 - x + 1$

For each problem below, graph the three parabolas using the same coordinate axes. Then compare the parabolas.

31. a) $y = x^2$ **b)** $y = 2x^2$ **c)** $y = \frac{1}{2}x^2$ **32. a)** $y = -x^2$ **b)** $y = -2x^2$ **c)** $y = -\frac{1}{2}x^2$

33. a) $y = x^2$ **b)** $y = x^2 - 2$ **c)** $y = x^2 + 2$ **34. a)** $y = x^2$ **b)** $y = (x - 2)^2$ **c)** $y = (x + 2)^2$

35. A particular frog leap follows a path given by the equation $y = 2x - \frac{1}{4}x^2$, where x and y are in feet and $0 \le x \le 8$. Graph this path.

36. A particular football punt follows a path given by the equation $y = 2x - \frac{1}{30}x^2$, where x and y are in yards and $0 \le x \le 60$. Graph this path.

37. A rancher plans to use 8 miles of fencing to construct a rectangular pasture. One side of the rancher's property lies along an existing fence and needs no additional fencing (see Figure 9.12). What should the dimensions of the pasture be to maximize its area? What is the maximum area?

38. The outside edges of a piece of sheet metal that is 16 inches wide are folded up to make a trough (see Figure 9.13). What should x be to maximize the cross-sectional area of the trough? What is the maximum area?

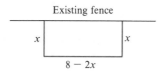

Figure 9.12

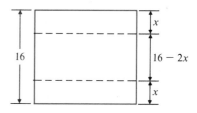

Figure 9.13

If we interchange x and y in the equation $y = ax^2 + bx + c$, we obtain $x = ay^2 + by + c$. The graph of the second equation is a parabola that opens right (if $a > 0$) or left (if $a < 0$). Use this information to graph each of the following parabolas.

† **39.** $x = y^2$ †**40.** $x = y^2 - 4$ †**41.** $x = y^2 - 4y - 5$ †**42.** $x = -y^2 + 6y$

CALCULATOR PROBLEMS

43. Find the vertex and intercepts of the parabola whose equation is $y = x^2 + 2.7x - 20.74$. Do not graph.

44. Suppose that the weekly profit P from manufacturing x rocking chairs is given by the equation $P = -0.68x^2 + 176.8x - 6807$. How many rocking chairs should be manufactured to maximize weekly profit? What is the maximum weekly profit?

9.2   *The Circle*

In Section 9.1 we learned to graph parabolas. In this section we learn to graph a different conic section — the circle. First, consider the circle with center at (h, k)

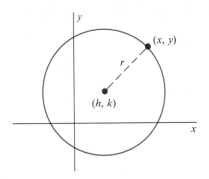

Figure 9.14

and radius r, as shown in Figure 9.14. If (x, y) is any point on the circle, then the distance between (x, y) and (h, k) is r. Using the distance formula given in Section 7.3, we can write

$$\sqrt{(x - h)^2 + (y - k)^2} = r.$$

If we square each side of this equation, we get the **standard form** for the equation of a circle.

Standard Form for the Equation of a Circle

An equation for the circle with center at (h, k) and radius r is

$$(x - h)^2 + (y - k)^2 = r^2.$$

EXAMPLE 1   Write an equation for the circle with center at $(1, 6)$ and radius 9.

Use the standard form for the equation of a circle with $(h, k) = (1, 6)$ and $r = 9$.

$$(x - 1)^2 + (y - 6)^2 = 9^2$$

Try Problem 1
$$(x - 1)^2 + (y - 6)^2 = 81 \blacktriangleleft$$

EXAMPLE 2   Determine the center and radius of the circle whose equation is $x^2 + y^2 = 9$.

This equation can be written as follows:

$$(x - 0)^2 + (y - 0)^2 = 3^2.$$

Try Problem 11   Therefore the center is at $(0, 0)$ and the radius is 3. $\blacktriangleleft$

EXAMPLE 3   Determine the center and radius of the circle whose equation is $(x - 5)^2 + (y + 2)^2 = 16$.

This equation can be written as follows:

$$(x - 5)^2 + [y - (-2)]^2 = 4^2.$$

Try Problem 21

Therefore the center is at $(5, -2)$ and the radius is 4. ◄

We can write the equation of Example 3 in a different form by expanding $(x - 5)^2$ and $(y + 2)^2$ and collecting terms on the left side.

$$(x - 5)^2 + (y + 2)^2 = 16$$
$$x^2 - 10x + 25 + y^2 + 4y + 4 = 16$$
$$x^2 + y^2 - 10x + 4y + 13 = 0$$

This is called the **general form** for the equation of a circle. To graph a circle whose equation is given in general form, we must first convert the equation to standard form by *completing the square* (refer to Section 6.2).

EXAMPLE 4   Graph $x^2 + y^2 - 6x + 4y - 12 = 0$.

Rearrange terms as follows:

$$(x^2 - 6x) + (y^2 + 4y) = 12.$$

Then complete the square on $x^2 - 6x$ by adding $(\frac{-6}{2})^2 = 9$ to each side. Complete the square on $y^2 + 4y$ by adding $(\frac{4}{2})^2 = 4$ to each side.

$$(x^2 - 6x + 9) + (y^2 + 4y + 4) = 12 + 9 + 4$$

Factor the two perfect-square trinomials on the left side.

$$(x - 3)^2 + (y + 2)^2 = 25$$

The graph is the circle with the center at $(3, -2)$ and radius 5, as shown in Figure 9.15.

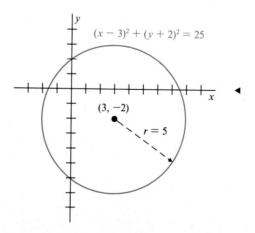

Try Problem 27

Figure 9.15

EXAMPLE 5   Write an equation for the circle with center at $(-2, -1)$ that passes through the point $(-3, 4)$.

Use the distance formula to find the radius r (see Figure 9.16).

$$r = \sqrt{[-3 - (-2)]^2 + [4 - (-1)]^2} = \sqrt{1 + 25} = \sqrt{26}$$

Substitute $(h, k) = (-2, -1)$ and $r = \sqrt{26}$ into the standard form of a circle and simplify.

$$[x - (-2)]^2 + [y - (-1)]^2 = (\sqrt{26})^2$$
$$(x + 2)^2 + (y + 1)^2 = 26$$

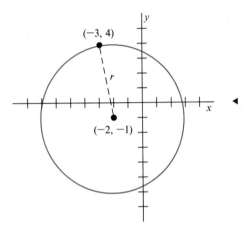

Figure 9.16

Try Problem 41

PROBLEM SET 9.2

Write an equation for the circle with the given center and radius.

1. Center at $(3, 5)$, radius 4
2. Center at $(1, 4)$, radius 25
3. Center at $(4, -10)$, radius 5
4. Center at $(9, -5)$, radius 6
5. Center at $(-\frac{1}{2}, -2)$, radius 1
6. Center at $(-3, -\frac{1}{3})$, radius 1
7. Center at $(0, 1)$, radius $\frac{1}{2}$
8. Center at $(2, 0)$, radius $\frac{1}{3}$
9. Center at $(0, 0)$, radius $\sqrt{2}$
10. Center at $(0, 0)$, radius $\sqrt{3}$

Determine the center and radius of each circle.

11. $x^2 + y^2 = 4$
12. $x^2 + y^2 = 16$
13. $3x^2 + 3y^2 = 3$
14. $5x^2 + 5y^2 = 5$

15. $4x^2 + 4y^2 = 25$

16. $9x^2 + 9y^2 = 16$

17. $x^2 + (y - 4)^2 = 16$

18. $x^2 + (y - 2)^2 = 4$

19. $(x - 3)^2 + y^2 = 81$

20. $(x - 5)^2 + y^2 = 100$

21. $(x - 1)^2 + (y + 6)^2 = 144$

22. $(x - 3)^2 + (y + 4)^2 = 64$

23. $(x + \frac{1}{4})^2 + (y + \frac{3}{4})^2 = \frac{1}{9}$

24. $(x + \frac{1}{5})^2 + (y + \frac{3}{5})^2 = \frac{1}{4}$

25. $(x - 8)^2 + (y - 8)^2 = 8$

26. $(x - 27)^2 + (y - 27)^2 = 27$

Graph each circle. Label the center and the radius.

27. $x^2 + y^2 - 4x + 6y - 12 = 0$

28. $x^2 + y^2 - 6x + 4y - 3 = 0$

29. $x^2 + y^2 - 8x - 10y + 40 = 0$

30. $x^2 + y^2 - 10x - 8y + 40 = 0$

31. $x^2 + y^2 + 2x - 2y - 7 = 0$

32. $x^2 + y^2 + 2x - 2y - 2 = 0$

33. $x^2 + y^2 + 12x + 20 = 0$

34. $x^2 + y^2 + 14x + 24 = 0$

35. $x^2 + y^2 - 4x = 0$

36. $x^2 + y^2 - 6x = 0$

37. $x^2 + y^2 = 1$

38. $x^2 + y^2 = 9$

39. An empty cylindrical water glass with a moist rim is inverted and placed so that it covers an equal amount of area in each of the four quadrants. If the diameter of the top of the glass is 3 inches, write an equation for the water mark left by the glass.

40. A goat is attached to a stake at the point $(2, -1)$ by a rope that is 9 feet long. If the goat walks clockwise while straining at the rope, write an equation for the goat's path.

41. Write an equation for the circle with center at $(-1, -6)$ that passes through the point $(4, -3)$.

42. Write an equation for the circle with center at $(-2, -4)$ that passes through the point $(5, -1)$.

Graph each equation.

† 43. $x^2 + y^2 = 0$

† 44. $9x^2 + 9y^2 - 6x + 12y + 1 = 0$

Graph each equation. Can you identify each curve?

† 45. $y = \sqrt{25 - x^2}$

† 46. $y = -\sqrt{25 - x^2}$

CALCULATOR PROBLEMS

Determine the center and radius of each circle.

47. $2.3x^2 + 2.3y^2 = 74.727$

48. $x^2 + y^2 + 6.4x - 3.9y - 4.02 = 0$

The Ellipse and the Hyperbola

In this section we learn to graph the remaining two conic sections — the ellipse and the hyperbola. To simplify matters somewhat, we consider only ellipses and hyperbolas that are centered at the origin.

THE ELLIPSE

The orbits of the planets and of manmade satellites are ellipses (see Figure 9.17), as are the cross-sections of many camshafts and gears. The Whispering Gallery in the Rotunda of the Capitol Building in Washington, D.C., uses its elliptical shape to bounce sound waves from a person whispering at one focus of the ellipse to a person listening at the other focus. In a similar manner, the elliptical reflector in a dentist's light concentrates the rays from a light placed at one focus to the patient's mouth at the other focus.

Any equation of the form

$$\frac{x^2}{a^2} + \frac{y^2}{b^2} = 1 \qquad (a \neq 0,\, b \neq 0)$$

represents an ellipse centered at the origin. To find the *x*-intercepts, set $y = 0$.

$$\frac{x^2}{a^2} + \frac{0^2}{b^2} = 1$$

$$\frac{x^2}{a^2} = 1 \qquad \text{Since } \frac{0^2}{b^2} = 0$$

$$x^2 = a^2 \qquad \text{Multiply by } a^2$$

$$x = \pm a \qquad \text{Extract roots}$$

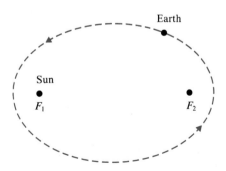

The Sun is at one focus
of the elliptical orbit.

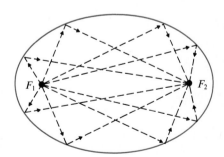

Sound or light is reflected
from one focus of an ellipse
to the other.

Figure 9.17

Therefore the x-intercepts are a and $-a$ (the square roots of the denominator of x^2/a^2). By setting $x = 0$, we can show that the y-intercepts are b and $-b$ (the square roots of the denominator of y^2/b^2).

EXAMPLE 1   Graph $\dfrac{x^2}{9} + \dfrac{y^2}{4} = 1$.

This represents an ellipse with x-intercepts 3 and -3 and y-intercepts 2 and -2. Draw a smooth oval-shaped curve through the four intercepts to produce the ellipse shown in Figure 9.18. ◄

Try Problem 1

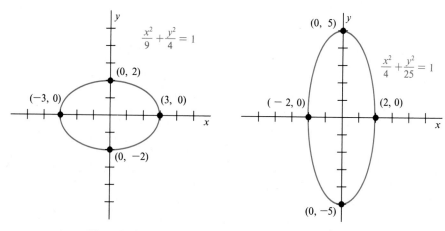

Figure 9.18   **Figure 9.19**

EXAMPLE 2   Graph $25x^2 + 4y^2 = 100$.

Divide each side by 100.

$$\frac{25x^2}{100} + \frac{4y^2}{100} = \frac{100}{100}$$

$$\frac{x^2}{4} + \frac{y^2}{25} = 1$$

This represents an ellipse with x-intercepts 2 and -2 and y-intercepts 5 and -5, as shown in Figure 9.19. ◄

Try Problem 11

EXAMPLE 3   Graph $x^2 + 16y^2 = 16$.

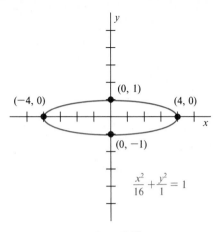

Figure 9.20

Divide each side by 16.

$$\frac{x^2}{16} + \frac{16y^2}{16} = \frac{16}{16}$$

$$\frac{x^2}{16} + \frac{y^2}{1} = 1$$

The graph is an ellipse with x-intercepts 4 and -4 and y-intercepts 1 and -1, as shown in Figure 9.20. ◄

Try Problem 13

We can summarize the procedure for graphing an ellipse as follows:

To Graph an Ellipse Centered at the Origin

1. Write the equation in the **standard form**

$$\frac{x^2}{a^2} + \frac{y^2}{b^2} = 1.$$

2. Locate the x-intercepts at a and $-a$. Locate the y-intercepts at b and $-b$.
3. Draw a smooth oval-shaped curve through the four intercepts.

THE HYPERBOLA

A hyperbola consists of two separate curves, called *branches*. Each branch looks like a parabola, but it is not. Hyperbolas are useful in solving problems involving optics, architecture, comets and meteors, market-area analysis, and the loran system of navigation for ships and aircraft.

Any equation of the form

$$\frac{x^2}{a^2} - \frac{y^2}{b^2} = 1 \qquad (a \neq 0, b \neq 0)$$

represents a hyperbola centered at the origin. The x-intercepts are a and $-a$. However, there are no y-intercepts. To see this, set $x = 0$.

$$\frac{0^2}{a^2} - \frac{y^2}{b^2} = 1$$

$$-\frac{y^2}{b^2} = 1$$

$$y^2 = -b^2 \qquad \text{Multiply by } -b^2$$

$$y = \pm\sqrt{-b^2} \qquad \text{Extract roots}$$

Since $-b^2$ is a negative number, $\sqrt{-b^2}$ is not a real number. Therefore the hyperbola has no y-intercepts.

EXAMPLE 4   Graph $\dfrac{x^2}{9} - \dfrac{y^2}{4} = 1$.

This represents the hyperbola with x-intercepts 3 and -3 shown in Figure 9.21. To help draw the hyperbola, we form a rectangle whose vertical sides intersect the x-axis at 3 and -3 (the square roots of 9 in $x^2/9$) and whose horizontal sides intersect the y-axis at 2 and -2 (the square roots of 4 in $y^2/4$). The extended diagonals of the rectangle are called **asymptotes.** The asymptotes help us to sketch the hyperbola accurately, since the two branches of the hyperbola approach, but never touch, the asymptotes. The rectangle and the asymptotes are dashed, since they are not part of the hyperbola itself. ◄

Try Problem 17

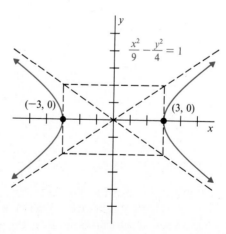

Figure 9.21

If the equation is of the form

$$\frac{y^2}{b^2} - \frac{x^2}{a^2} = 1 \qquad (a \neq 0,\, b \neq 0)$$

the hyperbola has y-intercepts b and $-b$, and no x-intercepts.

EXAMPLE 5   Graph $\dfrac{y^2}{4} - \dfrac{x^2}{9} = 1$.

This equation produces the same rectangle and asymptotes shown in Figure 9.21. In this case, however, the y-intercepts of the hyperbola are 2 and -2, and *Try Problem 19* there are no x-intercepts. The graph of this equation is shown in Figure 9.22. ◄

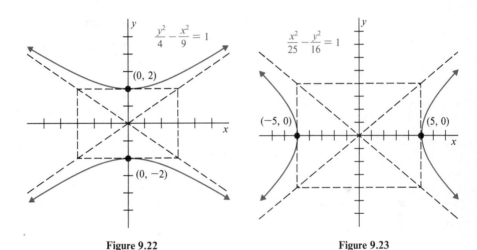

Figure 9.22   Figure 9.23

EXAMPLE 6   Graph $16x^2 - 25y^2 = 400$.

Divide each side by 400.

$$\frac{16x^2}{400} - \frac{25y^2}{400} = \frac{400}{400}$$

$$\frac{x^2}{25} - \frac{y^2}{16} = 1$$

Try Problem 25   The graph is the hyperbola with x-intercepts 5 and -5 shown in Figure 9.23. ◄

We can summarize the procedure for graphing a hyperbola as shown on the following page.

To Graph a Hyperbola Centered at the Origin

1. Write the equation in one of the **standard forms**

$$\frac{x^2}{a^2} - \frac{y^2}{b^2} = 1 \quad \text{or} \quad \frac{y^2}{b^2} - \frac{x^2}{a^2} = 1.$$

2. Draw a dashed rectangle whose vertical sides intersect the x-axis at a and $-a$, and whose horizontal sides intersect the y-axis at b and $-b$.
3. Draw two dashed asymptotes by extending the diagonals of the rectangle.
4. Determine the intercepts of the hyperbola.
5. Sketch each branch of the hyperbola so that it passes through an intercept and approaches the asymptotes.

PROBLEM SET 9.3

Graph each ellipse. Label the intercepts.

1. $\dfrac{x^2}{25} + \dfrac{y^2}{9} = 1$
2. $\dfrac{x^2}{16} + \dfrac{y^2}{9} = 1$
3. $\dfrac{x^2}{9} + \dfrac{y^2}{25} = 1$
4. $\dfrac{x^2}{9} + \dfrac{y^2}{16} = 1$

5. $\dfrac{x^2}{4} + \dfrac{y^2}{9} = 1$
6. $\dfrac{x^2}{4} + \dfrac{y^2}{16} = 1$
7. $\dfrac{x^2}{81} + \dfrac{y^2}{49} = 1$
8. $\dfrac{x^2}{64} + \dfrac{y^2}{36} = 1$

9. $\dfrac{x^2}{4} + \dfrac{y^2}{5} = 1$
10. $\dfrac{x^2}{4} + \dfrac{y^2}{7} = 1$
11. $25x^2 + 16y^2 = 400$
12. $4x^2 + 25y^2 = 100$

13. $x^2 + 9y^2 = 9$
14. $x^2 + 4y^2 = 4$
15. $x^2 + 4y^2 = 1$
16. $x^2 + 9y^2 = 1$

Graph each hyperbola. Label the intercepts.

17. $\dfrac{x^2}{16} - \dfrac{y^2}{9} = 1$
18. $\dfrac{x^2}{25} - \dfrac{y^2}{9} = 1$
19. $\dfrac{y^2}{9} - \dfrac{x^2}{16} = 1$
20. $\dfrac{y^2}{9} - \dfrac{x^2}{25} = 1$

21. $\dfrac{x^2}{4} - \dfrac{y^2}{4} = 1$
22. $\dfrac{x^2}{16} - \dfrac{y^2}{16} = 1$
23. $\dfrac{y^2}{4} - \dfrac{x^2}{4} = 1$
24. $\dfrac{y^2}{16} - \dfrac{x^2}{16} = 1$

25. $9x^2 - 64y^2 = 576$
26. $4x^2 - 49y^2 = 196$
27. $x^2 - y^2 = 1$
28. $y^2 - x^2 = 1$

Write an equation for each conic section graphed below.

29.

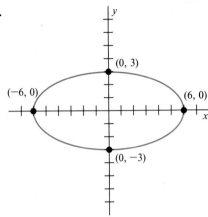

30.

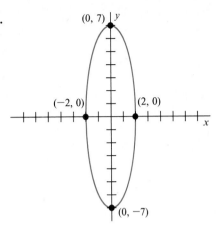

31.

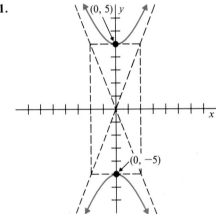

32.

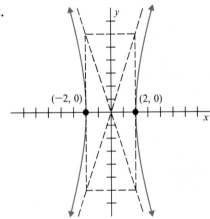

Identify each equation as representing a parabola, a circle, an ellipse, or a hyperbola. Do not graph.

33. $9x^2 + 9y^2 - 36 = 0$ **34.** $4x^2 + 4y^2 - 64 = 0$ **35.** $x^2 = 49 + y^2$ **36.** $y^2 = 81 + x^2$

37. $4y^2 = 100 - 25x^2$ **38.** $9x^2 = 900 - 100y^2$ **39.** $y - 4x^2 = 1$ **40.** $y + 9x^2 = 1$

Graph each equation. Do you see how your graph could be formed by intersecting a plane with a cone?

† 41. $y^2 - x^2 = 0$ **† 42.** $y^2 - 4x^2 = 0$

CALCULATOR PROBLEMS

Graph each conic section.

43. $14.44x^2 + 5.76y^2 = 83.1744$ **44.** $2.89x^2 - 20.25y^2 = 58.5225$

9.4   *Nonlinear Systems of Equations*

In the first three sections of the chapter, we learned to graph the four conic sections. Sometimes, however, we want to find the intersection point(s) of two or more conic sections. For example, suppose that seismic station *A* determines that the center of an earthquake (the epicenter) is 130 miles away. Then the epicenter must lie on a circle whose center is at station *A* and whose radius is 130 miles (see Figure 9.24a). If similar circles are drawn centered at two other seismic stations, then the intersection point of the three circles gives the location of the epicenter (see Figure 9.24b).

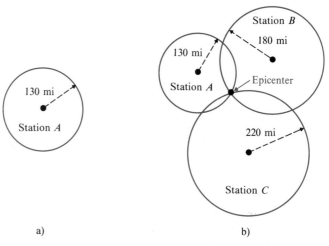

a)   b)

Figure 9.24

To find the intersection point(s) of two or more conic sections, we must solve a nonlinear system of equations. A **nonlinear system of equations** is a system that contains at least one nonlinear equation. We can solve a nonlinear system using the substitution method, the addition method, or a combination of the two. We illustrate the substitution method in Examples 1 and 2.

EXAMPLE 1   Solve the following system.
$$x^2 + y^2 = 25$$
$$2x + y = 10$$

Solve the linear equation for *y*.
$$2x + y = 10$$
$$y = 10 - 2x$$

Substitute $10 - 2x$ for y in the nonlinear equation $x^2 + y^2 = 25$.

$$x^2 + (10 - 2x)^2 = 25$$
$$x^2 + 100 - 40x + 4x^2 = 25 \qquad \text{Expand } (10 - 2x)^2$$
$$5x^2 - 40x + 75 = 0 \qquad \text{Write in standard form}$$
$$x^2 - 8x + 15 = 0 \qquad \text{Divide by 5}$$
$$(x - 3)(x - 5) = 0 \qquad \text{Factor}$$
$$x = 3 \quad \text{or} \quad x = 5$$

Substitute the x-values into the equation $y = 10 - 2x$ to find the corresponding y-values.

When $x = 3$, $\quad y = 10 - 2 \cdot 3$ $\qquad$ When $x = 5$, $\quad y = 10 - 2 \cdot 5$
$$y = 4. \qquad\qquad\qquad\qquad\qquad y = 0.$$

Therefore the solutions are $(3, 4)$ and $(5, 0)$. To check, each solution must satisfy *both* equations in the original system.

CHECK:

(3, 4)		(5, 0)	
$x^2 + y^2 = 25$	$2x + y = 10$	$x^2 + y^2 = 25$	$2x + y = 10$
$3^2 + 4^2 \stackrel{?}{=} 25$	$2 \cdot 3 + 4 \stackrel{?}{=} 10$	$5^2 + 0^2 \stackrel{?}{=} 25$	$2 \cdot 5 + 0 \stackrel{?}{=} 10$
$9 + 16 \stackrel{\checkmark}{=} 25$	$6 + 4 \stackrel{\checkmark}{=} 10$	$25 \stackrel{\checkmark}{=} 25$	$10 \stackrel{\checkmark}{=} 10$

The solutions to Example 1 correspond to the intersection points of the circle $x^2 + y^2 = 25$ and the line $2x + y = 10$, as shown in Figure 9.25.

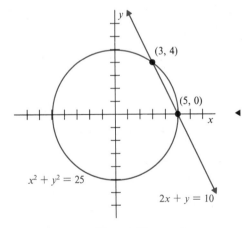

Try Problem 7

Figure 9.25

EXAMPLE 2 Solve the following system.

$$x^2 + y^2 = 16$$
$$y = x^2 - 4$$

Solve the second equation for x^2.

$$y = x^2 - 4$$
$$y + 4 = x^2$$

Substitute $y + 4$ for x^2 in the first equation.

$$(y + 4) + y^2 = 16$$
$$y^2 + y - 12 = 0$$
$$(y + 4)(y - 3) = 0$$
$$y = -4 \quad \text{or} \quad y = 3$$

Use $y = x^2 - 4$ to find the corresponding x-values.

$$\text{When } y = -4, \quad -4 = x^2 - 4$$
$$0 = x^2$$
$$0 = x.$$

$$\text{When } y = 3, \quad 3 = x^2 - 4$$
$$7 = x^2$$
$$\pm \sqrt{7} = x.$$

Try Problem 13 \ The solutions are $(0, -4)$, $(\sqrt{7}, 3)$, and $(-\sqrt{7}, 3)$. Check in the original system. ◄

The addition method is more appropriate for solving the next example.

EXAMPLE 3 Solve the following system.

$$3x^2 + 2y^2 = 116$$
$$x^2 - y^2 = 32$$

Multiply the second equation by 2 and add it to the first.

$$3x^2 + 2y^2 = 116$$
$$\underline{2x^2 - 2y^2 = 64}$$
$$5x^2 = 180$$
$$x^2 = 36$$
$$x = \pm 6$$

Use $x^2 - y^2 = 32$ to find the corresponding y-values.

When $x = 6$, $\quad 6^2 - y^2 = 32$	When $x = -6$, $\quad (-6)^2 - y^2 = 32$
$-y^2 = -4$	$-y^2 = -4$
$y^2 = 4$	$y^2 = 4$
$y = \pm 2.$	$y = \pm 2.$

The solutions are $(6, 2)$, $(6, -2)$, $(-6, 2)$, and $(-6, -2)$. Check in the original

Try Problem 19 \ system. ◄

A combination of the addition method and the substitution method offers the best approach to solving the next example.

EXAMPLE 4   Solve the following system.

$$x^2 + 2xy + y^2 = 16$$
$$x^2 + y^2 = 10$$

Multiply the second equation by -1 and add it to the first.

$$
\begin{array}{rcl}
x^2 + 2xy + y^2 &=& 16 \\
-x^2 \qquad\quad - y^2 &=& -10 \\
\hline
2xy \qquad\quad &=& 6
\end{array}
$$

$$y = \frac{6}{2x} \qquad \text{Divide by } 2x$$

$$y = \frac{3}{x} \qquad \text{Simplify}$$

Substitute $y = \frac{3}{x}$ into $x^2 + y^2 = 10$.

$$x^2 + \left(\frac{3}{x}\right)^2 = 10$$

$$x^2 + \frac{9}{x^2} = 10$$

$$x^4 + 9 = 10x^2 \qquad \text{Multiply by } x^2$$

$$x^4 - 10x^2 + 9 = 0$$

$$(x^2 - 1)(x^2 - 9) = 0 \qquad \text{Factor}$$

$$
\begin{array}{ccc}
x^2 - 1 = 0 & \text{or} & x^2 - 9 = 0 \\
x^2 = 1 & \text{or} & x^2 = 9 \\
x = \pm 1 & \text{or} & x = \pm 3
\end{array}
$$

Use $y = \frac{3}{x}$ to find the corresponding y-values.

When $x = 1$, $y = \frac{3}{1} = 3$.   When $x = -1$, $y = \frac{3}{-1} = -3$.

When $x = 3$, $y = \frac{3}{3} = 1$.   When $x = -3$, $y = \frac{3}{-3} = -1$.

The solutions are $(1, 3)$, $(-1, -3)$, $(3, 1)$, and $(-3, -1)$. Check in the original system. ◄

Try Problem 29

PROBLEM SET 9.4

Solve each system using the substitution method.

1. $y = x^2$
$y = 2x - 1$

2. $y = x^2$
$y = 4x - 4$

3. $y = x^2 + 2x + 1$
$y = x + 3$

4. $y = x^2 + 3x + 1$
$y = x + 4$

5. $y^2 - 8x^2 = 32$
$y = 4x$

6. $y^2 - 4x^2 = 45$
$y = 3x$

7. $x^2 + y^2 = 25$
$3x + y = 15$

8. $x^2 + y^2 = 25$
$2x + y = 5$

9. $x^2 + y^2 = 4$
    $x - 2y = 4$

10. $x^2 + y^2 = 16$
    $x - 2y = 8$

11. $x + y = 8$
    $xy = 16$

12. $x + y = 6$
    $xy = 9$

13. $x^2 + y^2 = 4$
    $y = x^2 - 2$

14. $x^2 + y^2 = 9$
    $y = x^2 - 3$

15. $x^2 + xy - y^2 = 19$
    $y - x = 3$

16. $x^2 + xy - y^2 = 31$
    $y - x = 2$

Solve each system using the addition method.

17. $x^2 + y^2 = 1$
    $x^2 - y^2 = 1$

18. $x^2 + y^2 = 16$
    $x^2 - y^2 = 16$

19. $3x^2 + 2y^2 = 140$
    $x^2 - y^2 = 20$

20. $3x^2 + 2y^2 = 200$
    $x^2 - y^2 = 60$

21. $x^2 + y^2 = 9$
    $9x^2 + 4y^2 = 36$

22. $x^2 + y^2 = 4$
    $4x^2 + 9y^2 = 36$

23. $4x^2 - 3y^2 = -11$
    $3x^2 + 2y^2 = 30$

24. $2x^2 - 3y^2 = 6$
    $3x^2 + 2y^2 = 35$

25. $xy - y^2 = 25$
    $2xy - y^2 = 75$

26. $xy - y^2 = 30$
    $3xy - y^2 = 162$

27. $4x^2 + 4y^2 = 25$
    $4x^2 + 2y = 13$

28. $9x^2 + 9y^2 = 13$
    $9x^2 + 3y = 7$

Use a combination of the addition method and the substitution method to solve each system.

29. $x^2 + 3xy + y^2 = 29$
    $x^2 + y^2 = 17$

30. $x^2 - 2xy + y^2 = 1$
    $x^2 + y^2 = 5$

31. $x^2 - xy + y^2 = 21$
    $x^2 + xy + y^2 = 31$

32. $x^2 + 2xy + y^2 = 25$
    $x^2 + xy + y^2 = 19$

Solve each system. The solutions contain complex numbers.

33. $5x^2 - y^2 = 16$
    $y = 3x$

34. $11x^2 - y^2 = 45$
    $y = 4x$

35. $3x^2 + 2y^2 = 2$
    $x^2 + y^2 = 9$

36. $2x^2 + 3y^2 = 19$
    $x^2 + y^2 = 5$

37. $x^2 - 3xy - y^2 = 17$
    $x^2 - y^2 = 8$

38. $x^2 + 2xy - y^2 = 7$
    $x^2 - y^2 = 3$

39. Find x and y given that the area of the triangle in Figure 9.26 is 3.

40. In Figure 9.27, the area of the triangle is 104, and the area of the rectangle is 247. Find x and y.

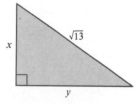

Figure 9.26

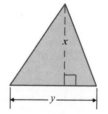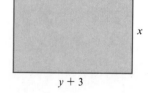

Figure 9.27

41. Determine the width and length of a rectangle whose perimeter is 20 feet and whose area is 24 square feet.

42. A rectangle has an area of 12 square feet and a diagonal of length 5 feet. Find its width and length.

CALCULATOR PROBLEMS

Solve each system.

43. $x + y = 11.06$
$\quad\quad xy = 28.7853$

44. $5.41x^2 + 6.84y^2 = 936.67$
$\quad\quad x^2 - y^2 = 1.75$

| 9.5 | *Second-Degree Inequalities and Systems of Inequalities* |

In Sections 9.1 through 9.3 we graphed second-degree equations in two variables, such as $x^2 + y^2 = 25$. In this section we graph **second-degree inequalities** in two variables, such as $x^2 + y^2 < 25$. We graph second-degree inequalities the same way we graphed linear inequalities in Section 7.5. Later in the section we graph **systems of inequalities.**

SECOND-DEGREE INEQUALITIES

EXAMPLE 1   Graph $x^2 + y^2 < 25$.

First graph the boundary curve $x^2 + y^2 = 25$. This is the circle with center at the origin and radius 5 shown in Figure 9.28(a). The circle is dashed because the original inequality sign is $<$ and not $\leq$. The boundary curve divides the plane into two regions—the region inside the circle and the region outside the circle. To determine which region represents the solution to $x^2 + y^2 < 25$, choose a

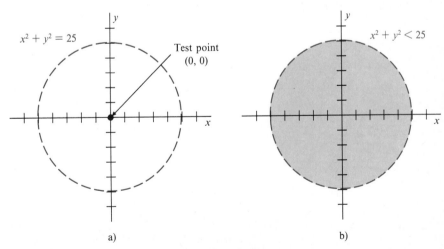

a)   b)

Figure 9.28

test point that is not on the boundary itself, say $(0, 0)$. Substitute the test point into the original inequality.

$$x^2 + y^2 < 25$$
$$(0)^2 + (0)^2 < 25$$
$$0 < 25 \qquad \text{True}$$

Try Problem 1

Since $(0, 0)$ satisfies the inequality $x^2 + y^2 < 25$, shade the region that contains $(0, 0)$ as shown in Figure 9.28(b). ◄

EXAMPLE 2 Graph $y \le x^2 - 1$.

Graph the boundary $y = x^2 - 1$. This is the parabola with intercepts at $(0, -1)$, $(-1, 0)$, and $(1, 0)$ shown in Figure 9.29(a). The parabola is solid (not dashed) because the original inequality symbol is $\le$ and not $<$. Substitute the test point $(0, 0)$ into the original inequality.

$$y \le x^2 - 1$$
$$0 \le (0)^2 - 1$$
$$0 \le -1 \qquad \text{False}$$

Since $(0, 0)$ does *not* satisfy $y \le x^2 - 1$, shade the region that does *not* contain $(0, 0)$ as shown in Figure 9.29(b).

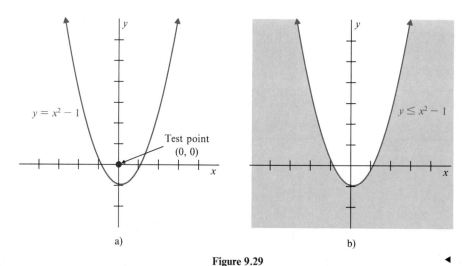

a)   b)

Try Problem 9

Figure 9.29 ◄

SYSTEMS OF INEQUALITIES

The solution to a system of equations is represented by the intersection points of the graphs of the equations. The solution to a system of inequalities is represented by the intersection points of the graphs of the inequalities.

EXAMPLE 3   Graph the following system.

$$x + y > 3$$
$$y \le x$$

First graph $x + y > 3$. This is the open half plane that lies above the line $x + y = 3$, as shown in Figure 9.30(a). Then graph $y \le x$ on the same axes. This is the closed half plane that lies on and below the line $y = x$, as shown in Figure 9.30(b). The intersection of the two half planes, shown as the shaded region in Figure 9.30(c), represents the solution to the system.

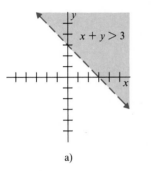

a)

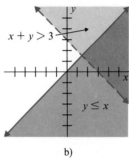

b)

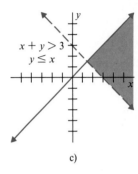

c)

Try Problem 21

Figure 9.30   ◀

EXAMPLE 4   Graph the following system.

$$x^2 + y^2 < 16$$
$$x - 2y < -4$$

Graph $x^2 + y^2 < 16$ as shown in Figure 9.31(a). On the same axes, graph $x - 2y < -4$ as shown in Figure 9.31(b). The intersection of the two shaded regions, shown in Figure 9.31(c), is the graph of the system.

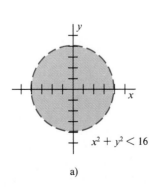

a)

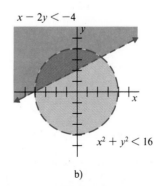

b)

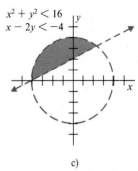

c)

Try Problem 25

Figure 9.31   ◀

EXAMPLE 5 Graph the following system.

$$x^2 - y^2 \geq 1$$
$$4x^2 + 9y^2 \leq 36$$

Graph $x^2 - y^2 \geq 1$ as shown in Figure 9.32(a). On the same axes, graph $\dfrac{x^2}{9} + \dfrac{y^2}{4} \leq 1$ as shown in Figure 9.32(b). The intersection of the two regions, shown in Figure 9.32(c), is the graph of the system.

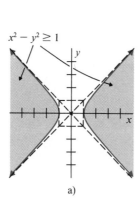

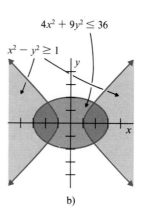

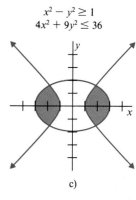

a)	b)	c)

Try Problem 35

Figure 9.32 ◄

PROBLEM SET 9.5

Graph each inequality.

1. $x^2 + y^2 < 4$

2. $x^2 + y^2 < 9$

3. $x^2 + y^2 > 4$

4. $x^2 + y^2 > 9$

5. $\dfrac{x^2}{9} + \dfrac{y^2}{4} \leq 1$

6. $\dfrac{x^2}{25} + \dfrac{y^2}{4} \leq 1$

7. $y > x^2$

8. $y < x^2$

9. $y \leq x^2 - 4$

10. $y \geq x^2 - 4$

11. $\dfrac{x^2}{9} - \dfrac{y^2}{25} \leq 1$

12. $\dfrac{x^2}{9} - \dfrac{y^2}{16} \leq 1$

13. $\dfrac{y^2}{9} - \dfrac{x^2}{9} \geq 1$

14. $\dfrac{y^2}{25} - \dfrac{x^2}{25} \geq 1$

15. $(x - 3)^2 + y^2 \geq 1$

16. $x^2 + (y - 4)^2 \geq 4$

Graph each system of inequalities.

17. $x \geq 2$
$\quad y \geq 1$

18. $x \leq -3$
$\quad y \leq -1$

19. $y < x + 3$
$\quad y > 3$

20. $y > x + 2$
$\quad y < 2$

21. $x + y > 1$
$\quad y \leq x$

22. $x - y < 3$
$\quad y \leq -x$

23. $x - y \leq 4$
$\quad x - y \geq 0$

24. $x + y \leq 4$
$\quad x + y \geq 0$

25. $x^2 + y^2 < 36$
$x - 2y < -6$

26. $x^2 + y^2 < 16$
$x + 2y < 4$

27. $x^2 + y^2 < 36$
$x - 2y > -6$

28. $x^2 + y^2 < 16$
$x + 2y > 4$

29. $x^2 + y^2 \leq 16$
$y < x^2$

30. $x^2 + y^2 \leq 25$
$y > x^2$

31. $y \leq x^2 - 1$
$y \geq -x^2 + 1$

32. $y \geq x^2 - 4$
$y \leq -x^2 + 4$

33. $x^2 + y^2 \leq 9$
$4x^2 + 9y^2 \geq 36$

34. $x^2 + y^2 \geq 4$
$4x^2 + 9y^2 \leq 36$

35. $x^2 - y^2 \geq 4$
$x^2 + 4y^2 \leq 16$

36. $x^2 - y^2 \leq 4$
$4x^2 + y^2 \geq 16$

37. $x^2 + y^2 \geq 1$
$-2 < x \leq 2$

38. $x^2 + y^2 \leq 25$
$-3 \leq y < 3$

CALCULATOR PROBLEMS

Graph each system.

39. $1.4x^2 + 1.4y^2 \geq 8.75$
$1.96x^2 + y^2 \leq 12.25$

40. $1.4x^2 + 1.4y^2 < 8.75$
$1.96x^2 + y^2 > 12.25$

CHAPTER 9 REVIEW

[9.1] *Graph each parabola. Label the intercepts and the vertex.*

1. $y = 2x^2$

2. $y = 3x^2 - 3$

3. $y = x^2 + 3x$

4. $y = 8 - 2x^2$

5. $y = x^2 - 5$

6. $y = -x^2 - 4x - 4$

7. $y = x^2 - x - 6$

8. $y = x^2 + x + 1$

9. A farmer plans to use 6 miles of fencing to construct a rectangular pasture. One side of the farmer's property lies along a straight river and needs no fence. What should the dimensions of the pasture be to maximize its area? What is the maximum area?

[9.2] *Write an equation for the circle with the given center and radius.*

10. Center at $(0, 0)$, radius 2

11. Center at $(2, -3)$, radius 3

Determine the center and radius of each circle.

12. $x^2 + (y - 2)^2 = 16$

13. $(x + 4)^2 + (y - 5)^2 = 5$

Graph each circle. Label the center and the radius.

14. $x^2 + y^2 = 16$

15. $x^2 + y^2 + 4x - 6y - 12 = 0$

16. $(x + 2)^2 + y^2 = \frac{1}{4}$

17. $x^2 + y^2 + 6x - 7 = 0$

18. Write an equation for the circle with center at $(-2, 1)$ that passes through the point $(2, -2)$.

[9.3] *Graph each ellipse. Label the intercepts.*

19. $\dfrac{x^2}{16} + \dfrac{y^2}{25} = 1$ **20.** $\dfrac{x^2}{49} + \dfrac{y^2}{16} = 1$ **21.** $25x^2 + y^2 = 25$

Graph each hyperbola. Label the intercepts.

22. $\dfrac{x^2}{36} - \dfrac{y^2}{25} = 1$ **23.** $\dfrac{x^2}{4} - y^2 = 1$ **24.** $y^2 - x^2 = 9$

Identify each equation as representing a parabola, a circle, an ellipse, or a hyperbola. Do not graph.

25. $4x^2 + 4y^2 = 1$ **26.** $9x^2 - 16y = 144$

27. $y^2 = 64 + x^2$ **28.** $y^2 = 100 - 2x^2$

Write an equation for each conic section graphed below.

29.

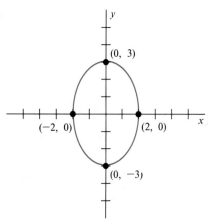

30.

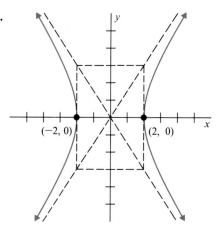

[9.4] *Solve each system using the substitution method.*

31. $\quad y = x^2$ **32.** $x^2 + y^2 = 25$ **33.** $\quad xy = -15$
$\quad\ \ y - 2x = 3$ $\quad\ 4x - 3y = 0$ $\qquad x + y = 2$

Solve each system using the addition method.

34. $x^2 + y^2 = 25$ **35.** $x^2 + 7y^2 = 16$ **36.** $4x^2 + y^2 = 17$
$\quad\ x^2 - y^2 = 7$ $\qquad x^2 + \ \ y^2 = 10$ $\qquad x^2 + y\ = 5$

Solve the following system. The solutions contain complex numbers.

37. $2x^2 + 3y^2 = 10$
$\quad\ x^2 + \ y^2 = 3$

38. Two squares have a combined perimeter of 68 inches. One square contains 17 square inches more than the other square. Find the dimensions of each square.

[9.5]   *Graph each inequality.*

39. $(x-2)^2 + y^2 < 4$ **40.** $y \geq x^2 - 2x$ **41.** $y^2 - x^2 \geq 9$ **42.** $\dfrac{x^2}{16} + y^2 < 1$

Graph each system of inequalities.

43. $y \geq -x$
$\quad\ y \leq x + 3$

44. $x^2 + y^2 < 16$
$\qquad\ y > x^2$

45. $x^2 + y^2 < 25$
$\quad\ x^2 + y^2 > 16$

46. $9x^2 + 16y^2 \leq 144$
$\qquad 16x^2 + 9y^2 \leq 144$

CHAPTER 9 TEST

Determine the center and radius of each circle.

1. $(x-6)^2 + (y-8)^2 = 25$

2. $(x+2)^2 + y^2 = 4$

Graph each conic section.

3. $y = x^2 + 1$ **4.** $y = -x^2 + 4x$ **5.** $\dfrac{x^2}{9} + \dfrac{y^2}{4} = 1$

6. $16x^2 + 9y^2 = 144$ **7.** $x^2 + y^2 + 4x - 6y - 3 = 0$ **8.** $x^2 + y^2 = 1$

9. $\dfrac{x^2}{16} - \dfrac{y^2}{25} = 1$ **10.** $\dfrac{y^2}{25} - \dfrac{x^2}{16} = 1$

Solve each system.

11. $x^2 + y^2 = 25$
$\qquad y = 2x - 5$

12. $x - y = 4$
$\qquad xy = 5$

13. $x - y = 15$
$\qquad x = 2y^2$

14. $x^2 + 4y^2 = 25$
$\qquad x^2 - y^2 = 5$

Graph each inequality.

15. $x^2 + y^2 < 9$

16. $y \leq x^2 - 4$

Graph each system of inequalities.

17. $\quad y > x$
$\quad x + y \leq 5$

18. $y^2 - x^2 \leq 9$
$\quad \dfrac{x^2}{4} + \dfrac{y^2}{9} \geq 1$

19. Write an equation for the circle with center at $(-3, 2)$ that passes through the point $(1, -1)$.

20. A farmer plans to use 40 feet of fencing to construct a rectangular pen for his pigs. One side of the pen is next to the barn and needs no fencing. What should the dimensions of the pen be to maximize its area? What is the maximum area?

CUMULATIVE REVIEW FOR CHAPTERS 7–9

[7]   *Find the distance between each pair of points.*

  1. (3, 5) and (6, 9)

  2. (−3, 4) and (1, −2)

Graph each equation. Label the x- and the y-intercept if they exist.

  3. $x - y = 4$ **4.** $2x + 9y = 18$ **5.** $y = 7$ **6.** $y = \frac{2}{3}x + 1$

Find the slope of the line passing through each pair of points.

  7. (−2, 5) and (3, −7)

  8. (3, −1) and (5, −4)

Find an equation for the line that passes through the given point and has the given slope. Then write the equation in standard form.

  9. (0, −3), $m = 5$

  10. (−2, 3), $m = \frac{3}{4}$

Determine the slope and the y-intercept of each line.

  11. $y = -2x - 1$

  12. $5x - 3y = 15$

Determine whether the lines with the given equations are parallel, perpendicular, or neither.

  13. $2x - 3y = 5$
      $4x - 6y = 5$

  14. $x - 2y = 3$
      $2x + \ y = 3$

Graph each inequality in the xy-plane.

  15. $2x - 5y \le 0$ **16.** $x \ge 4$ **17.** $3x - 2y > 6$ **18.** $|2x - 1| < 3$

  19. Find an equation for the line with y-intercept 2 that is perpendicular to $2x + 3y = 4$. Then write the equation in standard form.

[8]   *Solve the following system by graphing. Classify the system as independent, inconsistent, or dependent.*

  20. $x + \ y = 6$
      $x - 2y = -6$

Solve each system using the substitution method or the addition method.

  21. $x + y = 2$
      $x - y = 0$

  22. $3x - 4y = 2$
      $x = 4y + 2$

  23. $x + 2y = 5$
      $3x + 7y = 21$

  24. $\frac{x}{2} + \frac{y}{4} = 3$
      $\frac{x}{8} - \frac{y}{6} = \frac{5}{3}$

  25. The difference between two numbers is 11, and the sum is 21. Find the numbers.

  26. A collection of nickels and quarters totals $4.10. The total number of coins is 30. How many of each kind of coin are there?

Solve each system.

  27. $x + 4y - 2z = -3$
      $2x - 3y + 4z = 8$
      $2x + \ y + 2z = 2$

  28. $4x - \ y + \ z = 1$
      $3x - \ y - 5z = 5$
      $x + 2y + 7z = 7$

Evaluate each determinant.

29. $\begin{vmatrix} 5 & -2 \\ 4 & 3 \end{vmatrix}$

30. $\begin{vmatrix} 3 & 4 \\ 2 & -3 \end{vmatrix}$

31. $\begin{vmatrix} 3 & 0 & 5 \\ 2 & 1 & 5 \\ -2 & 0 & 2 \end{vmatrix}$

32. $\begin{vmatrix} 4 & 1 & 0 \\ -1 & 2 & -3 \\ 8 & 2 & 0 \end{vmatrix}$

Use Cramer's rule to solve each system.

33. $2x - 3y = 11$
    $x - 2y = 6$

34. $x - y \quad = 6$
    $y + z = 1$
    $x \quad - z = -1$

35. The sum of three numbers is 4. The first number equals the sum of the second number and twice the third. The sum of the second number and the third minus the first is 2. Find the numbers.

[9]   *Graph each parabola. Label the intercepts and the vertex.*

36. $y = x^2 - 4$

37. $y = x^2 + 2x - 8$

38. $y = x^2 + 2x + 3$

Graph each circle. Label the center and the radius.

39. $x^2 + (y + 2)^2 = 9$

40. $x^2 + 8x + y^2 - 6y = 0$

Graph each ellipse. Label the intercepts.

41. $4x^2 + 9y^2 = 36$

42. $\dfrac{x^2}{25} + \dfrac{y^2}{36} = 1$

Graph each hyperbola. Label the intercepts.

43. $\dfrac{y^2}{81} - \dfrac{x^2}{49} = 1$

44. $x^2 - y^2 = 25$

Solve each system.

45. $x - y = 5$
    $xy = 14$

46. $3x^2 + y^2 = 52$
    $2x^2 - 5y^2 = 12$

Graph each inequality.

47. $(x + 1)^2 + (y - 2)^2 < 4$

48. $y \geq x^2 + 7x + 12$

Graph each system of inequalities.

49. $y > x^2 - 4$
    $y < 2 - x$

50. $x^2 + 4y^2 \leq 36$
    $x^2 + y^2 \geq 9$

51. Write an equation for the circle with center at $(-2, 5)$ that passes through the point $(3, -7)$.

52. The sum of the squares of two positive numbers is 292, and the difference of the squares is 220. Find the numbers.

You have probably heard the expression, "Price is a function of demand."
Roughly speaking, this means that the price of an item and the demand
for that item are related in such a way that as the quantity demanded
changes, so does the price. In this chapter we state a precise definition of
a function, and we show how functions are used to express the
relationship between two variable quantities. We introduce the notation
we use to write a function, and we discuss several different types of
functions and their graphs. Once you understand the definition of a
function, you will learn to find the inverse of a function. The chapter
concludes with a discussion of variation—a special kind of function that
appears in many types of applied problems.

Functions occupy a central position in mathematics. It is important
that you understand how to recognize them, how to write them, and how
to use them.

10

Functions

10.1   *Definition of a Function*

Suppose you earn $4 an hour. Then your earnings are related to the number of hours that you work. One way to express this relationship is to write a set of ordered pairs. For example, the ordered pair (2, 8) represents the fact that if you work 2 hours, you will earn $8. The ordered pair (40, 160) represents the fact that if you work 40 hours, you will earn $160. In this way, a set of ordered pairs defines a **relation** between two variable quantities.

Definition   A **relation** is any set of ordered pairs. The set of all first coordinates is called the **domain** of the relation. The set of all second coordinates is called the **range.**

The set of ordered pairs $\{(1, 4), (-3, 2), (0, 7), (8, 4)\}$ is an example of a relation. The domain of the relation is $\{1, -3, 0, 8\}$. The range is $\{4, 2, 7\}$.

Our primary concern in this chapter is with a particular type of relation known as a **function.***

Definition   A **function** is a relation in which for each first coordinate there is exactly one second coordinate.

EXAMPLE 1   Determine whether the given relation is a function.

a) $\{(1, 4), (-3, 2), (0, 7), (8, 4)\}$
This is a relation in which for each first coordinate there is exactly one second coordinate. Therefore the relation *is* a function.

b) $\{(5, 6), (1, 3), (5, -9)\}$
This relation states that the first coordinate 5 corresponds to two different second coordinates, namely 6 and -9. Therefore this relation is *not* a function. ◄

Try Problem 1

CAUTION Every function is a relation, but some relations are not functions.

We can also use an equation to define a function. For example, if you earn $4 an hour, then your earnings y are a function of the number of hours x that you work according to the equation $y = 4x$. Since your earnings y depend upon the number of hours you work, we call y the **dependent variable.** On the other

---

* The term "function" was first defined by the German mathematician Lejeune Dirichlet (1805–1859).

hand, since the values of x may be chosen at will, we call x the **independent variable.**

$$y = 4x$$

Dependent   Independent
variable   variable

CAUTION Every equation in x and y defines a relation, but not all equations define y as a function of x.

EXAMPLE 2   Determine whether the given equation defines y as a function of x.

a) $y = x + 1$

This equation defines y as a function of x, since to each value of x there corresponds exactly one value of y.

b) $x^2 + y^2 = 25$

This equation does *not* define y as a function of x. For example, when $x = 4$,

$$4^2 + y^2 = 25$$
$$16 + y^2 = 25$$
$$y^2 = 9$$
$$y = \pm 3.$$

Try Problem 11

That is, there are *two* y-values that correspond to the x-value 4. ◄

We can determine whether a relation is a function by applying the **vertical-line test** to its graph.

Vertical-Line Test

A graph in the plane represents a function provided that no vertical line intersects the graph at more than one point.

The vertical-line test is illustrated in Figures 10.1 and 10.2. Note that no vertical line intersects the graph of $y = x + 1$ at more than one point. Therefore $y = x + 1$ defines y as a function of x. In Figure 10.2, however, the vertical line $x = 4$ intersects the graph of $x^2 + y^2 = 25$ at $(4, 3)$ and at $(4, -3)$. Therefore $x^2 + y^2 = 25$ does not define y as a function of x.

When a function is defined by an equation, its domain consists of all those real values of the independent variable that produce real values for the dependent variable, unless specified otherwise.

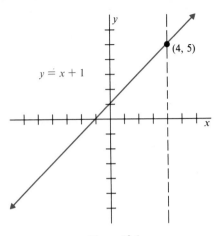

Figure 10.1   **Figure 10.2**

Try Problem 27

EXAMPLE 3   Find the domain of the function defined by $y = 2x + 7$.

If we substitute any real number for x, multiply by 2, and add 7, the result will be a real value for y. Therefore the domain consists of all real numbers. ◄

EXAMPLE 4   Find the domain of the function defined by $y = \dfrac{x}{x^2 - 6x + 5}$.

First, factor the denominator.

$$y = \frac{x}{(x - 1)(x - 5)}$$

If y is to be a real number, the denominator on the right side cannot be zero. Therefore the domain consists of all real numbers except 1 and 5. We shall

Try Problem 37

simply write $x \neq 1, 5$. ◄

EXAMPLE 5   Find the domain of the function defined by $y = \sqrt{2x + 7}$.

If y is to be a real number, the radicand $2x + 7$ must be nonnegative. That is,

$$2x + 7 \geq 0$$
$$2x \geq -7$$
$$x \geq -\tfrac{7}{2}.$$

Try Problem 43

Therefore the domain is $x \geq -\tfrac{7}{2}$. ◄

Finding the range of a function (or relation) from its equation is usually more difficult than finding the domain. However, we can determine both the domain and the range of a function (or relation) from its graph.

EXAMPLE 6   Find the domain and the range of the relation graphed in Figure 10.3. Determine whether the relation is a function.

From Figure 10.3, we see that the domain is $-4 \leq x \leq 4$, and the range is $-3 \leq y \leq 3$. The graph fails the vertical-line test, so the relation is not a function. ◄

Try Problem 47

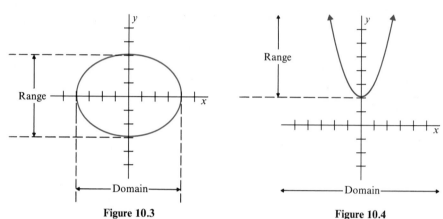

Figure 10.3   **Figure 10.4**

EXAMPLE 7   Find the domain and the range of the relation graphed in Figure 10.4. Determine whether the relation is a function.

The domain consists of all real numbers, and the range is $y \geq 2$. The graph passes the vertical-line test, so the relation is a function. ◄

Try Problem 49

EXAMPLE 8   The maximum pulse rate r that a healthy person aged 18 to 55 should attain while exercising is a function of the person's age a. That rate is determined by subtracting the person's age from 220. Write an equation that defines r as a function of a, and state the domain of the function.

Since r is determined by subtracting a from 220, we write

$$r = 220 - a.$$

The equation is true only for ages from 18 to 55; therefore the domain of the function is $18 \leq a \leq 55$. ◄

Try Problem 65

PROBLEM SET 10.1

State the domain and the range of each relation. Determine whether the relation is a function.

1. $\{(2, 5), (-6, 4), (0, 3), (9, 5)\}$

2. $\{(8, 1), (-5, 7), (10, 1), (0, 2)\}$

3. $\{(3, 9), (1, 7), (3, -8)\}$

4. $\{(6, 9), (4, 8), (4, -3)\}$

5. $\{(1, 1), (2, 2), (3, 3)\}$

6. $\{(0, 0), (-1, -1), (-2, -2)\}$

7. $\{(-5, 0), (-6, 0), (-7, 0)\}$

8. $\{(8, 11), (9, 11), (10, 11)\}$

9. $\{(4, 5), (5, 6), (6, 7)\}$

10. $\{(1, 2), (2, 3), (3, 4)\}$

Determine whether the given equation defines y as a function of x.

11. $y = x + 3$

12. $y = x + 2$

13. $2x + y = 6$

14. $3x + y = 5$

15. $y = \sqrt{x}$

16. $y = \sqrt{x - 1}$

17. $y = \pm\sqrt{x}$

18. $y = \pm\sqrt{x - 1}$

19. $y = \dfrac{1}{x - 5}$

20. $y = \dfrac{1}{x + 4}$

21. $x^2 + y^2 = 9$

22. $x^2 + y^2 = 16$

23. $y = x^2$

24. $y = x^3$

25. Does the inequality $y < x + 5$ define y as a function of x? Why or why not?

26. Does the inequality $y > x - 1$ define y as a function of x? Why or why not?

Find the domain of the function defined by each equation.

27. $y = 2x + 9$

28. $y = 3x + 8$

29. $y = x^2 - 4$

30. $y = x^2 - 9$

31. $y = \dfrac{8}{x - 1}$

32. $y = \dfrac{9}{x - 2}$

33. $y = \dfrac{1}{x^2 - 16}$

34. $y = \dfrac{1}{x^2 - 25}$

35. $y = \dfrac{1}{x^2 + 16}$

36. $y = \dfrac{1}{x^2 + 25}$

37. $y = \dfrac{x}{x^2 - 5x + 6}$

38. $y = \dfrac{x}{x^2 - 7x + 12}$

39. $y = \sqrt{x - 6}$

40. $y = \sqrt{x - 7}$

41. $y = \sqrt[3]{x - 6}$

42. $y = \sqrt[3]{x - 7}$

43. $y = \sqrt{2x + 9}$

44. $y = \sqrt{3x + 8}$

45. $y = \dfrac{x}{\sqrt{15 - 3x}}$

46. $y = \dfrac{x}{\sqrt{24 - 4x}}$

Find the domain and the range of each relation graphed below. Determine whether the relation is a function.

47.

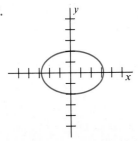

48.

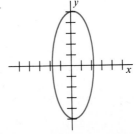

49.

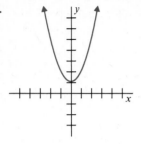

50.

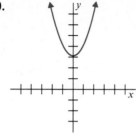

51.

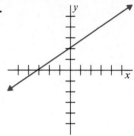

52.

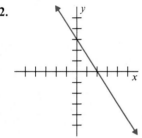

53.

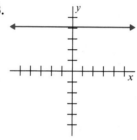

54.

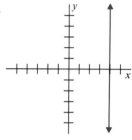

55.

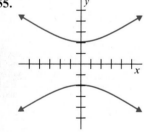

56.

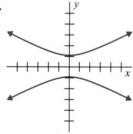

57.

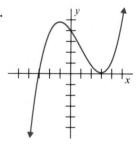

58.

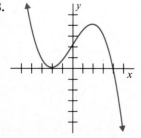

59.

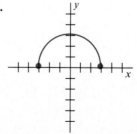

60.

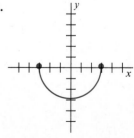

61. Write an equation that defines the area A of a square as a function of its side x. State the domain of the function.

62. Write an equation that defines the perimeter P of a square as a function of its side x. State the domain of the function.

63. If you earn \$6 an hour, write an equation that defines your earnings y as a function of the number of hours x that you work.

64. A sales representative earns \$250 per week plus a commission of 15% of her sales x. If y denotes her weekly earnings, write an equation that defines y as a function of x.

65. The target heart rate T that a healthy person aged 18 to 55 should maintain while exercising is 80% of the maximum heart rate (see Example 8). Write an equation that defines T as a function of age a, and state the domain of the function.

66. Suppose the price p of water skis is determined by subtracting three times the quantity demanded q from 600. Write an equation that defines p as a function of q, and state the domain of the function.

CALCULATOR PROBLEMS

Find the domain of each function.

67. $y = \dfrac{1}{2.56x^2 - 7.84}$

68. $y = \sqrt{1.550x - 99.014}$

10.2   *Function Notation*

To denote that y is a function of x, we write $y = f(x)$.* The expression "$f(x)$" is read "f of x". It does *not* mean f times x. Since y and $f(x)$ are equal, they can be used interchangably. That is, we can write $y = x^2$ or $f(x) = x^2$. The **function notation** $f(x) = x^2$ has certain advantages over the notation $y = x^2$. For example, using $y = x^2$ we would have to write

$$\text{when } x = 3, \quad y = 3^2 = 9.$$

Using $f(x) = x^2$, we would simply write

$$f(3) = 3^2 = 9.$$

Note that $f(3)$ is the range value that corresponds to the domain value 3. As such, $f(3)$ is called the **value of the function f at 3.**

$$f(3) = 9$$

Name of   Domain   Value of
function   value   function

---

* The notation $f(x)$ was introduced by the Swiss mathematician Leonhard Euler (1707–1783).

We can think of the function f as a "machine" that squares each number it is given. The inputs to the machine constitute the domain, and the outputs the range (see Figure 10.5).

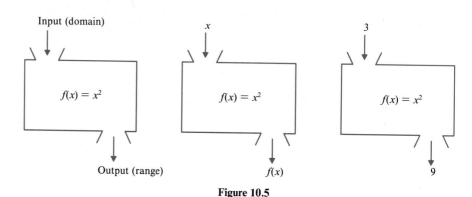

Figure 10.5

EXAMPLE 1   Suppose $f(x) = 3x^2 - 5x + 8$. Determine each of the following: **a)** $f(4)$,   **b)** $f(\frac{1}{3})$,   **c)** $f(-a)$.

a)   Replace x with 4.

$$f(x) = 3x^2 - 5x + 8$$
$$f(4) = 3(4)^2 - 5(4) + 8$$
$$= 48 - 20 + 8$$
$$= 36$$

b)   Replace x with $\frac{1}{3}$.

$$f(\tfrac{1}{3}) = 3(\tfrac{1}{3})^2 - 5(\tfrac{1}{3}) + 8$$
$$= \tfrac{1}{3} - \tfrac{5}{3} + \tfrac{24}{3}$$
$$= \tfrac{20}{3}$$

c)   Replace x with $-a$.

$$f(-a) = 3(-a)^2 - 5(-a) + 8$$
$$= 3a^2 + 5a + 8 \blacktriangleleft$$

Try Problem 13

Although we usually use the letter f to name a function, any letter may be used. Therefore each of the functions

$$g(x) = x^2, \qquad h(x) = x^2, \quad \text{and} \quad P(t) = t^2$$

is equivalent to the function $f(x) = x^2$, since each has the range value a^2 when the domain value is a.

EXAMPLE 2   Suppose $g(x) = 5$. Determine each of the following: **a)** $g(1)$, **b)** $g(5)$,   **c)** $g(0)$.

We can write this function as $g(x) = 0 \cdot x + 5$.

a) Replace x with 1.
$$g(1) = 0 \cdot 1 + 5 = 5$$

b) Replace x with 5.
$$g(5) = 0 \cdot 5 + 5 = 5$$

c) Replace x with 0.

Try Problem 25
$$g(0) = 0 \cdot 0 + 5 = 5 \blacktriangleleft$$

Figure 10.6 illustrates the results of Example 2.

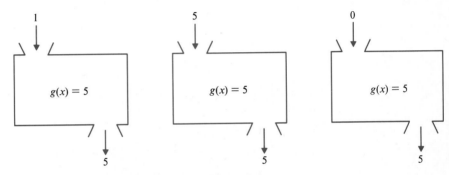

Figure 10.6

EXAMPLE 3   Suppose $f(x) = 3x - 5$ and $g(x) = x^2 + 4$. Determine each of the following: **a)** $f(1) + g(1)$,   **b)** $\dfrac{f(-1)}{g(0)}$,   **c)** $f(g(3))$.

a) Find $f(1)$ and $g(1)$, then add.
$$f(1) = 3(1) - 5 = -2$$
$$g(1) = 1^2 + 4 = 5$$
Therefore $f(1) + g(1) = -2 + 5 = 3$.

b) Find $f(-1)$ and $g(0)$, then divide.
$$\frac{f(-1)}{g(0)} = \frac{3(-1) - 5}{0^2 + 4} = \frac{-8}{4} = -2$$

c) Find $g(3)$, then find $f(g(3))$.
$$g(3) = 3^2 + 4 = 13$$

Try Problem 39
$$f(g(3)) = f(13) = 3(13) - 5 = 34 \blacktriangleleft$$

Figure 10.7 illustrates Example 3(c).

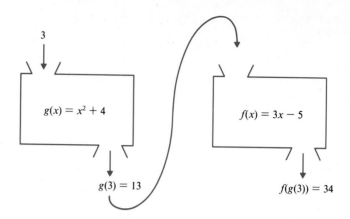

Figure 10.7

EXAMPLE 4   Suppose $f = \{(8, 9), (1, 2), (0, 3)\}$ and $g = \{(6, -4), (-7, 1), (4, -5)\}$. Determine each of the following: **a)** $f(8) - g(4)$,   **b)** $f(0) \cdot g(6)$,   **c)** $f(g(-7))$.

a)   Since $f(8) = 9$ and $g(4) = -5$, we have

$$f(8) - g(4) = 9 - (-5) = 14.$$

b)   Since $f(0) = 3$ and $g(6) = -4$, we have

$$f(0) \cdot g(6) = 3 \cdot (-4) = -12.$$

c)   Since $g(-7) = 1$, we have

Try Problem 43

$$f(g(-7)) = f(1) = 2. \blacktriangleleft$$

The expression

$$\frac{f(t) - f(a)}{t - a}$$

is called a difference quotient. If you go on to study calculus, you will need to know how to work with difference quotients.

EXAMPLE 5   Find $\dfrac{f(t) - f(a)}{t - a}$ for the function $f(x) = x^2 + 3$. Then simplify.

Since $f(t) = t^2 + 3$ and $f(a) = a^2 + 3$, we have

$$\frac{f(t) - f(a)}{t - a} = \frac{(t^2 + 3) - (a^2 + 3)}{t - a}$$

Try Problem 55

$$= \frac{t^2 - a^2}{t - a} = \frac{(t + a)(t - a)}{t - a} = t + a. \blacktriangleleft$$

PROBLEM SET 10.2

Suppose $f(x) = 2x + 1$. Determine each of the following.

1. $f(5)$ **2.** $f(4)$ **3.** $f(-5)$ **4.** $f(-4)$

5. $f(0)$ **6.** $f(1)$ **7.** $f(\frac{3}{2})$ **8.** $f(\frac{5}{2})$

9. $f(a)$ **10.** $f(b)$ **11.** $f(a+1)$ **12.** $f(b-1)$

Suppose $f(x) = 6x^2 - 4x + 3$. Determine each of the following.

13. $f(2)$ **14.** $f(3)$ **15.** $f(-2)$ **16.** $f(-3)$

17. $f(\frac{1}{3})$ **18.** $f(\frac{2}{3})$ **19.** $f(1)$ **20.** $f(0)$

21. $f(-a)$ **22.** $f(-b)$ **23.** $f(x+h)$ **24.** $f(x-h)$

Suppose $g(x) = 8$. Determine each of the following.

25. $g(1)$ **26.** $g(2)$ **27.** $g(8)$ **28.** $g(-8)$

29. $g(0)$ **30.** $g(\frac{1}{2})$ **31.** $g(a)$ **32.** $g(b)$

Suppose $f(x) = 3x + 2$ and $g(x) = x^2 - 5$. Determine each of the following.

33. $f(2) + g(2)$ **34.** $f(4) + g(4)$ **35.** $f(\frac{4}{3}) \cdot g(-1)$ **36.** $f(\frac{2}{3}) \cdot g(-2)$

37. $\dfrac{f(1)}{g(0)}$ **38.** $\dfrac{f(6)}{g(0)}$ **39.** $f(g(3))$ **40.** $f(g(1))$

41. $g(f(3))$ **42.** $g(f(1))$

Suppose $f = \{(7, 9), (0, -4), (12, 7), (2, 6), (-7, 4)\}$ and $g = \{(6, 7), (8, -8), (-8, 2), (4, 2), (0, -3)\}$. Determine each of the following.

43. $f(12) - g(4)$ **44.** $f(2) - g(8)$ **45.** $f(7) \cdot g(0)$ **46.** $f(0) \cdot g(6)$

47. $f(g(-8))$ **48.** $g(f(-7))$ **49.** $f(f(12))$ **50.** $g(g(8))$

51. $g(f(g(4)))$ **52.** $f(g(f(2)))$

Find $\dfrac{f(t) - f(a)}{t - a}$ for each function. Then simplify.

53. $f(x) = 3x + 6$ **54.** $f(x) = 5x + 1$ **55.** $f(x) = x^2 + 1$ **56.** $f(x) = x^2 + 5$

57. $f(x) = x^3 + 2$ **58.** $f(x) = x^3 + 4$ **59.** $f(x) = 4$ **60.** $f(x) = 7$

61. Suppose $f(x) = x^2$ and $g(x) = x + 1$. Find $f(g(a))$ and $g(f(a))$.

62. Suppose $f(x) = x^2 - 1$ and $g(x) = 3x$. Find $f(g(a))$ and $g(f(a))$.

63. A company can produce shower curtains at a cost of $4 per curtain plus a daily overhead of $275. If $C(x)$ denotes the total daily cost of producing x curtains, write a cost function for the curtains. Then find and interpret $C(0)$ and $C(235)$.

64. A manufacturer can produce small electric motors at a cost of $8 per motor plus a daily overhead of $350. If $C(x)$ denotes the total daily cost of producing x motors, write a cost function for the motors. Then find and interpret $C(0)$ and $C(125)$.

CALCULATOR PROBLEMS

65. Given $f(x) = 5.2x^2 - 7.4x + 11$, find each of the following.

    a) $f(1.6)$ **b)** $f(115)$

66. If $8500 is invested for 10 years, its value V is a function of the interest rate r (expressed as a decimal) according to the equation

$$V(r) = 8500(1 + r)^{10}.$$

Find and interpret $V(0.06)$ and $V(0.1375)$.

10.3   *Constant, Linear, and Quadratic Functions*

In this section we examine three specific types of functions and their graphs. They are the **constant function,** the **linear function,** and the **quadratic function.**

CONSTANT FUNCTIONS

The constant function is so named because it has the same range value for all domain values.

Definition

A **constant function** is a function that can be written in the form

$$f(x) = a,$$

where a is a constant.

The constant function $f(x) = a$ is equivalent to $y = a$. From Section 7.2, we know that the graph of this equation is a horizontal line through the point $(0, a)$.

EXAMPLE 1   Graph the constant function $f(x) = 5$.

This equation is equivalent to $y = 5$. The graph is the horizontal line through $(0, 5)$ shown in Figure 10.8. We can label the vertical axis as the y-axis or the $f(x)$-axis. ◂

Try Problem 1

LINEAR FUNCTIONS

A linear equation in two variables defines a linear function.

Definition

A **linear function** is a function that can be written in the form
$$f(x) = ax + b,$$
where a and b are constants and $a \neq 0$.

The linear function $f(x) = ax + b$ is equivalent to $y = ax + b$. From Section 7.4, we know that the graph of this equation is a line with slope a and y-intercept b.

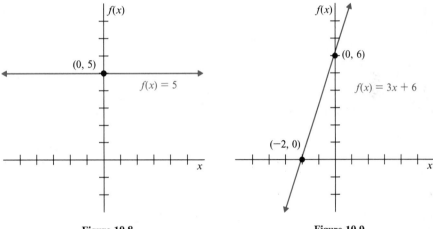

Figure 10.8   Figure 10.9

EXAMPLE 2   Graph the linear function $f(x) = 3x + 6$.

Try Problem 7

The graph is the line with slope 3 and y-intercept 6 shown in Figure 10.9. ◀

QUADRATIC FUNCTIONS

When $f(x)$ equals a polynomial of the second degree, the result is a quadratic function.

Definition

A **quadratic function** is a function that can be written in the form
$$f(x) = ax^2 + bx + c,$$
where a, b, and c are constants and $a \neq 0$.

The quadratic function $f(x) = ax^2 + bx + c$ is equivalent to the equation $y = ax^2 + bx + c$. From Section 9.1, we know that the graph of this equation is a U-shaped curve known as a parabola (see Figure 10.10). The vertex occurs at

The parabola $f(x) = ax^2 + bx + c$

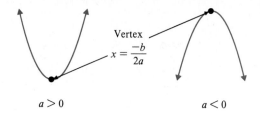

Figure 10.10

$x = -b/2a$. If $a > 0$, the parabola opens upward. If $a < 0$, the parabola opens downward. The y-intercept is c.

EXAMPLE 3   Graph the quadratic function $f(x) = x^2 + 4x - 5$.

The y-intercept is -5. To find the x-intercepts, set $f(x) = 0$.

$$0 = x^2 + 4x - 5$$
$$0 = (x - 1)(x + 5)$$
$$x = 1 \quad \text{or} \quad x = -5$$

The x-value of the vertex is

$$x = \frac{-b}{2a} = \frac{-4}{2(1)} = \frac{-4}{2} = -2.$$

The y-value of the vertex is

$$f(-2) = (-2)^2 + 4(-2) - 5 = 4 - 8 - 5 = -9.$$

Since $a = 1$ (a positive number), the parabola opens upward as shown in Figure 10.11. If we "fold" Figure 10.11 along the vertical line through its vertex, the

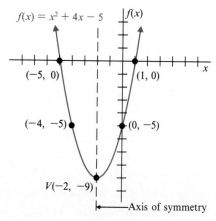

Figure 10.11

two halves of the parabola will coincide. Therefore this line is the axis of symmetry of the parabola. Using symmetry, we conclude that the point $(-4, -5)$ must be on the parabola, since the point $(0, -5)$ is on the parabola. ◄

Try Problem 13

EXAMPLE 4   Graph the quadratic function $f(x) = x^2 - 2x$.

The y-intercept is 0. To find the x-intercepts, set $f(x) = 0$.

$$0 = x^2 - 2x$$
$$0 = x(x - 2)$$
$$x = 0 \quad \text{or} \quad x = 2$$

The x-value of the vertex is

$$x = \frac{-b}{2a} = \frac{-(-2)}{2(1)} = \frac{2}{2} = 1.$$

The y-value of the vertex is

$$f(1) = 1^2 - 2(1) = 1 - 2 = -1.$$

The graph is the parabola shown in Figure 10.12.

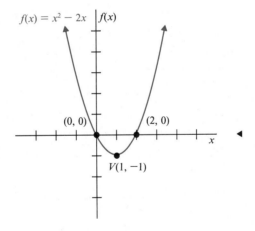

Try Problem 15

Figure 10.12

EXAMPLE 5   Graph the quadratic function $f(x) = -2x^2$.

The y-intercept is 0. To find the x-intercepts, set $f(x) = 0$.

$$0 = -2x^2$$
$$0 = x^2 \qquad \text{Divide by } -2$$
$$0 = x \qquad \text{Extract roots}$$

The x-value of the vertex is

$$x = \frac{-b}{2a} = \frac{-0}{2(-2)} = \frac{0}{-4} = 0.$$

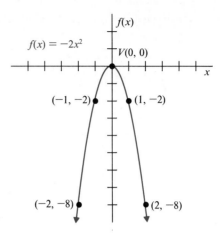

Figure 10.13

The y-value of the vertex is

$$f(0) = -2(0)^2 = -2(0) = 0.$$

Since $a = -2$ (a negative number), the parabola opens downward. The y-intercept, the x-intercept, and the vertex all occur at the same point $(0, 0)$. Therefore we graph some additional points to get the parabola shown in Figure 10.13. ◄

Try Problem 17

PROBLEM SET 10.3

Graph each constant function.

1. $f(x) = 3$ 2. $f(x) = 4$ 3. $f(x) = -2$

4. $f(x) = -1$ 5. $f(x) = \frac{1}{2}$ 6. $f(x) = 0$

Graph each linear function.

7. $f(x) = 2x + 2$ 8. $f(x) = 3x + 3$ 9. $f(x) = -x - 3$

10. $f(x) = -x - 2$ 11. $f(x) = \frac{2}{3}x$ 12. $f(x) = \frac{3}{4}x$

Graph each quadratic function.

13. $f(x) = x^2 + 2x - 3$ 14. $f(x) = x^2 + 6x + 5$ 15. $f(x) = x^2 - 6x$

16. $f(x) = x^2 - 4x$ 17. $f(x) = -3x^2$ 18. $f(x) = -x^2$

Identify as a constant function, a linear function, or a quadratic function. Then graph the function. Label all intercepts and vertices.

19. $f(x) = x$

20. $f(x) = 2x$

21. $f(x) = x^2$

22. $f(x) = 2x^2$

23. $f(x) = 1$

24. $f(x) = 2$

25. $g(x) = -x^2 + 2x$

26. $g(x) = -x^2 + 4x$

27. $6x - 2y = 12$

28. $4x - 2y = 8$

29. $y + x^2 = 4$

30. $y + x^2 = 9$

31. $2y + 3 = 0$

32. $2y + 7 = 0$

33. $y = t^2 - 4t + 4$

34. $y = t^2 - 2t + 1$

35. $f(t) = \frac{1}{2}t^2 + 1$

36. $f(t) = \frac{1}{2}t^2 + 2$

Determine each of the following based on Figure 10.14.

37. $f(0)$

38. $f(4)$

39. $f(-2)$

40. $f(5)$

41. $f(7)$

42. $f(-3)$

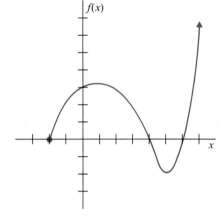

Figure 10.14

CALCULATOR PROBLEMS

Graph each function.

43. $f(x) = \dfrac{\sqrt{317}}{0.87\pi}$

44. $f(x) = 7x^2 - 2.8x + 1.1$

10.4   *Polynomial Functions and Special Functions*

The constant, linear, and quadratic functions of Section 10.3 all fall into a broader class of functions known as **polynomial functions.** A constant function is a polynomial function of degree 0, a linear function is a polynomial function of degree 1, and a quadratic function is a polynomial function of degree 2. To draw an accurate graph of a polynomial function of degree 3 or higher generally requires calculus techniques. We can usually get a good idea of the graph, however, by constructing a table of values, graphing the points from the table, and drawing a smooth curve through the points.

EXAMPLE 1   Graph $f(x) = x^3$.

Substitute values for x and calculate the corresponding values of $f(x)$.

x	-2	-1	$-\frac{1}{2}$	0	$\frac{1}{2}$	1	2
$f(x)$	-8	-1	$-\frac{1}{8}$	0	$\frac{1}{8}$	1	8

Graph these points and draw a smooth curve through them to produce the graph shown in Figure 10.15. ◄

Try Problem 1

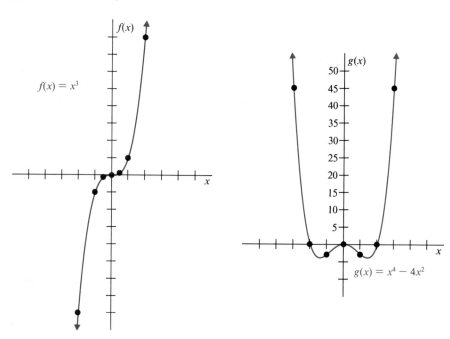

Figure 10.15   **Figure 10.16**

EXAMPLE 2   Graph $g(x) = x^4 - 4x^2$.

Construct the following table of values.

x	-3	-2	-1	0	1	2	3
$g(x)$	45	0	-3	0	-3	0	45

Graph these points and draw a smooth curve through them to produce the graph shown in Figure 10.16. To fit the graph on the page, we have changed the scale on the $g(x)$-axis. ◄

Try Problem 7

CAUTION The graph of a polynomial function does not contain any "gaps" or "corners."

SPECIAL FUNCTIONS

Certain special functions occur often enough to merit special attention. One such function is the **square-root function.**

Definition   The function

$$f(x) = \sqrt{x}$$

is called the **square-root function.**

EXAMPLE 3   Graph $f(x) = \sqrt{x}$.

Since the domain of this function is $x \geq 0$, we choose only nonnegative x-values when constructing the table of values.

x	0	1	2	3	4	9
$f(x)$	0	1	$\sqrt{2}$	$\sqrt{3}$	2	3

Try Problem 9   The graph is shown in Figure 10.17.

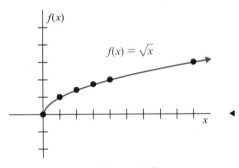

Figure 10.17

Another special function is the **absolute-value function.**

Definition   The function

$$f(x) = |x|$$

is called the **absolute-value function.**

EXAMPLE 4   Graph $f(x) = |x|$.

Construct the following table of values.

x	−3	−2	−1	0	1	2	3
f(x)	3	2	1	0	1	2	3

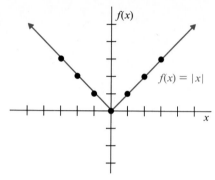

Figure 10.18

Try Problem 19

The graph consists of two half lines joined at their endpoints, as shown in Figure 10.18. ◄

When a function is defined by different equations over different parts of its domain, it is called a **piecewise-defined function.**

EXAMPLE 5   Graph the following piecewise-defined function.

$$h(x) = \begin{cases} 2 & \text{if } x > 1 \\ 2x & \text{if } -1 \le x \le 1 \\ -2 & \text{if } x < -1 \end{cases}$$

The graph of $h(x) = 2$ is a horizontal line through $(0, 2)$, but we graph only that part of the line for which $x > 1$ (see Figure 10.19a). The graph of $h(x) = 2x$ is a line with slope 2 and y-intercept 0, but we graph only that part of the line for which $-1 \le x \le 1$ (see Figure 10.19b). Finally, we graph that part of the horizontal line $h(x) = -2$ for which $x < -1$ (see Figure 10.19c). Combining these graphs produces the graph shown in Figure 10.20.

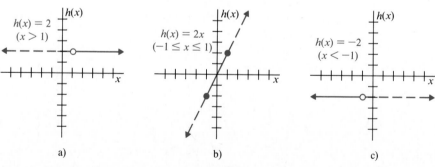

Figure 10.19

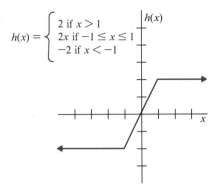

$$h(x) = \begin{cases} 2 \text{ if } x > 1 \\ 2x \text{ if } -1 \le x \le 1 \\ -2 \text{ if } x < -1 \end{cases}$$

Try Problem 31

Figure 10.20

Problem Solving

Blood Alcohol Level

The percentage $P(t)$ of alcohol in the average adult's bloodstream t hours after consuming 4 ounces of alcohol is approximately given by the function

$$P(t) = -0.0015t^3 + 0.106t.$$

This function is valid over the domain $0 \le t \le 8$. If a person is legally drunk when his or her blood alcohol level is greater than 0.10%, use the graph of this function to determine the approximate time interval over which a person who consumes 4 ounces of alcohol is legally drunk.

Using a calculator, construct the following table of values.

t	0	1	2	3	4	5	6	7	8
$P(t)$	0	.10	.20	.28	.33	.34	.31	.23	.08

Graph these points and draw a smooth curve through them to produce the graph shown in Figure 10.21. From the graph, we see that $P(t)$ is greater than 0.10 for values of t from approximately 1 hour to approximately 7.8 hours after consumption.

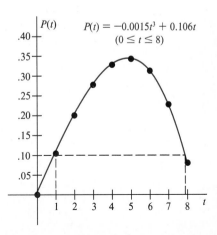

Figure 10.21

PROBLEM SET 10.4

Graph each polynomial function.

1. $f(x) = x^3 + 1$
2. $f(x) = x^3 - 1$
3. $f(x) = -x^3 + 2x$
4. $f(x) = x^3 - 2x$
5. $g(x) = x^4$
6. $g(x) = x^4 - 1$
7. $g(x) = x^4 - 9x^2$
8. $g(x) = -x^4 + 9x^2$

Graph each function.

9. $f(x) = \sqrt{x + 4}$
10. $f(x) = \sqrt{x + 1}$
11. $g(x) = -\sqrt{x + 4}$
12. $g(x) = -\sqrt{x + 1}$
13. $h(x) = \sqrt{x} + 2$
14. $h(x) = \sqrt{x} - 1$
15. $p(x) = \sqrt{x - 3}$
16. $p(x) = \sqrt{x - 4}$
17. $r(x) = \sqrt{-x}$
18. $r(x) = \sqrt{4 - x}$

Graph each function.

19. $f(x) = |x| - 4$
20. $f(x) = |x| + 1$
21. $g(x) = |x - 2|$
22. $g(x) = |x + 3|$
23. $h(x) = 2|x|$
24. $h(x) = \frac{1}{2}|x|$
25. $p(x) = -|x|$
26. $p(x) = -|x| + 5$

Graph each piecewise-defined function.

27. $f(x) = \begin{cases} 1 \text{ if } x \geq 0 \\ -1 \text{ if } x < 0 \end{cases}$

28. $f(x) = \begin{cases} 2 \text{ if } x \geq 0 \\ -2 \text{ if } x < 0 \end{cases}$

29. $g(x) = \begin{cases} -4 \text{ if } x > 0 \\ 0 \text{ if } x = 0 \\ 4 \text{ if } x < 0 \end{cases}$

30. $g(x) = \begin{cases} -5 \text{ if } x > 0 \\ 0 \text{ if } x = 0 \\ 5 \text{ if } x < 0 \end{cases}$

31. $h(x) = \begin{cases} 6 \text{ if } x > 2 \\ 3x \text{ if } -2 \leq x \leq 2 \\ -6 \text{ if } x < -2 \end{cases}$

32. $h(x) = \begin{cases} 2 \text{ if } x > 4 \\ \frac{1}{2}x \text{ if } -4 \leq x \leq 4 \\ -2 \text{ if } x < -4 \end{cases}$

33. $r(x) = \begin{cases} 2 \text{ if } x = 3 \\ 1 \text{ if } x \neq 3 \end{cases}$

34. $r(x) = \begin{cases} 3 \text{ if } x = 2 \\ 1 \text{ if } x \neq 2 \end{cases}$

35. Using Figure 10.21, determine the approximate time interval over which $P(t) > 0.20$.

36. Using Figure 10.21, determine the approximate time interval over which $P(t) > 0.25$.

37. A parking garage charges $2 for the first hour (or part of an hour) and $1 for each hour (or part of an hour) after the first (up to a maximum of 4 hours). Write the cost $C(x)$ of parking x hours as a piecewise-defined function. Then graph the function.

38. A parcel delivery service charges $5 for the first pound (or part of a pound) and $3 for each pound (or part of a pound) after the first (up to a maximum of 4 pounds). Write the cost $C(x)$ of delivering a package weighing x pounds as a piecewise-defined function. Then graph the function.

CALCULATOR PROBLEMS

Graph each function.

39. $f(x) = x^3 - x^2 - 8x + 6$

40. $g(x) = x^4 - x^3 - 7x^2 + x + 6$

10.5   *Inverse of a Function*

Consider the function f below.

$$f = \{(3, 5), (-1, 6), (-4, 0)\}$$

If we interchange the coordinates of each ordered pair, the resulting function is called the **inverse** of f, written f^{-1}. The symbol f^{-1} is read "f inverse".

$$f^{-1} = \{(5, 3), (6, -1), (0, -4)\}$$

Notice that the range of f becomes the domain of f^{-1}, and the domain of f becomes the range of f^{-1} (see Figure 10.22). The function f^{-1} "undoes" what the function f "does."

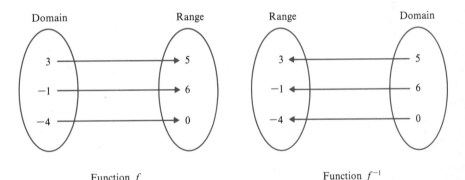

Function f Function f^{-1}

Figure 10.22

Definition

The **inverse** of a function f is the set of all ordered pairs of the form (b, a), where (a, b) belongs to f. The inverse of f is written f^{-1}.

CAUTION When x is a *number*, x^{-1} means $\frac{1}{x}$. When f is a *function*, f^{-1} means the inverse of f, *not* the reciprocal of f.

EXAMPLE 1   Find the inverse of the function $f = \{(4, 6), (-2, -1), (-5, 6), (7, -3)\}$.

Interchange the coordinates of each ordered pair to get f^{-1}.

$$f^{-1} = \{(6, 4), (-1, -2), (6, -5), (-3, 7)\}$$

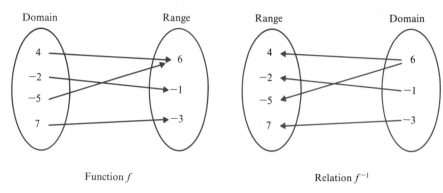

Function f Relation f^{-1}

Figure 10.23

Try Problem 5

Note that f^{-1} is not a function, since the first coordinate 6 corresponds to both 4 and -5 (see Figure 10.23). ◄

CAUTION Every function has an inverse, but the inverse may not be a function.

When a function is defined by an equation, we find its inverse by interchanging the independent and the dependent variables.

EXAMPLE 2 Given $f(x) = 2x - 4$, find $f^{-1}(x)$.
Replace $f(x)$ with y.

$$y = 2x - 4$$

Interchange x and y to get the equation that defines f^{-1}.

$$x = 2y - 4$$

Solve for the new y.

$$2y = x + 4$$
$$y = \tfrac{1}{2}x + 2$$

Replace the new y with $f^{-1}(x)$.

Try Problem 11

$$f^{-1}(x) = \tfrac{1}{2}x + 2 \blacktriangleleft$$

The graphs of $f(x) = 2x - 4$ and $f^{-1}(x) = \tfrac{1}{2}x + 2$ are shown in Figure 10.24. If we "fold" Figure 10.24 along the line $y = x$, the graphs of f and f^{-1} will coincide. That is, the graphs of f and f^{-1} are symmetric about the line $y = x$. In fact, the graphs of *any* function f and its inverse f^{-1} are symmetric about $y = x$. This occurs because when (a, b) belongs to f, then (b, a) belongs to f^{-1} (see Figure 10.25).

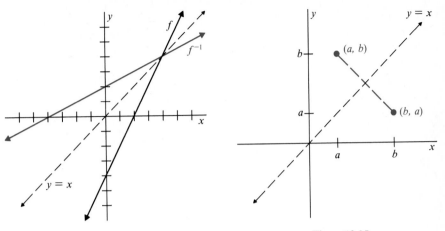

Figure 10.24   Figure 10.25

EXAMPLE 3   Given $f(x) = x^2 + 1$, find $f^{-1}(x)$. Then graph f and f^{-1} on the same axes.

$$y = x^2 + 1 \qquad \text{Replace } f(x) \text{ with } y$$
$$x = y^2 + 1 \qquad \text{Interchange } x \text{ and } y$$
$$\left.\begin{array}{l} y^2 = x - 1 \\ y = \pm\sqrt{x-1} \end{array}\right\} \text{Solve for the new } y$$
$$f^{-1}(x) = \pm\sqrt{x-1} \qquad \text{Replace the new } y \text{ with } f^{-1}(x)$$

The graphs of $f(x) = x^2 + 1$ and $f^{-1}(x) = \pm\sqrt{x-1}$ are shown in Figure 10.26. Note that they are symmetric about $y = x$. Also note that f^{-1} is not a function, since it fails the vertical-line test.

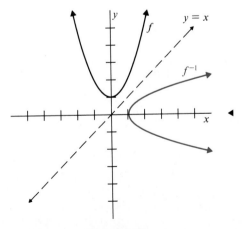

Try Problem 15

Figure 10.26

EXAMPLE 4   Find the inverse of $g(x) = x^3 - 5$.

$$y = x^3 - 5 \qquad \text{Replace } g(x) \text{ with } y$$
$$x = y^3 - 5 \qquad \text{Interchange } x \text{ and } y$$
$$\left. \begin{array}{l} y^3 = x + 5 \\ y = \sqrt[3]{x + 5} \end{array} \right\} \text{Solve for the new } y$$

Try Problem 23

$$g^{-1}(x) = \sqrt[3]{x + 5} \qquad \text{Replace the new } y \text{ with } g^{-1}(x) \blacktriangleleft$$

We can summarize the procedure for finding the inverse of a function as follows:

To Find $f^{-1}(x)$

1. Replace $f(x)$ with y.
2. Interchange x and y.
3. Solve the equation resulting from step 2 for the new y.
4. Replace the new y with $f^{-1}(x)$.

PROBLEM SET 10.5

Find the inverse of each function f. State whether f^{-1} is a function.

1. $f = \{(2, 5), (-1, 2), (-3, 0)\}$

2. $f = \{(3, 4), (-2, 3), (-1, 0)\}$

3. $f = \{(0, 0), (2, 2), (4, -4)\}$

4. $f = \{(3, 3), (-1, -1), (-5, 5)\}$

5. $f = \{(3, 6), (-4, -1), (5, -2), (-7, 6)\}$

6. $f = \{(0, -3), (2, 7), (-5, -1), (-4, 7)\}$

Find $f^{-1}(x)$ for each function. Graph f and f^{-1} on the same axes. State whether f^{-1} is a function.

7. $f(x) = x + 2$

8. $f(x) = x + 4$

9. $f(x) = 2x$

10. $f(x) = 3x$

11. $f(x) = 3x - 6$

12. $f(x) = 2x - 6$

13. $f(x) = -x + 3$

14. $f(x) = -x + 5$

15. $f(x) = x^2 + 2$

16. $f(x) = x^2 + 3$

17. $f(x) = x^2 + 2, x \geq 0$

18. $f(x) = x^2 + 3, x \geq 0$

19. $f(x) = 1$

20. $f(x) = 4$

Find the inverse of each function.

21. $f(x) = 2x^2 - 4$

22. $f(x) = 3x^2 - 9$

23. $g(x) = x^3 - 1$

24. $g(x) = x^3 - 2$

25. $h(x) = x$

26. $h(x) = -x$

27. $2x - 3y = 6$

28. $2x - 5y = 10$

Given f = {(− 1, 8), (5, 2), (3, 7), (− 4, 6)}, determine each of the following.

29. a) $f(5)$ **b)** $f^{-1}(2)$ **30. a)** $f(3)$ **b)** $f^{-1}(7)$

     c) $f^{-1}(f(-1))$ **d)** $f(f^{-1}(6))$ **c)** $f^{-1}(f(-4))$ **d)** $f(f^{-1}(8))$

31. Given $f(x) = 5x$, find $f^{-1}(x)$. Then determine each of the following.

     a) $f(4)$ **b)** $f^{-1}(20)$ **c)** $f^{-1}(f(a))$ **d)** $f(f^{-1}(a))$

32. Given $f(x) = x + 1$, find $f^{-1}(x)$. Then determine each of the following.

     a) $f(7)$ **b)** $f^{-1}(8)$ **c)** $f^{-1}(f(a))$ **d)** $f(f^{-1}(a))$

Given the graph of f, sketch f⁻¹ on the same axes.

33.

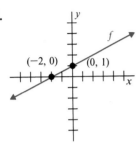

34.

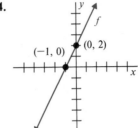

35.

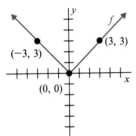

36.

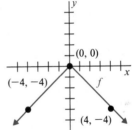

37.

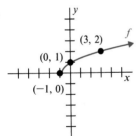

38.

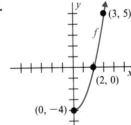

39.

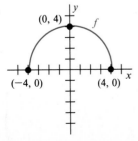

40.

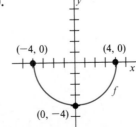

41.

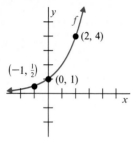

42.

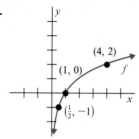

CALCULATOR PROBLEMS

43. Find $f^{-1}(x)$ if f is defined by $1.6x - 6.98y = 14.16$.

44. Given $f(x) = x^3$, find $f^{-1}(x)$. Then determine each of the following.

    a) $f(19)$ **b)** $f^{-1}(6859)$ **c)** $f^{-1}(9261)$ **d)** $f^{-1}(f(1.4))$

10.6   Variation

A particular type of function known as variation appears so often in applied problems that it deserves special attention. For example, the circumference C of a circle is a function of its diameter d according to the equation

$$C = \pi d.$$

This means that if the diameter is doubled, the circumference is also doubled. If the diameter is halved, the circumference is halved. We say the C **varies directly with d,** or C **is directly proportional to d.** The constant π is called the **variation constant,** or the **proportionality constant.**

Definition

The statement y **varies directly with** x means that

$$y = kx$$

for some nonzero constant k. We call k the **variation constant,** and we call $y = kx$ the **general variation function.**

We now illustrate how to find the variation constant.

EXAMPLE 1   Suppose y varies directly with x, and $y = 30$ when $x = 5$. Find the variation constant k.

The statement "y varies directly with x" means that
$$y = kx.$$
Since $y = 30$ when $x = 5$, we substitute and get
$$30 = k \cdot 5.$$
Solving for k gives

Try Problem 15a
$$k = \tfrac{30}{5} = 6. \blacktriangleleft$$

If we substitute the value of k into the general variation function, we obtain the **specific variation function.**

EXAMPLE 2 Write the specific variation function for Example 1.

The general variation function is
$$y = kx.$$
Substitute $k = 6$ to obtain the specific variation function.

Try Problem 15b
$$y = 6x \blacktriangleleft$$

We can use the specific variation function to find values for the dependent variable that correspond to given values of the independent variable.

EXAMPLE 3 Use the specific variation function of Example 2 to find y when $x = \tfrac{3}{2}$.

Substitute $x = \tfrac{3}{2}$ into $y = 6x$.

Try Problem 15c
$$y = 6 \cdot \tfrac{3}{2} = 9 \blacktriangleleft$$

Normally, a single variation problem consists of all the steps performed in Examples 1, 2, and 3.

EXAMPLE 4 Under normal conditions, the price of gold y is directly proportional to the price of silver x. If gold is $320 an ounce when silver is $8 an ounce, find the price of gold when silver is $11 an ounce.

Write the general variation function.
$$y = kx$$
Substitute $y = 320$ and $x = 8$.
$$320 = k \cdot 8$$
Solve for k.
$$k = \tfrac{320}{8} = 40$$
Substitute $k = 40$ into $y = kx$ to get the specific variation function.
$$y = 40x$$

Substitute $x = 11$ into $y = 40x$.

$$y = 40 \cdot 11 = 440$$

Try Problem 25   The price of gold will be $440 an ounce. ◄

Sometimes one quantity varies directly with the *square* of another quantity. For example, the area A of a circle is a function of its radius r according to the equation

$$A = \pi r^2.$$

This means that if the radius is doubled, the area is multiplied by *four* (see Figure 10.27). In other words, a pipe whose radius is twice the radius of another pipe can carry four times as much water as the first pipe. We say that A **varies directly with the square of r.**

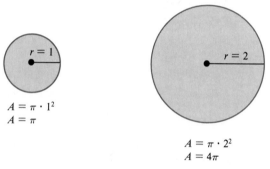

$$A = \pi \cdot 1^2$$
$$A = \pi$$

$$A = \pi \cdot 2^2$$
$$A = 4\pi$$

Figure 10.27

Definition

The statement y **varies directly with the nth power of x** means that

$$y = kx^n$$

for some nonzero constant k.

Note that $y = kx$ is the special case of $y = kx^n$ that occurs when $n = 1$.

Sometimes one quantity gets *smaller* as another quantity gets *larger*. For example, the resistance R of a length of wire is a function of the diameter d of the wire according to the equation

$$R = \frac{k}{d^2},$$

where k is a constant that is determined by such factors as the material used to make the wire and the temperature of the wire. Note that R gets smaller as d^2 gets larger. We say that R **varies inversely with the square of d.**

Definition

The statement **y varies inversely with the nth power of x** means that

$$y = \frac{k}{x^n}$$

for some nonzero constant k.

EXAMPLE 5   A wire 0.06 centimeter in diameter has a resistance of 1 ohm. Find the resistance of a wire of the same length and material but with a diameter of 0.03 centimeter.

Write the general variation function.

$$R = \frac{k}{d^2}$$

Since $R = 1$ when $d = 0.06$, we have

$$1 = \frac{k}{(0.06)^2}$$

$$1 = \frac{k}{0.0036}$$

$$k = 0.0036.$$

Hence for this particular wire the specific variation function is

$$R = \frac{0.0036}{d^2}.$$

Therefore when $d = 0.03$, we have

$$R = \frac{0.0036}{(0.03)^2} = \frac{0.0036}{0.0009} = 4.$$

Try Problem 31   The resistance is 4 ohms. ◂

Sometimes one quantity varies directly with the product of two or more other quantities. For example, the maximum uniformly distributed load L that a rectangular beam will support is a function of its width w and its depth d according to the equation

$$L = kwd^2.$$

We say that **L varies jointly with w and the square of d.**

Definition

The statement **y varies jointly with x and z** means that

$$y = kxz$$

for some nonzero constant k.

EXAMPLE 6   A beam 4 inches wide and 10 inches deep will support a load of 1200 pounds. Determine the load that a beam 2 inches wide and 8 inches deep, of the same length and material, will support.

$$L = kwd^2$$ Write the general variation function

$$1200 = k(4)(10)^2$$ Substitute $L = 1200$, $w = 4$, and $d = 10$

$$1200 = k \cdot 400$$

$$k = 3$$ Solve for k

$$L = 3wd^2$$ Write the specific variation function

$$L = 3(2)(8)^2$$ Substitute $w = 2$ and $d = 8$

$$L = 384$$ Simplify

Try Problem 33   The beam will support 384 pounds. ◄

We can summarize the procedure for solving a variation problem as follows:

To Solve a Variation Problem

1. Write the general variation function.
2. Substitute the original values of the variables into the general variation function.
3. Solve the equation resulting from step 2 for the variation constant k.
4. Substitute the value of k found in step 3 into the general variation function. This gives the specific variation function.
5. Use the specific variation function and the given value(s) of the independent variable(s) to find the value of the dependent variable.

PROBLEM SET 10.6

Write each statement in Problems 1–12 as a general variation function.

1. p varies directly with q.
2. r varies directly with s.
3. r is directly proportional to s^3.
4. p is directly proportional to q^2.
5. u varies inversely with v.
6. z varies inversely with w.
7. w is inversely proportional to the square of z.
8. v is inversely proportional to the cube of u.
9. y varies jointly with x and the cube of z.
10. y varies jointly with x and the square of z.
11. T varies directly with r and inversely with s.
12. T varies directly with r and inversely with the cube of s.

13. Newton's law of gravitation states that the gravitational force F between any two bodies varies jointly with their masses m_1 and m_2 and inversely with the square of the distance d between them. Write Newton's law as a general variation function.

14. A sociologist, Joseph Cavanaugh, found that the number of long-distance phone calls N between two cities varies directly with their populations p_1 and p_2 and inversely with the distance d between the two cities. Write Cavanaugh's finding as a general variation function.

Perform the following steps for Problems 15–20: **a)** *find the variation constant k,* **b)** *use k to write the specific variation function,* **c)** *find y when x = 6.*

15. y varies directly with x, and $y = 20$ when $x = 5$.

16. y varies directly with x, and $y = 40$ when $x = 5$.

17. y varies directly with x^2, and $y = 12$ when $x = 2$.

18. y varies directly with x^2, and $y = 18$ when $x = 3$.

19. y varies inversely with x, and $y = 8$ with $x = 5$.

20. y varies inversely with x, and $y = 7$ with $x = 4$.

21. Given that y varies jointly with x and z, and $y = 18$ when $x = 6$ and $z = 2$, find y when $x = 4$ and $z = 5$.

22. Given that y varies jointly with x and z, and $y = 16$ when $x = 4$ and $z = 3$, find y when $x = 6$ and $z = 5$.

23. Given that y varies jointly with x and the square of z, and $y = 48$ when $x = \frac{2}{3}$ and $z = 6$, find y when $x = 8$ and $z = \frac{1}{2}$.

24. Given that y varies jointly with x and the square of z, and $y = 96$ when $x = \frac{3}{4}$ and $z = 8$, find y when $x = 6$ and $z = \frac{1}{2}$.

25. Suppose the price of gold, y, is directly proportional to the price of silver, x. If gold is $350 an ounce when silver is $10 an ounce, find the price of gold when silver is $9 an ounce.

26. According to Hooke's law,* the distance y that a spring is stretched is directly proportional to the amount of force x exerted on the spring. Given that a force of 100 kilograms stretches a spring a distance of 4 centimeters, how far will a force of 175 kilograms stretch that same spring?

27. Ignoring air resistance, the distance d that an object falls varies directly with the square of the time t that it falls. If an object falls 64 feet in 2 seconds, how far will it fall in 4 seconds?

28. The pressure p in pounds per square foot of a wind varies directly with the square of the velocity v of the wind. If a 10-mph wind produces a pressure of 0.3 pound per square foot, what pressure will a 100-mph wind produce? What will be the total force on the side of a barn that is 40 feet by 20 feet?

29. Suppose the time T required to complete a task is inversely proportional to the number of people, N, working on the task. If it takes 12 people 15 hours to construct the stage for a rock concert, how long will it take 10 people?

30. Boyle's law** states that the volume V of a gas is inversely proportional to the pressure P. If a gas has a volume of 100 cubic inches when the pressure is 15 pounds per square inch, what is its volume when the pressure is 10 pounds per square inch?

31. A wire 0.06 centimeter in diameter has a resistance of 1 ohm. Find the resistance of a wire of the same length and material but with a diameter of 0.02 centimeter. (See Example 5.)

32. A wire 0.03 centimeter in diameter has a resistance of 2 ohms. Find the resistance of a wire of the same length and material but with a diameter of 0.01 centimeter. (See Example 5.)

---

* Named after the English physicist Robert Hooke (1635–1703).
** Named after the English chemist Robert Boyle (1627–1691).

33. A beam 2 inches wide and 10 inches deep will support a load of 800 pounds. Determine the load that a beam 6 inches wide and 8 inches deep, of the same length and material, will support. (See Example 6.)

34. A beam 4 inches wide and 5 inches deep will support a load of 600 pounds. Determine the load that a beam 3 inches wide and 12 inches deep, of the same length and material, will support. (See Example 6.)

†35. The distance d required to stop an automobile is directly proportional to the square of its speed s. How many times greater is the distance required to stop an auto traveling at 60 mph than an auto traveling at 30 mph?

†36. The illumination I from a light source is inversely proportional to the square of the distance d from the source. How many times greater is the illumination at 10 feet from the source than at 20 feet from the source?

CALCULATOR PROBLEMS

37. Given that y varies inversely with x^3, and $y = 2.5$ when $x = 7.4$, find y when $x = 2$.

38. Given that y varies jointly with x and the square of w, and $y = 32{,}076$ when $x = 16.2$ and $w = 15$, find y when $x = 23.4$ and $w = 25$.

CHAPTER 10 REVIEW

[10.1]   *State the domain and the range of each relation. Determine whether the relation is a function.*

1. $\{(1, 3), (2, 3), (-6, 1), (3, 2)\}$

2. $\{(0, 4), (4, 0), (0, 0)\}$

Determine whether the given equation defines y as a function of x.

3. $y = 2x - 1$

4. $y = \dfrac{1}{1 - x}$

5. $x = y^2$

6. $y = \sqrt{2 - x}$

7. $y = -x^3$

Find the domain of the function defined by each equation.

8. $y = 1 - 2x$

9. $y = \dfrac{x}{x^2 - 9}$

10. $y = \dfrac{\sqrt{1 - x}}{x - 1}$

†11. $y = \sqrt[3]{x + 3}$

12. $y = \dfrac{1}{x^2 + 1}$

Find the domain and the range of each relation graphed below. Determine whether the relation is a function.

13.

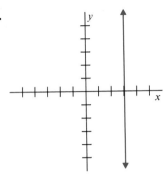

14.

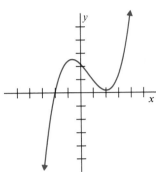

[10.2]   *Suppose $f(x) = 3x + 1$. Determine each of the following.*

15. $f(0)$ **16.** $f(-1)$ **17.** $f(a)$ **18.** $f(a - 1)$

Suppose $f(x) = 2x - 1$ and $g(x) = x^2 + 2$. Determine each of the following.

19. $g(2) - f(2)$ **20.** $\dfrac{f(0)}{g(0)}$ **21.** $g(2) \cdot f(\frac{1}{2})$

22. $f(f(2))$ **23.** $f(g(1))$ **24.** $g(f(1))$

Find $\dfrac{f(t) - f(a)}{t - a}$ for each function. Then simplify.

25. $f(x) = x^2 - 1$ **26.** $f(x) = 3$

27. A company can produce electric fans at a cost of $5 per fan plus a daily overhead of $375. If $C(x)$ denotes the total daily cost of producing x electric fans, write a cost function for the fans. Then find and interpret $C(0)$ and $C(100)$.

[10.3]   *Graph each function.*

28. $f(x) = -4$ **29.** $f(x) = 1$ **30.** $f(x) = x - 2$

31. $f(x) = -3x - 6$ **32.** $f(x) = -x^2 + 4x$ **33.** $f(x) = x^2$

Identify as a constant function, a linear function, or a quadratic function. Then graph the function. Label all intercepts and vertices.

34. $5x - 3y = 15$ **35.** $f(x) = x^2 - 6x + 9$ **36.** $g(t) = -t^2 + 4$ **37.** $f(x) = -3$

[10.4]   *Graph each polynomial function.*

38. $f(x) = -x^3$ **39.** $f(x) = x^3 - 4x$ **40.** $g(x) = 1 - x^4$ **41.** $f(x) = x^3 - 2x^2$

Graph each function.

42. $h(x) = \sqrt{2 - x}$ **43.** $h(x) = |x + 1|$ **44.** $g(x) = |x| - 3$

45. $p(x) = \sqrt{x} + 1$

46. $f(x) = \begin{cases} x \text{ if } x \geq 0 \\ 0 \text{ if } x < 0 \end{cases}$

[10.5] *Find the inverse of each function f. State whether f^{-1} is a function.*

47. $f = \{(2, 2), (3, 4), (-1, 3)\}$

48. $f = \{(-2, 3), (0, 1), (1, 3), (2, 2)\}$

Find $f^{-1}(x)$ for each function. Graph f and f^{-1} on the same axes. State whether f^{-1} is a function.

49. $f(x) = -2x$

50. $f(x) = 2x + 4$

51. $f(x) = x^2 + 4$

52. $f(x) = x^2 + 4, x \geq 0$

Find the inverse of each function.

53. $g(x) = x^3$

54. $h(x) = 3x^2 - 4$

55. Given $f(x) = 2x - 1$, find $f^{-1}(x)$. Then determine each of the following.
   a) $f^{-1}(-3)$ **b)** $f^{-1}(0)$ **c)** $f(f^{-1}(a))$ **d)** $f(-1)$

[10.6] *Write each statement as a general variation function.*

56. s varies directly with t.

57. y varies inversely with x.

58. A is inversely proportional to the square of x.

59. The circumference C of a circle varies directly with the radius r.

60. Given that y varies directly with x^2 and $y = 24$ when $x = 4$, find y when $x = 3$.

61. Given that z varies directly with x and inversely with y, and $z = 10$ when $x = 4$ and $y = 2$, find z when $x = 8$ and $y = 10$.

62. The volume V of a gas is inversely proportional to the pressure P. If a gas has a volume of 8 cubic inches when the pressure is 30 pounds per square inch, what is its volume when the pressure is 80 pounds per square inch?

CHAPTER 10 TEST

1. Determine whether the given equation defines y as a function of x.
   a) $y = \sqrt{x}$ **b)** $y = \frac{1}{x}$

2. State the domain of each function.
   a) $f(x) = \dfrac{x}{x - 2}$ **b)** $g(x) = \sqrt{3 - x}$

3. Consider the relation $\{(0, 3), (1, 4), (2, 2), (1, 1)\}$.
   a) State the domain.   **b)** State the range.
   c) Is the relation a function?   **d)** State the inverse.

4. Given $f(x) = 3 - x^2$, find each of the following.

   a) $f(0)$ **b)** $f(3)$ **c)** $f(a)$

Graph each function.

5. $f(x) = \sqrt{x + 2}$ **6.** $g(x) = |x - 2|$ **7.** $h(x) = 3x^2 - x^3$

8. $f(x) = -2$ **9.** $f(x) = 4 - 2x$ **10.** $g(x) = x^2 - 2x - 8$

11. $f(x) = \begin{cases} x \text{ if } x \geq 1 \\ 1 \text{ if } x < 1 \end{cases}$

Find the inverse of each function f. State whether f^{-1} is a function.

12. $f(x) = 3x + 4$ **13.** $f(x) = x^2 - 2$

14. Let $f(x) = 2 - 3x$ and $g(x) = x^2 + 1$. Determine each of the following.

   a) $f(2) + g(2)$ **b)** $f(g(-2))$

15. Find $\dfrac{f(t) - f(a)}{t - a}$ for $f(x) = x^2$. Then simplify.

16. Given $f(x) = x + 3$, find $f^{-1}(x)$. Then determine each of the following.

   a) $f^{-1}(1)$ **b)** $f^{-1}(f(a))$

17. Write each statement as a general variation function.

   a) y varies jointly with x and z.

   b) T varies directly with x and inversely with y.

18. Suppose y varies directly with x and $y = 12$ when $x = 8$. Find y when $x = 6$.

19. The distance d an object falls is directly proportional to the square of the time t that it falls. If an object falls 144 feet in 3 seconds, how far will it fall in 5 seconds?

20. The current i in a resistor is inversely proportional to the resistance R. If $i = 8$ amps when $R = 4$ ohms, find i when $R = 16$ ohms.

In this chapter we study two new functions—the exponential function and the logarithmic function. We study these two functions together because they are inverse functions; that is, one is formed from the other by interchanging the dependent and the independent variables. Exponential functions are used to calculate compound interest, the growth of bacteria, and the decay of a radioactive substance. Logarithmic functions are used to determine the pH of a solution, to measure the magnitude of an earthquake, and to compute the loudness of a sound wave. In Section 11.5, you will see how logarithms are used to solve equations that contain a variable in an exponent.

11

Exponential and Logarithmic Functions

11.1   *Exponential Functions*

Exponential functions are so named because the independent variable is in the exponent.

Definition

The **exponential function** with base a is a function of the form

$$f(x) = a^x,$$

where a is any positive constant except 1.

Note that if $a = 1$, the function $f(x) = a^x$ becomes $f(x) = 1^x = 1$, which is a constant function. That is why we eliminate $a = 1$ as a base. We eliminate zero and negative values of a for similar reasons.

EXAMPLE 1   Graph $f(x) = 2^x$.

This is the exponential function with base 2. Substitute values for x and compute the corresponding values of $f(x)$.

$$f(3) = 2^3 = 8 \qquad f(-3) = 2^{-3} = \frac{1}{2^3} = \frac{1}{8}$$

$$f(2) = 2^2 = 4 \qquad f(-2) = 2^{-2} = \frac{1}{2^2} = \frac{1}{4}$$

$$f(1) = 2^1 = 2 \qquad f(-1) = 2^{-1} = \frac{1}{2^1} = \frac{1}{2}$$

$$f(0) = 2^0 = 1$$

These calculations produce the following table of values.

x	-3	-2	-1	0	1	2	3
$f(x)$	$\frac{1}{8}$	$\frac{1}{4}$	$\frac{1}{2}$	1	2	4	8

Graph these points and draw a smooth curve through them to produce the graph shown in Figure 11.1. This graph is typical of graphs of $f(x) = a^x$, where $a > 1$.

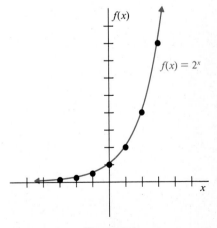

Figure 11.1

Try Problem 1

Although we have defined rational powers, such as 2^3 and $2^{1/2}$, we have *not* defined irrational powers, such as 2^{π}. A definition of irrational powers is beyond the scope of this text. Instead, we shall simply assume that 2^x is defined for all real values of x, to produce the graph without holes shown in Figure 11.1. Therefore the domain of $f(x) = 2^x$ consists of all real numbers. From the graph, we see that the range consists of all positive real numbers.

EXAMPLE 2   Graph $g(x) = (\frac{1}{2})^x$.

This is the exponential function with base $\frac{1}{2}$. Construct the following table of values.

x	-3	-2	-1	0	1	2	3
$g(x)$	8	4	2	1	$\frac{1}{2}$	$\frac{1}{4}$	$\frac{1}{8}$

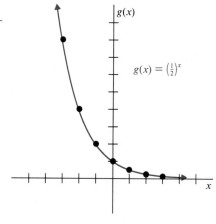

Graph these points and draw a smooth curve through them to produce the graph shown in Figure 11.2. This graph is typical of graphs of $g(x) = a^x$, where $0 < a < 1$.

$g(x) = (\frac{1}{2})^x$

Try Problem 3

Figure 11.2   ◄

Problem Solving

Radioactive Decay

The radioactive substance plutonium-239 is present in nuclear reactor wastes. It has a half-life of about 25,000 years. This means that it takes 25,000 years for a given quantity to decay to half of its original quantity. If Q is the quantity of 100 grams of plutonium-239 that remains after t years, write an equation that defines Q as a function of t. Then use this equation to find the quantity that remains after 100,000 years.

After 25,000 years, the quantity that remains is $100 \cdot \frac{1}{2} = 100(2)^{-1}$.
After 50,000 years, the quantity that remains is $100 \cdot \frac{1}{4} = 100(2)^{-2}$.
After 75,000 years, the quantity that remains is $100 \cdot \frac{1}{8} = 100(2)^{-3}$.
After t years, the quantity that remains is $100(2)^{-t/25,000}$.

Therefore an equation that defines Q as a function of t is

$$Q = 100(2)^{-t/25,000}.$$

To find the quantity that remains after 100,000 years, replace t with 100,000.

$$Q = 100(2)^{-100,000/25,000}$$
$$Q = 100(2)^{-4}$$
$$Q = 100(\tfrac{1}{16})$$
$$Q = 6.25$$

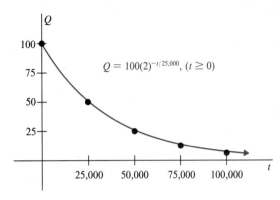

$$Q = 100(2)^{-t/25,000}, \ (t \geq 0)$$

Figure 11.3

That is, 6.25 grams will remain after 100,000 years. A graph of this function for $t \geq 0$ is shown in Figure 11.3.

PROBLEM SET 11.1

Graph each function.

1. $f(x) = 3^x$
2. $f(x) = 4^x$
3. $g(x) = (\tfrac{1}{3})^x$
4. $g(x) = (\tfrac{1}{4})^x$
5. $y = -3^x$
6. $y = -4^x$
7. $y = 3^{-x}$
8. $y = 4^{-x}$
9. $y = 2^x + 1$
10. $y = 2^x - 1$
11. $y = 2^{x+1}$
12. $y = 2^{x-1}$
13. $y = 4^{|x|}$
14. $y = 3^{|x|}$
15. $y = 4^{-|x|}$
16. $y = 3^{-|x|}$
17. $y = 5^{x/2}$
18. $y = 10^{x/2}$

19. A bacteria culture that contains 5 cells doubles its number each hour. This means that the number of bacteria B present after t hours is given by the function

$$B = 5(2)^t.$$

Find the number of bacteria present at each time below. Then graph the function for $t \geq 0$.
   a) $t = 0$ b) $t = 1$ c) $t = 2$ d) $t = 3$

20. The value of a comic book that you purchased for $2 triples each year. This means that the value after t years is given by the function

$$V = 2(3)^t.$$

Find the value of the comic book at each time below. Then graph the function for $t \geq 0$.
   a) $t = 0$ b) $t = 1$ c) $t = 2$ d) $t = 3$

21. The half-life of caffeine in the bloodstream is about 6 hours. This means that if you consume 100 milligrams

of caffeine (about one cup of coffee), the number of milligrams of caffeine C that will still be in your bloodstream t hours later is given by the function

$$C = 100(2)^{-t/6}.$$

Find the number of milligrams of caffeine that remains after each time below. Then graph the function for $t \geq 0$.

a) 0 hours   **b)** 6 hours   **c)** 12 hours   **d)** 18 hours

22. The substance radium-226, which is present in radioactive wastes, has a half-life of 1600 years. This means that if 100 grams of radium-226 are allowed to decay, the quantity Q that remains after t years is given by the function

$$Q = 100(2)^{-t/1600}.$$

Find the quantity that remains after each time below. Then graph the function for $t \geq 0$.

a) 0 years   **b)** 1600 years   **c)** 3200 years   **d)** 4800 years

CALCULATOR PROBLEMS

†**23.** The future value FV of a sequence of yearly payments, each of size p, is given by the formula

$$FV = p\left[\frac{(1+r)^t - 1}{r}\right],$$

where r is the interest rate, and t is the time in years. Suppose you make payments of $500 each year into your retirement account. What is the future value of your account in 50 years if the interest rate is 12%? What is the total amount that you paid into the account?

†**24.** The present value PV of a sequence of monthly payments, each of size p, is given by the formula

$$PV = p\left[\frac{1 - (1 + r/12)^{-12t}}{r/12}\right],$$

where r is the interest rate, and t is the time in years. Suppose you take out a home mortgage loan for $40,000 at 12% interest. Complete the following table by using the present value formula with $PV = \$40,000$.

Number of Years	Monthly Payment	Total Amount Repaid
20	$440	$105,600
30	?	?
40	?	?

11.2   *Logarithmic Functions*

Logarithms were originally invented* to simplify the tedious computations involved in astronomy. Today, such computations can be done with a calculator. Although logarithms are no longer used as a computational device, they are more important than ever. Logarithms are used today to describe many natural phenomena. As we will see in Section 11.5, logarithms are also invaluable in solving certain types of exponential equations.

---

* By the Scottish Baron John Napier (1550–1617).

Consider the exponential function

$$y = a^x.$$

Given values of x, we can compute the corresponding values of y. But suppose we were given values of y and asked to compute the corresponding values of x. Then y would be the independent variable, and x the dependent variable. Since it is customary to label the independent variable as x and the dependent variable as y, we interchange the two variables to get the inverse function

$$x = a^y.$$

To solve this equation for y, we introduce the following notation.

Definition

If a is any positive number except 1, then

$$y = \log_a x \quad \text{is equivalent to} \quad x = a^y.$$

The symbol "log" is an abbreviation for the word "logarithm." The notation $\log_a x$ is read "the logarithm of x to the base a." The equation $y = \log_a x$ defines the **logarithmic function** with base a. The relationship between exponential form and logarithmic form is illustrated below.

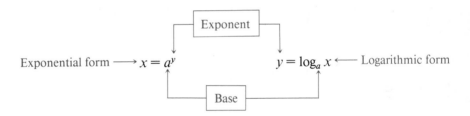

The examples and problems in this section will familiarize you with logarithmic notation. Your work will be made easier if you keep in mind that *a logarithm is an exponent.* More specifically,

$$\log_a x = \text{the exponent to which we raise } a \text{ to get } x.$$

Therefore

$$\log_2 8 = \text{the exponent to which we raise 2 to get 8}$$
$$\log_2 8 = 3.$$

EXAMPLE 1   Convert to logarithmic form: **a)** $9 = 3^2$,   **b)** $10^3 = 1000$, **c)** $6^{-1} = \frac{1}{6}$.

a) $9 = 3^2$ is equivalent to $\log_3 9 = 2$

b) $10^3 = 1000$ is equivalent to $\log_{10} 1000 = 3$

Try Problem 1   **c)** $6^{-1} = \frac{1}{6}$ is equivalent to $\log_6 \frac{1}{6} = -1$ ◄

EXAMPLE 2   Convert to exponential form: **a)** $\log_4 64 = 3$, **b)** $\log_{1/5} 25 = -2$,   **c)** $\log_9 3 = \frac{1}{2}$.

a) $\log_4 64 = 3$ is equivalent to $64 = 4^3$

b) $\log_{1/5} 25 = -2$ is equivalent to $25 = (\frac{1}{5})^{-2}$

Try Problem 17   **c)** $\log_9 3 = \frac{1}{2}$ is equivalent to $3 = 9^{1/2}$ ◄

We can solve equations of the form $y = \log_a x$ for any of the three variables y, a, or x by first writing the equation in exponential form. Then, if necessary, we apply the following property of the exponential function.

One-to-One Property of the Exponential Function

If a is any positive number except 1, then

$$a^x = a^y \quad \text{implies that} \quad x = y.$$

EXAMPLE 3   Find y if $y = \log_2 16$.

Convert to exponential form.

$$2^y = 16$$

Write each side with the same base.

$$2^y = 2^4$$

Use the one-to-one property of the exponential function to equate exponents.

$$y = 4$$

Check by replacing y with 4 in the original equation to get $4 = \log_2 16$. Since this *Try Problem 33* is equivalent to $2^4 = 16$, the solution checks. ◄

EXAMPLE 4   Find y if $y = \log_{1/2} 8$.

$$\begin{aligned} (\tfrac{1}{2})^y &= 8 & &\text{Convert to exponential form} \\ (2^{-1})^y &= 2^3 \\ 2^{-y} &= 2^3 \end{aligned} \Big\} \quad \text{Write each side with the same base}$$

$$\begin{aligned} -y &= 3 & &\text{Equate exponents} \end{aligned}$$

Try Problem 41 $\quad\quad y = -3 \quad$ Solve for y ◄

EXAMPLE 5   Find y if $y = \log_3 \sqrt{3}$.

$$3^y = \sqrt{3}$$
$$3^y = 3^{1/2}$$

Try Problem 47 $\quad\quad y = \frac{1}{2}$ ◄

EXAMPLE 6   Find x if $\log_{64} x = \frac{2}{3}$.
$$x = 64^{2/3}$$

Try Problem 53
$$x = (\sqrt[3]{64})^2 = (4)^2 = 16 \blacktriangleleft$$

EXAMPLE 7   Find a if $\log_a 49 = 2$.
$$a^2 = 49$$
$$a = 7 \quad \text{or} \quad a = -7$$

Try Problem 57
Eliminate $a = -7$ since a must be positive. Therefore $a = 7$. $\blacktriangleleft$

EXAMPLE 8   Graph $y = \log_2 x$.

This is the logarithmic function with base 2. It will be easier to construct a table of values if we convert to exponential form.
$$x = 2^y$$

Substitute values for y and compute the corresponding values of x.

When $y = 0$, $x = 2^0 = 1$.
When $y = 1$, $x = 2^1 = 2$.
When $y = 2$, $x = 2^2 = 4$, etc.

We obtain the following table of values.

x	$\frac{1}{8}$	$\frac{1}{4}$	$\frac{1}{2}$	1	2	4	8
y	-3	-2	-1	0	1	2	3

Graph these points and draw a smooth curve through them to produce the graph shown in Figure 11.4. This graph is typical of graphs of $y = \log_a x$, where $a > 1$.

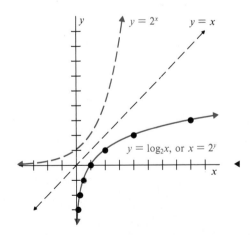

Figure 11.4

Try Problem 63

Note that the graph of $y = 2^x$ (drawn as a dashed curve in Figure 11.4) is symmetric to the graph of $y = \log_2 x$ about the line $y = x$. This symmetry occurs because the two functions are inverses of each other. From the graph of $y = \log_2 x$, we see that the domain is $x > 0$, whereas the range consists of all real numbers.

EXAMPLE 9   Graph $y = \log_{1/2} x$.

This is the logarithmic function with base $\frac{1}{2}$. Convert to the exponential form $x = (\frac{1}{2})^y$. Then assign values to y to get the following table of values.

x	$\frac{1}{8}$	$\frac{1}{4}$	$\frac{1}{2}$	1	2	4	8
y	3	2	1	0	-1	-2	-3

Graph these points and draw a smooth curve through them to produce the graph shown in Figure 11.5. This graph is typical of graphs of $y = \log_a x$, where $0 < a < 1$.

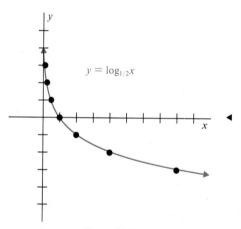

$y = \log_{1/2} x$

Try Problem 65

Figure 11.5

PROBLEM SET 11.2

Convert each equation to logarithmic form.

1. $25 = 5^2$
2. $49 = 7^2$
3. $3^4 = 81$
4. $2^5 = 32$
5. $3^{-1} = \frac{1}{3}$
6. $2^{-1} = \frac{1}{2}$
7. $10^2 = 100$
8. $10^3 = 1000$
9. $10^{-2} = 0.01$
10. $10^{-4} = 0.0001$
11. $6^1 = 6$
12. $8^1 = 8$
13. $4^{1/2} = 2$
14. $100^{1/2} = 10$
15. $t = b^s$
16. $v = b^u$

Convert each equation to exponential form.

17. $\log_2 16 = 4$

18. $\log_3 27 = 3$

19. $\log_5 125 = 3$

20. $\log_5 625 = 4$

21. $\log_{1/7} 49 = -2$

22. $\log_{1/9} 81 = -2$

23. $\log_{25} 5 = \frac{1}{2}$

24. $\log_{36} 6 = \frac{1}{2}$

25. $3 = \log_{1/2} \frac{1}{8}$

26. $3 = \log_{1/3} \frac{1}{27}$

27. $\log_{\sqrt{3}} 3 = 2$

28. $\log_{\sqrt{5}} 5 = 2$

29. $\log_{11} 1 = 0$

30. $\log_{12} 1 = 0$

31. $r = \log_b M$

32. $s = \log_b N$

Find y in each case.

33. $y = \log_2 8$

34. $y = \log_4 64$

35. $y = \log_2 \frac{1}{8}$

36. $y = \log_4 \frac{1}{64}$

37. $y = \log_{10} 10{,}000$

38. $y = \log_{10} 100{,}000$

39. $y = \log_{10} 0.1$

40. $y = \log_{10} 0.001$

41. $y = \log_{1/3} 27$

42. $y = \log_{1/3} 9$

43. $y = \log_{10} 1$

44. $y = \log_5 1$

45. $y = \log_{10} 10$

46. $y = \log_5 5$

47. $y = \log_5 \sqrt{5}$

48. $y = \log_7 \sqrt{7}$

49. $\log_a a^t = y$

50. $\log_b b^t = y$

Find x in each case.

51. $\log_9 x = 2$

52. $\log_4 x = 2$

53. $\log_8 x = \frac{2}{3}$

54. $\log_{27} x = \frac{2}{3}$

55. $\log_2 x = -5$

56. $\log_3 x = -4$

Find a in each case.

57. $\log_a 16 = 2$

58. $\log_a 64 = 2$

59. $\log_a \frac{1}{64} = 3$

60. $\log_a \frac{1}{125} = 3$

61. $\log_a 4 = \frac{1}{2}$

62. $\log_a 9 = \frac{1}{2}$

Graph each function.

63. $y = \log_3 x$

64. $y = \log_4 x$

65. $y = \log_{1/3} x$

66. $y = \log_{1/4} x$

67. Show that there is no real value for $\log_2 0$.

68. Show that there is no real value for $\log_2 (-4)$.

69. The magnitude M of an earthquake measured on the Richter scale* is given by the formula

$$M = \log_{10} \frac{I}{I_0},$$

where I is the intensity of the earthquake, and I_0 is the intensity of a "minimum" earthquake. Find the magnitude of an earthquake with the given intensity I.

a) $I = I_0$ (minimum earthquake)

b) $I = 10^{6.7} I_0$ (Los Angeles earthquake of 1971)

c) $I = 10^{8.3} I_0$ (San Francisco earthquake of 1906)

---

* Devised in 1935 by Charles Richter (1900–1985).

70. The loudness L of a sound wave in decibels* is given by the formula

$$L = 10 \log_{10} \frac{I}{I_0},$$

where I is the intensity of the wave, and I_0 is the intensity of a wave at the threshold of audibility. Find the loudness of a sound wave with the given intensity I.

a) $I = I_0$ (threshold of audibility)
b) $I = 10^{6.5} I_0$ (normal conversation)
c) $I = 10^{12} I_0$ (threshold of pain)

CALCULATOR PROBLEMS

Verify each equation. Then convert to logarithmic form.

71. $10^{0.4771} \approx 3$ **72.** $10^{1.6990} \approx 50$ **73.** $10^{2.5391} \approx 346$ **74.** $10^{3.3032} \approx 2010$

| 11.3 | *Properties of Logarithms* |

Since logarithms are exponents, we can use the laws of exponents to establish several important properties of logarithms. It is these properties that make logarithms so useful. The first property, the **product rule,** states that the logarithm of a product is the sum of the logarithms of the factors.

Product Rule for Logarithms

If a, M, and N are positive numbers and $a \neq 1$, then

$$\log_a M \cdot N = \log_a M + \log_a N.$$

We can prove the product rule as follows:

$$x = \log_a M \quad \text{is equivalent to} \quad M = a^x,$$
$$y = \log_a N \quad \text{is equivalent to} \quad N = a^y.$$

Apply the product rule for exponents.

$$M \cdot N = a^x \cdot a^y = a^{x+y}$$

---

* Named after Alexander Graham Bell (1847–1922).

Convert this equation to logarithmic form.

$$M \cdot N = a^{x+y} \quad \text{is equivalent to} \quad \log_a M \cdot N = x + y$$

Substitute for x and y to produce the desired result.

$$\log_a M \cdot N = \log_a M + \log_a N$$

EXAMPLE 1   Express as a sum of two logarithms: **a)** $\log_2 5 \cdot 7$, **b)** $\log_3 8x$,   **c)** $\log_{10} x \,(x + 2)$.

a) $\log_2 5 \cdot 7 = \log_2 5 + \log_2 7$

b) $\log_3 8x = \log_3 8 + \log_3 x$

Try Problem 1   **c)** $\log_{10} x(x + 2) = \log_{10} x + \log_{10} (x + 2)$ ◄

CAUTION The expression $\log_{10} (x + 2)$ cannot be simplified. It is *not* equal to $\log_{10} x + \log_{10} 2$.

The **quotient rule** for logarithms states that the logarithm of a fraction is the logarithm of the numerator minus the logarithm of the denominator.

Quotient Rule for Logarithms

If a, M, and N are positive numbers and $a \neq 1$, then

$$\log_a \frac{M}{N} = \log_a M - \log_a N.$$

The proof of the quotient rule is similar to the proof of the product rule.

EXAMPLE 2   Express as a difference of two logarithms: **a)** $\log_5 \dfrac{2}{7}$, **b)** $\log_2 \dfrac{3}{x}$,   **c)** $\log_{10} \dfrac{x - 1}{x}$.

a) $\log_5 \dfrac{2}{7} = \log_5 2 - \log_5 7$

b) $\log_2 \dfrac{3}{x} = \log_2 3 - \log_2 x$

Try Problem 9   **c)** $\log_{10} \dfrac{x - 1}{x} = \log_{10} (x - 1) - \log_{10} x$ ◄

CAUTION The quotient rule does *not* apply to expressions like $\dfrac{\log_5 6}{\log_5 2}$. This expression cannot be simplified.

The **power rule** for logarithms states that the logarithm of a number to a power is the power times the logarithm of the number.

Power Rule for Logarithms

If a and M are positive numbers and $a \neq 1$, then

$$\log_a M^r = r \log_a M$$

for any real number r.

EXAMPLE 3   Express as a product of a number and a logarithm:
a) $\log_4 x^3$, **b)** $\log_7 x^{-2}$, **c)** $\log_a \sqrt{x}$.

a) $\log_4 x^3 = 3 \log_4 x$

b) $\log_7 x^{-2} = -2 \log_7 x$

Try Problem 17   **c)** $\log_a \sqrt{x} = \log_a x^{1/2} = \frac{1}{2} \log_a x$ ◄

CAUTION The power rule does *not* apply to expressions like $(\log_4 x)^3$. This expression cannot be simplified.

Note what happens in the power rule when the base of the power is the same as the base of the logarithm.

$$\log_a a^t = t \log_a a = t \cdot 1 = t$$

Therefore we have the following special case of the power rule.

$$\log_a a^t = t$$

This means that $\log_2 2^3 = 3$, $\log_{10} 10^1 = 1$, and $\log_5 1 = \log_5 5^0 = 0$.

EXAMPLE 4   Expand and simplify: **a)** $\log_2 8x$, **b)** $\log_5 \frac{1}{x}$.

a) $\log_2 8x = \log_2 8 + \log_2 x$

$\qquad\qquad = \log_2 2^3 + \log_2 x$

$\qquad\qquad = 3 + \log_2 x$ $\qquad\qquad$ Since $\log_a a^t = t$

b) $\log_5 \frac{1}{x} = \log_5 1 - \log_5 x$

$\qquad\qquad = 0 - \log_5 x$

Try Problem 27 $\qquad\qquad = -\log_5 x$ ◄

You can also simplify Example 4(b) as follows:

$$\log_5 \frac{1}{x} = \log_5 x^{-1} = (-1) \log_5 x = -\log_5 x.$$

Sometimes we must use all three properties of logarithms to expand a logarithmic expression.

EXAMPLE 5   Expand $\log_3 \dfrac{x\sqrt{y}}{z^3}$.

$$\log_3 \dfrac{x\sqrt{y}}{z^3} = \log_3 x\sqrt{y} - \log_3 z^3 \qquad\qquad \text{Quotient rule}$$

$$= \log_3 x + \log_3 y^{1/2} - \log_3 z^3 \qquad \text{Product rule}$$

Try Problem 41

$$= \log_3 x + \dfrac{1}{2}\log_3 y - 3\log_3 z \qquad \text{Power rule} \quad \blacktriangleleft$$

We can combine a sum or difference of logarithms into a single logarithm by using the properties of logarithms in reverse. We can do this, however, only when the logarithms have the same base.

EXAMPLE 6   Express each part as a single logarithm with a coefficient of 1.

a) $6\log_5 x = \log_5 x^6$ Power rule

b) $\log_7 r + \log_7 s = \log_7 rs$ Product rule

Try Problem 53 **c)** $\log_{10} r - \log_{10} s = \log_{10} \dfrac{r}{s}$ Quotient rule ◂

EXAMPLE 7   Express $\frac{1}{3}\log_a x - \log_a y - \log_a z$ as a single logarithm with a coefficient of 1.

$$\dfrac{1}{3}\log_a x - \log_a y - \log_a z = \log_a x^{1/3} - (\log_a y + \log_a z)$$

$$= \log_a \sqrt[3]{x} - \log_a yz$$

Try Problem 61

$$= \log_a \dfrac{\sqrt[3]{x}}{yz} \quad \blacktriangleleft$$

PROBLEM SET 11.3

Express as a sum of two logarithms.

1. $\log_2 3\cdot 7$ **2.** $\log_3 2\cdot 5$ **3.** $\log_3 xy$ **4.** $\log_5 xy$

5. $\log_5 4x$ **6.** $\log_7 9x$ **7.** $\log_{10} x(x+1)$ **8.** $\log_{10} x(x+3)$

Express as a difference of two logarithms.

9. $\log_4 \frac{3}{7}$ **10.** $\log_2 \frac{5}{7}$ **11.** $\log_2 \frac{x}{y}$ **12.** $\log_4 \frac{x}{y}$

13. $\log_5 \frac{2}{x}$ **14.** $\log_5 \frac{6}{x}$ **15.** $\log_{10} \dfrac{x-4}{x}$ **16.** $\log_{10} \dfrac{x-9}{x}$

Express as a product of a number and a logarithm.

17. $\log_6 x^2$ **18.** $\log_6 x^4$ **19.** $\log_{10} x^{-3}$ **20.** $\log_{10} x^{-2}$

21. $\log_2 3^{1/2}$ **22.** $\log_2 5^{1/2}$ **23.** $\log_a \sqrt[3]{x}$ **24.** $\log_a \sqrt[4]{x}$

Expand and simplify.

25. $\log_{10} 3 \cdot 100$ **26.** $\log_{10} 7 \cdot 100$ **27.** $\log_3 27x$ **28.** $\log_4 64x$

29. $\log_2 \frac{16}{3}$ **30.** $\log_3 \frac{81}{2}$ **31.** $\log_4 \frac{1}{x}$ **32.** $\log_2 \frac{1}{x}$

Expand each expression. Simplify where possible.

33. $\log_2 x^3 y$ **34.** $\log_2 x^5 y$ **35.** $\log_2 8x^4$ **36.** $\log_2 16x^2$

37. $\log_3 \frac{x^2}{y}$ **38.** $\log_3 \frac{x^3}{y}$ **39.** $\log_3 \frac{x^5}{9}$ **40.** $\log_3 \frac{x^4}{27}$

41. $\log_4 \frac{x\sqrt{y}}{z^2}$ **42.** $\log_5 \frac{x\sqrt{y}}{z^4}$ **43.** $\log_5 \frac{x}{yz}$ **44.** $\log_4 \frac{x^2}{yz}$

45. $\log_5 \frac{x}{5y}$ **46.** $\log_4 \frac{x}{4y}$

Express as a single logarithm with a coefficient of 1.

47. $7 \log_5 x$ **48.** $8 \log_5 x$

49. $\frac{1}{2} \log_3 x$ **50.** $\frac{1}{3} \log_2 x$

51. $-\log_{10} x$ **52.** $-\log_{10} x$

53. $\log_2 r + \log_2 s$ **54.** $\log_3 r + \log_3 s$

55. $\log_3 r - \log_3 s$ **56.** $\log_2 r - \log_2 s$

57. $\log_5 x + \log_5 y + \log_5 z$ **58.** $\log_7 x + \log_7 y + \log_7 z$

59. $\log_{10} x + \log_{10} y - \log_{10} z$ **60.** $\log_{10} x - \log_{10} y + \log_{10} z$

61. $\frac{1}{2} \log_a x - \log_a y - \log_a z$ **62.** $\frac{1}{4} \log_a x - \log_a y - \log_a z$

63. $\log_a x - 2 \log_a y - \log_a z$ **64.** $\log_a x - \log_a y - 3 \log_a z$

65. $\log_{10} (x + 1) + \log_{10} (x - 1)$ **66.** $\log_{10} (x + 2) + \log_{10} (x - 2)$

67. $\log_{10} (x^2 + 7x + 12) - \log_{10} (x + 3)$ **68.** $\log_{10} (x^2 + 8x + 15) - \log_{10} (x + 5)$

69. Show that $\log_2 4 \cdot 8$ and $(\log_2 4)(\log_2 8)$ are not equal.

70. Show that $\log_2 (4 + 4)$ and $\log_2 4 + \log_2 4$ are not equal.

71. Show that $\dfrac{\log_3 9}{\log_3 3}$ and $\log_3 9 - \log_3 3$ are not equal.

72. Show that $\log_3 3^2$ and $(\log_3 3)^2$ are not equal.

† 73. Prove the quotient rule for logarithms.

† 74. Prove the power rule for logarithms.

CALCULATOR PROBLEMS

Given that $log_{10}\ 3 \approx 0.4771$ and $log_{10}\ 5 \approx 0.6990$, use the properties of logarithms to determine each of the following.

75. $\log_{10} 3 \cdot 5$

76. $\log_{10} \frac{3}{5}$

77. $\log_{10} 3^4$

78. $\log_{10} \sqrt{5}$

11.4   *Common Logarithms*

Since our number system is based on 10, logarithms with base 10 are more efficient in many situations than logarithms with other bases. A logarithm with base 10 is called a **common logarithm.** For simplicity, $\log_{10} x$ is written log x.

We can determine common logarithms of powers of 10 by using the rule $\log_a a^t = t$.

$$\begin{aligned}
\log 1000 &= \log 10^3 &&= 3 \\
\log 100 &= \log 10^2 &&= 2 \\
\log 10 &= \log 10^1 &&= 1 \\
\log 1 &= \log 10^0 &&= 0 \\
\log 0.1 &= \log 10^{-1} &&= -1 \\
\log 0.01 &= \log 10^{-2} &&= -2 \\
\log 0.001 &= \log 10^{-3} &&= -3
\end{aligned}$$

Note that numbers greater than 1 have common logarithms that are positive. Numbers between 0 and 1 have common logarithms that are negative.

To determine common logarithms of other positive numbers, we use either a table or a calculator. Appendix 5 contains a table of common logarithms. A calculator is faster, however, so we will use a calculator throughout this section. (Not all calculators operate the same way, so check your owner's manual.)

Let's begin by trying to get a feel for the value of a common logarithm. Suppose we want the common logarithm of 55. First, we note that

$$10 < 55 < 100.$$

Therefore we expect

$$\log 10 < \log 55 < \log 100.$$

That is, we expect

$$1 < \log 55 < 2.$$

Since 55 is halfway between 10 and 100, you might also expect that log 55 is halfway between 1 and 2. This would be the case if the logarithmic function were a linear function, but it's not.

EXAMPLE 1 Find log 55.

Press the following keys on your calculator.

Display

The number displayed is actually an approximate value of log 55. However, since most logarithms are approximations, it is customary to use the symbol = instead of the symbol ≈. Rounding off to four decimal places, we write log 55 = 1.7404. ◄

Try Problem 7

You can use your calculator to check the answer to Example 1 by showing that the equivalent exponential equation below is true.

$$10^{1.7404} = 55$$

To do this, press the following keys.

Display

If a number is too large or too small to be entered into your calculator in standard form, you must first write the number in scientific notation.

EXAMPLE 2 Find each logarithm: **a)** log 437,000,000,000, **b)** log 0.00000000437.

a) log 437,000,000,000 = log (4.37×10^{11}) Scientific notation

$= \log 4.37 + \log 10^{11}$ Product rule

$= 0.6405 + 11$

$= 11.6405$

b) log 0.00000000437 = log (4.37×10^{-9}) Scientific notation

$= \log 4.37 + \log 10^{-9}$ Product rule

$= 0.6405 + (-9)$

Try Problem 17

$= -8.3595$ ◄

You can avoid the product rule in Example 2(a) by entering 437,000,000,000 into your calculator in scientific notation.

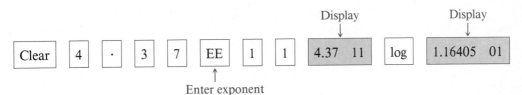

Enter exponent

This gives a value of $1.16405 \times 10^1 = 11.6405$. This method also works for Example 2(b).

Sometimes we want to reverse the procedure of finding the logarithm of a number. For example, since

$$\log 100 = 2,$$

we say that 2 is the logarithm of 100. On the other hand, we say that 100 is the **antilogarithm** of 2.

Definition

If $\log x = y$, then x is called the **antilogarithm** of y. This is abbreviated as $x = \text{antilog } y$.

EXAMPLE 3 Find x if $\log x = 2.49$.

There are two ways to do this on most calculators.

Method I Use the inverse of the log function.

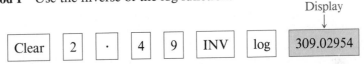

Method II Use the fact that $x = 10^{2.49}$.

Try Problem 29

In either case, the answer to three significant figures is $x = \text{antilog } 2.49 = 309$. ◄

You can check the solution to Example 3 by substituting it into the original equation. That is,

$$\log 309 = 2.49.$$

EXAMPLE 4 Find x if $\log x = 37.98$.

In this case, the display on your calculator will read as shown below.

$$\boxed{9.5499 \quad 37}$$

Try Problem 33

The antilogarithm of 37.98 is very large, so your calculator was forced to express it in scientific notation. That is, $x = $ antilog $37.98 = 9.55 \times 10^{37}$. ◄

EXAMPLE 5   Find x if $\log x = -0.158$.

Display

Try Problem 35

Therefore $x = $ antilog $(-0.158) = 0.695$. ◄

Problem Solving

pH of a Solution

Chemists define the pH (power of hydrogen) of a solution as

$$pH = -\log [H^+].$$

The symbol $[H^+]$ stands for the hydrogen ion concentration of the solution in moles per liter. On a scale of 0 to 14, pure distilled water has a pH of 7, acidic solutions have a pH of less than 7, and alkaline (base) solutions have a pH of greater than 7 (see Figure 11.6).

The pH Scale

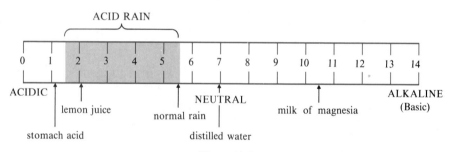

Figure 11.6

Acid rains have been measured that have hydrogen ion concentrations of 3.16×10^{-2} moles per liter. Determine the pH of this acid rain.

$$
\begin{aligned}
pH = -\log [H^+] &= -\log (3.16 \times 10^{-2}) && \text{Substitution} \\
&= -(\log 3.16 + \log 10^{-2}) && \text{Product rule} \\
&= -(0.5 + (-2)) \\
&= -(-1.5) \\
&= 1.5
\end{aligned}
$$

Generally, pH is expressed only to the nearest tenth, so the pH is 1.5.

PROBLEM SET 11.4

Find each common logarithm to four decimal places.

1. log 3

2. log 7

3. log 2.6

4. log 8.1

5. log 7.52

6. log 3.49

7. log 65

8. log 75

9. log 6500

10. log 7500

11. log 65,000

12. log 75,000

13. log 0.514

14. log 0.371

15. log 0.0514

16. log 0.0371

17. log 37,500,000,000

18. log 8,650,000,000

19. log 0.000000000375

20. log 0.0000000865

Find each antilogarithm x to three significant figures.

21. $\log x = 0.7443$

22. $\log x = 0.8293$

23. $\log x = 1.7443$

24. $\log x = 1.8293$

25. $\log x = 2.7443$

26. $\log x = 2.8293$

27. $\log x = 6.7443$

28. $\log x = 5.8293$

29. $\log x = 2.42$

30. $\log x = 2.73$

31. $\log x = 16.8$

32. $\log x = 17.6$

33. $\log x = 54.29$

34. $\log x = 86.94$

35. $\log x = -0.7126$

36. $\log x = -0.5717$

37. $\log x = -1.7126$

38. $\log x = -1.5717$

39. $\log x = -15.384$

40. $\log x = -17.684$

41. Why does your calculator register an error when you try to find log 0?

42. Why does your calculator register an error when you try to find log (-10)?

43. Determine the pH of milk of magnesia, given that it has a hydrogen ion concentration of 3.16×10^{-11} moles per liter.

44. Determine the pH of stomach acid, given that it has a hydrogen ion concentration of 6.31×10^{-2} moles per liter.

45. Find the hydrogen ion concentration of normal rainwater, given that it has a pH of 5.6. [*Hint:* Replace pH by 5.6 in the pH formula, multiply each side by -1, then find the antilogarithm.]

46. Find the hydrogen ion concentration of lemon juice, given that it has a pH of 2.1. (See the hint for Problem 45.)

†**47.** How many times greater is the hydrogen ion concentration of solution A than solution B, given that A has a pH of 5 and B has a pH of 6?

†**48.** Do Problem 47 if A has a pH of 4.

11.5   *Exponential and Logarithmic Equations*

EXPONENTIAL EQUATIONS

An equation that contains a variable in an exponent is called an **exponential equation.**

EXAMPLE 1   Solve $5^x = 25$.

Write both sides with the same base 5.

$$5^x = 5^2$$

Use the one-to-one property of the exponential function, given in Section 11.2, to equate exponents.

$$x = 2$$

Try Problem 1   The solution is 2. Check in the original equation. ◄

If we cannot write both sides of an exponential equation with the same base, we solve by taking the logarithm of each side. We use a common logarithm because we can find its value using a table or a calculator.

EXAMPLE 2   Solve $5^x = 7$.

Take the log of each side.

$$\log 5^x = \log 7$$

Use the power rule for logarithms to bring the exponent down as a multiplier.

$$x \log 5 = \log 7$$

Divide each side by log 5.

$$x = \frac{\log 7}{\log 5}$$

This is the exact value of x. You can approximate this answer using a table or a calculator as follows:

$$x = \frac{\log 7}{\log 5} \approx \frac{0.8451}{0.6990} \approx 1.21.$$

Try Problem 3   The solution is 1.21. Check in the original equation. ◄

CAUTION Remember that $\dfrac{\log 7}{\log 5}$ does *not* equal $\log 7 - \log 5$.

EXAMPLE 3   Solve $9^{x-1} = 27$.

$$9^{x-1} = 27$$
$$\left.\begin{array}{l}(3^2)^{x-1} = 3^3\\ 3^{2x-2} = 3^3\end{array}\right\}\quad \text{Write with the same base 3}$$
$$2x - 2 = 3 \qquad \text{Equate exponents}$$
$$\left.\begin{array}{l}2x = 5\\ x = \tfrac{5}{2}\end{array}\right\}\quad \text{Solve for } x$$

Try Problem 9   The solution is $\tfrac{5}{2}$. Check in the original equation. ◂

EXAMPLE 4   Solve $9^{x-1} = 15$.

$$\log 9^{x-1} = \log 15 \qquad \text{Take the log of each side}$$
$$(x - 1)\log 9 = \log 15 \qquad \text{Apply the power rule}$$
$$x - 1 = \frac{\log 15}{\log 9} \qquad \text{Divide by log 9}$$
$$x = \frac{\log 15}{\log 9} + 1 \qquad \text{Add 1}$$
$$\left.\begin{array}{l}x \approx \dfrac{1.1761}{0.9542} + 1\\[2mm] x \approx 2.23\end{array}\right\}\quad \text{Approximate}$$

Try Problem 11   The solution is 2.23. Check in the original equation. ◂

We can summarize the procedure for solving an exponential equation as follows:

To Solve an Exponential Equation

1. Write both sides with the same base, then equate exponents.
2. If step 1 fails, take the common logarithm of each side. Then use the power rule for logarithms to bring the variable exponent down as a multiplier.
3. Solve the equation resulting from step 1 or step 2.
4. Check your solution in the original equation.

LOGARITHMIC EQUATIONS

An equation that contains a logarithm of a variable quantity is called a **logarithmic equation.** One method for solving a logarithmic equation utilizes the following property.

One-to-One Property of the Logarithmic Function

If a, x, and y are positive and $a \neq 1$, then

$$\log_a x = \log_a y \quad \text{implies that} \quad x = y.$$

EXAMPLE 5 Solve $\log_7 (2x + 5) = \log_7 11$.

Both logarithms have the same base. Therefore apply the one-to-one property of the logarithmic function and get

$$2x + 5 = 11.$$

Then solve this equation.

$$2x = 6$$
$$x = 3$$

Try Problem 23

The solution is 3. Check in the original equation. ◄

EXAMPLE 6 Solve $\log (x + 2) + \log (x - 4) = \log 5x$.

Use the product rule for logarithms to write the left side as a single logarithm.

$$\log (x + 2)(x - 4) = \log 5x$$
$$\log (x^2 - 2x - 8) = \log 5x$$

Since both logarithms have base 10, apply the one-to-one property of the logarithmic function.

$$x^2 - 2x - 8 = 5x$$

Solve this quadratic equation by factoring.

$$x^2 - 7x - 8 = 0$$
$$(x - 8)(x + 1) = 0$$
$$x = 8 \quad \text{or} \quad x = -1$$

Try Problem 27

But $x = -1$ does not check since it causes both $\log (x - 4)$ and $\log 5x$ to equal $\log (-5)$, which is undefined. Therefore the only solution is 8. ◄

CAUTION The domain of any logarithmic function consists only of positive real numbers. Therefore we must always check the solutions to a logarithmic equation to make certain they do not cause a logarithm to be undefined.

If we cannot isolate a logarithm on each side of a logarithmic equation, we solve by converting to exponential form.

EXAMPLE 7 Solve $\log (x + 1) - \log (x - 2) = 1$.

Use the quotient rule for logarithms to write the left side as a single logarithm.

$$\log \frac{x+1}{x-2} = 1$$

Convert to exponential form. Remember, $\log = \log_{10}$.

$$\frac{x+1}{x-2} = 10^1$$

Solve for x.

$$x + 1 = 10x - 20 \qquad \text{Multiply by } x - 2$$
$$21 = 9x$$
$$x = \frac{21}{9} = \frac{7}{3}$$

Try Problem 39

The solution is $\frac{7}{3}$. Check in the original equation. ◄

We can summarize the procedure for solving a logarithmic equation as follows:

To Solve a Logarithmic Equation

1. Write each side as a single logarithm with the same base. Then apply the one-to-one property of the logarithmic function.
2. If step 1 fails, write one side as a single logarithm and the other side as a number. Then convert to exponential form.
3. Solve the equation resulting from step 1 or step 2.
4. Check your solution in the original equation. Discard any solution that causes a logarithm to be undefined.

The next example illustrates a practical application of an exponential equation.

EXAMPLE 8   How long will it take $100 to double if it is invested at 15%?

Recall the following formula from Section 2.1.

$$V = P(1 + r)^t$$

Substitute $V = 200$, $P = 100$, and $r = 0.15$. Then simplify.

$$200 = 100(1 + 0.15)^t$$
$$200 = 100(1.15)^t$$
$$2 = 1.15^t \qquad \text{Divide by 100}$$

Solve this exponential equation by taking the log of each side.

$$\log 2 = \log 1.15^t$$
$$\log 2 = t \log 1.15 \qquad \text{Power rule}$$
$$t = \frac{\log 2}{\log 1.15} \qquad \text{Divide by log 1.15}$$
$$t \approx 5 \qquad \text{Approximate}$$

Try Problem 41 \ It will take approximately 5 years for $100 to double at 15%. ◄

PROBLEM SET 11.5

Solve each equation. Round approximate answers to the nearest hundredth.

1. $2^x = 4$
2. $3^x = 9$
3. $2^x = 5$
4. $3^x = 11$
5. $3^{2x} = 81$
6. $2^{2x} = 16$
7. $3^{2x} = 13$
8. $2^{2x} = 7$
9. $4^{x+1} = 8$
10. $9^{x+1} = 27$
11. $4^{x+1} = 15$
12. $9^{x+1} = 14$
13. $5^{-x} = 125$
14. $5^{-x} = 625$
15. $5^{-x} = 100$
16. $5^{-x} = 1000$
17. $6^{3x-1} = 1$
18. $7^{3x+1} = 1$
19. $2^{x^2+3x} = \frac{1}{4}$
20. $3^{x^2-3x} = \frac{1}{9}$

Solve each equation.

21. $\log_4 5x = \log_4 15$
22. $\log_4 7x = \log_4 14$
23. $\log_7 (3x + 5) = \log_7 8$
24. $\log_5 (2x - 1) = \log_5 9$
25. $\log (2x - 4) = \log (x + 1)$
26. $\log (3x + 2) = \log (2x + 5)$
27. $\log (x + 3) + \log (x - 2) = \log 6x$
28. $\log (x + 4) + \log (x - 3) = \log 5x$

Solve each equation.

29. $\log_2 x = 5$
30. $\log_3 x = 4$
31. $\log (x - 1) = 3$
32. $\log (x + 1) = 2$
33. $\log_3 (x^2 - 8x) = 2$
34. $\log_2 (x^2 - 7x) = 3$
35. $\log_2 x - \log_2 3 = 1$
36. $\log_3 x - \log_3 2 = 1$
37. $\log x + \log (x - 15) = 2$
38. $\log x + \log (x - 21) = 2$
39. $\log (x + 2) - \log (x - 1) = 1$
40. $\log (x + 5) - \log (x - 1) = 1$

41. How long will it take $100 to double if it is invested at 10%? How long will it take $10,000 to double at 10%?

42. How long will it take $500 to double if it is invested at 12%? How long will it take $50,000 to double at 10%?

43. How long will it take an investment to triple if it earns 18%?

44. How long will it take an investment to triple if it earns 20%?

† **45.** A person is murdered in a room whose temperature is 70°. According to Newton's law of cooling, the temperature T of the body t hours after death is given by the formula

$$T = 70 + 28.6(0.97)^t.$$

The temperature of the body when it is discovered at 9 A.M. is 95°. At what time did the murder take place?

† **46.** The half-life of carbon-14 is 5700 years. This means that the quantity Q of carbon-14 that remains t years after an organism dies is given by the formula

$$Q = Q_o (2)^{-t/5700},$$

where Q_o is the original quantity. An old bone is unearthed and determined to contain only 30% of its original carbon-14 content. Determine the age of the bone.

11.6   *Change of Base, The Number e*

CHANGE OF BASE

Since there are an infinite number of possible bases for the logarithmic function, it would be impossible to construct a table or a calculator key for every base. We can find logarithms to any base, however, using the **change-of-base formula** below.

Change-of-Base Formula

If a, b, and x are positive numbers and neither a nor b is 1, then

$$\log_b x = \frac{\log_a x}{\log_a b}.$$

We can verify the change-of-base formula as shown below.

$$y = \log_b x$$
$$b^y = x \qquad \text{Convert to exponential form}$$
$$\log_a b^y = \log_a x \qquad \text{Take } \log_a \text{ of each side}$$
$$y \log_a b = \log_a x \qquad \text{Apply the power rule}$$
$$y = \frac{\log_a x}{\log_a b} \qquad \text{Divide by } \log_a b$$
$$\log_b x = \frac{\log_a x}{\log_a b} \qquad \text{Replace } y \text{ with } \log_b x$$

We can now find logarithms to any base simply by converting to common logarithms.

EXAMPLE 1   Find $\log_2 5$.

Use the change-of-base formula with $b = 2$, $x = 5$, and $a = 10$.

$$\log_2 5 = \frac{\log_{10} 5}{\log_{10} 2}$$

Find $\log_{10} 5$ and $\log_{10} 2$ using a calculator or a table.

Try Problem 1

$$\log_2 5 \approx \frac{0.6990}{0.3010} \approx 2.32 \blacktriangleleft$$

THE NUMBER e

The number that is used as a logarithmic base most often in calculus is the number e.* The number e is irrational, like π. An approximate value of e is 2.718. Since the base e appears naturally in many applied problems, a logarithm with base e is called a **natural logarithm.** For simplicity, $\log_e x$ is written $\ln x$.

We can determine the natural logarithm of any positive number using either a table or a calculator with a $\boxed{\ln x}$ key.

EXAMPLE 2   Find $\ln 1350$.

Press the following keys on your calculator.

Display

Try Problem 15   Therefore $\ln 1350 = 7.2079$. $\blacktriangleleft$

You can use your calculator to check the answer to Example 2 by showing that the equivalent exponential equation below is true.

$$e^{7.2079} = 1350$$

To do this, press the following keys.

Display

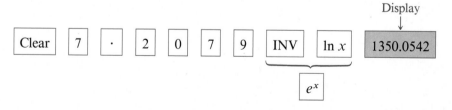

---

* Named by the Swiss mathematician Leonhard Euler (1707–1783).

Note that e^x is the inverse of the natural logarithmic function $\ln x$. The exponential function e^x appears in many applied problems. For example, if interest is compounded *continuously,* the value V of an account is given by the formula

$$V = Pe^{rt},$$

where P is the principal, r is the interest rate, and t is the time in years.

EXAMPLE 3 Determine the value of $1000 invested for 2 years at 8% compounded continuously.

Substitute $P = 1000$, $r = 0.08$, and $t = 2$ into the formula $V = Pe^{rt}$.

$$V = 1000e^{\,0.08(2)}$$
$$= 1000\, e^{0.16}$$

Find $e^{0.16}$ using the table in Appendix 6, or by pressing the following keys on your calculator.

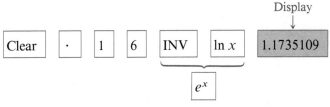

Therefore $e^{0.16} \approx 1.17351$, and

$$V \approx 1000(1.17351) = 1173.51.$$

Try Problem 41 \ The value is $1173.51. ◄

PROBLEM SET 11.6

Find each logarithm to the nearest hundredth.

1. $\log_2 3$
2. $\log_3 7$
3. $\log_5 632$
4. $\log_6 588$
5. $\log_{11} 5.14$
6. $\log_{11} 3.16$
7. $\log_{1/2} 9.1$
8. $\log_{1/2} 9.7$
9. $\log_3 1000$
10. $\log_2 100$
11. Express $\log_2 15$ in terms of $\log_9$.
12. Express $\log_3 25$ in terms of $\log_8$.
13. Express $\log_7 5$ in terms of $\log_5$.
14. Express $\log_5 11$ in terms of $\log_{11}$.

Find each natural logarithm to four decimal places.

15. ln 1260   **16.** ln 1480   **17.** ln 100   **18.** ln 1000

19. ln 47   **20.** ln 68   **21.** ln 5.13   **22.** ln 7.42

23. ln 0.00707   **24.** ln 0.00303

When a = e, the property $\log_a a^t = t$ becomes $\ln e^t = t$. Use this property to find each natural logarithm.

25. ln e   **26.** ln e^2 **27.** ln e^3 **28.** ln e^4

29. ln $\dfrac{1}{e}$ **30.** ln $\dfrac{1}{e^2}$ **31.** ln $\sqrt{e}$ **32.** ln $\sqrt[3]{e}$

Use the table in Appendix 6 or your calculator to find each power of e to three significant figures.

33. e^2 **34.** e^4 **35.** $e^{3.8}$ **36.** $e^{3.9}$

37. $e^{1.52}$ **38.** $e^{1.26}$ **39.** $e^{-0.3}$ **40.** $e^{-0.8}$

41. Determine the value of $2000 invested for 5 years at 8% compounded continuously.

42. Determine the value of $3000 invested for 10 years at 12% compounded continuously.

43. The atmospheric pressure P in pounds per square inch at an altitude of a miles is given by the formula

$$P = 14.7\, e^{-0.21a}.$$

Find the atmospheric pressure
a) at sea level,
b) at an altitude of 10 miles.

44. Use the formula given in Problem 43 to find the atmospheric pressure
a) on Mt. Everest, about 5.5 miles above sea level,
b) in Death Valley, about 0.05 miles below sea level.

†45. How long will it take an investment to double if it earns 10% compounded continuously?

†46. Use the change-of-base formula to show that
$$\log_b a = \frac{1}{\log_a b}.$$

CHAPTER 11 REVIEW

[11.1] *Graph each function.*

1. $f(x) = 5^x$ **2.** $g(x) = 5^{-x}$ **3.** $y = 3^x + 1$ **4.** $y = 2^{|x|}$

5. A bacteria culture that contains 10 cells triples its number each hour. This means that the number of bacteria B present after t hours is given by the function

$$B = 10(3)^t.$$

Find the number of bacteria present at each time below.
a) $t = 0$ b) $t = 1$ c) $t = 2$ d) $t = 3$

[11.2] *Convert each equation to logarithmic form.*

6. $3^5 = 243$ **7.** $3^{-3} = \frac{1}{27}$ **8.** $8^{2/3} = 4$ **9.** $10^{-5} = 0.00001$

Convert each equation to exponential form.

10. $\log_{1/2} \frac{1}{16} = 4$ **11.** $\log_7 1 = 0$ **12.** $\log_4 1024 = 5$ **13.** $\log_{32} 4 = \frac{2}{5}$

Find y in each case.

14. $y = \log_2 32$ **15.** $y = \log_3 \frac{1}{81}$ **16.** $y = \log_{10} 1000$ **17.** $y = \log_{10} 0.000001$

Find x in each case.

18. $\log_5 x = 4$ **19.** $\log_9 x = \frac{3}{2}$

Find a in each case.

20. $\log_a 144 = 2$ **21.** $\log_a 27 = -3$

Graph each function.

22. $y = \log_5 x$ **23.** $y = \log_{1/2} x$

[11.3] *Express as a sum or difference of two logarithms.*

24. $\log_4 5 \cdot 7$ **25.** $\log_2 \frac{9}{x}$ **26.** $\log_6 16x$ **27.** $\log_4 \frac{x}{x+1}$

Express as a product of a number and a logarithm.

28. $\log_5 x^6$ **29.** $\log_4 \sqrt[4]{x}$

Expand and simplify.

30. $\log_7 \frac{7}{x}$ **31.** $\log_4 \frac{3}{64}$ **32.** $\log_6 36x^2$

33. $\log_2 8xy^2$ **34.** $\log_{10} \frac{100x^2}{y^2}$ **35.** $\log_6 \frac{x^2}{6y}$

Express as a single logarithm with a coefficient of 1.

36. $7 \log_4 x$ **37.** $\log_3 a - 4 \log_3 b$

38. $\frac{1}{2} \log_a x - 3 \log_a y - 4 \log_a z$ **39.** $\log_{10} (x^2 + 8x + 7) - \log_{10} (x + 1)$

[11.4] *Find each common logarithm to four decimal places.*

40. $\log 7.3$ **41.** $\log 6.25$ **42.** $\log 34{,}100$ **43.** $\log 0.000452$

Find each antilogarithm x to three significant figures.

44. $\log x = 0.4983$ **45.** $\log x = 6.7372$ **46.** $\log x = 1.4407$

47. $\log x = -0.3958$ **48.** $\log x = -10.0022$ **49.** $\log x = -4.6860$

[11.5] *Solve each equation.*

50. $5^x = 125$

51. $5^x = 100$

52. $3^{x^2-5} = 9^{2x}$

53. $3^{x-1} = 7$

54. $\log_2 (2x + 1) = \log_2 9$

55. $\log (x + 10) - \log x = \log 6$

56. $\log_6 (x + 3) + \log_6 (x + 2) = 1$

57. $\log_3 x + \log_3 (x - 6) = 3$

58. How long will it take $1000 to triple if it is invested at 12% compounded annually?

[11.6] *Find each logarithm to the nearest hundredth.*

59. $\log_7 4$ **60.** $\log_{12} 226$ **61.** $\log_{1/2} 6.2$ **62.** $\log_4 1000$

63. Express $\log_3 21$ in terms of $\log_7$.

Find each natural logarithm to four decimal places.

64. $\ln 10$ **65.** $\ln 37$ **66.** $\ln 0.001$ **67.** $\ln \sqrt[4]{e}$

Use the table in Appendix 6 or your calculator to find each power of e to three significant figures.

68. $e^{2.4}$ **69.** $e^{-0.5}$

70. Determine the value of $5000 invested for four years at 9% compounded continuously.

CHAPTER 11 TEST

Convert to exponential form.

1. $\log_5 25 = 2$

2. $\log 0.01 = -2$

Convert to logarithmic form.

3. $6^3 = 216$

4. $9^{1/2} = 3$

Graph each function.

5. $y = 2^{-x}$

6. $y = \log_4 x$

Find the unknown in each case.

7. $\log_a 32 = 5$

8. $\log_4 x = 3$

9. $y = \log_{10} 100$

10. $y = \log_2 \frac{1}{16}$

Expand and simplify.

11. $\log_2 2y$

12. $\log_7 \frac{49}{xy}$

Express as a single logarithm with a coefficient of 1.

13. $\log x + \frac{1}{2} \log y$

14. $3 \log_5 x - \log_5 y$

Solve each equation.

15. $2^{2x-1} = 8$

16. $\log (5x - 3) = \log 2x$

17. $4^{x+1} = 9$

18. $\log_8 (x + 2) + \log_8 x = 1$

19. $\log_5 (x + 12) - \log_5 x = 1$

Find each common logarithm to four decimal places.

20. $\log 248$

21. $\log 0.00017$

Find each antilogarithm x to three significant figures.

22. $\log x = 4.8488$

23. $\log x = -3.3526$

24. Express $\log_2 9$ in terms of $\log_7$.

25. Determine the value of $4000 invested for 3 years at 10% compounded continuously. Use the formula $V = Pe^{rt}$.

Many real-world problems are best described using a particular type of function known as a sequence. In this chapter, we examine the two most basic types of sequences—arithmetic sequences and geometric sequences. You will learn to find any term of such a sequence without writing the entire sequence. The indicated sum of a sequence is called a series. You will learn to find the sum of a series without actually adding the terms. The chapter concludes with a discussion of the binomial theorem. This theorem allows us to expand expressions of the form $(a + b)^n$ without actually multiplying. As you can see, this is a chapter of shortcuts.

12

Sequences and Series

12.1	*Sequences and Series*

SEQUENCES

Suppose a radio station announces that it will award $5 to the first person who correctly identifies a certain "mystery voice." Moreover, for each day the mystery voice remains unidentified, an additional $5 will be added to the prize. If $a(n)$ denotes the size of the award on the nth day, then

$$a(n) = 5n.$$

For example, the size of the award on the 3rd day is

$$a(3) = 5 \cdot 3 = 15.$$

The following numbers represent the size of the award on each successive day.

$$5, 10, 15, 20, 25, \ldots$$

The function $a(n) = 5n$ is valid only when n is a positive integer. That is, $a(2.5)$ and $a(-3)$ have no meaning. A function whose domain consists of only positive integers is called a **sequence function,** or simply a **sequence.** The range values of a sequence function are called the **terms** of the sequence.

Definition

An **infinite sequence** is a function whose domain is the entire set of positive integers. A **finite sequence** is a function whose domain is the first k positive integers.

A finite sequence has a last term, whereas an infinite sequence does not. For example, the *finite* sequence $a(n) = 5n$ with domain $n = 1, 2, 3, 4$ is written

$$5, 10, 15, 20.$$

The *infinite* sequence $a(n) = 5n$ with domain $n = 1, 2, 3, 4, \ldots$ is written

$$5, 10, 15, 20, \ldots.$$

It is customary to name sequence functions with the letters a, b, and c, as opposed to the usual f, g, and h. It is also customary to use the letter n as the independent variable rather than x. Finally, we denote the value of a sequence function as a_n (read "a sub n") rather than $a(n)$. We can summarize this notation as shown below.

$$a_1 = \text{first term of the sequence}$$
$$a_2 = \text{second term of the sequence}$$
$$\vdots$$
$$a_n = n\text{th term of the sequence}$$

The nth term of the sequence is also called the **general term,** since from it we can obtain all the terms of the sequence. To conserve space, the terms of a sequence are usually written horizontally as shown below.

$$a_1, a_2, a_3, \ldots, a_n, \ldots$$

EXAMPLE 1   Find the first four terms of the sequence whose general term is $a_n = 2n - 1$.

Replace n with 1, 2, 3, and 4.

$$a_1 = 2(1) - 1 = 1$$
$$a_2 = 2(2) - 1 = 3$$
$$a_3 = 2(3) - 1 = 5$$
$$a_4 = 2(4) - 1 = 7$$

Try Problem 1

Therefore the first four terms are 1, 3, 5, 7. ◄

EXAMPLE 2   Find the first four terms of the sequence whose general term is $a_n = \dfrac{(-1)^n}{n + 1}$.

Replace n with 1, 2, 3, and 4.

$$a_1 = \frac{(-1)^1}{1 + 1} = -\frac{1}{2}$$
$$a_2 = \frac{(-1)^2}{2 + 1} = \frac{1}{3}$$
$$a_3 = \frac{(-1)^3}{3 + 1} = -\frac{1}{4}$$
$$a_4 = \frac{(-1)^4}{4 + 1} = \frac{1}{5}$$

Try Problem 11

The first four terms are $-\frac{1}{2}, \frac{1}{3}, -\frac{1}{4}, \frac{1}{5}$. ◄

Sometimes we must reverse the procedure. That is, given some of the terms of a sequence, we must find a formula for the general term a_n. There are no rules for finding the general term; we simply inspect the given terms and try to identify a pattern. Don't be afraid to guess at a formula for the general term. Often an incorrect guess will lead to the correct formula.

EXAMPLE 3   Find the general term a_n for the sequence 1, 4, 9, 16, 25, ...

Observe that each term is the square of the term number. That is,

$$a_1 = 1 = 1^2$$
$$a_2 = 4 = 2^2$$
$$a_3 = 9 = 3^2,$$

Try Problem 19

and so on. Therefore the general term is $a_n = n^2$. ◄

SERIES

Consider the finite sequence below.

$$5, 10, 15, 20$$

The indicated sum of this sequence, written

$$5 + 10 + 15 + 20,$$

is called a **series.**

Definition   The indicated sum of a finite sequence is called a **finite series.** The indicated sum of an infinite sequence is called an **infinite series.**

In Sections 12.1 – 12.3 we deal with finite series only. In Section 12.4 we will study a particular type of infinite series.

We can write a series in a more compact form, called **summation notation,** using the symbol $\sum$ (the Greek letter sigma) along with the general term of the corresponding sequence. For example, since the general term of the sequence 5, 10, 15, 20 is $5n$, we can write the corresponding series $5 + 10 + 15 + 20$ as $\sum_{n=1}^{4} 5n$. To evaluate $\sum_{n=1}^{4} 5n$, replace n by 1, 2, 3, and 4. Then add the resulting terms.

$$\sum_{n=1}^{4} 5n = 5(1) + 5(2) + 5(3) + 5(4)$$
$$= 5 + 10 + 15 + 20$$
$$= 50$$

The expression $\sum_{n=1}^{4} 5n$ is read "the summation of $5n$ as n goes from 1 to 4." The letter n is called the **index** of summation. The number 1 is the **lower limit** of summation, and 4 is the **upper limit** of summation.

EXAMPLE 4   Expand $\sum_{n=1}^{3} (n^2 - n)$ and simplify.

Replace n with 1, 2, and 3. Then add.

$$\sum_{n=1}^{3} (n^2 - n) = (1^2 - 1) + (2^2 - 2) + (3^2 - 3)$$
$$= 0 + 2 + 6$$
$$= 8 \blacktriangleleft$$

Try Problem 37

EXAMPLE 5   Expand $\sum_{n=1}^{5} \dfrac{x^n}{n}$.

Replace n (*not* x) with 1, 2, 3, 4, and 5.

$$\sum_{n=1}^{5} \frac{x^n}{n} = \frac{x^1}{1} + \frac{x^2}{2} + \frac{x^3}{3} + \frac{x^4}{4} + \frac{x^5}{5}$$

Try Problem 43

$$= x + \frac{x^2}{2} + \frac{x^3}{3} + \frac{x^4}{4} + \frac{x^5}{5} \blacktriangleleft$$

We now reverse the procedure and convert from expanded form to summation notation. Generally, there is more than one answer to this type of problem.

EXAMPLE 6 Write $4 + 7 + 10 + 13 + 16 + 19$ in summation notation.

The difference between any two successive terms is 3. This suggests writing $3n$. To obtain the first term 4 when $n = 1$, write $a_n = 3n + 1$. Therefore

$$4 + 7 + 10 + 13 + 16 + 19 = \sum_{n=1}^{6} (3n + 1).$$

Try Problem 45

You may verify that the answer can also be written as $\sum_{n=0}^{5} (3n + 4)$. $\blacktriangleleft$

CAUTION You must use parentheses when writing $\sum_{n=1}^{6} (3n + 1)$. The expression

$\sum_{n=1}^{6} 3n + 1$ means $\left(\sum_{n=1}^{6} 3n \right) + 1.$

PROBLEM SET 12.1

Find the first four terms of the sequence whose general term is given. Then find the tenth term.

1. $a_n = 3n - 1$ **2.** $a_n = 3n + 1$ **3.** $a_n = n^2 + n$ **4.** $a_n = n^2 - n$

5. $a_n = \dfrac{n - 1}{n}$ **6.** $a_n = \dfrac{n + 1}{n}$ **7.** $a_n = n + \dfrac{1}{n}$ **8.** $a_n = n - \dfrac{1}{n}$

9. $a_n = 6$ **10.** $a_n = 8$ **11.** $a_n = \dfrac{(-1)^n}{n}$ **12.** $a_n = \dfrac{(-1)^n}{n + 2}$

13. $a_n = 2^n$ **14.** $a_n = 3^n$

Find the general term a_n for each sequence.

15. $2, 4, 6, 8, 10, \ldots$ **16.** $3, 6, 9, 12, 15, \ldots$ **17.** $1, 2, 3, 4, 5, \ldots$

18. $1, 3, 5, 7, 9, \ldots$ **19.** $1, 8, 27, 64, 125, \ldots$ **20.** $1, 16, 81, 256, 625, \ldots$

21. $1, \frac{1}{4}, \frac{1}{9}, \frac{1}{16}, \frac{1}{25}, \ldots$ **22.** $1, \frac{1}{8}, \frac{1}{27}, \frac{1}{64}, \frac{1}{125}, \ldots$ **23.** $-1, 1, -1, 1, -1, \ldots$

24. $1, -1, 1, -1, 1, \ldots$ **25.** $2, -4, 6, -8, 10, \ldots$ **26.** $-3, 6, -9, 12, -15, \ldots$

Expand and simplify.

27. $\displaystyle\sum_{n=1}^{6} 4n$

28. $\displaystyle\sum_{n=1}^{7} 10n$

29. $\displaystyle\sum_{n=1}^{6} 4$

30. $\displaystyle\sum_{n=1}^{7} 10$

31. $\displaystyle\sum_{n=2}^{5} (2n+1)$

32. $\displaystyle\sum_{n=2}^{6} (2n-1)$

33. $\displaystyle\sum_{n=2}^{5} 2n+1$

34. $\displaystyle\sum_{n=2}^{6} 2n-1$

35. $\displaystyle\sum_{n=0}^{4} (n^2-3n+4)$

36. $\displaystyle\sum_{n=0}^{4} (n^2-4n+5)$

37. $\displaystyle\sum_{n=1}^{3} (n^3-n)$

38. $\displaystyle\sum_{n=1}^{3} (n^3+n)$

39. $\displaystyle\sum_{n=1}^{4} \frac{1}{n}$

40. $\displaystyle\sum_{n=0}^{3} \frac{1}{n+1}$

41. $\displaystyle\sum_{n=1}^{3} (-n)^n$

42. $\displaystyle\sum_{n=0}^{3} (-n)^{n+1}$

43. $\displaystyle\sum_{n=1}^{5} \frac{x^{n+1}}{n+1}$

44. $\displaystyle\sum_{n=1}^{5} \frac{x^{2n}}{2n}$

Write each series in summation notation.

45. $2+5+8+11+14+17$

46. $1+5+9+13+17+21$

47. $5+8+11+14+17+20$

48. $5+9+13+17+21+25$

49. $3+6+9+12$

50. $2+4+6+8$

51. $3+9+27+81$

52. $2+4+8+16$

53. $\frac{1}{2}+\frac{2}{3}+\frac{3}{4}+\frac{4}{5}$

54. $\frac{3}{4}+\frac{4}{5}+\frac{5}{6}+\frac{6}{7}$

55. $2+5+10+17+26$

56. $0+3+8+15+24$

57. $1+x+x^2+x^3+x^4$

58. $x+x^2+x^3+x^4+x^5$

59. Show that $\displaystyle\sum_{n=1}^{5} n$ and $\displaystyle\sum_{k=1}^{5} k$ represent the same value.

60. Show that $\displaystyle\sum_{n=1}^{5} n^2$ and $\displaystyle\sum_{k=1}^{5} k^2$ represent the same value.

61. An automobile costs $10,000 when new. If it depreciates 10% each year, write a sequence that describes its value for each of the first five years.

62. Your starting salary is $10,000. If you receive a 10% raise each year, write a sequence that describes your salary for each of the first five years.

†63. Determine the fifth term of a sequence if $a_1 = 3$, and $a_n = 2a_{n-1}$ when $n \geq 2$.

†64. The Fibonacci sequence* below describes, among other things, the reproductive behavior of rabbits. Try to determine its pattern.

$$1, 1, 2, 3, 5, 8, 13, 21, 34, \ldots$$

CALCULATOR PROBLEMS

65. Suppose the rate of inflation is 6% per year. If a loaf of bread costs $0.89 today, write a sequence that describes the cost of a loaf of bread for each of the next ten years.

---

* Named after Leonardo Fibonacci, who introduced it in 1202.

66. An old legend has it that the King of Persia invited the inventor of the game of chess to name his own reward. What the inventor requested didn't seem like much. He asked for one grain of wheat for the 1st square of the chessboard, two grains for the 2nd square, four grains for the 3rd square, eight for the 4th, and so on for all 64 squares. Write a series, first in expanded form and then in summation notation, that describes the total amount of wheat requested. Then calculate the amount of wheat needed for

a) the first 10 squares.   **b)** the last (64th square).

(Incidentally, the total amount of wheat for all 64 squares would cover the entire state of California with a layer over a foot deep!)

12.2   *Arithmetic Sequences and Series*

ARITHMETIC SEQUENCES

We will now examine the two most basic types of sequences in more detail. We study arithmetic sequences in this section and geometric sequences in Section 12.3.

Definition   An **arithmetic sequence** is a sequence in which each term after the first is obtained by adding a fixed constant to the preceding term. This fixed constant is called the **common difference** and is denoted by the letter d.

The sequence

$$3, 7, 11, 15, 19, \ldots$$

is an arithmetic sequence, since each term after the first is obtained by adding 4 to the preceding term. The next two terms in the sequence are 23 and 27.

Given an arithmetic sequence, we can determine the common difference by subtracting any term in the sequence from the term that follows it.

EXAMPLE 1   Find the common difference for the arithmetic sequence 10, 2, $-6, -14, -22, \ldots$. Then write the next two terms.

To find d, subtract 10 from 2 (or 2 from -6, etc.).

$$d = 2 - 10 = -8$$

The common difference is -8. Therefore the next two terms are

$$a_6 = -22 + (-8) = -30,$$

Try Problem 7
$$a_7 = -30 + (-8) = -38. \blacktriangleleft$$

An arithmetic sequence is completely defined when its first term a_1 and its common difference d are known. For then we can write the first few terms as shown here.

$$a_1$$
$$a_2 = a_1 + d$$
$$a_3 = a_2 + d = (a_1 + d) + d = a_1 + 2d$$
$$a_4 = a_3 + d = (a_1 + 2d) + d = a_1 + 3d$$

This suggests the following formula.

General Term of an Arithmetic Sequence
$$a_n = a_1 + (n - 1)d$$

This formula has four variables. They are a_n, a_1, n, and d. If any three of the variables are known, we can use the formula to find the fourth.

EXAMPLE 2   Find the eleventh term of the arithmetic sequence with first term 6 and common difference 7.

Substitute $n = 11$, $a_1 = 6$, and $d = 7$ into the formula for a_n.

$$a_n = a_1 + (n - 1)d$$
$$a_{11} = 6 + (11 - 1)7$$
$$a_{11} = 6 + (10)7$$
$$a_{11} = 76$$

Try Problem 13   The eleventh term is 76. ◄

CAUTION The variables n and a_n are *not* the same. The variable n represents the term number, whereas a_n represents the value of the nth term.

EXAMPLE 3   Find the fortieth term of the arithmetic sequence $\frac{1}{2}, \frac{7}{6}, \frac{11}{6}, \frac{5}{2}, \ldots$

First find d.

$$d = \frac{11}{6} - \frac{7}{6}$$
$$= \frac{4}{6}$$
$$= \frac{2}{3}$$

Then substitute $n = 40$, $a_1 = \frac{1}{2}$, and $d = \frac{2}{3}$ into the formula for a_n.

$$a_n = a_1 + (n-1)d$$

$$a_{40} = \frac{1}{2} + (40 - 1)\frac{2}{3}$$

$$a_{40} = \frac{1}{2} + (39)\frac{2}{3}$$

$$a_{40} = \frac{1}{2} + 26$$

$$a_{40} = \frac{53}{2}$$

Try Problem 21

The fortieth term is $\frac{53}{2}$. ◀

EXAMPLE 4   Find the common difference for the arithmetic sequence with first term 3 and sixth term 28.

Since we know the value of a_6, write

$$a_n = a_1 + (n-1)d$$
$$a_6 = a_1 + (6-1)d$$
$$a_6 = a_1 + 5d.$$

Then substitute $a_6 = 28$ and $a_1 = 3$ and solve for d.

$$28 = 3 + 5d$$
$$5d = 25$$
$$d = 5$$

Try Problem 23

The common difference is 5. ◀

ARITHMETIC SERIES

The indicated sum of an arithmetic sequence is called an **arithmetic series**. If we denote the sum of the first n terms as S_n, then

First term	Second term	Third term		Next-to-last term	Last term

$$S_n = a_1 + (a_1 + d) + (a_1 + 2d) + \cdots + (a_n - d) + a_n.$$

To develop a formula for S_n, write the same series in reverse order.

$$S_n = a_n + (a_n - d) + (a_n - 2d) + \cdots + (a_1 + d) + a_1$$

Then add the two equations as shown below.

$$S_n = a_1 \qquad + (a_1 + d) + (a_1 + 2d) + \cdots + (a_n - d) + a_n$$
$$S_n = a_n \qquad + (a_n - d) + (a_n - 2d) + \cdots + (a_1 + d) + a_1$$
$$\overline{2S_n = (a_1 + a_n) + (a_1 + a_n) + (a_1 + a_n) + \cdots + (a_1 + a_n) + (a_1 + a_n)}$$

Since the expression $(a_1 + a_n)$ appears n times, write this last equation as

$$2S_n = n(a_1 + a_n).$$

Divide each side by 2 to produce the following formula.

Sum of the First n Terms of an Arithmetic Series

$$S_n = \frac{n}{2}(a_1 + a_n)$$

EXAMPLE 5   Find the sum of the first ten terms of the arithmetic series with first term 4 and tenth term 31.

Substitute $n = 10$, $a_1 = 4$, and $a_{10} = 31$ into the formula for S_n.

$$S_n = \frac{n}{2}(a_1 + a_n)$$

$$S_{10} = \frac{10}{2}(4 + 31) = 5(35)$$

$$S_{10} = 175$$

Try Problem 27   The sum of the first ten terms is 175. ◄

Problem Solving

Totaling a Series of Fines

A contract specifies that construction of an office building must be completed by June 1. If not, the general contractor will be fined $100 for the first late day, $150 for the second late day, $200 for the third late day, and so on. If the contractor is two weeks late, what is the total fine?

The total fine is the sum of the first 14 terms of the following arithmetic series.

$$100 + 150 + 200 + 250 + \cdots$$

First find a_{14}.

$$a_n = a_1 + (n - 1)d$$
$$a_{14} = 100 + (14 - 1)50 = 100 + (13)50$$
$$a_{14} = 750$$

Then find S_{14}.

$$S_n = \frac{n}{2}(a_1 + a_n)$$

$$S_{14} = \frac{14}{2}(100 + 750) = 7(850)$$

$$S_{14} = 5950$$

The total fine is $5950.

PROBLEM SET 12.2

Identify each sequence as arithmetic or nonarithmetic. For those that are arithmetic, find the common difference and write the next two terms.

1. $1, 2, 3, 4, 5, \ldots$

2. $1, 3, 5, 7, 9, \ldots$

3. $3, 9, 15, 21, 27, \ldots$

4. $4, 11, 18, 25, 32, \ldots$

5. $-6, -2, 2, 6, 10, \ldots$

6. $-12, -4, 4, 12, 20, \ldots$

7. $8, 3, -2, -7, -12, \ldots$

8. $9, 3, -3, -9, -15, \ldots$

9. $1, \frac{3}{2}, 2, \frac{5}{2}, 3, \ldots$

10. $1, \frac{4}{3}, \frac{5}{3}, 2, \frac{7}{3}, \ldots$

11. $1, \frac{1}{2}, \frac{1}{3}, \frac{1}{4}, \frac{1}{5}, \ldots$

12. $1, 4, 9, 16, 25, \ldots$

Find the indicated unknown for each arithmetic sequence.

13. $a_1 = 4, d = 6; a_{11} = ?$

14. $a_1 = 2, d = 5; a_{21} = ?$

15. $a_1 = 4, d = 6; a_{110} = ?$

16. $a_1 = 2, d = 5; a_{111} = ?$

17. $a_1 = 4, d = 6; a_n = ?$

18. $a_1 = 2, d = 5; a_n = ?$

19. $10, 7, 4, 1, \ldots; a_{15} = ?$

20. $15, 11, 7, 3, \ldots; a_{25} = ?$

21. $\frac{1}{5}, \frac{9}{5}, \frac{17}{5}, 5, \ldots; a_{101} = ?$

22. $\frac{1}{2}, 3, \frac{11}{2}, 8, \ldots; a_{99} = ?$

23. $a_1 = 9, a_5 = 25; d = ?$

24. $a_1 = 7, a_6 = 22; d = ?$

25. $a_1 = -1, a_7 = 11; d = ?$

26. $a_1 = -2, a_8 = 19; d = ?$

Find the indicated sum for each arithmetic series.

27. $a_1 = 5, a_{10} = 32; S_{10} = ?$

28. $a_1 = 6, a_{10} = 15; S_{10} = ?$

29. $a_n = 3n + 4; S_{20} = ?$

30. $a_n = 2n + 3; S_{20} = ?$

31. $a_1 = \frac{1}{3}, d = \frac{2}{3}; S_{10} = ?$

32. $a_1 = \frac{1}{2}, d = \frac{1}{3}; S_{10} = ?$

33. $3 + 11 + 19 + \cdots; S_{18} = ?$

34. $2 + 11 + 20 + \cdots; S_{19} = ?$

35. Evaluate $\sum_{k=1}^{2000} (4k + 1)$.

36. Evaluate $\sum_{k=1}^{2000} (6k - 1)$.

37. What term in the arithmetic sequence 8, 13, 18, . . . is 88?

38. What term in the arithmetic sequence 6, 10, 14, . . . is 58?

39. Find the sum of the first 100 positive integers.

40. Find the sum of the first 200 positive integers.

41. Heather saved $100 in January, $125 in February, $150 in March, and so on. How much did she save for the year?

42. A disgruntled baseball player is told to report to spring training or be fined $50 for the first late day, $60 for the second late day, $70 for the third late day, and so on. If the player reports three weeks late, what is the total fine?

43. An auditorium has 31 rows of seats. The back row contains 76 seats and each row has two fewer seats than the row behind it.
a) How many seats are in the first row?
b) How many seats are in the auditorium?

44. Bottles are stacked in 23 rows so that the bottom row contains 97 bottles, and each row contains three fewer bottles than the row below it.
a) How many bottles are in the top row?
b) How many bottles are there in all?

† **45.** Find the general term of the arithmetic sequence with fourth term 10 and twelfth term 26.

† **46.** Combine the formulas for S_n and a_n to show that the formula for the sum of the first n terms of an arithmetic series can also be written as

$$S_n = \frac{n}{2}[2a_1 + (n-1)d].$$

CALCULATOR PROBLEMS

47. Find a_{86} for the arithmetic sequence with $a_1 = 7.863$ and $d = 4.795$.

48. Your grandfather gives you 1¢ on the day of your birth, 2¢ on your second day of life, 3¢ on your third day, and so on. How much money have you received by your eighteenth birthday? (Assume 1 year = 365 days.)

12.3   *Geometric Sequences and Series*

GEOMETRIC SEQUENCES

In an arithmetic sequence, each term after the first is obtained by adding a fixed constant to the preceding term. In a geometric sequence, we *multiply* each term by a fixed constant to obtain the next term.

Definition

A **geometric sequence** is a sequence in which each term after the first is obtained by multiplying the preceding term by a fixed constant. This fixed constant is called the **common ratio** and is denoted by the letter r.

The sequence

$$3, 6, 12, 24, \ldots$$

is a geometric sequence, since each term after the first is obtained by multiplying the preceding term by 2. The next two terms in the sequence are 48 and 96.

Given a geometric sequence, we can determine the common ratio by dividing any term in the sequence by the term that precedes it.

EXAMPLE 1   Find the common ratio for the geometric sequence $1, \frac{1}{2}, \frac{1}{4}, \frac{1}{8}, \ldots$ Then write the next two terms.

To find r, divide $\frac{1}{2}$ by 1 (or $\frac{1}{4}$ by $\frac{1}{2}$, etc.).

$$r = \frac{1}{2} \div 1 = \frac{1}{2}$$

Therefore the next two terms are

$$a_5 = \frac{1}{8} \cdot \frac{1}{2} = \frac{1}{16},$$

Try Problem 3

$$a_6 = \frac{1}{16} \cdot \frac{1}{2} = \frac{1}{32}. \blacktriangleleft$$

A geometric sequence is completely defined when its first term a_1 and its common ratio r are known. For then we can write the first few terms as shown below.

$$a_1$$
$$a_2 = a_1 r$$
$$a_3 = a_2 r = (a_1 r)r = a_1 r^2$$
$$a_4 = a_3 r = (a_1 r^2)r = a_1 r^3$$

This suggests the following formula.

General Term of a Geometric Sequence

$$a_n = a_1 r^{n-1}$$

This formula contains the four variables a_n, a_1, r, and n. If any three of the variables are known, we can use the formula to find the fourth.

EXAMPLE 2   Find the seventh term of the geometric sequence with first term 10 and common ratio 3.

Substitute $n = 7$, $a_1 = 10$, and $r = 3$ into the formula for a_n.

$$a_n = a_1 r^{n-1}$$
$$a_7 = 10(3)^{7-1}$$
$$a_7 = 10(3)^6$$
$$a_7 = 10(729)$$
$$a_7 = 7290$$

Try Problem 15

The seventh term is 7290. ◄

CAUTION Remember that n is the term number, and a_n is the value of the nth term.

EXAMPLE 3   Find the fifth term of the geometric sequence $\frac{1}{4}, -\frac{1}{6}, \frac{1}{9}, \ldots$

First find r.

$$r = -\frac{1}{6} \div \frac{1}{4} = -\frac{1}{6} \cdot \frac{4}{1} = -\frac{2}{3}$$

Then substitute $n = 5$, $a_1 = \frac{1}{4}$, and $r = -\frac{2}{3}$ into the formula for a_n.

$$a_n = a_1 r^{n-1}$$

$$a_5 = \frac{1}{4}\left(-\frac{2}{3}\right)^{5-1}$$

$$a_5 = \frac{1}{4}\left(-\frac{2}{3}\right)^{4}$$

$$a_5 = \frac{1}{4}\left(\frac{16}{81}\right)$$

$$a_5 = \frac{4}{81}$$

Try Problem 21

The fifth term is $\frac{4}{81}$. ◀

EXAMPLE 4 Find the common ratio for the geometric sequence with first term 5 and fourth term 0.135.

Since we know the value of a_4, write

$$a_n = a_1 r^{n-1}$$
$$a_4 = a_1 r^{4-1}$$
$$a_4 = a_1 r^3.$$

Then substitute $a_4 = 0.135$ and $a_1 = 5$ and solve for r.

$$0.135 = 5r^3$$
$$0.027 = r^3 \qquad \text{Divide by 5}$$
$$r = 0.3 \qquad \text{Take the cube root}$$

Try Problem 25

The common ratio is 0.3. ◀

GEOMETRIC SERIES

The indicated sum of a geometric sequence is called a **geometric series.** If we denote the sum of the first n terms as S_n, then

First term	Second term	Third term	Fourth term		Next-to-last term	Last term
↓	↓	↓	↓		↓	↓

$$S_n = a_1 + a_1 r + a_1 r^2 + a_1 r^3 + \cdots + a_1 r^{n-2} + a_1 r^{n-1}.$$

To develop a formula for S_n, multiply each side by $-r$.

$$-rS_n = -a_1 r - a_1 r^2 - a_1 r^3 - \cdots - a_1 r^{n-1} - a_1 r^n$$

Then add the two equations as shown below.

$$S_n = a_1 + a_1 r + a_1 r^2 + a_1 r^3 + \cdots + a_1 r^{n-1}$$
$$-rS_n = \quad - a_1 r - a_1 r^2 - a_1 r^3 - \cdots - a_1 r^{n-1} - a_1 r^n$$
$$\overline{S_n - rS_n = a_1 \qquad\qquad\qquad\qquad\qquad\qquad\quad - a_1 r^n}$$

Factor out S_n on the left side and a_1 on the right side.

$$(1 - r)S_n = a_1(1 - r^n)$$

Divide each side by $1 - r$ to produce the following formula.

Sum of the First n Terms of a Geometric Series

$$S_n = \frac{a_1(1 - r^n)}{1 - r} \qquad (r \neq 1)$$

EXAMPLE 5   Find the sum of the first four terms of the geometric series with first term 6 and common ratio 5.

Substitute $n = 4$, $a_1 = 6$, and $r = 5$ into the formula for S_n.

$$S_n = \frac{a_1(1 - r^n)}{1 - r}$$

$$S_4 = \frac{6(1 - 5^4)}{1 - 5}$$

$$S_4 = \frac{6(1 - 625)}{-4}$$

$$S_4 = 936$$

Try Problem 31

The sum of the first four terms is 936. ◄

Problem Solving

Totaling Your First Ten Years' Income
Suppose your first year's salary is $20,000, and you are guaranteed a 7% raise each year. Find your total income for the first ten years.
    Observe the following pattern.

first year's salary = $20,000
second year's salary = first year's salary + raise

$$= 20,000 + 20,000(0.07)$$
$$= 20,000(1 + 0.07) \qquad \text{Factor out 20,000}$$
$$= 20,000(1.07)$$

third year's salary = second year's salary + raise

$$= 20,000(1.07) + 20,000(1.07)(0.07)$$
$$= 20,000(1.07)(1 + 0.07) \qquad \text{Factor out 20,000(1.07)}$$
$$= 20,000(1.07)(1.07)$$
$$= 20,000(1.07)^2$$

Therefore the sum of your yearly salaries forms the following geometric series.

$$20{,}000 + 20{,}000(1.07) + 20{,}000(1.07)^2 + 20{,}000(1.07)^3 + \cdots$$

Your total income for the first ten years is the sum of the first ten terms.

$$S_n = \frac{a_1(1 - r^n)}{1 - r}$$

$$S_{10} = \frac{20{,}000(1 - 1.07^{10})}{1 - 1.07}$$

$$S_{10} \approx \frac{20{,}000(1 - 1.96715)}{-0.07} \qquad \text{From a calculator}$$

$$S_{10} \approx 276{,}329$$

Your total income for the first ten years is approximately \$276,329.

PROBLEM SET 12.3

Identify each sequence as geometric or nongeometric. For those that are geometric, find the common ratio and write the next two terms.

1. $1, 2, 4, 8, 16, \ldots$

2. $1, 3, 9, 27, 81, \ldots$

3. $1, \frac{1}{3}, \frac{1}{9}, \frac{1}{27}, \ldots$

4. $1, \frac{1}{4}, \frac{1}{16}, \frac{1}{64}, \ldots$

5. $1, 5, 10, 15, 20, \ldots$

6. $1, 2, 4, 6, 8, \ldots$

7. $-5, -25, -125, \ldots$

8. $-2, -4, -8, \ldots$

9. $1, 4, 9, 16, 25, \ldots$

10. $1, \frac{1}{2}, \frac{1}{3}, \frac{1}{4}, \frac{1}{5}, \ldots$

11. $6, -6, 6, -6, 6, \ldots$

12. $7, -7, 7, -7, 7, \ldots$

13. $-2, 1, -\frac{1}{2}, \frac{1}{4}, \ldots$

14. $-3, 1, -\frac{1}{3}, \frac{1}{9}, \ldots$

Find the indicated unknown for each geometric sequence.

15. $a_1 = 2, r = 3; a_4 = ?$

16. $a_1 = 3, r = 2; a_7 = ?$

17. $a_1 = 2, r = 3; a_n = ?$

18. $a_1 = 3, r = 2; a_n = ?$

19. $4, 2, 1, \ldots; a_6 = ?$

20. $9, 3, 1, \ldots; a_5 = ?$

21. $\frac{1}{3}, -\frac{1}{2}, \frac{3}{4}, \ldots; a_4 = ?$

22. $\frac{1}{2}, -\frac{1}{3}, \frac{2}{9}, \ldots; a_5 = ?$

23. $-6, -12, -24, \ldots; a_5 = ?$

24. $-4, -8, -16, \ldots; a_4 = ?$

25. $a_1 = 5, a_5 = 0.008; r = ?$

26. $a_1 = 5, a_4 = 0.32; r = ?$

27. $1, 1, 1, 1, \ldots; a_{56} = ?$

28. $1, 1, 1, 1, \ldots; a_{67} = ?$

29. $1, -1, 1, -1, \ldots; a_n = ?$

30. $-1, 1, -1, 1, \ldots; a_n = ?$

Find the indicated sum for each geometric series.

31. $a_1 = 7, r = 4; S_5 = ?$

32. $a_1 = 9, r = 5; S_4 = ?$

33. $a_1 = 40, r = \frac{1}{2}; S_6 = ?$

34. $a_1 = 80, r = \frac{1}{2}; S_7 = ?$

35. $2 + (-6) + 18 + \cdots; S_7 = ?$

36. $3 + (-6) + 12 + \cdots; S_9 = ?$

37. $1 + (-1) + 1 + (-1) + \cdots$; $S_{50} = ?$ $S_{51} = ?$

38. $-1 + 1 + (-1) + 1 + \cdots$; $S_{100} = ?$ $S_{101} = ?$

39. Evaluate $\sum_{k=1}^{10} 16(\frac{3}{2})^k$.

40. Evaluate $\sum_{k=1}^{10} 81(\frac{2}{3})^k$.

41. A tank contains 500 gallons of water. With each stroke a pump removes $\frac{1}{5}$ of the water in the tank. How much water remains after 4 strokes?

42. A rubber ball is dropped on a hard surface from a height of 80 feet. On each rebound it bounces $\frac{3}{4}$ as high as on its previous bounce. How high does it bounce on its fourth bounce?

†43. Find the general term of the geometric sequence with third term 15 and sixth term 120.

†44. Combine the formula for S_n and a_n to show that the formula for the sum of the first n terms of a geometric series can also be written as

$$S_n = \frac{a_1 - a_n r}{1 - r} \qquad (r \neq 1).$$

CALCULATOR PROBLEMS

45. Suppose your first year's salary is $10,000, and you are guaranteed an 8% raise each year. Find your salary for the fifteenth year. Find your total income for the first fifteen years.

46. What will a car that costs $12,000 today cost in 20 years if inflation is 9% per year for that period?

12.4 *Infinite Geometric Series*

The Greek philosopher Zeno was famous for his paradoxes.* One such paradox went something like this: Suppose Achilles (the swiftest runner of his day) gives a tortoise a 100-yard head start in a race. If we assume that Achilles runs ten times as fast as the tortoise, then while Achilles runs the first 100 yards, the tortoise runs 10 yards. While Achilles runs the 10 yards already traveled by the tortoise, the tortoise runs an additional 1 yard. While Achilles runs that 1 yard, the tortoise runs an additional 1/10 yard, and so on. Reasoning this way it seems that Achilles never does catch the tortoise, though we know in fact that he would.

The sum of the distances run by the tortoise form the **infinite geometric series** below.

$$10 + 1 + 0.1 + 0.01 + \cdots$$

But how can we find the sum of an infinite number of terms? First, recall that the sum of the first n terms of a geometric series is given by the following formula.

$$S_n = \frac{a_1(1 - r^n)}{1 - r}$$

_____

* Zeno lived around the year 450 B.C.

But if $|r| < 1$, then r^n approaches zero as n grows larger. For example, in the tortoise series $r = 0.1$. The table below shows that $r^n = (0.1)^n$ approaches zero as n grows larger.

n	$r^n = (0.1)^n$
1	$r^1 = (0.1)^1 = 0.1$
5	$r^5 = (0.1)^5 = 0.00001$
10	$r^{10} = (0.1)^{10} = 0.0000000001$

Therefore if n grows large without bound, r^n approaches zero, and the right side of the formula for S_n approaches the fraction below.

$$\frac{a_1(1 - 0)}{1 - r} = \frac{a_1}{1 - r}$$

We summarize this discussion by stating the following formula.

Infinite Geometric Series

If $|r| < 1$, the sum S of the infinite geometric series with first term a_1 and common ratio r is given by the formula

$$S = \frac{a_1}{1 - r}.$$

If $|r| \geq 1$, the sum does not exist.

EXAMPLE 1   Find the total distance run by the tortoise before being caught by Achilles.

We must find the sum of the infinite geometric series below.

$$10 + 1 + 0.1 + 0.01 + \cdots$$

To do this, substitute $a_1 = 10$ and $r = 0.1$ into the formula for S.

$$S = \frac{a_1}{1 - r} = \frac{10}{1 - 0.1} = \frac{10}{0.9} = \frac{100}{9}$$

Try Problem 9

The tortoise runs $\frac{100}{9} = 11\frac{1}{9}$ yards. ◄

EXAMPLE 2   Find the sum of the infinite geometric series $9 - 6 + 4 - \frac{8}{3} + \cdots$.

First write the series as a sum.

$$9 + (-6) + 4 + \left(-\frac{8}{3}\right) + \cdots .$$

Then find r.

$$r = \frac{-6}{9} = -\frac{2}{3}$$

Since $|r| < 1$, substitute $a_1 = 9$ and $r = -\frac{2}{3}$ into the formula for S.

$$S = \frac{a_1}{1-r} = \frac{9}{1-\left(-\frac{2}{3}\right)} = \frac{9}{1+\frac{2}{3}} = \frac{9}{\frac{5}{3}} = \frac{27}{5}$$

Try Problem 11 ⟍   The sum is $\frac{27}{5}$. ◂

We can use the formula for the sum of an infinite geometric series to write a repeating decimal as a ratio of two integers.

EXAMPLE 3   Write $0.\overline{45}$ as a ratio of two integers.

First write $0.\overline{45}$ as follows:

$$0.\overline{45} = 0.45 + 0.0045 + 0.000045 + \cdots.$$

Note that the right side is the infinite geometric series with $a_1 = 0.45$ and $r = 0.01$. Since $|r| < 1$, we have

$$S = \frac{a_1}{1-r} = \frac{0.45}{1-0.01} = \frac{0.45}{0.99} = \frac{45}{99} = \frac{5}{11}.$$

Try Problem 23 ⟍   Therefore $0.\overline{45} = \frac{5}{11}$. This answer can be checked by division. ◂

PROBLEM SET 12.4

Find the sum, if it exists, of each infinite geometric series.

1. $a_1 = 16$, $r = \frac{1}{5}$
2. $a_1 = 12$, $r = \frac{2}{3}$
3. $a_1 = 16$, $r = 2$
4. $a_1 = 12$, $r = 3$
5. $a_1 = 4$, $r = -\frac{1}{3}$
6. $a_1 = 3$, $r = -\frac{1}{2}$
7. $1 + \frac{1}{2} + \frac{1}{4} + \frac{1}{8} + \cdots$
8. $1 + \frac{1}{3} + \frac{1}{9} + \frac{1}{27} + \cdots$
9. $10 + 3 + 0.9 + \cdots$
10. $10 + 7 + 4.9 + \cdots$
11. $16 - 12 + 9 - \frac{27}{4} + \cdots$
12. $25 - 20 + 16 - \frac{64}{5} + \cdots$
13. $\frac{1}{9} - \frac{1}{6} + \frac{1}{4} - \cdots$
14. $\frac{1}{16} - \frac{1}{12} + \frac{1}{9} - \cdots$
15. $1 + 1 + 1 + 1 + \cdots$
16. $1 - 1 + 1 - 1 + \cdots$

Find the indicated unknown for each infinite geometric series.

17. $S = 8$, $r = \frac{1}{4}$; $a_1 = ?$
18. $S = 18$, $r = \frac{1}{6}$; $a_1 = ?$
19. $S = 12$, $a_1 = 2$; $r = ?$
20. $S = 21$, $a_1 = 6$; $r = ?$

Write each repeating decimal as a ratio of two integers.

21. $0.\overline{3}$
22. $0.\overline{6}$
23. $0.\overline{36}$
24. $0.\overline{18}$
25. $0.\overline{07}$
26. $0.\overline{08}$
27. $0.\overline{123}$
28. $0.\overline{471}$

29. $0.\overline{9}$ **30.** $9.\overline{9}$ **31.** $5.\overline{7}$ **32.** $7.\overline{8}$

33. Evaluate $\displaystyle\sum_{n=1}^{\infty} 8(\tfrac{2}{3})^n$.

34. Evaluate $\displaystyle\sum_{n=1}^{\infty} 5(\tfrac{3}{4})^n$.

35. A ball is dropped on a hard surface from a height of 10 feet. On each rebound it bounces $\tfrac{3}{5}$ as high as its previous height. What is the total distance the ball travels before coming to rest?

36. On the first swing, the bob of a pendulum travels an arc of length 45 centimeters. The arc on each subsequent swing is $\tfrac{4}{5}$ the length of the previous arc. What is the total distance the bob travels before coming to rest?

CALCULATOR PROBLEMS

37. Find the sum of the infinite geometric series $5.64 + 4.794 + 4.0749 + \cdots$.

38. Find the sum S of the infinite geometric series $1 + \tfrac{1}{2} + \tfrac{1}{4} + \tfrac{1}{8} + \cdots$. Then use the formula for S_n from Section 12.3 to calculate S_5, S_{10}, and S_{15}, and note that these sums approach S.

12.5   *The Binomial Theorem*

When a binomial of the form $a + b$ is raised to a power, the result can be thought of as a series. To see the pattern of the series, consider the following expansions of $(a + b)^n$.

$$(a+b)^0 = 1$$
$$(a+b)^1 = a+b$$
$$(a+b)^2 = a^2 + 2ab + b^2$$
$$(a+b)^3 = a^3 + 3a^2b + 3ab^2 + b^3$$
$$(a+b)^4 = a^4 + 4a^3b + 6a^2b^2 + 4ab^3 + b^4$$
$$(a+b)^5 = a^5 + 5a^4b + 10a^3b^2 + 10a^2b^3 + 5ab^4 + b^5$$

Based on the expansions above, we draw the following conclusions.

Expansion of $(a + b)^n$

1. There are $n + 1$ terms.
2. The exponents on a start with n and decrease to 0.
3. The exponents on b start with 0 and increase to n.
4. The sum of the exponents on a and b in every term is n.

All we need now is a method for determining the coefficients. One method involves constructing an array of numbers called **Pascal's triangle.***

Pascal's Triangle

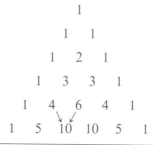

$$
\begin{array}{ccccccccccccc}
 & & & & & & 1 & & & & & & \\
 & & & & & 1 & & 1 & & & & & \\
 & & & & 1 & & 2 & & 1 & & & & \\
 & & & 1 & & 3 & & 3 & & 1 & & & \\
 & & 1 & & 4 & & 6 & & 4 & & 1 & & \\
 & 1 & & 5 & & 10 & & 10 & & 5 & & 1 &
\end{array}
$$

The sides of Pascal's triangle consist of 1's. To find a number in the interior of the triangle simply add the two adjacent numbers in the row above. For example, $4 + 6 = 10$.

Note that the coefficient in the expansion of $(a + b)^0$ is given by the 1st row of Pascal's triangle, the coefficients of $(a + b)^1$ by the 2nd row, the coefficients of $(a + b)^2$ by the 3rd row, and so on.

EXAMPLE 1 Expand $(a + b)^6$.

The coefficients in the expansion of $(a + b)^6$ are given by the 7th row of Pascal's triangle.

$$1 \quad 6 \quad 15 \quad 20 \quad 15 \quad 6 \quad 1$$

Use these coefficients, along with the four rules given earlier, to write the expansion.

Try Problem 1

$$(a + b)^6 = a^5 + 6a^5b + 15a^4b^2 + 20a^3b^3 + 15a^2b^4 + 6ab^5 + b^6 \blacktriangleleft$$

Although Pascal's triangle provides a simple method for determining the coefficients of a binomial expansion, it is impractical for large values of n. More important, it does not give us a formula for the expansion of $(a + b)^n$ in terms of a, b, and n. Such a formula is essential in proving some of the theorems we encounter in higher mathematics. Before we state such a formula, we introduce a notational shorthand called **factorial notation.**

Definition

If n is a positive integer, then $n!$ (read "n factorial") is given by

$$n! = n(n - 1)(n - 2) \cdots 3 \cdot 2 \cdot 1.$$

In addition, we define $0! = 1$.

---

* Named after the French mathematician and philosopher Blaise Pascal (1623–1662).

Examples of this definition are shown below.

$$4! = 4 \cdot 3 \cdot 2 \cdot 1 = 24$$
$$3! = 3 \cdot 2 \cdot 1 = 6$$
$$2! = 2 \cdot 1 = 2$$
$$1! = 1$$
$$0! = 1$$

We now use factorial notation to define the **binomial coefficient** of n over r.

Definition

The expression $\binom{n}{r}$ is called the **binomial coefficient** of n over r and is given by

$$\binom{n}{r} = \frac{n!}{r!(n-r)!}.$$

EXAMPLE 2   Calculate $\binom{5}{2}$.

Try Problem 3

$$\binom{5}{2} = \frac{5!}{2!(5-2)!} = \frac{5!}{2!3!} = \frac{5 \cdot 4 \cdot \cancel{3 \cdot 2 \cdot 1}}{(2 \cdot 1)(\cancel{3 \cdot 2 \cdot 1})} = \frac{5 \cdot 4}{2 \cdot 1} = 10 \blacktriangleleft$$

CAUTION The expressions $(5-2)!$ and $5! - 2!$ are *not* the same. Note that $(5-2)! = 3! = 6$, but $5! - 2! = 120 - 2 = 118$.

Actually the binomial coefficient of Example 2 can be calculated more efficiently as follows:

$$\binom{5}{2} = \frac{5!}{2!(5-2)!} = \frac{5 \cdot 4 \cdot \cancel{3!}}{(2 \cdot 1)\cancel{3!}} = 10.$$

We use this technique in the next example.

EXAMPLE 3   Calculate each binomial coefficient: **a)** $\binom{4}{0}$,   **b)** $\binom{4}{1}$, **c)** $\binom{4}{2}$,   **d)** $\binom{4}{3}$,   **e)** $\binom{4}{4}$.

a) $\binom{4}{0} = \dfrac{4!}{0!(4-0)!} = \dfrac{4!}{1 \cdot 4!} = 1$ **b)** $\binom{4}{1} = \dfrac{4!}{1!(4-1)!} = \dfrac{4 \cdot 3!}{1 \cdot 3!} = 4$

c) $\binom{4}{2} = \dfrac{4!}{2!(4-2)!} = \dfrac{4 \cdot 3 \cdot 2!}{(2 \cdot 1)2!} = 6$ **d)** $\binom{4}{3} = \dfrac{4!}{3!(4-3)!} = \dfrac{4 \cdot 3!}{3! 1!} = 4$

Try Problem 13   **e)** $\binom{4}{4} = \dfrac{4!}{4!(4-4)!} = \dfrac{4!}{4!0!} = 1 \blacktriangleleft$

Note that the values of the binomial coefficients $\binom{4}{0}, \binom{4}{1}, \binom{4}{2}, \binom{4}{3}$, and $\binom{4}{4}$ are the coefficients in the expansion of $(a + b)^4$. We use this fact to state the **binomial theorem.**

The Binomial Theorem

If n is a positive integer, then

$$(a + b)^n = \binom{n}{0} a^n + \binom{n}{1} a^{n-1}b + \binom{n}{2} a^{n-2}b^2 + \cdots + \binom{n}{n} b^n.$$

EXAMPLE 4   Expand $(x + 2)^4$ using the binomial theorem.

Apply the binomial theorem with $a = x$, $b = 2$, and $n = 4$.

$$(x + 2)^4 = \binom{4}{0} x^4 + \binom{4}{1} x^3(2) + \binom{4}{2} x^2(2)^2 + \binom{4}{3} x(2)^3 + \binom{4}{4} 2^4$$

$$= \frac{4!}{0!4!} x^4 + \frac{4!}{1!3!} x^3(2) + \frac{4!}{2!2!} x^2(4) + \frac{4!}{3!1!} x(8) + \frac{4!}{4!0!} (16)$$

$$= 1x^4 + 4x^3(2) + 6x^2(4) + 4x(8) + 1(16)$$

$$= x^4 + 8x^3 + 24x^2 + 32x + 16 \blacktriangleleft$$

Try Problem 17

EXAMPLE 5   Expand $(2m - 3)^5$ using the binomial theorem.

Write $(2m - 3)^5$ in the form $(a + b)^n$.

$$(2m - 3)^5 = ((2m) + (-3))^5$$

Apply the binomial theorem with $a = 2m$, $b = -3$, and $n = 5$.

$$(2m + (-3))^5 = \binom{5}{0} (2m)^5 + \binom{5}{1} (2m)^4(-3) + \binom{5}{2} (2m)^3(-3)^2$$

$$+ \binom{5}{3} (2m)^2(-3)^3 + \binom{5}{4} (2m)(-3)^4 + \binom{5}{5} (-3)^5$$

$$= \frac{5!}{0!5!} (32m^5) + \frac{5!}{1!4!} (16m^4)(-3) + \frac{5!}{2!3!} (8m^3)(9)$$

$$+ \frac{5!}{3!2!} (4m^2)(-27) + \frac{5!}{4!1!} (2m)(81) + \frac{5!}{5!0!} (-243)$$

$$= 1(32m^5) + 5(16m^4)(-3) + 10(8m^3)(9)$$

$$+ 10(4m^2)(-27) + 5(2m)(81) + 1(-243)$$

$$= 32m^5 - 240m^4 + 720m^3 - 1080m^2 + 810m - 243$$

Note that the expansion of a binomial of the form $(a - b)^n$ produces terms with alternating signs. $\blacktriangleleft$

Try Problem 25

We can also use the binomial theorem to find a particular term in an expansion without finding the other terms.

The rth term in the expansion of $(a + b)^n$

$$r\text{th term} = \binom{n}{r-1} a^{n-(r-1)} b^{r-1}$$

EXAMPLE 6   Find the middle term in the expansion of $\left(x + \dfrac{y}{4}\right)^8$.

Since there are $8 + 1 = 9$ terms in the expansion, the middle term is the 5th term. Therefore substitute $r = 5$, $n = 8$, $a = x$, and $b = \frac{y}{4}$ into the formula for the rth term.

$$\begin{aligned}
5\text{th term} &= \binom{8}{4} x^{8-4} \left(\frac{y}{4}\right)^4 \\
&= \frac{8!}{4!4!} x^4 \left(\frac{y^4}{4^4}\right) \\
&= 70x^4 \left(\frac{y^4}{256}\right) \\
&= \frac{35}{128} x^4 y^4 \blacktriangleleft
\end{aligned}$$

Try Problem 41

PROBLEM SET 12.5

1. Expand $(a + b)^7$. Use Pascal's triangle to determine the coefficients.

2. Expand $(a + b)^8$. Use Pascal's triangle to determine the coefficients.

Calculate each binomial coefficient.

3. $\dbinom{6}{2}$

4. $\dbinom{7}{2}$

5. $\dbinom{5}{3}$

6. $\dbinom{7}{5}$

7. $\dbinom{11}{8}$

8. $\dbinom{10}{7}$

9. $\dbinom{8}{1}$

10. $\dbinom{9}{1}$

11. $\dbinom{8}{0}$

12. $\dbinom{9}{0}$

13. $\dbinom{15}{14}$

14. $\dbinom{20}{19}$

Expand each binomial using the binomial theorem.

15. $(x + y)^3$ **16.** $(x + y)^5$ **17.** $(x + 3)^4$ **18.** $(x + 5)^4$

19. $(x - y)^5$ **20.** $(x - y)^3$ **21.** $(2x + y)^4$ **22.** $(3x + y)^4$

23. $(a - 5)^3$ **24.** $(a - 2)^5$ **25.** $(3m - 2)^5$ **26.** $(3m - 4)^5$

27. $(x + \frac{y}{2})^4$ **28.** $(x + \frac{y}{3})^4$ **29.** $(t^2 + 1)^6$ **30.** $(t^2 - 1)^6$

Use the binomial theorem to write the first three terms in each expansion.

31. $(a + b)^{20}$ **32.** $(a + b)^{30}$ **33.** $(x + 10)^{15}$

34. $(x + 10)^{25}$ **35.** $(t^2 - 1)^{12}$ **36.** $(t^2 + 1)^{13}$

Find the indicated term in each expansion.

37. $(a + b)^9$; 4th term   **38.** $(a + b)^{11}$; 5th term   **39.** $(p - 2q)^{14}$; 6th term

40. $(p - 2q)^{16}$; 4th term   **41.** $\left(x + \frac{y}{2}\right)^8$; middle term   **42.** $\left(x + \frac{y}{3}\right)^{10}$; middle term

CALCULATOR PROBLEMS

43. Expand $(x - 2.5)^4$.

44. Factorials increase very rapidly. To see this, find
   a) 0!   **b)** 7!   **c)** 11!   **d)** 69!   **e)** 70!

CHAPTER 12 REVIEW

[12.1]   *Find the first four terms of the sequence whose general term is given. Then find the tenth term.*

1. $a_n = 2n + 1$ **2.** $a_n = \dfrac{n - 2}{n + 2}$ **3.** $a_n = (-1)^n 2^n$

Find the general term a_n for each sequence.

4. 4, 8, 12, 16, . . .   **5.** $\frac{2}{3}, \frac{4}{5}, \frac{6}{7}, \frac{8}{9}, \ldots$ **6.** $1, -\frac{1}{8}, \frac{1}{27}, -\frac{1}{64}, \ldots$

Expand and simplify.

7. $\displaystyle\sum_{n=1}^{6} 5n$ **8.** $\displaystyle\sum_{n=0}^{4} (n^2 - 2n + 5)$ **9.** $\displaystyle\sum_{n=1}^{5} \frac{1}{2^{n-1}}$

Write each series in summation notation.

10. $1 + 3 + 5 + 7 + 9 + 11 + 13$ **11.** $6 + 11 + 16 + 21$

12. $1 + \frac{1}{4} + \frac{1}{16} + \frac{1}{64} + \frac{1}{256}$

[12.2] *For each arithmetic sequence, find the common difference and write the next two terms.*

13. $3, 7, 11, 15, 19, \ldots$ **14.** $-6, -1, 4, 9, 14, \ldots$ **15.** $4, 3\frac{1}{2}, 3, 2\frac{1}{2}, 2, \ldots$

Find the indicated unknown for each arithmetic sequence.

16. $a_1 = 5, d = 2; a_7 = ?$ **17.** $a_1 = 4, d = -3; a_{10} = ?$ **18.** $a_1 = 15, a_{11} = -25; d = ?$

Find the indicated sum for each arithmetic series.

19. $a_1 = 7, a_{14} = 85; S_{14} = ?$ **20.** $a_1 = 6, d = 4; S_{12} = ?$

21. $4 + 7 + 10 + 13 + \cdots ; S_{20} = ?$

22. A woman accepts a position at a salary of $30,000 for the first year with an increase in salary of $2000 for each year thereafter. Find the woman's total earnings if she works at the position for fifteen years.

[12.3] *For each geometric sequence, find the common ratio and write the next two terms.*

23. $2, 8, 32, 128, \ldots$ **24.** $5, -10, 20, -40, \ldots$ **25.** $1, \frac{2}{3}, \frac{4}{9}, \frac{8}{27}, \frac{16}{81}, \ldots$

Find the indicated unknown for each geometric sequence.

26. $a_1 = 3, r = \frac{1}{3}; a_6 = ?$ **27.** $a_1 = 5, r = -\frac{2}{3}; a_5 = ?$ **28.** $a_1 = 3, a_5 = 48; r = ?$

Find the indicated sum for each geometric series.

29. $a_1 = 3, r = 2; S_6 = ?$ **30.** $25 + 5 + 1 + \cdots ; S_4 = ?$

31. A ball is dropped from a height of 90 feet. On each rebound it bounces $\frac{2}{3}$ of the height from which it last fell. How high does it bounce on its fifth bounce?

[12.4] *Find the sum, if it exists, of each infinite geometric series.*

32. $a_1 = 9, r = \frac{1}{3}$ **33.** $a_1 = 5, r = -0.2$

34. $16 + 4 + 1 + \cdots$ **35.** $1 - \frac{1}{2} + \frac{1}{4} - \frac{1}{8} + \cdots$

Find the indicated unknown for each infinite geometric series.

36. $S = 2, r = \frac{1}{2}; a_1 = ?$ **37.** $S = 6, a_1 = 2; r = ?$

Write each repeating decimal as a ratio of two integers.

38. $0.\overline{12}$ **39.** $2.\overline{7}$

40. Suppose a golf ball rebounds $\frac{3}{4}$ of the distance it falls. Find the total distance the golf ball will travel before coming to rest if it is dropped from a height of 20 feet.

[12.5] *Calculate each binomial coefficient.*

41. $\binom{7}{4}$ **42.** $\binom{11}{10}$ **43.** $\binom{13}{0}$ **44.** $\binom{8}{2}$

Expand each binomial using the binomial theorem.

45. $(x + y)^6$ **46.** $(3m + 2)^4$ **47.** $(t^2 - 2)^5$ **48.** $(x - \frac{y}{2})^3$

49. Use the binomial theorem to write the first three terms in the expansion of $(x + 5)^{10}$.

50. Find the fifth term of $(a - 2b)^7$.

CHAPTER 12 TEST

1. Calculate each binomial coefficient.

   a) $\binom{6}{5}$

   b) $\binom{5}{3}$

2. For the arithmetic sequence 4.5, 5.2, 5.9, 6.6, . . . , find the common difference and write the next two terms.

3. For the geometric sequence 20, 10, 5, . . . , find the common ratio and write the next two terms.

Expand and simplify.

4. $\sum\limits_{n=1}^{4} n(n + 1)$ **5.** $\sum\limits_{n=1}^{5} (5n - 3)$

Write each series in summation notation.

6. $3 + 6 + 9 + 12 + 15$ **7.** $\frac{3}{2} + \frac{4}{3} + \frac{5}{4} + \frac{6}{5}$

Find the general term a_n for each sequence.

8. $5, 7, 9, \ldots$ **9.** $1, -4, 9, -16, \ldots$

Find the first four terms of the sequence whose general term is given. Then find the tenth term.

10. $a_n = \dfrac{2n}{n + 2}$ **11.** $a_n = 4n - 3$

Find the indicated unknown for each arithmetic sequence or series.

12. $a_1 = 3, d = 2; a_8 = ?$ **13.** $a_1 = 17, a_6 = -3; d = ?$ **14.** $5 + 9 + 13 + \cdots; S_{10} = ?$

Find the indicated unknown for each geometric sequence or series.

15. $a_1 = 9, r = \frac{1}{3}; a_5 = ?$ **16.** $6 + 3 + \frac{3}{2} + \cdots; S_4 = ?$

17. Write $0.\overline{24}$ as a ratio of two integers.   **18.** Expand $(x - 2)^4$ using the binomial theorem.

19. Find the fourth term of $(a + 3b)^9$.

20. The first swing of a pendulum is 24 inches. The second swing is 20 inches, and the lengths of the successive swings form a geometric sequence. Find the total distance traveled by the pendulum before it comes to rest.

CUMULATIVE REVIEW FOR CHAPTERS 10–12

[10] *Determine whether the given equation defines y as a function of x.*

1. $y = 3x + 4$

2. $x = y^2 + 1$

Find the domain of the function defined by each equation.

3. $y = \dfrac{1}{(x + 2)(x - 3)}$

4. $y = \sqrt{x + 6}$

Graph each function. Label all intercepts.

5. $f(x) = x^2 - 5x$

6. $h(x) = -2$

7. $f(x) = 4 - 2x$

8. $g(x) = 4x^2 - x^4$

9. $f(x) = |x - 1|$

10. $g(x) = \sqrt{x + 2}$

11. $h(x) = \begin{cases} 3, & \text{if } x \geq 0 \\ 0, & \text{if } x < 0 \end{cases}$

12. Suppose $f(x) = \sqrt{2x + 1}$. Determine each of the following.

a) $f(0)$

b) $f(4)$

c) $f(t + h)$

d) $f(f(x))$

13. If $f(x) = 5x - 2$, find $f^{-1}(x)$.

14. Find $\dfrac{f(t) - f(a)}{t - a}$ for $f(x) = x^3 + 3$. Then simplify.

Write each statement as a general variation function.

15. x varies directly with the square of y and inversely with the cube of z.

16. The volume V of a gas varies directly with its absolute temperature T and inversely with its pressure P.

17. Given that y varies inversely with the square of x, and $y = 4$ when $x = 6$, find y when $x = 12$.

18. The load L that a wooden beam can carry varies with the square of the depth d of the beam. If a beam 5 inches deep can carry a load of 2000 pounds, determine how much could be carried by a beam that is 8 inches deep.

[11] *Convert each equation to logarithmic form.*

19. $2^5 = 32$

20. $3^t = 100$

Convert each equation to exponential form.

21. $\log_2 512 = 9$

22. $\log_{16} 8 = \frac{3}{4}$

23. Express $\log_{12} 8$ in terms of $\log_5$.

Find y in each case.

24. $y = \log 100$

25. $y = \log_4 8$

26. $y = \log_5 1$

27. $y = \ln e^6$

Graph each function.

28. $y = 6^x$ **29.** $f(x) = 3^{|x|}$ **30.** $y = \log_4 x$ **31.** $y = \log_{1/2} x$

Solve each equation.

32. $\log (x + 1) - \log x = 1$ **33.** $2^{2x+3} = 128$

34. $2^{3x} = 4^{2x-1}$ **35.** $\log_3 (5 + x) + \log_3 (5 - x) = 2$

36. Write $\log \dfrac{x^5}{y^3 z^4}$ in terms of log x, log y, and log z.

Express as a single logarithm with a coefficient of 1.

37. $3 \ln x + \frac{1}{2} \ln (x + 1)$ **38.** $2 \log x - 4 \log y$

Find each logarithm to four decimal places.

39. $\log 452$ **40.** $\log 0.000000963$ **41.** $\ln 16$ **42.** $\ln 0.02$

Find each antilogarithm x to three significant figures.

43. $\log x = 2.5527$ **44.** $\log x = -3.1605$ **45.** $\ln x = 0.95$ **46.** $\ln x = 4.8$

47. If the present population of a city is 100,000 and it is increasing at the rate of 9% a year, how long will it be before the population is 150,000? Use $V = P(1 + r)^t$ with $V = 150,000$, $P = 100,000$, and $r = 0.09$.

48. The Dutch bought the island of Manhattan from American Indians in 1626 for goods valued at $24. If the Indians had invested their $24 in an investment that paid 5% compounded annually, how much would their investment be worth in 1986? Use $V = P(1 + r)^t$.

[12]   *Find the first four terms of the sequence whose general term is given. Then find the tenth term.*

49. $a_n = \dfrac{(n + 1)^2}{n}$ **50.** $a_n = \dfrac{3n}{n + 3}$

Find the general term a_n for each sequence.

51. $\frac{1}{4}, \frac{2}{9}, \frac{3}{16}, \frac{4}{25}, \ldots$ **52.** $1, -\frac{1}{3}, \frac{1}{5}, -\frac{1}{7}, \ldots$

Write each series in summation notation.

53. $3 + 9 + 27 + 81 + 243$ **54.** $7 + 11 + 15 + 19$

For each arithmetic sequence, find the common difference and write the next two terms.

55. $2, 6, 10, 14, \ldots$ **56.** $-2, -\frac{3}{2}, -1, -\frac{1}{2}, \ldots$

For each geometric sequence, find the common ratio and write the next two terms.

57. $2, \frac{1}{2}, \frac{1}{8}, \frac{1}{32}, \ldots$ **58.** $1, -3, 9, -27, \ldots$

Find the indicated sum.

59. $a_1 = 10$, $d = 3$; $S_{20} = ?$ **60.** $a_1 = 4$, $r = \frac{1}{4}$; $S_4 = ?$ **61.** $\sum\limits_{n=1}^{\infty} \left(\frac{2}{3}\right)^n$

Calculate each binomial coefficient.

62. $\dbinom{13}{11}$ **63.** $\dbinom{7}{0}$

Expand each binomial using the binomial theorem.

64. $(x - y)^8$ **65.** $(x + 2y)^5$

Find the indicated term in each expansion.

66. $(x^2 + y)^{14}$; 12th term   **67.** $(2a - b)^8$; middle term

68. The length of the first arc traversed by a pendulum is 15 cm, and the length of each succeeding arc is $\frac{3}{5}$ of the preceding one. Find the total distance traveled by the pendulum before it comes to rest.

Appendixes

Appendix 1: Sets

A **set*** is a collection of objects. The objects are the **elements,** or **members,** of the set. We enclose the elements of a set in braces { } and separate them with commas. Therefore if S is the set whose elements are the first three natural numbers, then

$$S = \{1, 2, 3\}.$$

The set $\{1, 2, 3\}$ is a **finite** set because it has a limited number of elements. The set of all natural numbers N is an **infinite** set, since it has an unlimited number of elements.

$$N = \{1, 2, 3, 4, 5, \ldots\}$$

The three dots mean "and so on."

The set that contains no elements is called the **empty set.** We use the symbol $\emptyset$ to represent the empty set.

CAUTION Write $\emptyset$ to represent the empty set. Do *not* write $\{\emptyset\}$.

To signify that 2 is an element of the set $S = \{1, 2, 3\}$, we write

$$2 \in S \quad \text{or} \quad 2 \in \{1, 2, 3\}.$$

To signify that 4 is not an element of the set $S = \{1, 2, 3\}$, we write

$$2 \notin S \quad \text{or} \quad 2 \notin \{1, 2, 3\}.$$

We describe a set either by listing its elements or by stating a rule that its elements obey. For example, using the listing method, we describe the set of natural numbers between 5 and 13 as

$$\{6, 7, 8, 9, 10, 11, 12\}.$$

Using the rule method, we describe the same set as

$$\{x | x \text{ is a natural number between 5 and 13}\}.$$

We read this last expression as "the set of all elements x such that x is a natural number between 5 and 13."

Two sets are **equal** if they have exactly the same elements. Therefore

$$\{1, 2, 3\} = \{3, 1, 2\}.$$

On the other hand

$$\{1, 2, 3\} \neq \{1, 2, 3, 4\},$$

where $\neq$ means "is not equal to."

---

* Sets were introduced by the German mathematician Georg Cantor (1845–1918).

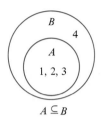

$$A \subseteq B$$

Figure A.1

Set A is a **subset** of set B, written $A \subseteq B$, if every element of A is also an element of B. Therefore

$$\{1, 2, 3\} \subseteq \{1, 2, 3, 4\},$$

since every element of $\{1, 2, 3\}$ is also an element of $\{1, 2, 3, 4\}$. A diagram of this relationship is shown in Figure A.1.*

On the other hand,

$$\{1, 2, 3, 4\} \nsubseteq \{1, 2, 3\},$$

where $\nsubseteq$ means "is not a subset of." Note that 4 is an element of $\{1, 2, 3, 4\}$, but 4 is not an element of $\{1, 2, 3\}$.

Since every element of set A is an element of A, set A is a subset of itself. Also, since the empty set has no elements, it cannot have any elements that are not elements of A. Therefore the empty set is a subset of A. We can summarize this discussion as follows:

For any set A,

$$A \subseteq A \text{ and } \emptyset \subseteq A.$$

CAUTION Do not confuse the symbols $\in$ and $\subseteq$. The symbol $\in$ is used between an element and a set. The symbol $\subseteq$ is used between two sets. For example, $2 \in \{1, 2, 3\}$, but $\{2\} \subseteq \{1, 2, 3\}$.

The **union** of set A and set B, written $A \cup B$, is the set of all elements that are either in A or in B, or in both A and B. For example,

$$\{1, 3, 5\} \cup \{2, 3, 4\} = \{1, 2, 3, 4, 5\}.$$

Even though 3 is an element of both sets, we do not list 3 twice in our answer. A Venn diagram of this operation is shown in Figure A.2.

---

* Diagrams that depict sets as regions are called **Venn diagrams,** after the English logician John Venn (1834–1923).

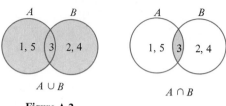

$$A \cup B$$

Figure A.2

$$A \cap B$$

Figure A.3

The **intersection** of set A and set B, written $A \cap B$, is the set of all elements common to A and B. For example,

$$\{1, 3, 5\} \cap \{2, 3, 4\} = \{3\}.$$

A Venn diagram of this operation is shown in Figure A.3.

PROBLEM SET A.1

1. Is $\{2, 4, 6, 8\}$ a finite set or an infinite set?
2. Is $\{2, 4, 6, 8, \ . \ . \ .\}$ a finite set or an infinite set?

True or false.

3. $5 \in \{5, 6, 7\}$
4. $8 \in \{5, 6, 7\}$
5. $4 \in \{x|x \text{ is an even natural number}\}$
6. $13 \in \{y|y \text{ is a multiple of 5}\}$

Write each set using the listing method.

7. $\{x|x \text{ is a natural number less than 5}\}$
8. $\{y|y \text{ is a natural number greater than 7}\}$
9. $\{t|t \text{ is a natural number between 1 and 6}\}$
10. $\{p|p \text{ is a natural number between 3 and 4}\}$

Write each set using the rule method.

11. $\{4, 5, 6, 7, 8\}$
12. $\{1, 2, 3, 4, 5\}$
13. $\{10, 11, 12, 13, \ . \ . \ .\}$
14. $\{7, 9, 11, 13, \ . \ . \ .\}$

True or false.

15. $\{3, 4, 7\} \subseteq \{1, 3, 4, 7\}$
16. $\{1, 3, 4, 7\} \subseteq \{3, 4, 7\}$
17. $\{5, 8, 9\} \subseteq \{8, 9, 5\}$
18. $\emptyset \subseteq \{2, 5, 6\}$
19. $\{x|x \text{ is an odd natural number}\} \subseteq \{x|x \text{ is an even natural number}\}$

20. $\{y|y$ is a multiple of $4\} \subseteq \{y|y$ is a multiple of $2\}$

21. If $A \subseteq B$ and $B \subseteq C$, then $A \subseteq C$.

22. If $A \subseteq B$ and $B \subseteq A$, then $A = B$.

Find all subsets of each set.

23. $\emptyset$ **24.** $\{1\}$ **25.** $\{1, 2\}$ **26.** $\{1, 2, 3\}$

If $A = \{1, 2, 4\}$, $B = \{2, 3, 5, 7\}$, and $C = \{6, 3, 5, 8\}$, find each of the following.

27. $A \cup B$ **28.** $B \cap C$ **29.** $(A \cap B) \cup C$ **30.** $B \cup (A \cap C)$

31. $A \cup A$ **32.** $A \cap A$ **33.** $\emptyset \cup A$ **34.** $\emptyset \cap A$

Appendix 2: Metric System

Early units of measure were often based on parts of the human body. A foot was the length of a man's foot, an inch was the distance from the tip of the thumb to the middle of the first joint, and a mile was 2000 paces. Later, in medieval England, King Henry I decreed that an *ell* (roughly a yard) should be the distance from the tip of his nose to the end of his outstretched thumb. Such units were not very precise, but they did serve a purpose for many years when more accurate scientific measurement was not in great demand. These and other units of measure were organized to form the **English** (or **British**) system of measurement. Because England was dominant in commerce and trade, the English system became the one that was most widely used.

In the English system, length can be measured in inches, feet, yards, or miles. Weight can be measured in ounces, pounds, or tons. Volume can be measured in pints, quarts, or gallons. Over the years these measures have been made more precise and standardized, but conversions between the units themselves have remained awkward. For example, small distances can be measured in inches or feet. Since 1 foot = 12 inches, changing from one of these units to the other requires either multiplying or dividing by 12.

In spite of its drawbacks, the English system remained popular for many years. But there was still no single, coordinated system acceptable world-wide. Then, in 1790, the French Academy of Science created a simple, scientific standard of weights and measures called the **metric system.** Since this system is based on multiples of ten, converting from one unit in the metric system to another can be accomplished simply by moving the decimal point. Therein lies one of the major advantages of the metric system over the English system.

One by one, nations have been adopting the metric system as their standard of measurement. It is significant to note that no nation that has adopted the metric system has abandoned it. The United States is now the only major industrial nation in the world that has not officially gone metric. It is almost as if all of the countries in the world are speaking one language and the United States is speaking another. It has been estimated that the United States could save millions of dollars per year in international trade alone by converting to the metric system.

Although the U.S. has not officially gone metric, the system itself is not a complete stranger to its citizens. Olympic track and field and swimming events are measured using the metric system. Athletes speak of the 100 meter dash, the 10,000 meter run, and the 200 meter breaststroke. Many of us have 8 millimeter home movie projectors or 35 millimeter cameras. We may glide down the slopes on skis that are 185 centimeters long, or puff on a 100 millimeter cigarette. Many hospitals are now maintaining patient statistics such as height, weight, and body temperature in metric units, and metric units have long been used for dispensing prescription medicines. Radio and TV stations now report temperature readings in both Fahrenheit and Celsius degrees. Almost all scientific research is done using the metric system, and industries such as electronics, shipbuilding, and even some auto industries have jumped onto the metric bandwagon.

Let's take a look at what we need to know to work successfully with the metric system.

$$
\text{Three Standards} \begin{cases} \text{meter (m):} & \text{length} \\ \text{liter } (\ell)\text{:} & \text{volume} \\ \text{gram (g):} & \text{mass (weight)} \end{cases}
$$

$$
\text{Six Prefixes} \begin{cases} \text{kilo:} & 1000 \\ \text{hecto:} & 100 \\ \text{deca:} & 10 \\ \text{deci:} & \frac{1}{10} \\ \text{centi:} & \frac{1}{100} \\ \text{milli:} & \frac{1}{1000} \end{cases}
$$

Also, the Celsius scale, which was discussed in Section 3.2, is used for measuring temperature.

To get a feel for the standards in the metric system, we observe that a meter is roughly 3 inches longer than a yard, a liter is just slightly larger than a quart, a gram is about the weight of half of a dime, and an outside temperature of 30° Celsius would send some people scurrying to the beach.

An obvious question that now arises is, "Why were these specific standards chosen?" First, the temperature scale was chosen so that 0° Celsius coincided with the freezing point of water and 100° Celsius coincided with the boiling point of water. The original meter was one ten-millionth of the distance between the North Pole and the equator (measured along the meridian passing through Paris), and the original gram was the weight of one cubic centimeter (a centimeter is one hundredth of a meter) of water. A liter is the volume contained in a cube that is ten centimeters on each side.

Other units in the metric system, for measuring quantities that are either too large or too small to be measured efficiently using one of the three standards, are created by using one of the six prefixes in conjunction with one of the three standards. The three prefixes that are most commonly used are kilo, centi, and milli. The prefixes hecto, deca, and deci are primarily used in scientific measurements. Thus a kilogram (kg) is 1000 grams (about 2.2 pounds), a centimeter (cm) is $\frac{1}{100}$ of a meter (about the width of a large paper clip), and a milliliter (mℓ) is $\frac{1}{1000}$ of a liter ($\frac{1}{5}$ of a teaspoon).

Conversions within the metric system involve multiplying or dividing by an appropriate power of ten. Hence a conversion can be made simply by moving the decimal point.

EXAMPLE 1   Convert 3 meters to centimeters.

Since 1 m = 100 cm,

$$3 \text{ m} = 3 \times 100 \text{ cm} = 300 \text{ cm.} \blacktriangleleft$$

EXAMPLE 2   Convert 55 kilograms to grams.

Since 1 kg = 1000 g,

$$55 \text{ kg} = 55 \times 1000 \text{ g} = 55,000 \text{ g.} \blacktriangleleft$$

EXAMPLE 3   Convert 1600 milliliters to liters.

Since $1 \text{ m}\ell = \dfrac{1}{1000} \ell$,

$$1600 \text{ m}\ell = \frac{1600}{1000} \ell = 1.6 \, \ell. \blacktriangleleft$$

EXAMPLE 4   Convert 2.8 centimeters to millimeters.

Since 1 cm = 10 mm,

$$2.8 \text{ cm} = 2.8 \times 10 \text{ mm} = 28 \text{ mm.} \blacktriangleleft$$

Eventually, we should learn to think directly in the metric system without using the English system as a crutch. That is, if someone says that they live 33 kilometers away, we should have an idea of just how far that is. Is it a half-hour drive or a five-hour drive? Is a room 7 meters by 5 meters a small bathroom or a large living room? Is a man who weighs 115 kilograms the size of a defensive tackle or a thoroughbred jockey? Such familiarity only comes with experience. For the time being, we shall have to be content with making conversions between the familiar English system and the somewhat unfamiliar metric system in order to place metric measurements in their proper perspective. In doing so, we shall be on our way to our ultimate goal, which is to "think metric." The conversion table provided on next page will enable us to make these conversions.

EXAMPLE 5   Convert 5 yards to meters.

Yards to meters is an *English to metric* conversion. From the table we see that yards can be converted to meters by multiplying by 0.9144.

$$5 \text{ yards} = 5 \times 0.9144 \text{ m} = 4.572 \text{ m} \blacktriangleleft$$

EXAMPLE 6   Convert 32 kilograms to pounds.

This is a *metric to English* conversion. From the table we see that we need to multiply by 2.205.

$$32 \text{ kg} = 32 \times 2.205 \text{ pounds} = 70.56 \text{ pounds} \blacktriangleleft$$

Metric to English			English to Metric		
From	*To*	*Multiply by*	*From*	*To*	*Multiply by*
meters	yards	1.094	yards	meters	0.9144
meters	feet	3.281	feet	meters	0.3048
meters	inches	39.37	inches	meters	0.0254
centimeters	inches	0.3937	inches	centimeters	2.540
kilometers	miles	0.6214	miles	kilometers	1.609
grams	ounces	0.0353	ounces	grams	28.35
kilograms	pounds	2.205	pounds	kilograms	0.4536
liters	quarts	1.057	quarts	liters	0.9464

PROBLEM SET A.2

Make the following conversions within the metric system.

1. 5 m to cm
2. 18 m to mm
3. 7.8 m to cm
4. 16 km to m
5. 5200 m to km
6. 6200 cm to m
7. 45 mm to m
8. 17.1 cm to mm
9. 5 mm to cm
10. 6 kg to g
11. 15,000 g to kg
12. 1.62 kg to g
13. 8 ℓ to mℓ
14. 500 mℓ to ℓ
15. 37,000 mℓ to ℓ
16. 2.93 ℓ to mℓ

Use the table above to make the conversions indicated.

17. 8 yards to m
18. 90 kg to pounds
19. 3 m to feet
20. 350 miles to km
21. 11 ounces to g
22. 23 g to ounces
23. 45 inches to m
24. 200 feet to m
25. 7 inches to cm
26. 5 quarts to ℓ
27. 10 cm to inches
28. 2 m to inches
29. 850 pounds to kg
30. 30 ℓ to quarts
31. 13 m to yards
32. 200 km to miles

†33. Convert 55,000 sq km to square miles.

†34. Convert 7200 cu cm to cubic inches.

Rectangle

Area
$$A = \ell w$$
Perimeter
$$P = 2\ell + 2w$$

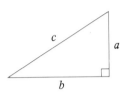

Right triangle

$$c^2 = a^2 + b^2$$

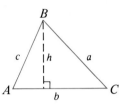

Triangle

Area
$$A = \tfrac{1}{2}bh$$
Perimeter
$$P = a + b + c$$
$$\angle A + \angle B + \angle C = 180°$$

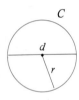

Circle

Area
$$A = \pi r^2$$
Circumference
$$C = 2\pi r$$

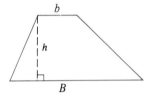

Trapezoid

Area
$$A = \tfrac{1}{2}h(b + B)$$

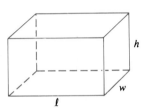

Parallelepiped

Volume
$$V = \ell wh$$
Surface area
$$S = 2(\ell w + \ell h + wh)$$

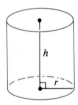

Right circular cylinder

Volume
$$V = \pi r^2 h$$
Lateral surface area
$$S = 2\pi rh$$
Total surface area
$$T = 2\pi r(h + r)$$

Right circular cone

Volume
$$V = \tfrac{1}{3}\pi r^2 h$$
Lateral surface area
$$S = \pi r\ell$$
Total surface area
$$T = \pi r(\ell + r)$$

Sphere

Volume
$$V = \tfrac{4}{3}\pi r^3$$
Surface area
$$S = 4\pi r^2$$

Appendix 4: Square Roots

n	$\sqrt{n}$	n	$\sqrt{n}$	n	$\sqrt{n}$	n	$\sqrt{n}$
1	1.000	51	7.141	101	10.05	151	12.29
2	1.414	52	7.211	102	10.10	152	12.33
3	1.732	53	7.280	103	10.15	153	12.37
4	2.000	54	7.348	104	10.20	154	12.41
5	2.236	55	7.416	105	10.25	155	12.45
6	2.449	56	7.483	106	10.30	156	12.49
7	2.646	57	7.550	107	10.34	157	12.53
8	2.828	58	7.616	108	10.39	158	12.57
9	3.000	59	7.681	109	10.44	159	12.61
10	3.162	60	7.746	110	10.49	160	12.65
11	3.317	61	7.810	111	10.54	161	12.69
12	3.464	62	7.874	112	10.58	162	12.73
13	3.606	63	7.937	113	10.63	163	12.77
14	3.742	64	8.000	114	10.68	164	12.81
15	3.873	65	8.062	115	10.72	165	12.85
16	4.000	66	8.124	116	10.77	166	12.88
17	4.123	67	8.185	117	10.82	167	12.92
18	4.243	68	8.246	118	10.86	168	12.96
19	4.359	69	8.307	119	10.91	169	13.00
20	4.472	70	8.367	120	10.95	170	13.04
21	4.583	71	8.426	121	11.00	171	13.08
22	4.690	72	8.485	122	11.05	172	13.11
23	4.796	73	8.544	123	11.09	173	13.15
24	4.899	74	8.602	124	11.14	174	13.19
25	5.000	75	8.660	125	11.18	175	13.23
26	5.099	76	8.718	126	11.22	176	13.27
27	5.196	77	8.775	127	11.27	177	13.30
28	5.292	78	8.832	128	11.31	178	13.34
29	5.385	79	8.888	129	11.36	179	13.38
30	5.477	80	8.944	130	11.40	180	13.42
31	5.568	81	9.000	131	11.45	181	13.45
32	5.657	82	9.055	132	11.49	182	13.49
33	5.745	83	9.110	133	11.53	183	13.53
34	5.831	84	9.165	134	11.58	184	13.56
35	5.916	85	9.220	135	11.62	185	13.60
36	6.000	86	9.274	136	11.66	186	13.64
37	6.083	87	9.327	137	11.70	187	13.67
38	6.164	88	9.381	138	11.75	188	13.71
39	6.245	89	9.434	139	11.79	189	13.75
40	6.325	90	9.487	140	11.83	190	13.78
41	6.403	91	9.539	141	11.87	191	13.82
42	6.481	92	9.592	142	11.92	192	13.86
43	6.557	93	9.644	143	11.96	193	13.89
44	6.633	94	9.695	144	12.00	194	13.93
45	6.708	95	9.747	145	12.04	195	13.96
46	6.782	96	9.798	146	12.08	196	14.00
47	6.856	97	9.849	147	12.12	197	14.04
48	6.928	98	9.899	148	12.17	198	14.07
49	7.000	99	9.950	149	12.21	199	14.11
50	7.071	100	10.000	150	12.25	200	14.14

Appendix 5: Common Logarithms

x	0	1	2	3	4	5	6	7	8	9
1.0	0.0000	0.0043	0.0086	0.0128	0.0170	0.0212	0.0253	0.0294	0.0334	0.0374
1.1	0.0414	0.0453	0.0492	0.0531	0.0569	0.0607	0.0645	0.0682	0.0719	0.0755
1.2	0.0792	0.0828	0.0864	0.0899	0.0934	0.0969	0.1004	0.1038	0.1072	0.1106
1.3	0.1139	0.1173	0.1206	0.1239	0.1271	0.1303	0.1335	0.1367	0.1399	0.1430
1.4	0.1461	0.1492	0.1523	0.1553	0.1584	0.1614	0.1644	0.1673	0.1703	0.1732
1.5	0.1761	0.1790	0.1818	0.1847	0.1875	0.1903	0.1931	0.1959	0.1987	0.2014
1.6	0.2041	0.2068	0.2095	0.2122	0.2148	0.2175	0.2201	0.2227	0.2253	0.2279
1.7	0.2304	0.2330	0.2355	0.2380	0.2405	0.2430	0.2455	0.2480	0.2504	0.2529
1.8	0.2553	0.2577	0.2601	0.2625	0.2648	0.2672	0.2695	0.2718	0.2742	0.2765
1.9	0.2788	0.2810	0.2833	0.2856	0.2878	0.2900	0.2923	0.2945	0.2967	0.2989
2.0	0.3010	0.3032	0.3054	0.3075	0.3096	0.3118	0.3139	0.3160	0.3181	0.3201
2.1	0.3222	0.3243	0.3263	0.3284	0.3304	0.3324	0.3345	0.3365	0.3385	0.3404
2.2	0.3424	0.3444	0.3464	0.3483	0.3502	0.3522	0.3541	0.3560	0.3579	0.3598
2.3	0.3617	0.3636	0.3655	0.3674	0.3692	0.3711	0.3729	0.3747	0.3766	0.3784
2.4	0.3802	0.3820	0.3838	0.3856	0.3874	0.3892	0.3909	0.3927	0.3945	0.3962
2.5	0.3979	0.3997	0.4014	0.4031	0.4048	0.4065	0.4082	0.4099	0.4116	0.4133
2.6	0.4150	0.4166	0.4183	0.4200	0.4216	0.4232	0.4249	0.4265	0.4281	0.4298
2.7	0.4314	0.4330	0.4346	0.4362	0.4378	0.4393	0.4409	0.4425	0.4440	0.4456
2.8	0.4472	0.4487	0.4502	0.4518	0.4533	0.4548	0.4564	0.4579	0.4594	0.4609
2.9	0.4624	0.4639	0.4654	0.4669	0.4683	0.4698	0.4713	0.4728	0.4742	0.4757
3.0	0.4771	0.4786	0.4800	0.4814	0.4829	0.4843	0.4857	0.4871	0.4886	0.4900
3.1	0.4914	0.4928	0.4942	0.4955	0.4969	0.4983	0.4997	0.5011	0.5024	0.5038
3.2	0.5051	0.5065	0.5079	0.5092	0.5105	0.5119	0.5132	0.5145	0.5159	0.5172
3.3	0.5185	0.5198	0.5211	0.5224	0.5237	0.5250	0.5263	0.5276	0.5289	0.5302
3.4	0.5315	0.5328	0.5340	0.5353	0.5366	0.5378	0.5391	0.5403	0.5416	0.5428
3.5	0.5441	0.5453	0.5465	0.5478	0.5490	0.5502	0.5514	0.5527	0.5539	0.5551
3.6	0.5563	0.5575	0.5587	0.5599	0.5611	0.5623	0.5635	0.5647	0.5658	0.5670
3.7	0.5682	0.5694	0.5705	0.5717	0.5729	0.5740	0.5752	0.5763	0.5775	0.5786
3.8	0.5798	0.5809	0.5821	0.5832	0.5843	0.5855	0.5866	0.5877	0.5888	0.5899
3.9	0.5911	0.5922	0.5933	0.5944	0.5955	0.5966	0.5977	0.5988	0.5999	0.6010
4.0	0.6021	0.6031	0.6042	0.6053	0.6064	0.6075	0.6085	0.6096	0.6107	0.6117
4.1	0.6128	0.6138	0.6149	0.6160	0.6170	0.6180	0.6191	0.6201	0.6212	0.6222
4.2	0.6232	0.6243	0.6253	0.6263	0.6274	0.6284	0.6294	0.6304	0.6314	0.6325
4.3	0.6335	0.6345	0.6355	0.6365	0.6375	0.6385	0.6395	0.6405	0.6415	0.6425
4.4	0.6435	0.6444	0.6454	0.6464	0.6474	0.6484	0.6493	0.6503	0.6513	0.6522
4.5	0.6532	0.6542	0.6551	0.6561	0.6571	0.6580	0.6590	0.6599	0.6609	0.6618
4.6	0.6628	0.6637	0.6646	0.6656	0.6665	0.6675	0.6684	0.6693	0.6702	0.6712
4.7	0.6721	0.6730	0.6739	0.6749	0.6758	0.6767	0.6776	0.6785	0.6794	0.6803
4.8	0.6812	0.6821	0.6830	0.6839	0.6848	0.6857	0.6866	0.6875	0.6884	0.6893
4.9	0.6902	0.6911	0.6920	0.6928	0.6937	0.6946	0.6955	0.6964	0.6972	0.6981
5.0	0.6990	0.6998	0.7007	0.7016	0.7024	0.7033	0.7042	0.7050	0.7059	0.7067
5.1	0.7076	0.7084	0.7093	0.7101	0.7110	0.7118	0.7126	0.7135	0.7143	0.7152
5.2	0.7160	0.7168	0.7177	0.7185	0.7193	0.7202	0.7210	0.7218	0.7226	0.7235
5.3	0.7243	0.7251	0.7259	0.7267	0.7275	0.7284	0.7292	0.7300	0.7308	0.7316
5.4	0.7324	0.7332	0.7340	0.7348	0.7356	0.7364	0.7372	0.7380	0.7388	0.7396

Common Logarithms (continued)

x	0	1	2	3	4	5	6	7	8	9
5.5	0.7404	0.7412	0.7419	0.7427	0.7435	0.7443	0.7451	0.7459	0.7466	0.7474
5.6	0.7482	0.7490	0.7497	0.7505	0.7513	0.7520	0.7528	0.7536	0.7543	0.7551
5.7	0.7559	0.7566	0.7574	0.7582	0.7589	0.7597	0.7604	0.7612	0.7619	0.7627
5.8	0.7634	0.7642	0.7649	0.7657	0.7664	0.7672	0.7679	0.7686	0.7694	0.7701
5.9	0.7709	0.7716	0.7723	0.7731	0.7738	0.7745	0.7752	0.7760	0.7767	0.7774
6.0	0.7782	0.7789	0.7796	0.7803	0.7810	0.7818	0.7825	0.7832	0.7839	0.7846
6.1	0.7853	0.7860	0.7868	0.7875	0.7882	0.7889	0.7896	0.7903	0.7910	0.7917
6.2	0.7924	0.7931	0.7938	0.7945	0.7952	0.7959	0.7966	0.7973	0.7980	0.7987
6.3	0.7993	0.8000	0.8007	0.8014	0.8021	0.8028	0.8035	0.8041	0.8048	0.8055
6.4	0.8062	0.8069	0.8075	0.8082	0.8089	0.8096	0.8102	0.8109	0.8116	0.8122
6.5	0.8129	0.8136	0.8142	0.8149	0.8156	0.8162	0.8169	0.8176	0.8182	0.8189
6.6	0.8195	0.8202	0.8209	0.8215	0.8222	0.8228	0.8235	0.8241	0.8248	0.8254
6.7	0.8261	0.8267	0.8274	0.8280	0.8287	0.8293	0.8299	0.8306	0.8312	0.8319
6.8	0.8325	0.8331	0.8338	0.8344	0.8351	0.8357	0.8363	0.8370	0.8376	0.8382
6.9	0.8388	0.8395	0.8401	0.8407	0.8414	0.8420	0.8426	0.8432	0.8439	0.8445
7.0	0.8451	0.8457	0.8463	0.8470	0.8476	0.8482	0.8488	0.8494	0.8500	0.8506
7.1	0.8513	0.8519	0.8525	0.8531	0.8537	0.8543	0.8549	0.8555	0.8561	0.8567
7.2	0.8573	0.8579	0.8585	0.8591	0.8597	0.8603	0.8609	0.8615	0.8621	0.8627
7.3	0.8633	0.8639	0.8645	0.8651	0.8657	0.8663	0.8669	0.8675	0.8681	0.8686
7.4	0.8692	0.8698	0.8704	0.8710	0.8716	0.8722	0.8727	0.8733	0.8739	0.8745
7.5	0.8751	0.8756	0.8762	0.8768	0.8774	0.8779	0.8785	0.8791	0.8797	0.8802
7.6	0.8808	0.8814	0.8820	0.8825	0.8831	0.8837	0.8842	0.8848	0.8854	0.8859
7.7	0.8865	0.8871	0.8876	0.8882	0.8887	0.8893	0.8899	0.8904	0.8910	0.8915
7.8	0.8921	0.8927	0.8932	0.8938	0.8943	0.8949	0.8954	0.8960	0.8965	0.8971
7.9	0.8976	0.8982	0.8987	0.8993	0.8998	0.9004	0.9009	0.9015	0.9020	0.9025
8.0	0.9031	0.9036	0.9042	0.9047	0.9053	0.9058	0.9063	0.9069	0.9074	0.9079
8.1	0.9085	0.9090	0.9096	0.9101	0.9106	0.9112	0.9117	0.9122	0.9128	0.9133
8.2	0.9138	0.9143	0.9149	0.9154	0.9159	0.9165	0.9170	0.9175	0.9180	0.9186
8.3	0.9191	0.9196	0.9201	0.9206	0.9212	0.9217	0.9222	0.9227	0.9232	0.9238
8.4	0.9243	0.9248	0.9253	0.9258	0.9263	0.9269	0.9274	0.9279	0.9284	0.9289
8.5	0.9294	0.9299	0.9304	0.9309	0.9315	0.9320	0.9325	0.9330	0.9335	0.9340
8.6	0.9345	0.9350	0.9355	0.9360	0.9365	0.9370	0.9375	0.9380	0.9385	0.9390
8.7	0.9395	0.9400	0.9405	0.9410	0.9415	0.9420	0.9425	0.9430	0.9435	0.9440
8.8	0.9445	0.9450	0.9455	0.9460	0.9465	0.9469	0.9474	0.9479	0.9484	0.9489
8.9	0.9494	0.9499	0.9504	0.9509	0.9513	0.9518	0.9523	0.9528	0.9533	0.9538
9.0	0.9542	0.9547	0.9552	0.9557	0.9562	0.9566	0.9571	0.9576	0.9581	0.9586
9.1	0.9590	0.9595	0.9600	0.9605	0.9609	0.9614	0.9619	0.9624	0.9628	0.9633
9.2	0.9638	0.9643	0.9647	0.9652	0.9657	0.9661	0.9666	0.9671	0.9675	0.9680
9.3	0.9685	0.9689	0.9694	0.9699	0.9703	0.9708	0.9713	0.9717	0.9722	0.9727
9.4	0.9731	0.9736	0.9741	0.9745	0.9750	0.9754	0.9759	0.9763	0.9768	0.9773
9.5	0.9777	0.9782	0.9786	0.9791	0.9795	0.9800	0.9805	0.9809	0.9814	0.9818
9.6	0.9823	0.9827	0.9832	0.9836	0.9841	0.9845	0.9850	0.9854	0.9859	0.9863
9.7	0.9868	0.9872	0.9877	0.9881	0.9886	0.9890	0.9894	0.9899	0.9903	0.9908
9.8	0.9912	0.9917	0.9921	0.9926	0.9930	0.9934	0.9939	0.9943	0.9948	0.9952
9.9	0.9956	0.9961	0.9965	0.9969	0.9974	0.9978	0.9983	0.9987	0.9991	0.9996

EXAMPLE 1 $\log 3.17 = 0.5011$ ◄

EXAMPLE 2 $\log 3170 = \log (3.17 \times 10^3)$
$$= \log 3.17 + \log 10^3$$
$$= 0.5011 + 3$$
$$= 3.5011 \blacktriangleleft$$

EXAMPLE 3 $\log 0.0317 = \log (3.17 \times 10^{-2})$
$$= \log 3.17 + \log 10^{-2}$$
$$= \log 3.17 + \log 10^{-2}$$
$$= 0.5011 + (-2)$$
$$= -1.4989 \blacktriangleleft$$

Appendix 6: Powers of e

x	e^x	e^{-x}	x	e^x	e^{-x}	x	e^x	e^{-x}
0.00	1.00000	1.00000	0.40	1.49182	0.67032	0.80	2.22554	0.44933
0.01	1.01005	0.99005	0.41	1.50682	0.66365	0.81	2.24791	0.44486
0.02	1.02020	0.98020	0.42	1.52196	0.65705	0.82	2.27050	0.44043
0.03	1.03045	0.97045	0.43	1.53726	0.65051	0.83	2.29332	0.43605
0.04	1.04081	0.96079	0.44	1.55271	0.64404	0.84	2.31637	0.43171
0.05	1.05127	0.95123	0.45	1.56831	0.63763	0.85	2.33965	0.42741
0.06	1.06184	0.94176	0.46	1.58407	0.63128	0.86	2.36316	0.42316
0.07	1.07251	0.93239	0.47	1.59999	0.62500	0.87	2.38691	0.41895
0.08	1.08329	0.92312	0.48	1.61607	0.61878	0.88	2.41090	0.41478
0.09	1.09417	0.91393	0.49	1.63232	0.61263	0.89	2.43513	0.41066
0.10	1.10517	0.90484	0.50	1.64872	0.60653	0.90	2.45960	0.40657
0.11	1.11628	0.89583	0.51	1.66529	0.60050	0.91	2.48432	0.40252
0.12	1.12750	0.88692	0.52	1.68203	0.59452	0.92	2.50929	0.39852
0.13	1.13883	0.87810	0.53	1.69893	0.58860	0.93	2.53451	0.39455
0.14	1.15027	0.86936	0.54	1.71601	0.58275	0.94	2.55998	0.39063
0.15	1.16183	0.86071	0.55	1.73325	0.57695	0.95	2.58571	0.38674
0.16	1.17351	0.85214	0.56	1.75067	0.57121	0.96	2.61170	0.38289
0.17	1.18530	0.84366	0.57	1.76827	0.56553	0.97	2.63794	0.37908
0.18	1.19722	0.83527	0.58	1.78604	0.55990	0.98	2.66446	0.37531
0.19	1.20925	0.82696	0.59	1.80399	0.55433	0.99	2.69123	0.37158
0.20	1.22140	0.81873	0.60	1.82212	0.54881	1.00	2.71828	0.36788
0.21	1.23368	0.81058	0.61	1.84043	0.54335	1.01	2.74560	0.36422
0.22	1.24608	0.80252	0.62	1.85893	0.53794	1.02	2.77319	0.36059
0.23	1.25860	0.79453	0.63	1.87761	0.53259	1.03	2.80107	0.35701
0.24	1.27125	0.78663	0.64	1.89648	0.52729	1.04	2.82922	0.35345
0.25	1.28403	0.77880	0.65	1.91554	0.52205	1.05	2.85765	0.34994
0.26	1.29693	0.77105	0.66	1.93479	0.51685	1.06	2.88637	0.34646
0.27	1.30996	0.76338	0.67	1.95424	0.51171	1.07	2.91538	0.34301
0.28	1.32313	0.75578	0.68	1.97388	0.50662	1.08	2.94468	0.33960
0.29	1.33643	0.74826	0.69	1.99372	0.50158	1.09	2.97427	0.33622
0.30	1.34986	0.74082	0.70	2.01375	0.49659	1.10	3.00417	0.33287
0.31	1.36343	0.73345	0.71	2.03399	0.49164	1.11	3.03436	0.32956
0.32	1.37713	0.72615	0.72	2.05443	0.48675	1.12	3.06485	0.32628
0.33	1.39097	0.71892	0.73	2.07508	0.48191	1.13	3.09566	0.32303
0.34	1.40495	0.71177	0.74	2.09594	0.47711	1.14	3.12677	0.31982
0.35	1.41907	0.70469	0.75	2.11700	0.47237	1.15	3.15819	0.31664
0.36	1.43333	0.69768	0.76	2.13828	0.46767	1.16	3.18993	0.31349
0.37	1.44773	0.69073	0.77	2.15977	0.46301	1.17	3.22199	0.31037
0.38	1.46228	0.68386	0.78	2.18147	0.45841	1.18	3.25437	0.30728
0.39	1.47698	0.67706	0.79	2.20340	0.45384	1.19	3.28708	0.30422

Powers of e (continued)

x	e^x	e^{-x}	x	e^x	e^{-x}	x	e^x	e^{-x}
1.20	3.32012	0.30119	1.60	4.95303	0.20190	4.0	54.598	0.01832
1.21	3.35348	0.29820	1.61	5.00281	0.19989	4.1	60.340	0.01657
1.22	3.38719	0.29523	1.62	5.05309	0.19790	4.2	66.686	0.01500
1.23	3.42123	0.29229	1.63	5.10387	0.19593	4.3	73.700	0.01357
1.24	3.45561	0.28938	1.64	5.15517	0.19398	4.4	81.451	0.01228
1.25	3.49034	0.28650	1.65	5.20698	0.19205	4.5	90.017	0.01111
1.26	3.52542	0.28365	1.66	5.25931	0.19014	4.6	99.484	0.01005
1.27	3.56085	0.28083	1.67	5.31217	0.18825	4.7	109.947	0.00910
1.28	3.59664	0.27804	1.68	5.36556	0.18637	4.8	121.510	0.00823
1.29	3.63279	0.27527	1.69	5.41948	0.18452	4.9	134.290	0.00745
1.30	3.66930	0.27253	1.70	5.47395	0.18268	5.0	148.41	0.00674
1.31	3.70617	0.26982	1.71	5.52896	0.18087	5.1	164.02	0.00610
1.32	3.74342	0.26714	1.72	5.58453	0.17907	5.2	181.27	0.00552
1.33	3.78104	0.26448	1.73	5.64065	0.17728	5.3	200.34	0.00499
1.34	3.81904	0.26185	1.74	5.69734	0.17552	5.4	221.41	0.00452
1.35	3.85743	0.25924	1.75	5.75460	0.17377	5.5	244.69	0.00409
1.36	3.89619	0.25666	1.80	6.04965	0.16530	5.6	270.43	0.00370
1.37	3.93535	0.25411	1.85	6.35982	0.15724	5.7	298.87	0.00335
1.38	3.97490	0.25158	1.90	6.68589	0.14957	5.8	330.30	0.00303
1.39	4.01485	0.24908	1.95	7.02869	0.14227	5.9	365.04	0.00274
1.40	4.05520	0.24660	2.0	7.3891	0.13534	6.0	403.43	0.00248
1.41	4.09596	0.24414	2.1	8.1662	0.12246	6.1	445.86	0.00224
1.42	4.13712	0.24171	2.2	9.0250	0.11080	6.2	492.75	0.00203
1.43	4.17870	0.23931	2.3	9.9742	0.10026	6.3	544.57	0.00184
1.44	4.22070	0.23693	2.4	11.0232	0.09072	6.4	601.85	0.00166
1.45	4.26311	0.23457	2.5	12.1825	0.08208	6.5	665.14	0.00150
1.46	4.30596	0.23224	2.6	13.4637	0.07427	6.6	735.10	0.00136
1.47	4.34924	0.22993	2.7	14.8797	0.06721	6.7	812.41	0.00123
1.48	4.39295	0.22764	2.8	16.4446	0.06081	6.8	897.85	0.00111
1.49	4.43710	0.22537	2.9	18.1741	0.05502	6.9	992.27	0.00101
1.50	4.48169	0.22313	3.0	20.086	0.04979	7.0	1096.6	0.00091
1.51	4.52673	0.22091	3.1	22.198	0.04505	7.5	1808.0	0.00055
1.52	4.57223	0.21871	3.2	24.533	0.04076	8.0	2981.0	0.00034
1.53	4.61818	0.21654	3.3	27.113	0.03688	8.5	4914.8	0.00020
1.54	4.66459	0.21438	3.4	29.964	0.03337	9.0	8103.1	0.00012
1.55	4.71147	0.21225	3.5	33.115	0.03020	9.5	13360	0.00007
1.56	4.75882	0.21014	3.6	36.598	0.02732	10.0	22026	0.00005
1.57	4.80665	0.20805	3.7	40.447	0.02472	10.5	36316	0.00003
1.58	4.85496	0.20598	3.8	44.701	0.02237	11.0	59874	0.00002
1.59	4.90375	0.20393	3.9	49.402	0.02024	11.5	98716	0.00001

Answers

CHAPTER 1

Problem Set 1.1, pp. 7–9

1. $x + 8$ **3.** $3 \cdot y$ **5.** $\frac{3}{5} \cdot z$ **7.** $25 \cdot t$
9. $s - 4$ **11.** $2.35r$ dollars   **13.** $3(m + 2)$
15. $21\frac{7}{8}x + 33y$ dollars   **17.** 123 lb
19. 21   **21.** 76   **23.** 17   **25.** 15
27. 1   **29.** 4   **31.** $\frac{9}{8}$ **33.** $\frac{7}{6}$ **35.** $\frac{19}{20}$
37. 0   **39.** 16   **41.** 14   **43.** 45
45. 48   **47.** 19.5   **49.** 13.49   **51.** 2
53. $\frac{11}{17}$ **55.** 37   **57.** 14   **59.** 15
61. $\$2\frac{11}{24}$ **63.** \$94.50   **65.** $9\frac{5}{6}$ yd
67. $25(\$18.50 + \$3.50) = \$550$;
     $25(\$18.50) + 25(\$3.50) = \$550$
69. $5x + 15$ **71.** $12y + 24$ **73.** $5z + 4$

75. $2a + 2b + 2c$ **77.** $6v + 9$ **79.** $b + c$
81. $x + 4$ **83.** $x + 3$ **85.** $13(r + s + t)$
87. $\dfrac{((n + 5)2 - 4)3}{6} - n = 3$

89. $\dfrac{a}{b} + \dfrac{c}{d} = \dfrac{ad}{bd} + \dfrac{bc}{bd} = \dfrac{ad + bc}{bd}$

91. a) 0.018   **b)** 8.232

Problem Set 1.2, pp. 15–17

1. $+4$ strokes, -6 strokes   **3.** $+2\frac{7}{8}, -4\frac{3}{8}$
5. $+7$ yd, -3 yd   **7.** $0.5 = 50\%$
9. $6.75 = 675\%$ **11.** $0.625 = 62.5\%$
13. $0.22\overline{2} = 22.\overline{2}\%$ **15.** $0.36\overline{36} = 36.\overline{36}\%$

17. $0.68\overline{18} = 68.\overline{18}\%$ **19.** $-3.8 = -380\%$
21. $0 = 0\%$ **23.** Undefined
25. a) 6, 203 **b)** 6, 203, $+0$
  c) $6, 203, +0, -10$ **d)** All of them
27. $1.06, 1.\overline{06}, \frac{8}{5}, 1.\overline{6}$
29. 8.27, infinite number
31. a) $\frac{17}{1000}$ **b)** $\frac{547}{100}$ **33.** 1
35. a) $\sqrt{64}, 3.14, -\frac{15}{7}, 0.\overline{47}$
  b) $\pi, 0.919919991\ldots, \sqrt{10}$
37. True **39.** True **41.** False
43. True **45.** True **47.** $8 + 6$
49. $7 \cdot (-2)$ **51.** $(3 + 4) + 5$
53. $x + (4 + (-4))$ **55.** $(9 \cdot 6) \cdot \frac{1}{3}$
57. $(\frac{1}{2} \cdot 2)x$ **59.** 13 **61.** 5 **63.** 0
65. 1 **67.** $(-2)5 + (-2)(-6)$
69. $(3 + 4)x$ **71.** $6 \div 2 \neq 2 \div 6$
73. $(10 - 4) - 1 \neq 10 - (4 - 1)$ **75.** $n + 1$
77. $m - 2$ **79.** $0.43r$ **81.** $0.0875s$
83. $0.2t$ **85. a)** Even **b)** Odd **87.** 97%, 3%
89. a) $0.\overline{909}$ **b)** 0 **c)** Undefined **d)** Undefined

Problem Set 1.3, pp. 22–24

1.

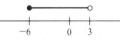

3. $-5 < -3$
5. $-101 < -100$ **7.** $0 > -1$ **9.** $-6 < 3$
11. $\pi < \frac{22}{7}$ **13.** $-\sqrt{2} < -1.4$ **15.** $\frac{3}{4} > \frac{5}{7}$
17. $-\frac{3}{4} < -\frac{5}{7}$ **19.** $x > 0$ **21.** $y \geq 0$
23. $z \leq 7$ **25.** $r + 5 > 14$ **27.** $8 < s < 9$
29. $-4 \leq t < 19$ **31.** $v \geq 18$ yr
33. 105 lb $\leq w \leq 112$ lb
35. 135 mi $\leq d \leq 165$ mi
37. $\$26.18 \leq p < \$32$
39. Four points

41. Infinite set of points

43. Open half line **45.** Closed half line

47. Open half line **49.** Closed interval

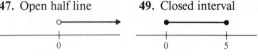

51. Open interval **53.** Half-open interval

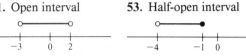

55. Half-open interval

57. $\{x \mid x \geq -2 \text{ and } x \text{ is an integer}\}$
59. $\{x \mid x > -2\}$ **61.** $\{x \mid -1 \leq x < 3\}$
63. 13 **65.** 9 **67.** $3\frac{1}{3}$ **69.** $\sqrt{2}$
71. 0 **73.** π **75.** -25 **77.** -99
79. 7 **81.** Undefined **83.** 0 **85.** $\frac{1}{2}$
87. False, for example, let $a = 3$ and $b = -3$
89. False, for example, let $a = -5$ and $b = 2$
91. Have each occupant add 1 to their room
  number and move to that room. This leaves
  room 1 unoccupied
93. a) $\frac{947}{2221} < \frac{773}{1811}$ **b)** $\sqrt{23} + \sqrt{29} > \sqrt{21} + \sqrt{31}$

Problem Set 1.4, pp. 29–31

1. -7 **3.** -4 **5.** 14 **7.** 13
9. -10 **11.** -20 **13.** -0.5 **15.** 1.37
17. $-\frac{14}{5}$ **19.** $\frac{13}{24}$ **21.** -1122
23. Down 2.25 points **25.** $1°$ **27.** -14
29. 5 **31.** 4 **33.** -4 **35.** -12
37. -14 **39.** $-\frac{1}{4}$ **41.** -6 **43.** 9
45. $-\frac{13}{28}$ **47.** $\frac{1}{3}$ **49.** -2.11 **51.** -8.9
53. -983 **55.** $\$-5.29$ **57.** $5°$
59. 13 **61.** 22 **63.** -28 **65.** $-\frac{31}{36}$
67. $\frac{12}{17}$ **69.** 3 **71.** -21 **73.** -3
75. 4 **77.** 15
79. If $a = -3$ and $b = 2$, then
  $|a + b| = |-3 + 2| = |-1| = 1$,
  but $|a| + |b| = |-3| + |2| = 3 + 2 = 5$
81. Down 15.4 points

Problem Set 1.5, pp. 34–36

1. -21 **3.** -20 **5.** 30 **7.** -17
9. 0 **11.** -66 **13.** $-\frac{1}{8}$ **15.** $\frac{1}{9}$

17. -27 **19.** 1   **21.** -618.87
23. $166   **25.** 3   **27.** 5   **29.** -4
31. -1 **33.** -5 **35.** 0   **37.** 23
39. -4.5 **41.** $-\frac{16}{9}$ **43.** $\frac{2}{9}$
45. Undefined   **47.** -2 yd per carry
49. $\frac{7}{9}$ **51.** -2 **53.** 2   **55.** $\frac{1}{6}$
57. $-\dfrac{x}{13}$ **59.** $\frac{1}{7}$ **61.** 7   **63.** $\frac{8}{3}$
65. $-\frac{5}{14}$ **67.** 1   **69.** $-\dfrac{q}{p}$ **71.** 22
73. 1   **75.** 36   **77.** -354 **79.** 2
81. 2   **83.** 6   **85.** $-\frac{23}{19}$ **87.** $-\frac{2}{5}$
89. $-\frac{32}{5}$ **91.** -3 **93.** $-\frac{23}{2}$
95. a) m and n have like signs
   b) m and n have unlike signs
97. a) -21.2 **b)** -23.6

Problem Set 1.6, pp. 41–42

1. Identity   **3.** Identity   **5.** Identity
7. Contradiction   **9.** Conditional
11. Identity   **13.** Conditional
15. Identity   **17.** Identity
19. Contradiction   **21.** Identity   **23.** 4
25. 8   **27.** 2   **29.** $\frac{8}{3}$ **31.** 5   **33.** 6
35. -3 **37.** -2 **39.** -4 **41.** 0
43. -12 **45.** -7 **47.** 48   **49.** 4
51. -6 **53.** $\frac{3}{4}$ **55.** -1 **57.** 0
59. 0   **61.** -15 **63.** 12   **65.** 1
67. $-\frac{11}{2}$ **69.** $x - 17 = 29, x = 46$
71. $\frac{2}{5}y = 7, y = \frac{35}{2}$ **73.** $6t + 13 = 61, t = 8$
75. $0.12r = 11.4, r = 95$
77. $110 + 0.30m = 305, m = 650$ mi
79. a) 3.613 **b)** -1.311

Chapter 1 Review, pp. 42–45

1. $4x - 9$ **2.** $5x + 10y$ **3.** 14   **4.** 45
5. $85.50   **6.** $10\frac{5}{6}$ ft   **7.** $12x + 8$
8. $3y + 8$ **9.** $b + c$ **10.** $x + 3$
11. $0.125 = 12.5\%$ **12.** $0.45\overline{45} = 45.\overline{45}\%$
13. $0 = 0\%$ **14.** Undefined   **15.** $\sqrt{25}$
16. $\sqrt{25}, 0, -4$
17. $\sqrt{25}, 2\frac{1}{3}, 0, -4, 0.\overline{37}, 1.2, -\frac{10}{3}$

18. $\sqrt{5}, 0.8282282228 \ldots, \pi$ **19.** $\frac{59}{10}$
20. $\frac{7}{1000}$ **21.** $4 + 1$ **22.** 6   **23.** 0
24. $(\frac{1}{3} \cdot 3)x$ **25.** yx **26.** 1   **27.** 9
28. $(-7 + 7) + x$ **29.** $n + 3$ **30.** $0.08r$
31.

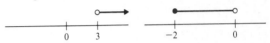

32. $-89 > -99$ **33.** $\frac{9}{16} < \frac{7}{12}$
34. $s \le 55$ mph   **35.** $1 \le t < 6$
36. Open half line   **37.** Half-open interval

38. 8   **39.** Undefined   **40.** 6
41. -12.8 **42.** $-\frac{2}{9}$ **43.** -4 **44.** -13
45. -8 **46.** -2.79 **47.** $\frac{1}{4}$ **48.** -13
49. 4   **50.** 1   **51.** -6 **52.** $-\frac{1}{10}$
53. 12.125   **54.** -4 **55.** -6.5 **56.** $\frac{8}{15}$
57. -19 **58.** $\frac{1}{5}$ **59.** $-\frac{7}{23}$ **60.** -24
61. -2 **62.** 4   **63.** $-\frac{1}{30}$ **64.** -9
65. Conditional   **66.** Contradiction
67. Identity   **68.** 81   **69.** 8   **70.** 0
71. $\frac{1}{2}$ **72.** -1 **73.** 9
74. $-x + 13 = -2, x = 15$ **75.** $\frac{3}{4}y = 12, y = 16$

Chapter 1 Test, pp. 45–46

1. a) $2x - 5$ **b)** $0.06y$ **2. a)** $\frac{17}{10}$ **b)** $\frac{3}{100}$
3. a) $10x + 15$ **b)** $4y + 10$
4. a) -8 **b)** $\frac{3}{8}$
5. a) $0.875 = 87.5\%$ **b)** $1.83\overline{3} = 183.\overline{3}\%$
6. a) $(x + y)$ **b)** $7(x + 4)$
7. a) $A > 10$ **b)** $8 \le C \le 14$
8. a) 4 **b)** $-7, 4$ **c)** $-7, 4, \frac{3}{4}, 0.\overline{5}$ **d)** $\sqrt{8}$
9. a) x **b)** $6 + 0$ **c)** $(\frac{2}{3} \cdot \frac{3}{2})y$ **d)** $(-4 + 4) + p$
10.   **11.**

12. -6 **13.** 6   **14.** 10   **15.** -2
16. -9 **17.** $\frac{1}{12}$ **18.** 4   **19.** $\frac{1}{3}$ **20.** 0
21. -2 **22.** -11 **23.** -6
24. $11\frac{11}{12}$ gal   **25.** $x = 64$

CHAPTER 2

Problem Set 2.1, pp. 51–52

1. Base = 8, exponent = 5
3. Base = -4, exponent = 2
5. Base = 4, exponent = 2
7. Base = $2x$, exponent = 3
9. Base = x, exponent = 3
11. Base = $-t$, exponent = 4
13. Base = t, exponent = 4
15. Base = $r + 2$, exponent = 7
17. Base = ab, exponent = $n + 1$

19. Base = $\dfrac{a}{b}$, exponent = $2n$

21. $4^2 = 16$ and $2^4 = 16$
23. $(-5)^2 = 25$ and $-5^2 = -25$
25. $(-10)^3 = -1000$ and $-10^3 = -1000$
27. $(-3)^4 = 81$ and $-3^4 = -81$
29. $(2 \cdot 4)^3 = 512$ and $2 \cdot 4^3 = 128$
31. $8(-\frac{3}{2})^4 = \frac{81}{2}$ and $-8(\frac{3}{2})^4 = -\frac{81}{2}$
33. $(2 + 4)^2 = 36$ and $2^2 + 4^2 = 20$
35. $(5 - 1)^2 = 16$ and $5^2 - 1^2 = 24$
37. $(2^3)^2 = 64$ and $2^{(3^2)} = 512$
39. $(x + 5)^3$
41. $19ab^4 - a^3b^2$
43. $(3r)^4$

45. $\left(\dfrac{m}{7}\right)^5$

47. a) \$22,000   **b)** \$24,200   **c)** \$26,620
d) \$29,282

49. a)

	Question 1
Choice 1	T
Choice 2	F

b)

	Question 1	Question 2
Choice 1	T	T
Choice 2	T	F
Choice 3	F	T
Choice 4	F	F

c)

	Question 1	Question 2	Question 3
Choice 1	T	T	T
Choice 2	T	T	F
Choice 3	T	F	T
Choice 4	T	F	F
Choice 5	F	T	T
Choice 6	F	T	F
Choice 7	F	F	T
Choice 8	F	F	F

$2^{10} = 1024$ ways, 2^n ways
51. a) 0.3   **b)** 3020.3   **c)** 2,377,764

Problem Set 2.2, pp. 57–58

1. x^7 **3.** 3^5 **5.** 5^{12} **7.** $2r^{11}$ **9.** t^{n+1}
11. $(y + 3)^{15}$ **13.** y^3 **15.** 6^2 **17.** $4t^4$
19. a^3b^4 **21.** r^{n+4} **23.** $3m - 1$
25. x^8 **27.** 2^{15} **29.** b^{30n} **31.** r^{2q+2}
33. $27p^3$ **35.** $49m^2n^2$ **37.** $16a^4$

39. $9^n t^n$ **41.** $\dfrac{9}{25}$ **43.** $\dfrac{-125}{z^3}$ **45.** $18x^3$

47. $\dfrac{9x^4}{y^6}$ **49.** $125p^{12}$ **51.** $-54s^6t^4$

53. $-27p^3q^{12}$ **55.** $-3a^4b^9$ **57.** $\dfrac{128m^9}{5b^{32}}$

59. $-5184x^{19}$ **61.** $\dfrac{16x^{34}}{25y^{10}}$ **63.** $30a^4b^4c^{15}$

65. $-512r^{44}s^{48}$ **67.** $-\dfrac{27n^{16}}{2m^6}$ **69.** $\dfrac{4w^2}{81}$

71. a) 1   **b)** 1   **c)** 6   **d)** 1   **e)** -1
73. 2^{360}
75. a) $3 \cdot 2^4 = 48$ cells   **b)** $3 \cdot 2^{24}$ cells
77. $(ab)^n = \underbrace{(ab)(ab)(ab) \cdots (ab)}_{n \text{ factors of } ab}$

$\qquad = \underbrace{a \cdot a \cdots a}_{n \text{ factors of } a} \cdot \underbrace{b \cdot b \cdots b}_{n \text{ factors of } b}$

$\qquad = a^n b^n$
79. a) 1   **b)** 1   **c)** Undefined   **d)** 1

Problem Set 2.3, pp. 61–62

1. Coefficient = 6, degree = 5
3. Coefficient = 1, degree = 2
5. Coefficient = 2, degree = 1
7. Coefficient = -1, degree = 4
9. Coefficient = 9, degree = 0
11. Coefficient = $-\frac{2}{3}$, degree = 1
13. $x^2 + 4x + 7$, degree = 2, trinomial
15. $x^2 - 3x + 9$, degree = 2, trinomial
17. $x + 2$, degree = 1, binomial
19. $5y + 3$, degree = 1, binomial
21. $r^2 - 9$, degree = 2, binomial
23. $-p^2 + 1$, degree = 2, binomial
25. $100t^{99}$, degree = 99, monomial
27. $6m^4 + m^2 - 1$, degree = 4, trinomial
29. $q^5 - q^2 - 7q + 4$, degree = 5
31. $5x$, degree = 1, monomial
33. 5, degree = 0, monomial
35. 8   37. -2 39. 10   41. 98
43. -5 45. -24 47. 3   49. 19   51. 4
53. $h = -16t^2 + 48t + 64$

    a) 96 ft   b) 96 ft   c) 64 ft

    d) 0 ft; maximum
        height = 100 ft

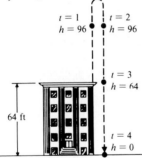

55. a) $h = -16t^2 + 48t$ b) $h = -16t^2 + 64$
57. 42.256

Problem Set 2.4, pp. 66–67

1. Like   3. Like   5. Unlike   7. Unlike
9. Like   11. Unlike   13. Unlike
15. Unlike   17. $10x$ 19. $13y^2$ 21. $5z$
23. $-3rs$ 25. $9q$ 27. $-3x^2y^2$
29. $15m^3 + 2m^2$ 31. $-c^2d + 7cd^2$
33. $3x - 6$ 35. $12x - 24$ 37. $-10y + 2$
39. $-r - 8$ 41. $t - 2$ 43. $3s^2 + 9s + 3$

45. $-y^2 + y - 5$ 47. $8x^2 + 7x + 11$
49. $8x^3 + 3x^2 - x - 2$
51. $15m^2 - 3mn + 6n^2$ 53. $6x^2 + 8x + 10$
55. $4x^3 + 5x^2 + x - 8$ 57. $5r^3 + 12s^3$
59. $24p^3 - 24p^2 - 21p + 75$ 61. $26b - 7c$
63. 73   65. $17u + 20$ 67. $-2t + 18$
69. $-m + n$ 71. $4x^3y^3$ 73. $20a + 7$
75. $98a^2 + 20b$ 77. $-48a^3$ 79. $3n + 3$
81. $65m + 360$ 83. $1.1p$ dollars
85. $66.014k^2 + 6.3755k - 111.55$

Problem Set 2.5, pp. 73–74

1. $12x$ 3. $12x^2$ 5. $5y^2$ 7. $28y^9$
9. $6r^6s^2$ 11. $-9r^5s^7$ 13. $40xy$
15. $x^2 + 6x$ 17. $8x^3 + 20x^2$
19. $5r^4 - 15r^3 + 5r$
21. $3r^4 + 6r^3s - 18r^2s^2$
23. $-m^3 + m^2n - mn^2$
25. $s^5t^2 - s^4t^3 - s^2t^4$ 27. $x^2 + 6x + 8$
29. $5x^3 - 17x^2 + 18x + 18$
31. $12r^4 - 32r^3 + 51r^2 - 41r + 7$
33. $b^3 + 3b^2 - 10b - 24$
35. $8p^3 + 36p^2 + 54p + 27$
37. $y^2 + 4y - 12$
39. $4x^3 - 11x^2 + 17x + 5$
41. $6m^4 - 29m^3 + 35m^2 + 4m - 10$
43. $2a^3 - 13a^2b - 28ab^2 - 6b^3$
45. $4m^2 + 5mn - 6n^2 + m + 13n - 5$
47. $x^2 + 7x + 12$ 49. $x^2 - 5x - 66$
51. $3r^2 + 5r - 2$ 53. $20s^2 - 23s + 6$
55. $12p^2 - 8pq - 15q^2$ 57. $t^4 - 13t^2 + 36$
59. $10a^4 - a^2b - 3b^2$ 61. $y^6 + 7y^3z - 8z^2$
63. $x^{2n} + 28x^n + 187$ 65. $x^2 + 8x + 16$
67. $9y^2 - 12y + 4$ 69. $p^2 - 49$
71. $25r^2 - 36s^2$ 73. $16m^2 + 24mn + 9n^2$
75. $t^{4k} - 1$ 77. $-m^4 + 6m^2$
79. $-5t^2 - 17t$ 81. $11p^2q - 7pq - 6p$
83. $n^3 + 6n^2 + 8n$ 85. 66 games
87. $n(n - 1)$ games
89. $24.64x^2 - 6.367xy - 44.84y^2$

Problem Set 2.6, pp. 79–80

1. $3 \cdot 7$ 3. $2^2 \cdot 3$ 5. $2 \cdot 3^2 \cdot 5$
7. $2^3 \cdot 3^2 \cdot 7$ 9. 2   11. 9   13. 4

15. 15   **17.** $3x^2$ **19.** 9   **21.** 6
23. 1   **25.** $5m^3n^3$ **27.** $6(x + 3)$
29. $13(a + b)$ **31.** $7(4m - 5n)$
33. Prime   **35.** $5(x^2 + 1)$
37. $4x^2(2x + 3)$ **39.** $a(a^4 + 1)$
41. $4a(4a - 3b)$ **43.** $-4p^2(p^3 + 3p - 4)$
45. $15rs(3r^2s^3 - 2rs^2 - 4)$
47. $11z^{51}(9z^4 + z^2 - 3)$
49. $-5r^2s^2t^2(2rt^2 + 5)$ **51.** $\pi abc(a + b - c)$
53. $(3m - 2n)(5x + y)$ **55.** $(x^2 + 9)^2$
57. $z^{n+1}(z^2 + 1)$ **59.** $(x + 3)(x - 3)$
61. $(2x + 3)(2x - 3)$
63. $(5m + 4n)(5m - 4n)$ **65.** Prime
67. $(6rs + 7)(6rs - 7)$ **69.** $(v - 1)(v + 5)$
71. $(a + b - c)(a - b + c)$
73. $(5x + 4)(-x + 6)$ **75.** $5(x + 2)(x - 2)$
77. $r(r + 5)(r - 5)$ **79.** $(t^2 + 4)(t + 2)(t - 2)$
81. $10(x^2 + y^2)(x + y)(x - y)$
83. $(x^2y^2 + 16)(xy + 4)(xy - 4)$
85. $77rs(1 + rs)(1 - rs)$
87. $(m^3 + 2)(m^3 - 2)$
89. $(1 + 9x^2)(1 + 3x)(1 - 3x)$
91. $25pq^2(p^2 + 1)(p + 1)(p - 1)$
93. $(85 + 15)(85 - 15) = (100)(70) = 7000$
95. $\pi(R + r)(R - r)$ **97.** $283t(t^4 - 7t^2 + 19)$

Problem Set 2.7, pp. 86–87

1. $(x + 3)(x + 5)$ **3.** $(x - 1)(x - 6)$
5. $(y + 3)(y - 1)$ **7.** $(y + 5)(y - 2)$
9. $(r - 6)(r + 3)$ **11.** $(r + 9)(r - 2)$
13. Prime   **15.** $(p - 5)(p - 7)$
17. $(t + 11)(t - 1)$ **19.** $-(t + 8)(t - 2)$
21. $(x + 3y)(x + y)$ **23.** $(x - 4y)(x + 3y)$
25. $(r + 7s)(r - 3s)$ **27.** $(p - 7q)(p - 2q)$
29. $(2x + 5)(x + 1)$ **31.** $(3x - 2)(x - 1)$
33. $(2y - 1)(y + 3)$ **35.** $(11y + 17)(y - 1)$
37. $(3m - 2)(m + 5)$ **39.** $(3x + 1)(2x - 3)$
41. $(5r + 2)(3r - 4)$ **43.** Prime
45. $(9s + 2)(s - 4)$ **47.** $(5x + 2)(x + 3)$
49. $-(2x - 3)(x + 4)$ **51.** $(2p - 3)(9p + 4)$
53. $(2x - y)(x - y)$ **55.** $(3x + 2y)(x + y)$
57. $(3m + 5n)(m - 2n)$
59. $(9p + 8q)(4p - 5q)$ **61.** $(x + 3)^2$
63. $(z - 1)^2$ **65.** $(2t + 1)^2$ **67.** $(2y - 5)^2$
69. $(r + 6)^2$ **71.** $(m + 4n)^2$
73. $(3r - 4s)^2$ **75.** $(2st + 3)^2$

77. $2(x + 5)(x - 2)$ **79.** $3(x^2 + x + 1)$
81. $10y(x + 13)(x + 1)$
83. $6mn^2(3m - 7n)(m + n)$
85. $(z^2 + 2)(z^2 + 3)$ **87.** $(2t^2 + 3)(3t^2 - 1)$
89. $w(w^2 + 1)(w + 2)(w - 2)$
91. $(4r^2 + 3s^2)(r + 3s)(r - 3s)$
93. **a)** $(3x - 5)^2 - 8(3x - 5) + 12$
$\qquad = 9x^2 - 54x + 77$
$\qquad = (3x - 7)(3x - 11)$
$\quad$ **b)** $u^2 - 8u + 12 = (u - 2)(u - 6)$
$\qquad = ((3x - 5) - 2)((3x - 5) - 6)$
$\qquad = (3x - 7)(3x - 11)$
95. $89x + 113$

Problem Set 2.8, p. 92

1. $(x - 3)(x^2 + 3x + 9)$
3. $(y + 2)(y^2 - 2y + 4)$
5. $(z + 1)(z^2 - z + 1)$
7. $(2 - r)(4 + 2r + r^2)$
9. $(5x + y)(25x^2 - 5xy + y^2)$
11. $(3s - 5t)(9s^2 + 15st + 25t^2)$
13. $(pq + 1)(p^2q^2 - pq + 1)$
15. $(7u + 6v)(49u^2 - 42uv + 36v^2)$
17. $(x^2 - 4y)(x^4 + 4x^2y + 16y^2)$
19. $(x^3 + 2)(x^6 - 2x^3 + 4)$
21. $((m - 4) - 2)((m - 4)^2 + 2(m - 4) + 4) =$
$\qquad (m - 6)(m^2 - 6m + 12)$
23. $((p + q) + (p - q))((p + q)^2 - (p + q)(p - q) +$
$\qquad (p - q)^2) = 2p(p^2 + 3q^2)$
25. $(x^2 + 3)(x + 7)$ **27.** $(a + 2)(b + 13)$
29. $(r + 3)(s - 1)$ **31.** $(5x + 2)(x + y)$
33. $(v - 1)(u - 1)$ **35.** $(s + 6t)(s + 8t^2)$
37. $(a + 3 + b)(a + 3 - b)$
39. $(x + 1 + 2y)(x + 1 - 2y)$
41. $(a + b + 4)(a - b - 4)$
43. $(2r - 5s + 3t)(2r - 5s - 3t)$
45. $(y + 1)(x + 3)(x - 3)$
47. $(y^2 - 5)(2y + 1)$
49. $(q^2 + 9)(p + 4)(p - 4)$
51. $(t + 5)(t - 2)(t^2 + 2t + 4)$
53. $(t + s)^2(t^2 - st + s^2)$
55. $2(x + 2)(x^2 - 2x + 4)$
57. $3(10x - 3)(100x^2 + 30x + 9)$
59. $5s(2r + 3)(4r^2 - 6r + 9)$
61. $3(4v^3 - 3)(v + 1)(v^2 - v + 1)$
63. $(3x^{4n} + 4y^{2m})(9x^{8n} - 12x^{4n}y^{2m} + 16y^{4m})$

65. $(2k - 1)(4k^2 + 2k + 1)(2k + 1)(4k^2 - 2k + 1)$
67. $(a + b)(a^2 - ab + b^2)$
$\quad = a^3 - a^2b + ab^2 + a^2b - ab^2 + b^3$
$\quad = a^3 + b^3$
69. $(13y - 17)(169y^2 + 221y + 289)$

Chapter 2 Review, pp. 93–94

1. Base $= 8$, exponent $= 4$
2. Base $= -6$, exponent $= 2$
3. Base $= 6$, exponent $= 2$
4. Base $= 7x$, exponent $= n$ **5.** 625
6. 54   **7.** 16   **8.** -16 **9.** $(x + 2)^3$
10. $5a^2b - ab^3$ **11.** \$5832   **12.** x^9
13. 5^{12} **14.** t^8 **15.** z^9 **16.** ab^4
17. 1   **18.** $32p^5$ **19.** 4^{18} **20.** a^{2n}
21. 7   **22.** $25a^2b^2$ **23.** $12y^4$
24. $-72a^{18}b^{17}c^7$ **25.** $400m^{20}$ **26.** $16x^4y$

27. $\dfrac{8s^2}{5r^6}$ **28.** Coefficient $= 5$, degree $= 4$

29. Coefficient $= 1$, degree $= 3$
30. Coefficient $= -1$, degree $= 1$
31. Coefficient $= 6$, degree $= 0$
32. $6x^2 + 4x + 3$, degree $= 2$, trinomial
33. $2x - 5$, degree $= 1$, binomial
34. $7p^3 + p^2 - p + 8$, degree $= 3$ **35.** 10
36. -15 **37.** $\frac{3}{2}$ **38.** $12x - 15$
39. $-t^2 + 4t - 2$ **40.** $6x^2 - 3x + 4$
41. $3y^3 - 9y^2 + y + 10$
42. $6p^3 - 12p^2 - 20p + 10$ **43.** $10t + 21$
44. a **45.** $12y^3 + 15y^2$
46. $a^2 + 8a + 12$
47. $8r^4 - 6r^3 - 17r^2 + 39r - 30$
48. $4x^2 + 12xy + 9y^2$
49. $6m^4 - 34m^3 + 40m^2 + 3m - 5$
50. $a^3 + 2a^2b - 5ab^2 + 12b^3$ **51.** $16t$
52. $9r^2(2r^3 - 4r + 3)$ **53.** $(3x - y)(5a + b)$
54. $(3t + 5)(3t - 5)$
55. $(2a + 11b)(2a - 11b)$
56. $((x - 1) + 2)((x - 1) - 2) = (x + 1)(x - 3)$
57. $(x^2 + 25)(x + 5)(x - 5)$
58. $7(p^2 + q^2)(p + q)(p - q)$
59. $xy(xy + 4)(xy - 4)$ **60.** $(z - 4)(z + 2)$
61. $(r + 7s)(r - 4s)$ **62.** $(p - 4)^2$
63. $(y - 7)(y - 1)$ **64.** $(3t - 1)(t + 5)$
65. $(3x - 2y)^2$ **66.** $(2m - 3)(m + 4)$
67. $5(x^2 - x + 2)$ **68.** $y^2(3x - 4)(2x - 3)$

69. $(t^2 + 9)(t + 2)(t - 2)$
70. $(p + 3)(p^2 - 3p + 9)$ **71.** $(r + 2)(s + 1)$
72. $(2a + 5b)(4a^2 - 10ab + 25b^2)$
73. $(m + 7)(m - 1)(m^2 + m + 1)$
74. $(x - 1 + y)(x - 1 - y)$
75. $(x + 3)(x + 2)(x - 2)$
76. $2(t - 4)(t^2 + 4t + 16)$
77. $3(z + 2)(z^2 - 2z + 4)(z - 1)(z^2 + z + 1)$

Chapter 2 Test, p. 95

1. a) 81 **b)** 50 **c)** 64 **d)** -64
2. a) 7^{14} **b)** $12x^3$ **c)** a^3b **d)** 1
3. a) Coefficient $= 2$, degree $= 1$
   b) Coefficient $= -1$, degree $= 3$
4. $6m^2 - m + 4$ **5.** $6p^3 - 6p^2 + 6p + 18$

6. $-54a^{10}b^{18}c^5$ **7.** $\dfrac{8s}{3r^2}$ **8.** -1

9. $15x^3 - 7x^2 - 8x + 16$ **10.** $3a$
11. $p^2 + 2p - 48$ **12.** $25x^2 + 10xy + y^2$
13. $3r^4 + 2r^3 - 23r^2 + 41r - 28$
14. $(3x + 7)(3x - 7)$ **15.** $(y + 2)(y - 6)$
16. $(r + 3s)^2$ **17.** $(2z + 5)(2z - 3)$
18. $(a - 3)(a^2 + 3a + 9)$
19. $6p^2(2p^3 + 3p - 4)$ **20.** $(r + 2)(s + 5)$
21. $(t^2 + 4)(t + 2)(t - 2)$
22. $y(6x + 5)(x - 2)$
23. $5(m + 2)(m^2 - 2m + 4)$
24. $(x - 4 + y)(x - 4 - y)$ **25.** \$4840

CHAPTER 3

Problem Set 3.1, pp. 101–102

1. 3   **3.** 3   **5.** 7   **7.** -3 **9.** 0
11. 1   **13.** -2 **15.** $-\frac{7}{2}$ **17.** -5
19. -3 **21.** 1   **23.** -3 **25.** 0
27. 0   **29.** -4 **31.** 8   **33.** -30
35. $\frac{1}{8}$ **37.** -21 **39.** 6   **41.** 15
43. $\frac{25}{9}$ **45.** -6 **47.** -14
49. Conditional   **51.** Identity
53. Contradiction   **55.** Contradiction
57. Identity   **59.** 7   **61.** 18   **63.** $\frac{5}{2}$
65. -5 **67.** $n = -\frac{7}{2}$ **69.** \$35   **71.** 97
73. 4.04

Problem Set 3.2, pp. 106–109

1. $l = 5$ **3.** $h = 4$ **5.** $r = \frac{1}{20}$
7. $a = 90$ **9.** $h = 3.5$ **11.** 62 mph

13. $l = \dfrac{P - 2w}{2}$ **15.** $F = \frac{9}{5}C + 32$

$l = 11\ m$ $F = 14°$

17. $y = 3x - 12$ **19.** $h = \dfrac{3A}{B}$

21. $c = 2s - a - b$ **23.** $y = \frac{1}{2}x - 2$

25. $r = \dfrac{P - V}{P}$ **27.** $P = \dfrac{V}{1 - r}$

29. $a = \dfrac{2(s - vt)}{t^2}$ **31.** $y = \dfrac{7x + 1}{2}$

33. $b^2 = c^2 - a^2$ **35.** $c^2 = \dfrac{E}{m}$

37. $n = \dfrac{A - a}{d} + 1$ **39.** $l = \dfrac{T}{\pi r} - h$

41. $x = \dfrac{3 + ab}{a + 2}$ **43.** 73.1 cm²

45. 7850 ft²   **47.** 47.1 ft³   **49.** 2750 ft
51. 30 mi   **53.** 4400 ft²
55. It doesn't matter.   **57.** 56°F
59. $5364.20

Problem Set 3.3, pp. 112–114

1. 17, 85   **3.** 116, 118   **5.** 23, 24, 25
7. 11, 12, 13   **9.** No solution
11. 75 yd by 120 yd   **13.** 26°, 57°, 97°
15. 5 m by 8 m   **17.** Karen is 12, Kathy is 7
19. Stacey is 12, her sister is 24
21. 14 nickels, 9 dimes, 7 quarters
23. 19 nickels, 21 dimes
25. 51 correct, 24 incorrect
27. a) 1995 **b)** 1983   **29.** 6.114

Problem Set 3.4, pp. 117–119

1. $4500 at 8%, $3000 at 12%   **3.** $3000
5. 140 adult, 360 student
7. 54 g of 20% alloy, 18 g of 28% alloy
9. 1 ℓ **11.** 2 qt

13. 33 sec for P-wave, 60 sec for S-wave, 165 mi
15. 4.5 hr   **17.** 20 min
19. 6 mph, 8 mph   **21.** 19.2 mi
23. 12,500   **25.** 250   **27.** 18 min
29. $19,500 at 8.31%, $17,950 at 13.78%

Problem Set 3.5, pp. 124–126

1. $x < 4$
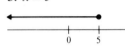

3. $x \le 5$
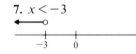

5. $x > 3$
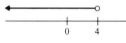

7. $x < -3$
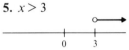

9. $x > -5$
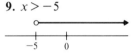

11. $x < 2$
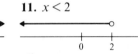

13. $y \ge -1$
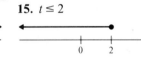

15. $t \le 2$
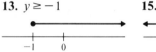

17. $m \ge 0$

19. $x < 4$

21. $y \ge 2$ **23.** $z \ge 0$ **25.** $m > 6$
27. $r < -2\frac{1}{3}$ **29.** $p > -5$ **31.** $q \le 0$
33. $x \le -3$ **35.** $y \le -2$ **37.** $t < -1$
39. $v \ge 2$ **41.** $u < 16$ **43.** $m \le 12$
45. $3 < x < 6$

47. $-2 \le r \le 3$

49. $-3 \le s < 0$
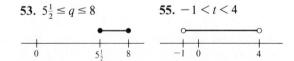

51. $-2\frac{1}{3} \le y < -1$

53. $5\frac{1}{2} \le q \le 8$

55. $-1 < t < 4$

57. Identity   **59.** Conditional

61. Contradiction　**63.** $-2 < x < 2$
65. $14° < F < 77°$　**67.** $48 \le x < 98$
69. a) More than 500　b) More than 625
　c) More than 800
71. $x < -6.375$

Problem Set 3.6, pp. 131–132

1. $4, -4$　**3.** $2, -2$　**5.** $13, -1$
7. $7, -2$　**9.** $3, -3$　**11.** $6, -\frac{15}{2}$
13. No solution　**15.** $6, -4$　**17.** $1, -\frac{1}{9}$
19. $\frac{3}{2}$　**21.** 0　**23.** $-4 < x < 4$
25. $-3 < x < 3$　**27.** $3 < y < 5$
29. $-5 \le z \le -1$　**31.** $-3 \le k \le 3$
33. $-4 < r < 3$　**35.** $-4 \le s \le \frac{7}{2}$
37. $x > 4$ or $x < -4$　**39.** $x > 2$ or $x < -2$
41. $r > 1$ or $r < -5$　**43.** $s \ge 2$ or $s \le -\frac{5}{3}$
45. $t \le -2$ or $t \ge \frac{9}{2}$　**47.** $t > 8$ or $t < -8$
49. All real numbers　**51.** $4, -\frac{2}{3}$
53. $-1 < y < 4$　**55.** $r > -3$ or $r < -4$
57. $3, 1$　**59.** $-\frac{1}{2}$　**61.** $q \le 5$ or $q \ge 11$
63. $\frac{7}{3}, -3$　**65.** $-\frac{1}{2} \le x \le 2$　**67.** 0
69. No solution　**71.** All real numbers
73. All real numbers except 2
75. No solution　**77.** 2
79. a) $|x| = 2$　b) $|x| < 2$　c) $|x| > 2$
81. $5, -13$　**83.** $-6 \le x \le 6$
85. $x > 1$ or $x < -6$　**87.** 5 hr $< t < 20$ hr
89. $2.09375, 2.03125$

Chapter 3 Review, pp. 133–134

1. 7　**2.** -2　**3.** 2　**4.** -2　**5.** 3
6. -3　**7.** 5　**8.** $\frac{32}{11}$　**9.** $-\frac{3}{2}$　**10.** 2
11. Contradiction　**12.** Identity
13. $n = 4$　**14.** \$440　**15.** $l = 27$
16. $r = 0.15$　**17.** $F = -13$　**18.** $y = 4x - 3$
19. $h = \dfrac{V}{\pi r^2}$　**20.** $d = \dfrac{A - a}{n - 1}$
21. $a = 2s - b - c$　**22.** 54 mph
23. 62.8 cm³　**24.** 1650 ft　**25.** 17, 68
26. 19, 20, 21　**27.** 17 ft by 44 ft
28. Adam is 12, his brother is 6
29. 8 dimes, 4 quarters
30. 211 student, 189 adult
31. 13 g of 22% alloy, 39 g of 30% alloy

32. $7\frac{1}{2}$ hr　**33.** 180
34. $x > 4$　**35.** $x \ge -2$

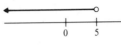

36. $t < 5$　**37.** $y \le 3$

38. $-4 \le x \le 3$　**39.** $0 < t \le 2$

40. $-2 < q \le \frac{5}{2}$　**41.** $-2 < t < 6$

42. $m > 2$　**43.** $p > 0$　**44.** $x \ge 2$
45. $y \le 5$　**46.** $m \le 0$　**47.** $t \ge \frac{2}{5}$
48. $-3 < x < 5$　**49.** $2, -2$
50. $1 < z < 4$　**51.** $t < -1$ or $t > \frac{5}{2}$
52. $6, -3$　**53.** $3, 1$　**54.** $m \le 3$ or $m \ge 5$
55. $-7 \le x \le 8$　**56.** 0
57. All real numbers　**58.** No solution
59. All real numbers　**60.** No solution
61. $-3, -11$　**62.** $-6 < y < 1$

Chapter 3 Test, p. 135

1. $w = 23$　**2.** $r = 0.25$　**3.** $r = \dfrac{I}{Pt}$
4. $y = 3x - 2$　**5.** Identity
6. Contradiction　**7.** $-\frac{3}{2}$　**8.** -2
9. 24　**10.** 2　**11.** $\frac{13}{4}$　**12.** $\frac{16}{7}$
13. $x > 4$　**14.** $x \le 9$

15. $-1 < y < 1$　**16.** $-5 < m \le 3$

17. $r \ge 1$　**18.** $y > 9$　**19.** $3, -3$
20. $-4, -2$　**21.** $y > -1$ or $y < -7$
22. $-1 \le r \le 4$　**23.** 19, 57

24. $9\frac{3}{4}$ ft by $34\frac{1}{4}$ ft
25. 25 g of 25% alloy, 75 g of 45% alloy

Cumulative Review for Chapters 1–3, pp. 136–138

1. 12 **2.** 28 **3.** 0 **4.** -32 **5.** ba
6. 1 **7.** $2a + 2b$ **8.** $(-3 + 3) + t$
9. $\frac{9}{13} > \frac{11}{16}$ **10.** $-63 > -73$
11. Half-open interval **12.** Open half line

13. $-\frac{1}{3}$ **14.** -6.1 **15.** -8.1 **16.** $\frac{3}{7}$
17. -8 **18.** $-\frac{1}{2}$ **19.** $\frac{9}{16}$ **20.** -5
21. 49 **22.** 4 **23.** $10\frac{1}{12}$ ft **24.** 24
25. -16 **26.** x^9 **27.** 1 **28.** a^5

29. $10m^5$ **30.** t^{16} **31.** $8x^3y^3$ **32.** $\frac{8}{b^3}$

33. $3x^3y^5$ **34.** -2 **35.** $6x^2 + 2x - 7$
36. $t^3 + t^2 - 9t + 4$ **37.** $x^2 + 9x + 20$
38. $9a^2 - 12ab + 4b^2$ **39.** $6a^3 - 7a^2b + b^3$
40. $6r^4 + 7r^3 + 9r^2 + 10r - 12$
41. $6x(2x^3 - 3x - 5)$ **42.** $(4t + 5)(4t - 5)$
43. $(m - 5n)(m - n)$ **44.** $(2a - 3b)^2$
45. $(3r + 2s)(9r^2 - 6rs + 4s^2)$
46. $(a + 2 + b)(a + 2 - b)$ **47.** 4 **48.** 3
49. 4 **50.** 11 **51.** $-\frac{1}{2}$ **52.** -4

53. $y = \frac{3}{4}x - 3$ **54.** $w = \dfrac{A}{2\pi p}$

55. $V = 1568$ **56.** \$670 **57.** 21, 42
58. 7 nickels, 21 dimes, 42 quarters
59. 15 g of 18% alloy, 33 g of 26% alloy
60. $13\frac{1}{3}$ mi
61. $x \le 5$ **62.** $y < 2$

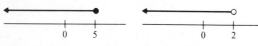

63. $y > -1$ **64.** $p \ge 2$

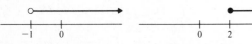

65. $m \ge -7$ **66.** $t > -\frac{1}{3}$

67. $-1 < x < 1$ **68.** $-\frac{4}{3} < q \le 3$

69. $t < -3$ or $t > 0$ **70.** $-\frac{1}{3}$, 3
71. No solution **72.** All real numbers
73. $-6 < y < 3$

CHAPTER 4

Problem Set 4.1, pp. 145–146

1. $x = 0$ **3.** $y = 1$ **5.** $p = -4$
7. Defined for all values **9.** $\frac{5}{9}$

11. Cannot be simplified **13.** $\dfrac{4b^2}{7}$

15. $\dfrac{-3}{c^2}$ **17.** $\dfrac{1}{3x^3y}$ **19.** $\dfrac{1}{6}$ **21.** $\dfrac{p}{q}$

23. $\dfrac{1}{2r - s}$ **25.** $\dfrac{5}{4t - 3}$ **27.** $\dfrac{1}{(w + 8)^4}$

29. $\dfrac{x + y}{x - y}$ **31.** $\dfrac{3(t - 3)}{2t}$ **33.** $\dfrac{3p + 8q}{p + 2q}$

35. 1 **37.** -1 **39.** $\dfrac{1}{m^2 - 2m + 4}$

41. $\dfrac{b + c}{b - c}$ **43.** $\dfrac{c^2 + 3cd + 9d^2}{4 - d}$ **45.** $\dfrac{1}{r + 5}$

47. $\dfrac{4q}{q^2 + 2}$ **49.** 20 **51.** $7x^7$

53. $r^2 + 9r$ **55.** $5m^2$ **57.** $6s^2 + 6s$
59. $3q^2 + 18q + 24$ **61.** $p^2 + 9p + 14$
63. $60x^2$ **65.** $2t^2 - 5t - 3$
67. $w^3 - w^2 + w$ **69.** $108°$
71. a) 668 vibrations/sec **b)** 605 vibrations/sec

Problem Set 4.2, pp. 149–151

1. $\dfrac{x}{5}$ **3.** $\dfrac{2y}{35x^3}$ **5.** $\dfrac{2r^2}{15}$ **7.** $\dfrac{a^2}{b}$

9. $-\dfrac{1}{15r}$ **11.** $\dfrac{2b^6}{5a^2c^4}$ **13.** $\dfrac{30q}{7}$

15. $\dfrac{9(p - 3)}{p^2}$ **17.** $\dfrac{3m^2}{10n}$ **19.** $\dfrac{y^2}{xy + 1}$

21. $\dfrac{5r(r+3)}{9(r-5)}$ **23.** $\dfrac{1}{8s^2}$ **25.** $\dfrac{m-4}{m-2}$

27. $\dfrac{1}{(x+3)(x^2+3x+9)}$ **29.** $\dfrac{c+d}{3c-d}$

31. $\dfrac{3(q-2)^2}{5(q+2)^2}$ **33.** $\dfrac{3w+4}{4w+3}$ **35.** $\dfrac{u(u-v)}{5}$

37. $s+t$ **39.** $\dfrac{1}{t^2-st+s^2}$ **41.** $\dfrac{b-a}{x(x+y)}$

43. $\dfrac{4u^2-6u+9}{u-3v}$ **45.** $\dfrac{z^2}{z-1}$

47. $R=\dfrac{(675-0.5x)x}{x+4}$ **49.** $\dfrac{99}{59r-73s}$

Problem Set 4.3, pp. 157–158

1. $\dfrac{2}{3r^2}$ **3.** $\dfrac{k^2-k+1}{k^3}$ **5.** 1 **7.** -1

9. $\dfrac{3}{x-3}$ **11.** $m-3n$ **13.** $\dfrac{x^2+9}{(x+3)(x-3)}$

15. $\dfrac{s^2}{s-6}$ **17.** $\dfrac{1}{c-d}$

19. a) $15a$ **b)** $\dfrac{8}{15a}$ **21. a)** b^2 **b)** $\dfrac{2b-1}{b^2}$

23. a) $9c^2$ **b)** $\dfrac{4c+21}{9c^2}$

25. a) $360x^3y^2$ **b)** $\dfrac{20x^2+27y}{360x^3y^2}$

27. a) k **b)** $\dfrac{k^2+1}{k}$ **29. a)** $p+3$ **b)** $\dfrac{3p}{p+3}$

31. a) $q(q+2)$ **b)** $\dfrac{4}{q}$

33. a) $(h-10)(h-5)$ **b)** $\dfrac{5h}{(h-10)(h-5)}$

35. a) $(r-4)(r+1)$ **b)** $\dfrac{4(2r-3)}{(r-4)(r+1)}$

37. a) $5(s+10)$ **b)** $-\dfrac{2}{5}$

39. a) $(x+5)(x-5)$ **b)** $\dfrac{2(x+10)}{(x+5)(x-5)}$

41. a) $2(y+9)(y-9)$ **b)** $\dfrac{-1}{2(y-9)}$

43. a) $m(m-2)^2$ **b)** $\dfrac{-12}{m(m-2)^2}$

45. a) $a(a-b)(a+b)$ **b)** $\dfrac{a^3+ab-b^2}{a(a-b)(a+b)}$

47. a) $(k+4)(k-3)(k+2)$

b) $\dfrac{2(k+3)}{(k+4)(k-3)(k+2)}$

49. a) $2(x+3)(x-3)$ **b)** $\dfrac{x+4}{2(x+3)}$

51. a) $(t-5)(t+3)$ **b)** $\dfrac{t-2}{t-5}$

53. a) $(a+1)^2$ **b)** $\dfrac{a^2}{(a+1)^2}$

55. a) $r-1$ **b)** $\dfrac{r^2-r-4}{1-r}$

57. a) $(s-8)^2(s+8)$ **b)** $\dfrac{2s^2}{(s-8)^2(s+8)}$

59. a) $(x+y)(x^2-xy+y^2)$

b) $\dfrac{x^2+y^2}{(x+y)(x^2-xy+y^2)}$

61. a) $(2c-d)(2c+d)$ **b)** $\dfrac{2(4c+3d)}{2c+d}$

63. a) $m(m+1)^2(m+4)$ **b)** $\dfrac{-3m^2+2m-1}{m(m+1)^2(m+4)}$

65. $\dfrac{2n+1}{n(n+1)}$ **67.** $\dfrac{Mg(2a-R)}{Ra}$

69. $\dfrac{5,382,941}{2,837,407x^5}$

Problem Set 4.4, pp. 162–163

1. a) $\dfrac{8}{3}$ **b)** $\dfrac{1}{6}$ **3.** $\dfrac{19}{14}$ **5.** $\dfrac{45}{4}$ **7.** $\dfrac{3(x+1)}{x-1}$

9. $\dfrac{p-2}{p}$ **11.** $\dfrac{s-r}{s+r}$ **13.** $\dfrac{x}{y}$

15. $\dfrac{4m-1}{m}$ **17.** $-7(a+b)$ **19.** $\dfrac{c-3}{18}$

21. $\dfrac{2p-1}{p-5}$ **23.** $\dfrac{d+8}{d+1}$ **25.** xy

27. $\dfrac{1}{t(t^2 - t + 1)}$ **29.** $\dfrac{m^2 n^2}{m^2 - 3mn + 9n^2}$

31. $\dfrac{uv(u + v)}{u^2 + uv + v^2}$ **33.** $\dfrac{q}{q - 2}$ **35.** $\dfrac{k + 1}{k + 3}$

37. $\dfrac{2w}{w^2 + 1}$ **39.** $\dfrac{x^2 - x + 1}{x}$ **41.** 2

43. $4\frac{4}{5}$ mph   **45.** 3.03

Problem Set 4.5, pp. 168–170

1. 20   **3.** -6 **5.** $-\frac{3}{2}$ **7.** 6   **9.** 1
11. $\frac{3}{5}$ **13.** -7 **15.** 6   **17.** $\frac{11}{2}$ **19.** 0
21. 3   **23.** -18 **25.** -8 **27.** 10
29. -24 **31.** No solution   **33.** -2
35. No solution   **37.** $\frac{1}{2}$ **39.** 0   **41.** 2

43. -3 **45.** $P = \dfrac{pvT}{Vt}$ **47.** $r = \dfrac{2S - a}{S}$

49. $R_2 = \dfrac{RR_1}{R_1 - R}$ **51.** $y = \dfrac{1}{x + 2}$

53. $r = \dfrac{d}{1 - dt}$ **55.** $x = \dfrac{3 + 3y}{y - 1}$

57. $x = -\frac{1}{3}$ **59.** -8360

Problem Set 4.6, pp. 174–175

1. 4 cm   **3.** 7.5 ohm   **5.** 2   **7.** 6, 18
9. 5, 7   **11.** $1\frac{1}{5}$ hr   **13.** 9 min
15. 2 mph   **17.** Freight 50 mph, express 80 mph
19. 7.17 cm

Problem Set 4.7, pp. 180–181

1. $4x^2 + 2x - 3$ **3.** $3x - 5 + \dfrac{2}{x}$

5. $y^2 + y - 1$ **7.** $a^2 - 2a + \dfrac{5}{3} - \dfrac{3}{a^2}$

9. $3q + \dfrac{2}{p} - \dfrac{1}{3pq}$

11. $3x^2 y^2 z^4 + x^2 yz^3 + 2xy^2 z^2$

13. $8r^3 s^3 + \dfrac{5s}{4} - 1$ **15.** 4   **17.** $3m - 2$

19. $2r + 5$ **21.** $x^2 + 2x + 3$

23. $5y^2 + y + 3$ **25.** $x^2 - x + 5 - \dfrac{30}{4x + 3}$

27. $3p^2 - p + 4 + \dfrac{6}{3p + 1}$ **29.** $t^2 + 3$

31. $2r + 7 + \dfrac{10r}{r^2 - r - 3}$ **33.** $3s^2 + s - 3$

35. $4w^2 + 5w + 3 + \dfrac{9w - 8}{5w^2 - 3w}$

37. $4m - 13$ **39.** $x^3 - 3x^2 + 9x - 27$

41. $AC = 7.5 + \dfrac{1500}{x}$ **43.** $k = -6$

45. $7.45p^4 - 1.3p^2 + \dfrac{0.55}{p}$

Problem Set 4.8, pp. 184–185

1. $3x + 10 + \dfrac{21}{x - 2}$ **3.** $3x - 2 + \dfrac{5}{x + 2}$

5. $x + 5$ **7.** $2x^2 + 5x + 8 + \dfrac{13}{x - 3}$

9. $2x^2 + 3x + 6 + \dfrac{14}{x - 4}$

11. $x^2 + 2x - 7 + \dfrac{25}{x + 3}$

13. $y^2 + 2y + 2 - \dfrac{2}{y - 1}$

15. $4m^2 - 2m + 2 - \dfrac{9}{m + 1}$

17. $5x^2 + 4 + \dfrac{3}{x + 2}$

19. $r^2 + 4r + 10 + \dfrac{50}{r - 5}$

21. $p^3 + p^2 + 5p + 5 + \dfrac{7}{p - 3}$

23. $2t^3 - 5t^2 + 6t + 1$

25. $x^3 - x^2 + 2x - 8 - \dfrac{10}{x + 4}$

27. $2y^3 - 2y^2 - 3y + 4 - \dfrac{3}{y + 1}$

29. $x^3 - x^2 + x - 10 + \dfrac{53}{x+5}$

31. $m^4 - 4m^3 + 11m^2 - 27m + 54 - \dfrac{112}{m+2}$

33. $x^2 + x + 1$
35. $x^4 - 2x^3 + 4x^2 - 8x + 16$
37. $4y^2 + 2y + 6$

39. $3x^2 + 106x + 1206 + \dfrac{121}{x-29}$

Chapter 4 Review, pp. 185–187

1. $x = 0$ **2.** $y = 2$ **3.** $\frac{2}{3}$

4. Cannot be simplified

5. 6   **6.** $\dfrac{5}{3t-4}$

7. $\dfrac{-5}{4y^2}$ **8.** $\dfrac{b-a}{b+a}$ **9.** $m^2 - 2mn + 4n^2$

10. $\dfrac{6p-5q}{3p-q}$ **11.** $6xy^2$ **12.** $t^2 + 2t$

13. $3s^2 - 3s$ **14.** $p^2 - 2p - 15$ **15.** $\dfrac{x^3}{3}$

16. $\dfrac{a^4}{b^5}$ **17.** $\dfrac{25q}{3}$ **18.** $\dfrac{p^4}{2(p-3)}$

19. $\dfrac{x-4}{x-2}$ **20.** $c(c-d)$

21. $(x-2)(x^2-2x+4)$ **22.** 1   **23.** $\dfrac{1}{4x^2}$

24. a) $8a^2$ **b)** $\dfrac{5a+12}{8a^2}$

25. a) $80x^2y^3$ **b)** $\dfrac{15y^2-4x}{80x^2y^3}$

26. a) $p+1$ **b)** $\dfrac{-p}{p+1}$

27. a) $(h-8)(h-2)$ **b)** $\dfrac{6h}{(h-8)(h-2)}$

28. a) $2(y+5)(y-5)$ **b)** $\dfrac{y^2+5y-4}{2(y+5)(y-5)}$

29. a) $a(a+b)(a-b)$ **b)** $\dfrac{a^2+ab+b^2}{a(a+b)(a-b)}$

30. a) $x(x+y)(x-y)$ **b)** $\dfrac{4y}{(x+y)(x-y)}$

31. a) $(k+3)(k-2)(k+4)$
b) $\dfrac{k^2+5k+3}{(k+3)(k-2)(k+4)}$

32. a) $r+1$ **b)** $\dfrac{r^2}{r+1}$

33. a) $(y+2)(y-2)$ **b)** $\dfrac{y+6}{y-2}$ **34.** $\dfrac{y+x}{y-x}$

35. $\dfrac{b+1}{a+1}$ **36.** $\dfrac{3m-1}{m}$ **37.** $\dfrac{4p+1}{2p+1}$

38. $\dfrac{1}{x(x^2+3x+9)}$ **39.** $\dfrac{m^2+mn+n^2}{mn(m+n)}$

40. $\frac{1}{2}$ **41.** -1 **42.** 6   **43.** $-\frac{2}{3}$

44. 0   **45.** No solution   **46.** -1

47. -2 **48.** $v = \dfrac{PVt}{pT}$ **49.** $r = \dfrac{S-a}{S-l}$

50. 7   **51.** $3\frac{3}{13}$ hr
52. Car 50 mph, plane 380 mph
53. $x^2 - 3x + 4$ **54.** $2a^2 + 3b - 9ab^2c$
55. $x - 5$ **56.** $3y^2 - 4y + 2$

57. $3x + 2 + \dfrac{9}{2x-3}$ **58.** $2y^2 - y - 1 + \dfrac{2}{2y+1}$

59. $3x + 2 + \dfrac{1}{2x^2-3x+2}$

60. $x^4 + x^3 + x^2 + x + 1$ **61.** $a + 2$
62. $t^2 - 4t + 3$

63. $2x^3 + 4x^2 + 8x + 29 + \dfrac{55}{x-2}$

64. $4y^2 - 6y + 5 - \dfrac{4}{y+4}$

65. $x^3 - 2x^2 + 4x - 8$

66. $x^2 + 5x + 7 - \dfrac{3}{x-3}$

Chapter 4 Test, pp. 187–188

1. $3x$ **2.** $x^2 + x$ **3.** 5   **4.** $\dfrac{t+2}{t+3}$

5. x **6.** $\dfrac{p-4}{p-2}$ **7.** 2   **8.** $\dfrac{b^2}{7a^3}$

9. $\dfrac{5a+4}{6a^2}$ **10.** $\dfrac{1}{x(x-2)}$

11. $\dfrac{x^2+5}{(x-1)(x+2)}$ **12.** $\dfrac{3x+2}{x-1}$

13. $\dfrac{y-6}{y-3}$ **14.** 2 **15.** -2 **16.** -1

17. 0 **18.** $T=\dfrac{PVt}{pv}$ **19.** $t=\dfrac{d-r}{dr}$

20. $3t^2-4t+2$

21. $x^2-3x+3+\dfrac{2x+5}{x^2-2}$ **22.** y^2+y-6

23. $x^2+x+2+\dfrac{8}{x-3}$ **24.** 5 **25.** $3\frac{3}{4}$ hr

CHAPTER 5

Problem Set 5.1, pp. 194–195

1. $\frac{1}{8}$ **3.** $\frac{1}{16}$ **5.** $-\frac{1}{8}$ **7.** $\frac{1}{16}$ **9.** $-\frac{1}{7}$

11. $\frac{1}{2}$ **13.** $\frac{7}{12}$ **15.** $\dfrac{6}{x^2}$ and $\dfrac{1}{36x^2}$

17. 125 and 250 **19.** $\dfrac{3r^2}{4}$ and $48r^2$

21. $\frac{9}{4}$ and -64 **23.** 2 and $\frac{1}{2}$ **25.** x^7

27. $\dfrac{1}{s^7}$ **29.** 1000 **31.** $\dfrac{1}{m^7}$ **33.** m^6

35. y^8 **37.** $\dfrac{1}{10^{10}}$ **39.** $\dfrac{1}{p^{15}}$ **41.** $\dfrac{p^4}{q^8}$

43. $\dfrac{q^{60}}{p^{30}}$ **45.** $\dfrac{4u^6}{v^4}$ **47.** $\dfrac{b^{12}}{64a^6}$ **49.** $\dfrac{y^7}{x^6}$

51. $\dfrac{1}{9x^2y^4}$ **53.** $\dfrac{2b}{5a}$ **55.** $\dfrac{ay^3}{bx^2}$ **57.** $\dfrac{9}{a^3}$

59. $\dfrac{16x^6z^4}{25y^4}$ **61.** $-\dfrac{32y^{22}}{x^{17}}$ **63.** $\dfrac{5b^9c^{10}}{8a^{23}}$

65. $\dfrac{1}{(a+b)^2}$ **67.** $\dfrac{c^2+1}{c}$ **69.** $\dfrac{d^2-1}{d^2+1}$

71. $\dfrac{1}{t^2+t}$ **73.** $-\dfrac{1}{rs}$ **75.** $\dfrac{xy}{y-x}$

77. a) 1.51 **b)** 31.01 **c)** 1.82

Problem Set 5.2, pp. 198–200

1. 60,000 **3.** 55,000,000 **5.** 0.004
7. 0.000019 **9.** $-817,000,000$
11. -0.000795 **13.** 9.007 **15.** 63.32
17. 0.101 **19.** 6×10^6 **21.** 1.4×10^5
23. 5×10^{-3} **25.** 7.2×10^{-4}
27. 8.09×10^{11} **29.** 3.084×10^{-11}
31. 1×10^7 **33.** 1×10^{-5}
35. 5.59×10^0 **37.** -2.751×10^2
39. 4.307×10 **41.** 1.19×10^{-1}
43. 3,000,000,000,000,000,000,000,000,000,000
45. 0.000 000 000 000 000 000 000 0016
47. 6.3×10^7 **49.** 8×10 **51.** 4×10
53. 1.6×10^2 **55.** 3×10^3 **57.** 6×10^{-17}
59. 2.73×10^{21} **61.** 1.2×10^3
63. 3.675×10 **65.** 6.5×10^{-11}
67. 1.62×10^{10} **69.** 6.6×10^{21} tons
71. 500 sec or $8\frac{1}{3}$ min
73. a) 3.12×10^{11} **b)** 1.70×10^{-22}

Problem Set 5.3, pp. 203–204

1. 6 **3.** -6 **5.** ±6
7. Not a real number **9.** 5 **11.** -5
13. -5 **15.** 5 **17.** 2 **19.** -2
21. Not a real number **23.** 3 **25.** -3
27. $\frac{2}{5}$ **29.** $\pm\frac{1}{9}$ **31.** 1
33. Not a real number **35.** 0 **37.** 1.1
39. 5 **41.** 4 **43.** 4 **45.** x **47.** x
49. x^3 **51.** x^2 **53.** x^3 **55.** $7x$
57. $4x$ **59.** $3y^2$ **61.** $2p^5$ **63.** $13b^2c^8$
65. $rs^{10}t^{20}$ **67.** $-2a^3b^6$ **69.** $6a^{11}b^{22}$
71. $\frac{3}{4}p^2q^2$ **73.** $(s+t)^9$ **75.** $(v+w)^2$
77. $r+2$ **79.** $p+q$ **81.** $4r^{3n}s^{2m}$
83. 40 ft/sec
85. a) 199 **b)** -17 **c)** 23 **d)** 1.26

Problem Set 5.4, pp. 208–209

1. 4 **3.** -4 **5.** Not a real number
7. 2 **9.** -2 **11.** 3 **13.** 1 **15.** 0
17. $\frac{7}{11}$ **19.** $-\frac{1}{5}$ **21.** $\sqrt{3x}$ **23.** $3\sqrt{x}$
25. $\sqrt[3]{x+6}$ **27.** $\sqrt{x^2+4}$ **29.** 9
31. $\frac{8}{125}$ **33.** 4 **35.** -32 **37.** $\sqrt[3]{y^2}$
39. $10\sqrt[4]{y^3}$ **41.** $\sqrt[3]{4r^2s^2}$ **43.** $\sqrt[5]{(r+3)^2}$

45. $\frac{1}{12}$ **47.** $\frac{1}{8}$ **49.** $\frac{1}{125}$ **51.** $\frac{1}{4}$
53. $-\frac{1}{3125}$ **55.** $\frac{3}{2}$ **57.** x **59.** $x^{3/2}$

61. $y^{1/3}$ **63.** y^2 **65.** $p^{1/2}$ **67.** $\frac{1}{p}$

69. $\frac{1}{r^9}$ **71.** $\frac{r^{1/4}}{s^{1/2}}$ **73.** $64r^9s^{10}$

75. $4p^4q^6$ **77.** $\frac{m^2}{n^3}$ **79.** $a^2b^3c^6$

81. $-\frac{3a^2}{b^5c}$ **83.** $\frac{p^3q^4}{2}$ **85.** $\frac{x^{29/2}}{y^9}$

87. $w + w^{3/2}$ **89.** $y^3 - y^{7/2}$ **91.** $z - 1$
93. 480 units   **95.** $x^{1/2}(x^{1/2} + 1)$
97. a) 3162.28   **b)** 2.03   **c)** 0.68

Problem Set 5.5, pp. 213–214

1. $2\sqrt{3}$ **3.** $5\sqrt{2}$ **5.** $10\sqrt{5}$
7. Cannot be simplified   **9.** $2\sqrt[3]{9}$
11. $2\sqrt[4]{3}$ **13.** $3x\sqrt{5}$ **15.** $24y\sqrt{y}$

17. $4p^3q\sqrt[3]{q}$ **19.** $2m^2\sqrt[4]{5m^3}$ **21.** $\frac{2\sqrt{5}}{9}$

23. $-\frac{10\sqrt{2}}{13}$ **25.** $\frac{7\sqrt{x}}{12}$ **27.** $\frac{3r\sqrt{3r}}{8}$

29. $\frac{3t\sqrt[3]{t^2}}{10}$ **31.** $\frac{\sqrt[5]{x}}{y^2}$ **33.** $\sqrt[4]{3}$ **35.** $\sqrt[6]{x}$

37. $\sqrt[8]{6}$ **39.** $\sqrt{z}$ **41.** $\sqrt{5}$ **43.** $\sqrt[3]{6xy^2}$

45. $3\sqrt{11}$ **47.** $a^2b^2\sqrt{b}$ **49.** $\frac{\sqrt{2}}{5}$

51. $-2\sqrt{15}$ **53.** $\sqrt[12]{x}$ **55.** $3\sqrt{3}$

57. $3\sqrt[3]{2}$ **59.** $\frac{y\sqrt{42}}{5}$ **61.** $5\sqrt[4]{2}$

63. $10\sqrt{7}$ **65.** $c^3d^4\sqrt{cd}$ **67.** $\sqrt{3x}$

69. $\frac{xy^2\sqrt{y}}{2z^2}$ **71.** $\frac{\sqrt{p}}{q}$ **73.** $11rs^3\sqrt{2r}$

75. $2t^2\sqrt[3]{2}$ **77.** $-3k^4\sqrt[3]{k^2}$ **79.** $5w\sqrt[4]{w}$

81. $\frac{2\sqrt[3]{2}}{5}$ **83.** $-\frac{m^5\sqrt[4]{m}}{4}$ **85.** $\sqrt{2}$

87. $5\sqrt[4]{k}$ **89.** $2a^3b^6c^4$ **91.** $a = 16, b = 9$
93. $10\sqrt{21} \approx 46$ mi
95. a) $\sqrt{2401 \cdot 4096} = \sqrt{2401} \cdot \sqrt{4096}$

b) $\sqrt{\dfrac{2401}{4096}} = \dfrac{\sqrt{2401}}{\sqrt{4096}}$

Problem Set 5.6, pp. 217–218

1. $\sqrt{15}$ **3.** $\sqrt{10x}$ **5.** 2   **7.** 6   **9.** $5\sqrt{5}$
11. y^2 **13.** $y^2\sqrt{y}$ **15.** $\sqrt{14m}$
17. $3x\sqrt{2}$ **19.** $5y\sqrt{2}$ **21.** $3\sqrt{11z}$
23. $8k\sqrt{k}$ **25.** $x + 1$ **27.** $rs^3\sqrt{s}$
29. $9\sqrt{15t}$ **31.** $6a^2\sqrt{a}$ **33.** $2p$
35. $3\sqrt[3]{4q}$ **37.** $rs^2\sqrt[3]{18rs}$ **39.** $2\sqrt[4]{ab}$
41. 3×10^4 **43.** 5   **45.** 7   **47.** $\sqrt{6}$

49. $11\sqrt{a}$ **51.** $\frac{1}{3y}$ **53.** $2m^2\sqrt{3}$

55. $12\sqrt{c}$ **57.** $ab^2c^3\sqrt{5}$ **59.** $\frac{9r}{s^2}$ **61.** $3t$

63. $\sqrt{7pq}$ **65.** $\frac{2r^3}{5s}$ **67.** 2×10^4

69. $\sqrt[6]{243}$ **71.** $\sqrt[4]{x^3}$ **73.** $\sqrt[4]{6}$ **75.** $\sqrt[3]{k}$
77. $\sqrt[4]{a^2b}$ **79.** $4^{1/2} = 2, 3^{1/2} = \sqrt{3}$
81. $\sqrt[3]{3375} \sqrt[3]{2744} = \sqrt[3]{3375 \cdot 2744}$

Problem Set 5.7, pp. 220–221

1. $8\sqrt{7}$ **3.** $13\sqrt{6}$
5. Cannot be simplified
7. Cannot be simplified   **9.** $5\sqrt{x}$
11. $4\sqrt{2y}$ **13.** Cannot be simplified
15. $-3\sqrt{11}$ **17.** $6\sqrt{m} + 5m$ **19.** $9a\sqrt{5}$
21. $3\sqrt{3}$ **23.** $5\sqrt{5}$ **25.** $9\sqrt{2}$
27. $-4\sqrt{7}$ **29.** $\sqrt{2}$ **31.** $29\sqrt{10}$
33. $3\sqrt{6x}$ **35.** $-\sqrt{3a}$ **37.** $7r\sqrt{11r}$
39. $4\sqrt[3]{3}$ **41.** $2\sqrt[3]{y}$ **43.** $5\sqrt[3]{3p}$
45. $-7\sqrt[3]{2}$ **47.** $5\sqrt[4]{t}$ **49.** $5\sqrt{3} + 20$
51. $12\sqrt{10} - 2\sqrt{5}$ **53.** $\sqrt{6} + \sqrt{15}$
55. $15\sqrt{2} + 5$ **57.** $x - x\sqrt{y}$
59. $30\sqrt{2} + 24$ **61.** $2 - \sqrt[3]{6}$
63. $\sqrt{6} + 5\sqrt{3} + \sqrt{2} + 5$ **65.** 3
67. $7 - \sqrt{14} - \sqrt{21} + \sqrt{6}$
69. $\sqrt{22} + \sqrt{33} + \sqrt{10} + \sqrt{15}$
71. $3 + \sqrt{6} + 4\sqrt{15} + 4\sqrt{10}$ **73.** $t^2 - 5$
75. $s - 4$ **77.** $24\sqrt{5} + 4\sqrt{10} - 27\sqrt{2} - 9$
79. $9 + 4\sqrt{5}$ **81.** $10 + 2\sqrt{21}$
83. $x + 2\sqrt{xy} + y$ **85.** $x - 2\sqrt{x + 4} + 5$

87. $3x + 1 + 2\sqrt{2x^2 + x}$ **89.** $7 + 5\sqrt{2}$
91. 34.4

Problem Set 5.8, pp. 225–226

1. $\dfrac{\sqrt{5}}{5}$ **3.** $\dfrac{\sqrt{10}}{2}$ **5.** $3\sqrt{2}$ **7.** $\dfrac{\sqrt{6}}{3}$

9. $\dfrac{3\sqrt{2}}{4}$ **11.** $\dfrac{2\sqrt{3x}}{x}$ **13.** $\dfrac{4\sqrt{5}}{5}$ **15.** $5\sqrt[3]{9}$

17. $4\sqrt[3]{2}$ **19.** $\dfrac{\sqrt{10}}{5}$ **21.** $\dfrac{\sqrt{3r}}{r}$ **23.** $\dfrac{\sqrt{70t}}{10t}$

25. $\dfrac{3x\sqrt{2y}}{2y}$ **27.** $\dfrac{\sqrt[3]{6p^2}}{3p}$ **29.** $\sqrt{2} - 1$

31. $2 + \sqrt{3}$ **33.** $\sqrt{6} + 2$ **35.** $\dfrac{5\sqrt{5} + 5}{2}$

37. $\dfrac{3\sqrt{5} - 1}{44}$ **39.** $4\sqrt{5} - 4\sqrt{3}$

41. $\dfrac{\sqrt{26} - 2}{11}$ **43.** $\dfrac{5 + \sqrt{21}}{2}$

45. $-6\sqrt{3} - 7\sqrt{2}$ **47.** $\dfrac{16x - 12\sqrt{xy}}{16x - 9y}$

49. $\sqrt{m} - 2$ **51.** $\dfrac{2\sqrt{5}}{15}$ **53.** $\dfrac{3\sqrt{2}}{2}$

55. $\dfrac{7\sqrt{3}}{6}$ **57.** $\dfrac{11\sqrt{2}}{2}$ **59.** $\dfrac{16\sqrt{10m}}{5}$

61. $\dfrac{y\sqrt{x} + x\sqrt{y}}{xy}$ **63.** 0.30

Problem Set 5.9, pp. 230–231

1. $3i$ **3.** $6i$ **5.** $\frac{1}{2}i$ **7.** $\sqrt{2}i$ **9.** $2\sqrt{3}i$

11. $35i$ **13.** $-10i$ **15.** $\dfrac{\sqrt{3}}{2}i$

17. a) True   b) True   c) True
19. $10 + 9i$ **21.** $-5 + 3i$ **23.** 2
25. $-20i$ **27.** $-8 + 3i$ **29.** $25 + i$
31. $8i$ **33.** $7 + i$ **35.** 0   **37.** $7 - 5i$
39. $11 - 9i$ **41.** $-12i$ **43.** $-4 - 9i$
45. $-2 + 23i$ **47.** $27 + 8i$
49. $-12 + 34i$ **51.** $-21 + 57i$ **53.** 41
55. -12 **57.** $6 + 10i$ **59.** $-45 + 30i$

61. $5 - 12i$ **63.** $-1 + 2\sqrt{2}i$
65. $-2 + 2i$ **67.** $4 + 4i$ **69.** $2 + i$
71. $2 - 5i$ **73.** $3 + 4i$ **75.** $\frac{10}{13} + \frac{24}{13}i$
77. $\frac{3}{17} - \frac{5}{17}i$ **79.** $2 - \frac{4}{5}i$ **81.** -5
83. $-\frac{1}{8}i$ **85.** $3 - 4i$
87. $(2i)^2 = 4i^2 = -4, (-2i)^2 = 4i^2 = -4$
89. $(a + bi) + (a - bi) = 2a$
91. $76.933 + 27.9341i$

Chapter 5 Review, pp. 231–233

1. $\frac{1}{9}$ **2.** $-\frac{1}{27}$ **3.** $\dfrac{r^2}{2}$ **4.** x^5 **5.** y^4

6. $\dfrac{x^3 + 1}{x^2}$ **7.** $\dfrac{a^4}{b^6}$ **8.** $-rs$ **9.** 610,000

10. 0.000302   **11.** 3.4×10^7
12. 2.53×10^{-6} **13.** 3.84×10^8
14. 1.65×10^{-15} **15.** 2   **16.** -3
17. $-2x^2y^3$ **18.** $(a + b)^2$ **19.** $3a^3$
20. ± 11 **21.** Not a real number
22. $5x^ny^{3m}$ **23.** 27   **24.** $\frac{5}{7}$ **25.** $\frac{1}{216}$
26. $x^3\sqrt{x}$ **27.** $4\sqrt[3]{y^2}$ **28.** $\sqrt{y + 2}$ **29.** x

30. $8p^6q^{12}$ **31.** $\dfrac{x^{3/2}}{y^2}$ **32.** $y^4 - y^{9/2}$

33. $2\sqrt{5}$ **34.** $3\sqrt[3]{2}$ **35.** $\dfrac{2a^2\sqrt{2a}}{11}$

36. $\dfrac{3x\sqrt[3]{x}}{5}$ **37.** $2x\sqrt{2}$ **38.** $a^2b\sqrt[3]{b}$

39. $\sqrt[12]{x}$ **40.** $4\sqrt[4]{a}$ **41.** $\dfrac{xy^2\sqrt{xy}}{z^3}$

42. $5\sqrt{2}$ **43.** y^4 **44.** $7x\sqrt{2}$ **45.** xy^2
46. $15a^3\sqrt{a}$ **47.** $2\sqrt[5]{ab}$ **48.** 7

49. $2\sqrt{5}$ **50.** $\dfrac{1}{7a}$ **51.** $2x$ **52.** $a^2b^4c\sqrt[3]{7}$

53. $5pq\sqrt{pq}$ **54.** $\sqrt{2}$ **55.** $\sqrt[6]{a}$ **56.** $7\sqrt{3}$
57. $5\sqrt{x}$ **58.** $\sqrt[3]{2y}$ **59.** $-2\sqrt{5}$
60. $2\sqrt{7} - 2$ **61.** $x - y$ **62.** $2 - \sqrt[3]{4}$
63. $t^2 + 2\sqrt{3}t + 3$ **64.** $5 + 2\sqrt{6}$

65. $18\sqrt{5} + 4\sqrt{15} - 15\sqrt{3} - 10$ **66.** $\dfrac{2\sqrt{3}}{3}$

67. $\dfrac{\sqrt{15}}{5}$ **68.** $4\sqrt[3]{4}$ **69.** $\dfrac{3 + \sqrt{5}}{4}$

70. $6 - \sqrt{35}$ **71.** $\dfrac{3\sqrt{2} + 2\sqrt{3}}{6}$

72. $10 + 4i$ **73.** $-2i$ **74.** $8 + 2i$
75. $14 + 5i$ **76.** $-3 + 12i$ **77.** $\frac{10}{13} - \frac{15}{13}i$
78. -4 **79.** $\frac{1}{2} - \frac{1}{2}i$

Chapter 5 Test, p. 234

1. a) 1.6×10^5 **b)** 7.69×10^{-4} **2.** $\frac{1}{81}$

3. $-\frac{1}{125}$ **4.** $\dfrac{b^2}{4a^4}$ **5.** $\dfrac{xy}{x+y}$ **6.** -4

7. $\dfrac{3xy}{5}$ **8.** $\dfrac{2a}{b^2}$ **9.** $\dfrac{2a\sqrt{3a}}{b^2}$ **10.** 11

11. $\frac{1}{9}$ **12.** $\dfrac{y^2}{x^4}$ **13.** $mn\sqrt[3]{m^2}$ **14.** 30

15. $\frac{1}{5}$ **16.** $\sqrt{x}$ **17.** -2 **18.** $10\sqrt{3}$

19. $6 + \sqrt{6}$ **20.** $2\sqrt[3]{25}$ **21.** $\dfrac{3 - \sqrt{5}}{4}$

22. $15 + 8i$ **23.** $4 + 3i$ **24.** $22 - 3i$
25. $\frac{14}{13} - \frac{5}{13}i$

CHAPTER 6

Problem Set 6.1, pp. 240–242

1. $2, -6$ **3.** $-\frac{1}{2}, -4$ **5.** $-7, 7$ **7.** 8
9. $0, -3$ **11.** $1, 4$ **13.** $-7, 3$
15. $-4, \frac{2}{3}$ **17.** $-8, 2$ **19.** -5 **21.** 2
23. $-3, \frac{1}{2}$ **25.** $0, 17$ **27.** $0, 5$ **29.** $\frac{3}{2}, \frac{5}{3}$
31. $-2, 2$ **33.** $-4, 4$ **35.** $-6, 6$
37. $\frac{3}{4}$ **39.** $-\frac{2}{3}, \frac{2}{3}$ **41.** $0, \frac{1}{2}$ **43.** $0, 1$
45. 0 **47.** $-\sqrt{3}, \sqrt{3}$ **49.** $-2\sqrt{3}, 2\sqrt{3}$
51. $-12, 10$ **53.** $-\frac{1}{3}$ **55.** $0, \frac{2}{7}$
57. $-\sqrt{5}, \sqrt{5}$ **59.** $x = -2a, 2a$
61. $x = 0, 2a$ **63.** $x = 3a$

65. $x = -a, 5a$ **67.** $x = -\dfrac{5a}{3}, \dfrac{a}{2}$

69. $a = \pm\sqrt{c^2 - b^2}$ **71.** $4\sqrt{3} \approx 6.9$ ft per sec
73. 7 sec   **75.** $x = -1$ or $x = 6$
77. $x^2 - 8x + 15 = 0$ **79.** $x^2 - 7x = 0$
81. $0, 79.32$

Problem Set 6.2, pp. 244–245

1. 16 **3.** 9 **5.** 1 **7.** $\frac{9}{4}$ **9.** $\frac{1}{4}$
11. $\frac{1}{25}$ **13.** $1, 5$ **15.** $-8, 4$
17. $1 \pm \sqrt{2}$ **19.** $-4 \pm 3\sqrt{3}$ **21.** $-2, 3$

23. $\dfrac{-3 \pm \sqrt{5}}{2}$ **25.** $1, 9$ **27.** $-1, 5$

29. -1 **31.** $0, -6$ **33.** $3 \pm \sqrt{6}$
35. $2 \pm \sqrt{7}$ **37.** $-5 \pm 2\sqrt{3}$ **39.** $4 \pm 3\sqrt{2}$

41. $\dfrac{-1 \pm \sqrt{5}}{2}$ **43.** $-6 \pm 4\sqrt{2}$

45. $\dfrac{-1 \pm \sqrt{7}}{2}$ **47.** $-\frac{1}{2}, \frac{3}{2}$ **49.** $\dfrac{-1 \pm \sqrt{17}}{4}$

51. $\dfrac{3 \pm 2\sqrt{3}}{3}$ **53.** $\dfrac{-1}{3}, 1$

55. $x = 1 + \sqrt{5}$ **57.** $x = a \pm 1$
59. $1190, 4188$

Problem Set 6.3, pp. 250–251

1. $-2, 1$ **3.** $-1, \frac{1}{3}$ **5.** 2 **7.** $0, 2$

9. ± 1 **11.** $\dfrac{-1 \pm \sqrt{13}}{2}$ **13.** $\dfrac{3 \pm \sqrt{33}}{2}$

15. $\dfrac{-7 \pm 3\sqrt{5}}{2}$ **17.** $2 \pm \sqrt{3}$

19. $-6 \pm \sqrt{11}$ **21.** $\dfrac{1 \pm \sqrt{17}}{4}$ **23.** $\dfrac{1 \pm \sqrt{3}}{2}$

25. $\dfrac{2 \pm \sqrt{19}}{3}$ **27.** $\dfrac{2 \pm \sqrt{3}}{2}$ **29.** $\dfrac{-6 \pm \sqrt{6}}{5}$

31. $\dfrac{1 \pm \sqrt{3}}{3}$ **33.** $\dfrac{2 \pm \sqrt{7}}{3}$ **35.** $\dfrac{2 \pm \sqrt{22}}{6}$

37. $\dfrac{-6 \pm \sqrt{46}}{2}$ **39.** $\dfrac{-2\sqrt{3}}{3}, \dfrac{\sqrt{3}}{3}$

41. $x = 1 + \sqrt{5}$

43. $\dfrac{5 - \sqrt{7}}{2} \approx 1.2$ sec, $\dfrac{5 + \sqrt{7}}{2} \approx 3.8$ sec

45. $x = 0$ or $x = 440$ **47.** $1.83, -2.17$

Problem Set 6.4, pp. 255–257

1. $6, 7$ **3.** $2, \frac{5}{2}$ **5.** $5, 6$

7. 2, 4 or $-2, -4$ 9. $-13, -11, -9$
11. Width 7 m, length 25 m   13. 2 ft
15. 15 in. by 15 in.   17. 6 in., 8 in., 10 in.
19. $2\sqrt{2}$ ft   21. Ship A 15 mph, ship B 20 mph
23. $x = 12$ 25. 29.7 ft

Problem Set 6.5, pp. 261–263

1. 16   3. 5   5. 42   7. 49   9. 19
11. No solution   13. 0   15. 6   17. 2
19. No solution   21. 100
23. No solution   25. 2   27. 4   29. $-\frac{7}{4}$
31. 1, -1 33. $\sqrt{2}, -\sqrt{2}$ 35. 2
37. $\frac{1}{4}$, 1   39. 9   41. -15, 3   43. 5
45. 2   47. 1   49. 9   51. 4, 20
53. 1   55. 64   57. 7   59. -29
61. -3 63. 5, -5 65. 6400   67. $A = s^2$

69. $V = \dfrac{\pi r^2 h}{2}$ 71. $V = e^3$ 73. $a = \sqrt{c^2 - b^2}$

75. $x = 7$ 77. $\dfrac{128}{\pi^2}$ ft   79. 8.92

Problem Set 6.6, pp. 267–268

1. 1, -1 3. $-2, -4$ 5. $-6, \frac{3}{4}$
7. $-\frac{1}{2}, 2$ 9. 1, 2   11. $-\frac{3}{5}, 2$
13. -3, 1   15. -5, 2   17. 2
19. 1, -1 21. 1, -1, 2, -2
23. $\sqrt{2}, -\sqrt{2}$, 3, -3 25. $\frac{3}{2}, -\frac{3}{2}$, 1, -1
27. 16, 81   29. -1, 27   31. $-\frac{1}{5}$, 6
33. Old 12 hr, new 6 hr
35. Car 60 mph, truck 50 mph   37. 1 mph

39. $r = \sqrt{\dfrac{Gm_1 m_2}{F}}$ 41. 1.11, -1.11, 0.65, -0.65

Problem Set 6.7, pp. 270–272

1. $\pm 2i$ 3. $\pm i$ 5. $\pm 8i$ 7. $\pm\sqrt{2}i$
9. $\pm\sqrt{5}i$ 11. $\pm 2\sqrt{3}i$ 13. $\pm 3\sqrt{5}i$

15. $\pm\frac{1}{2}i$ 17. $\pm\dfrac{2\sqrt{2}}{5}i$ 19. $-1 \pm i$

21. $3 \pm 2i$ 23. $2 \pm 3i$ 25. $\dfrac{-1 \pm i}{2}$

27. $\dfrac{1 \pm \sqrt{3}i}{2}$ 29. $1 \pm \sqrt{2}i$ 31. $\pm 6i, \pm 1$

33. $\pm\sqrt{2}, \pm i$ 35. $\pm 2i, \pm 5i$
37. Two rational solutions
39. Two rational solutions
41. Two irrational solutions
43. One rational solution
45. Two complex solutions
47. One rational solution   49. $k = 9$
51. $k = \pm 8$ 53. $x = 1 \pm i$ 55. Never
57. $\pm 79i$

Problem Set 6.8, pp. 277–278

1. $x < -3$ or $x > 2$ 3. $-3 < x < 2$

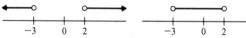

5. $-3 \le x \le 2$ 7. $x \le -3$ or $x \ge 2$

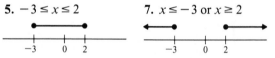

9. $y \le 2$ or $y \ge 4$ 11. $z < -6$ or $z > -1$

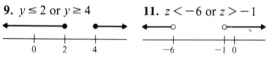

13. $-2 < x < 2$ 15. $0 < x < 4$

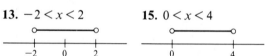

17. $\frac{1}{2} < p < 5$ 19. $y < -2$ or $y > \frac{3}{4}$

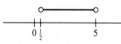
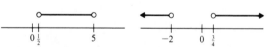

21. $m \le -4$ or $m \ge 0$ 23. $t < -5$ or $t > 5$

25. $0 \le r \le 1$ 27. $x \le -\frac{1}{3}$ or $x \ge \frac{1}{3}$

29. $-\sqrt{2} \le y \le \sqrt{2}$ 31. $m \le -6$ or $m \ge 6$

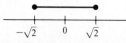

33. $0 \le p \le 9$ **35.** $1 < y < 3$
37. $y < -\frac{1}{3}$ or $y > \frac{5}{2}$
39. $x \le 1 - \sqrt{3}$ or $x \ge 1 + \sqrt{3}$
41. $\dfrac{-2 - \sqrt{6}}{2} \le r \le \dfrac{-2 + \sqrt{6}}{2}$ **43.** $x \ne 2$
45. All real numbers   **47.** No solution
49. $x = 2$ **51.** All real numbers
53. No solution   **55.** All real numbers
57. No solution   **59.** All real numbers
61. No solution   **63.** $2 \le x \le 4$ or $x \ge 6$
65. $p < -3$ or $0 < p < 2$ **67.** $1 \le x \le 4$
69. $-39 < x < 73$

54. $t \le 0$ or $t \ge 9$ **55.** $x \le -\frac{5}{3}$ or $x \ge \frac{5}{3}$

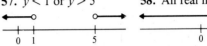

56. No real solution
57. $y < 1$ or $y > 5$ **58.** All real numbers

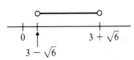

59. $3 - \sqrt{6} < x < 3 + \sqrt{6}$

Chapter 6 Review, pp. 278–280

1. $-4, 3$ **2.** $0, 5$ **3.** $-5, -2$
4. $-3, 2$ **5.** $\frac{7}{2}, -\frac{2}{3}$ **6.** $\frac{4}{5}, -\frac{4}{5}$ **7.** $0, \frac{4}{3}$
8. $3, -3$ **9.** $x = 0, x = 3b$
10. $x = -2a, x = 3a$ **11.** 16 **12.** $\frac{1}{4}$
13. $-2, 8$ **14.** $-\frac{8}{3}, 2$ **15.** $-3, -5$
16. $1 \pm \sqrt{3}$ **17.** $\dfrac{-1 \pm \sqrt{5}}{2}$ **18.** $\dfrac{9 \pm 2\sqrt{21}}{3}$
19. $-6, 4$ **20.** 3 **21.** $\dfrac{-1 \pm \sqrt{17}}{4}$
22. $-\frac{3}{4}, \frac{1}{2}$ **23.** $-2, \frac{3}{2}$ **24.** $\dfrac{1 \pm 2\sqrt{3}}{2}$
25. $\dfrac{3 \pm \sqrt{57}}{2}$ **26.** $\frac{1}{2}$ **27.** $5, 7$ or $-7, -5$
28. Width 5 m, length 14 m   **29.** 1 ft
30. 8 in., 15 in., 17 in.   **31.** 26
32. No solution   **33.** 1   **34.** 6   **35.** -2
36. $5, -5$ **37.** $g = \dfrac{v^2}{2R}$ **38.** $l = \dfrac{8t^2}{\pi^2}$
39. $-1, \frac{1}{3}$ **40.** $-1, 6$ **41.** $2, -2, 5, -5$
42. $1, -1, \sqrt{3}, -\sqrt{3}$ **43.** $-27, 1$
44. Jim 20 days, Bill 30 days
45. Moped 30 mph, bicycle 20 mph
46. $\pm 5i$ **47.** $\pm \frac{3}{2}i$ **48.** $\dfrac{3 \pm \sqrt{7}i}{4}$
49. $6 \pm i$ **50.** $\pm \sqrt{5}i, \pm 1$
51. Two complex solutions
52. Two rational solutions   **53.** $k = \pm 6$

Chapter 6 Test, p. 280

1. $-6, 3$ **2.** $\frac{9}{2}, -\frac{9}{2}$ **3.** $0, \frac{6}{5}$ **4.** 10
5. $\dfrac{-1 \pm \sqrt{17}}{2}$ **6.** $-4 \pm 2\sqrt{7}$ **7.** $-3 \pm \sqrt{6}$
8. $\pm i$ **9.** $\dfrac{-2 \pm i}{2}$ **10.** $\frac{25}{4}$
11. $x < -3$ or $x > 1$ **12.** $-2 \le x \le 4$

13. 7 **14.** $\sqrt{5}, -\sqrt{5}, 3, -3$ **15.** $-2, \frac{3}{5}$
16. $-2, 4$ **17.** 5 **18.** 13
19. 5 in., 12 in., 13 in.   **20.** 4 mph

Cumulative Review for Chapters 4–6, pp. 281–283

1. $\dfrac{x + 3y}{2}$ **2.** $\dfrac{x + 2}{x - 3}$ **3.** $\dfrac{n^2 + 5n - 1}{n + 5}$
4. $\dfrac{t^2 + 4t + 6}{(t + 2)(t - 2)}$ **5.** $\dfrac{x - 3}{x - 9}$ **6.** 1
7. $\dfrac{p}{p - 1}$ **8.** $x - y$ **9.** $-\frac{1}{6}$ **10.** -5
11. No solution   **12.** -6 **13.** $f = \dfrac{xn}{x + n}$
14. $2ab^2 + 6ab - 1$ **15.** $x^3 - 3x + 2 + \dfrac{6x - 4}{x^2 + 4}$
16. $3x^3 - 2x^2 + 5x - 2 - \dfrac{1}{2x + 1}$

17. $x^5 - 2x^4 + 3x^2 - 6x + 14 - \dfrac{27}{x + 2}$

18. $11\frac{1}{9}$ min

19. Freight 25 mph, express 45 mph   **20.** $\frac{1}{16}$

21. $\frac{1}{8}$ **22.** a **23.** p^2 **24.** $\dfrac{ay^3}{bx^2}$

25. $\dfrac{x^9}{8}$ **26.** $\dfrac{9b^8}{a^{18}}$ **27.** $\dfrac{a^2b^2}{a^2 + b^2}$

28. 9.63×10^{-7} **29.** 6.4×10^5

30. -3 **31.** $\frac{5}{2}$ **32.** $4x^3$ **33.** $r + 5$

34. $\dfrac{a^4}{b^3}$ **35.** $\dfrac{2b^2}{c^6x}$ **36.** $w^{1/2} + 1$ **37.** 9

38. $\sqrt[4]{a}$ **39.** $5x\sqrt{x}$ **40.** $2q\sqrt[3]{9}$ **41.** $\dfrac{1}{5m}$

42. $-\sqrt{3}$ **43.** $3\sqrt{6t}$ **44.** $6\sqrt{3} - 2$
45. -5 **46.** $\sqrt[3]{9}$ **47.** $3\sqrt{5} + 3\sqrt{3}$
48. $4 - 5i$ **49.** $7 - i$ **50.** $\frac{9}{13} - \frac{6}{13}i$
51. $-\frac{3}{2}i$ **52.** $-4, \frac{3}{2}$ **53.** $0, 3$
54. $9, -9$ **55.** $3i, -3i$ **56.** $-2 \pm \sqrt{10}$

57. $\dfrac{-3 \pm \sqrt{2}}{2}$ **58.** $\frac{1}{3}, 1$ **59.** $-4 \pm \sqrt{17}$

60. $\frac{1}{16}$ **61.** $x = 0, x = 5d$ **62.** $-\frac{2}{3}, \frac{1}{2}$
63. 0 **64.** $-2, 8$ **65.** $\pm\sqrt{3}i, \pm 3$
66. ± 2 **67.** Two rational solutions
68. $0 < x < 3$ **69.** $x \le -1$ or $x \ge 2$

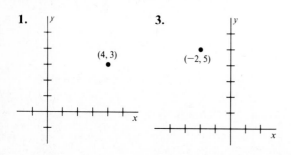

70. 3 in., 4 in., 5 in.   **71.** 2 ft

CHAPTER 7

Problem Set 7.1, pp. 287–288

5.

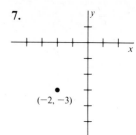

7.

9.   **11.**

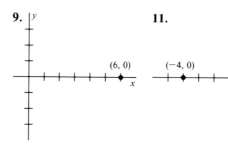

13.   **15.**

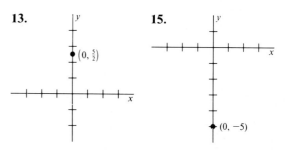

1.

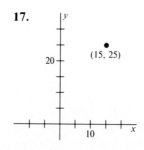

3.

17.   **19.**

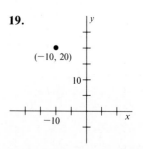

21.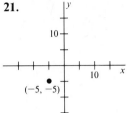

23.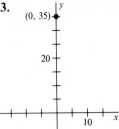

55. a)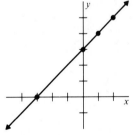

b)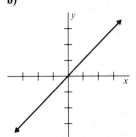

25.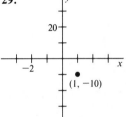

27.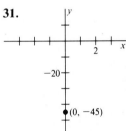

57. $97.\overline{2}\%$

Problem Set 7.2, pp. 293–294

1. Linear   **3.** Nonlinear   **5.** Nonlinear
7. Nonlinear

9.

x	y
0	3
1	4
2	5
−3	0

11.

x	y
−1	−4
0	−2
1	0
$\frac{5}{2}$	3

29.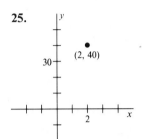

31.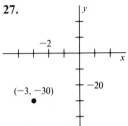

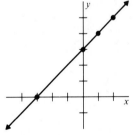

33. Quadrant IV   **35.** Quadrant II
37. Quadrant III   **39.** On the positive *x*-axis
41. $\frac{5}{36}$ **43.** $\frac{1}{12}$ **45.** $\frac{1}{6}$ **47.** 1   **49.** 0
51. 7

13.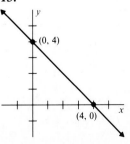

15.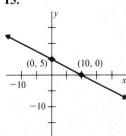

53. a)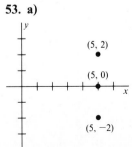

b)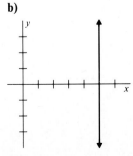

17.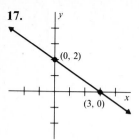

19.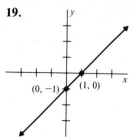

21.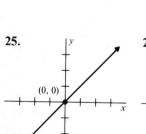

23.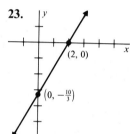

25.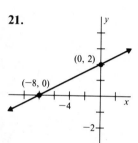

27.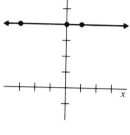

29.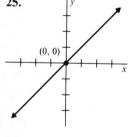

31.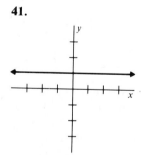

33.

x	y
−3	4
0	4
1	4
4	4

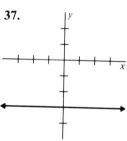

35.

x	y
3	−2
3	0
3	3
3	4

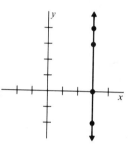

37.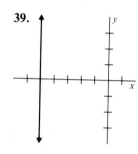

39.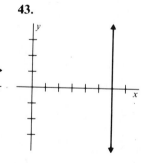

41.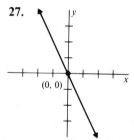

43.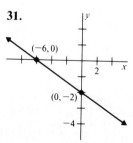

45.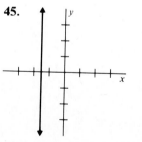

47.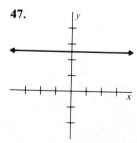

49.

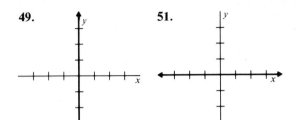

51.

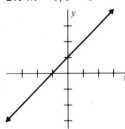

53.

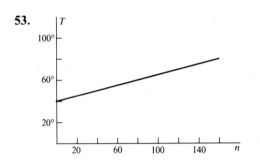

55. a) Yes   **b)** No

45. $\frac{3}{5}$ **47.** -2 **49.** 0

51. Undefined

53. $AC = \sqrt{17}$, $BC = \sqrt{17}$

55. $x = 1$ or $x = 5$

57. a) $-\frac{3}{4}$ **b)** $\frac{4}{3}$

59. a) 0 **b)** Undefined

61. Slope of both $\overline{AB}$ and $\overline{CD}$ is $\frac{3}{2}$; slope of both $\overline{AC}$ and $\overline{BD}$ is $-\frac{4}{3}$; it is not a rectangle since adjacent sides are not perpendicular

63. $t = -3, 5$ **65.** 1300 sq ft   **67.** 10.7

Problem Set 7.4, pp. 307–309

1. $3x + y = 11$ **3.** $2x - y = -19$

5. $x - 4y = 7$ **7.** $x - y = 4$

9. $5x + 6y = 60$ **11.** $y = 2$ **13.** $x = 1$

15. $3x - y = 16$ **17.** $x + y = 0$

19. $x + 3y = -25$ **21.** $y = 7$

23. $x = -8$

25. $\sqrt{3}x - 2y = -\sqrt{3}$

Problem Set 7.3, pp. 300–303

1. 5   **3.** 20   **5.** $\sqrt{13}$ **7.** $\sqrt{26}$ **9.** $3\sqrt{2}$

11. $2\sqrt{17}$ **13.** $2\sqrt{34}$ **15.** 1.3   **17.** $\frac{5}{12}$

19. $a^2 + b^2$ **21.** 3   **23.** 7   **25.** 2

27. -6 **29.** $\frac{3}{4}$ **31.** 0   **33.** Undefined

35. $\frac{11}{67}$ **37.** $-\frac{11}{4}$ **39.** $-\frac{1}{b}$

41. a) 3   **b)** 1   **43. a)** 0 **b)** 0

   c) $\frac{1}{3}$ **d)** -3 **c)** Undefined

   e) -1 **f)** $-\frac{1}{3}$ **d)** Undefined

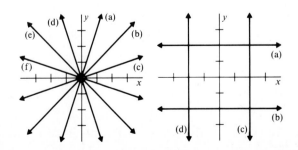

27. $m = 1, b = 1$ **29.** $m = 2, b = 6$

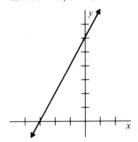

31. $m = -4, b = 2$ **33.** $m = \frac{2}{3}, b = -2$

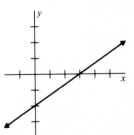

35. $m = -\frac{1}{2}$, $b = \frac{3}{2}$ **37.** $m = \frac{3}{4}$, $b = 0$

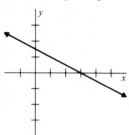   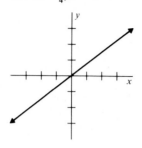

51. Parallel   **53.** Perpendicular   **55.** Neither
57. a) $10x + 15y = 6$ **b)** $15x - 10y = -4$
59. a) $2x + 5y = 14$ **b)** $5x - 2y = 35$
61. $2.88x - y = 6.5848$

Problem Set 7.5, pp. 314–315

1. 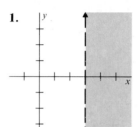   **3.**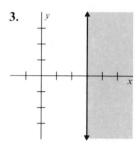

39. $m = 0$, $b = 3$ **41.** $y = x + 3$

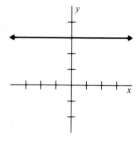

5. 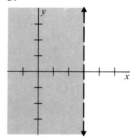   **7.**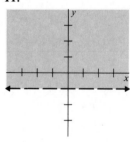

43. $y = -2x - 4$
45. Slope is undefined, no y-intercept

47.

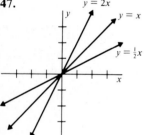

$y = 2x$
$y = x$
$y = \frac{1}{2}x$

9. 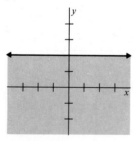   **11.**

49.

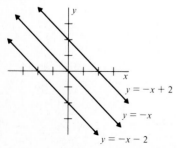

$y = -x + 2$
$y = -x$
$y = -x - 2$

13. 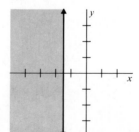   **15.**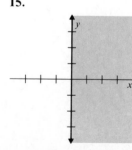

17.

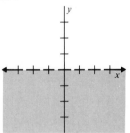

19.

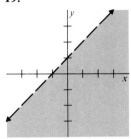

33.

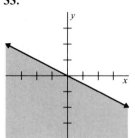

35.

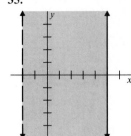

21.

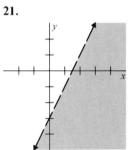

23.

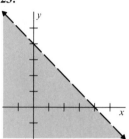

37.

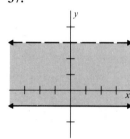

39.

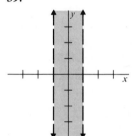

25.

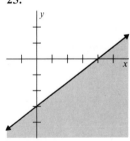

27.

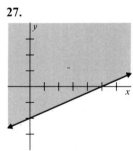

41.

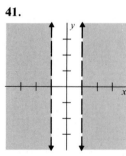

43.

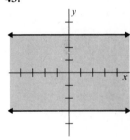

29.

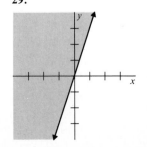

31.

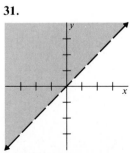

45.

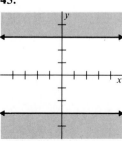

47.

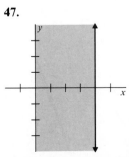

49.

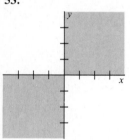

51. $4x + 8y \le 24$

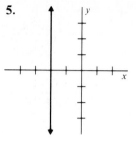

5.

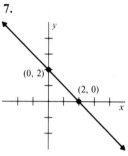

6.

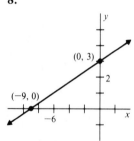

53.

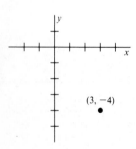

55.

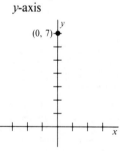

7.

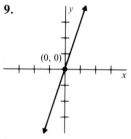

8.

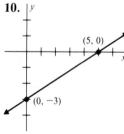

Chapter 7 Review, pp. 315–317

1. Quadrant IV

2. On the positive
y-axis

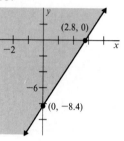

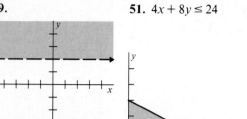

9.

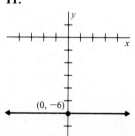

10.

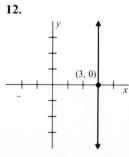

3. Quadrant III

4. On the negative
x-axis

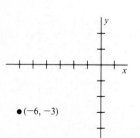

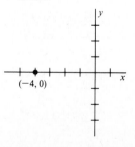

11.

12.

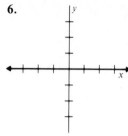

13.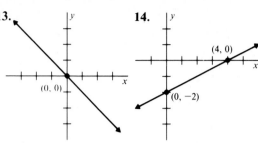

14. (labels: (4, 0), (0, −2))

15. 13 **16.** $\sqrt{41}$ **17.** 3 **18.** $-\frac{2}{3}$

19. Undefined **20.** $\frac{3}{2}$ **21.** 3 **22.** $-\frac{4}{3}$

23. Perpendicular

24. Slope of both $\overline{AB}$ and $\overline{CD}$ is $-\frac{1}{6}$; slope of both $\overline{AD}$ and $\overline{BC}$ is $-\frac{7}{2}$

25. $x + y = -1$ **26.** $3x - 4y = -26$

27. $y = 4$ **28.** $x = 4$ **29.** $x - y = -6$

30. $9x - 8y = -6$ **31.** $x + y = 0$

32. $m = 2, b = -3$ **33.** $m = -3, b = 4$

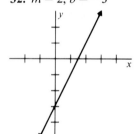 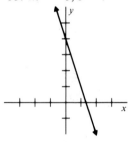

34. $m = -\frac{3}{5}, b = 3$ **35.** $m = 0, b = -2$

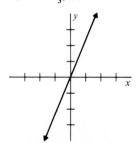 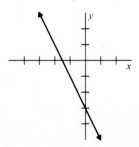

36. $m = \frac{7}{3}, b = 0$ **37.** $m = -2, b = -3$

38. $y = -\frac{3}{2}x + 3$ **39.** $y = 3$

40. $y = 2x + 4$ **41.** $y = \frac{1}{3}x + \frac{8}{3}$

42. Perpendicular **43.** Parallel

44. 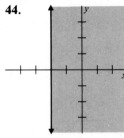 **45.**

46. 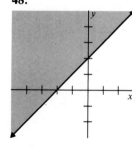 **47.**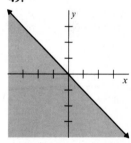

48. **49.**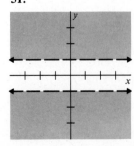

50. 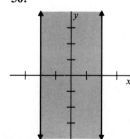 **51.**

52.

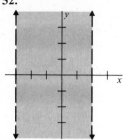

Chapter 7 Test, pp. 317–318

1.

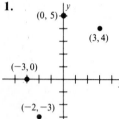

2.

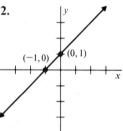

3.

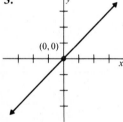

4.

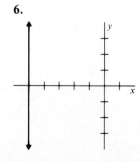

5.

6.

7.

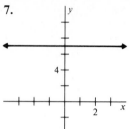

8. $\sqrt{34}$
9. $\frac{1}{2}$ **10.** $3x - y = -8$ **11.** $3x - 2y = 6$
12. $3x - 2y = 1$

13.

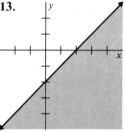

14.

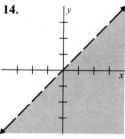

15.

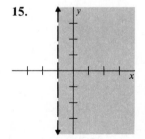

16.

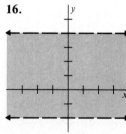

17. $m = -\frac{3}{2}$, $b = 6$ **18.** $x = 4$
19. $y = -\frac{3}{2}x - 3$ **20.** Perpendicular

CHAPTER 8

Problem Set 8.1, pp. 323–324

1. Yes   **3.** No   **5.** No
7. Independent   **9.** Independent

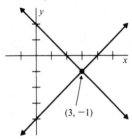

11. Independent   **13.** Independent

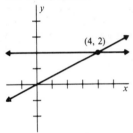

15. Inconsistent   **17.** Dependent

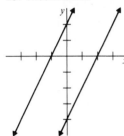

19. Independent   **21.** $x + y = 4$
                        $x - y = 6$

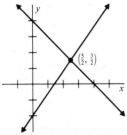

23. Yes

Problem Set 8.2, pp. 328–329

1. $(1, 1)$ **3.** $(3, 2)$ **5.** $(4, 2)$
7. $(-1, 0)$ **9.** $(-4, 5)$ **11.** $(2, -1)$
13. $(\frac{1}{2}, \frac{2}{3})$
15. Any ordered pair that satisfies $y = 4x$ is a
solution
17. No solution   **19.** $(-\frac{6}{7}, \frac{11}{7})$ **21.** $(2, 1)$
23. $(-1, -\frac{2}{3})$ **25.** $(-4, 0)$ **27.** $(1, 3)$
29. $(-\frac{1}{4}, -\frac{3}{4})$
31. Any ordered pair that satisfies $-x + 2y = 3$ is a
solution
33. No solution   **35.** $(5, 6)$
37. $x = 16, y = 7$ **39.** $x = 8, y = 5$

41. $(\frac{1}{2}, 1)$ **43.** $\left(\dfrac{2}{a}, 4\right)$ **45.** $(4.40, -2.80)$

Problem Set 8.3, pp. 334–335

1. Marigold cost $3, geranium cost $4
3. 7 oz macaroni, 3 oz tuna
5. 10 teddy bears, 15 dolls
7. Square 12 cm sides, triangle 16 cm sides
9. Speed of boat 9 mph, speed of current 2 mph
11. 83
13. Mickey 16 years old, Minnie 6 years old
15. Glenn 5 cards, Jon 7 cards
17. ($12.50, 175)   **19.** ($45, 375)
21. $67,500 at 9.42%, $48,100 at 13.65%

Problem Set 8.4, pp. 340–342

1. No   **3.** Yes   **5.** $(3, 2, -2)$
7. $(2, 1, 1)$ **9.** $(1, 2, 3)$ **11.** $(-3, 5, 4)$
13. $(-2, 0, \frac{1}{2})$ **15.** $(5, -6, 7)$
17. $(8, -1, -4)$ **19.** $(-\frac{9}{5}, \frac{3}{5}, -2)$
21. $(-4, -1, -2)$ **23.** No unique solution
25. No unique solution   **27.** $(0, 0, 0)$
29. No unique solution
31. No unique solution   **33.** $(11, 9, -13)$
35. $(6, 3, 1)$ **37.** $(-4, 1, 7)$
39. Muffins $2, pies $3, cake $5
41. Width of rectangle 10 cm, length of rectangle
65 cm, side of triangle 25 cm
43. Test 1 71, test 2 75, test 3 82
45. Curly 10 bananas, Larry 12 bananas, Moe 19
bananas
47. 574   **49.** $(29, -9, -14)$

Problem Set 8.5, pp. 346–347

1. 7   **3.** −15   **5.** 0   **7.** 16   **9.** −25
11. −2   **13.** $\frac{1}{5}$ **15.** 1   **17.** 0

19. $x^2 - y^2$ **21.** $\begin{vmatrix} 5 & 6 \\ 8 & 9 \end{vmatrix}$

23. $\begin{vmatrix} 4 & 5 \\ 7 & 8 \end{vmatrix}$ **25.** $\begin{vmatrix} 2 & 3 \\ 8 & 9 \end{vmatrix}$ **27.** $\begin{vmatrix} 1 & 3 \\ 4 & 6 \end{vmatrix}$

29. 33   **31.** −68

33. $-3\begin{vmatrix} -6 & -1 \\ -3 & 2 \end{vmatrix} + 0\begin{vmatrix} 4 & -1 \\ 1 & 2 \end{vmatrix} + 2\begin{vmatrix} 4 & -6 \\ 1 & -3 \end{vmatrix} = 33$

35. 40   **37.** −28   **39.** 5   **41.** 55
43. 0   **45.** $-3a + 3b - c$ **47.** −1
49. $x = 6$ **51.** $x = -5$ **53.** $x = \frac{1}{2}, -4$
55. a) 0 **b)** 0 **c)** 0; Conclusion: If any row of a
   3×3 determinant consists of zeros, the value
   of the determinant is zero
57. −96

Problem Set 8.6, pp. 351–352

1. $(1, 2)$ **3.** $(6, -5)$ **5.** $(\frac{1}{3}, -\frac{7}{2})$
7. $(3, 0)$ **9.** No unique solution
11. $(-3, 4)$ **13.** $(3, -1, 2)$
15. $(-2, 0, 4)$ **17.** $(5, -6, \frac{3}{4})$
19. $(0, 0, 1)$ **21.** No unique solution
23. $(0, 0, 0)$ **25.** $(8, 4)$

Chapter 8 Review, pp. 352–353

1. Independent   **2.** Inconsistent; no
                                solution

3. Dependent; any solution of $2x - 3y = 6$ is a
   solution

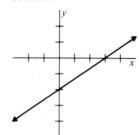

4. $(3, 3)$ **5.** $(-1, 1)$ **6.** $(-3, -2)$
7. $(3, -2)$ **8.** $(2, 2)$ **9.** $(5, 1)$
10. $(-3, 5)$ **11.** $(0, 0)$
12. Inconsistent system; no solution
13. Dependent system; any solution of $2x + 3y = 2$
   is a solution
14. $x = 7, y = 2$ **15.** 86
16. Ballpoint pen $3, fountain pen $7
17. 17 cars, 21 trucks
18. Mother 24 years old, daughter 4 years old
19. $(6, 3, 0)$ **20.** $(1, 3, 2)$ **21.** $(2, -5, 1)$
22. No unique solution   **23.** $(3, 2, 1)$
24. No unique solution   **25.** 176
26. −2   **27.** 6   **28.** $\frac{7}{16}$ **29.** 10
30. 45   **31.** 0   **32.** $x = -1$ **33.** $x = 6$
34. $(-2, 3)$ **35.** $(3, 2)$ **36.** $(-2, \frac{1}{5})$
37. $(2, -3, 7)$ **38.** $(\frac{1}{2}, -1, 3)$ **39.** $(\frac{1}{2}, \frac{1}{3}, -\frac{1}{2})$

Chapter 8 Test, pp. 353–354

1. Inconsistent   **2.** Dependent   **3.** Yes
4. 8   **5.** 0
6.   **7.**

8. $(2, -1)$
9. Inconsistent system; no solution

10. $(5, 6)$ **11.** $(\frac{1}{2}, -\frac{1}{2})$ **12.** 3
13. -10 **14.** $x = 3$ **15.** $(2, -1, 2)$
16. $(-\frac{6}{5}, \frac{13}{15}, \frac{19}{5})$ **17.** $(2, -3)$ **18.** $(1, 2, -3)$
19. Speed of boat 12.5 mph, speed of current 1.5 mph
20. Magazine $1, book $3

13.
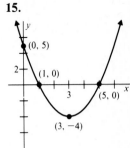

15.
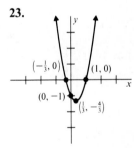

CHAPTER 9

Problem Set 9.1, pp. 363–364

1.
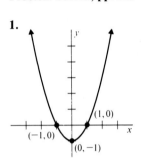

3.
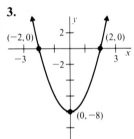

17.
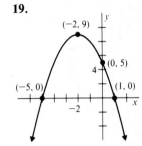

19.
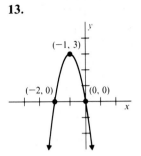

5.
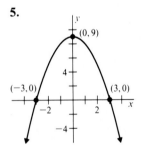

7.
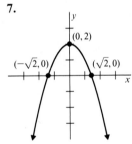

21.
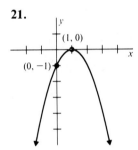

23.
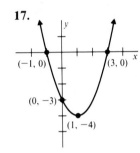

9.
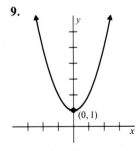

11.
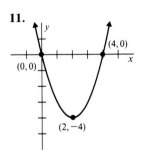

25.
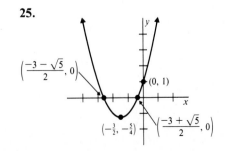

27.

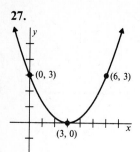

29.

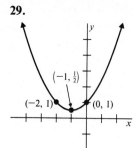

31.

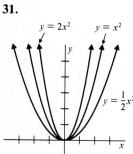

33.

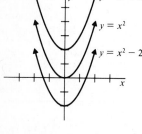

35.

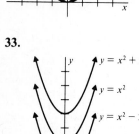

37. 2 mi × 4 mi, maximum area = 8 mi²

39.

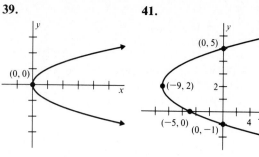

41.

43. (0, −20.74), (−6.1, 0), (3.4, 0),
(−1.35, −22.5625)

Problem Set 9.2, pp. 367–368

1. $(x - 3)^2 + (y - 5)^2 = 16$
3. $(x - 4)^2 + (y + 10)^2 = 25$
5. $(x + \frac{1}{2})^2 + (y + 2)^2 = 1$
7. $x^2 + (y - 1)^2 = \frac{1}{4}$ **9.** $x^2 + y^2 = 2$
11. $C(0, 0), r = 2$ **13.** $C(0, 0), r = 1$
15. $C(0, 0), r = \frac{5}{2}$ **17.** $C(0, 4), r = 4$
19. $C(3, 0), r = 9$ **21.** $C(1, -6), r = 12$
23. $C(-\frac{1}{4}, -\frac{3}{4}), r = \frac{1}{3}$ **25.** $C(8, 8), r = 2\sqrt{2}$

27.

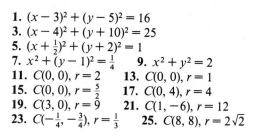

29.

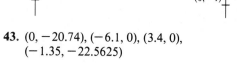

31.

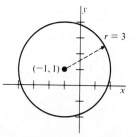

33.

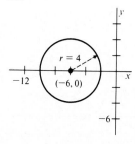

35.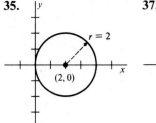

37.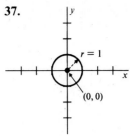

9.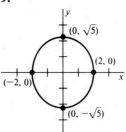

11.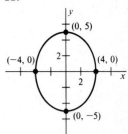

39. $x^2 + y^2 = 2.25$ **41.** $(x + 1)^2 + (y + 6)^2 = 34$

43.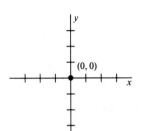

45. The top half of the circle $x^2 + y^2 = 25$

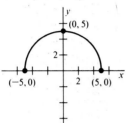

13.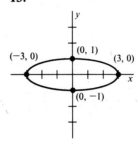

15.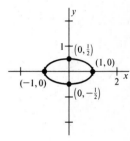

47. $C(0, 0)$, $r = 5.7$

17.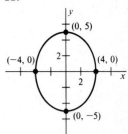

19.

Problem Set 9.3, pp. 374–375

1.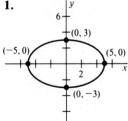

3.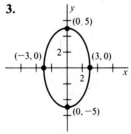

17.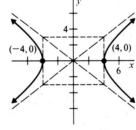

19.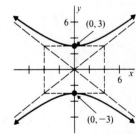

21.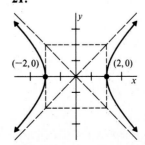

23.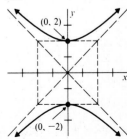

5.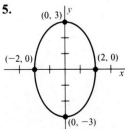

7.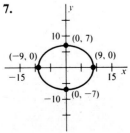

25.

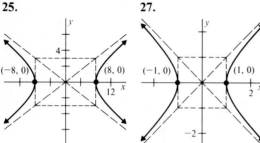

27.

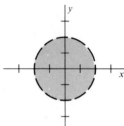

29. $\dfrac{x^2}{36} + \dfrac{y^2}{9} = 1$ **31.** $\dfrac{y^2}{25} - \dfrac{x^2}{4} = 1$

33. Circle   **35.** Hyperbola   **37.** Ellipse
39. Parabola

41.

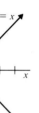

43.

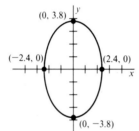

Problem Set 9.4, pp. 379–381

1. $(1, 1)$ **3.** $(-2, 1), (1, 4)$
5. $(-2, -8), (2, 8)$ **7.** $(4, 3), (5, 0)$
9. $(\frac{8}{5}, -\frac{6}{5}), (0, -2)$ **11.** $(4, 4)$
13. $(0, -2), (-\sqrt{3}, 1), (\sqrt{3}, 1)$
15. $(-4, -1), (7, 10)$ **17.** $(-1, 0), (1, 0)$
19. $(-6, -4), (6, -4), (-6, 4), (6, 4)$
21. $(0, 3), (0, -3)$
23. $(-2, -3), (2, -3), (-2, 3), (2, 3)$
25. $(-10, -5), (10, 5)$
27. $(-2, -\frac{3}{2}), (2, -\frac{3}{2}), (-\frac{3}{2}, 2), (\frac{3}{2}, 2)$
29. $(4, 1), (-4, -1), (1, 4), (-1, -4)$
31. $(5, 1), (-5, -1), (1, 5), (-1, -5)$
33. $(-2i, -6i), (2i, 6i)$
35. $(-4i, 5), (4i, -5), (-4i, -5), (4i, 5)$
37. $(3, -1), (-3, 1), (i, 3i), (-i, -3i)$
39. $x = 2, y = 3$ or $x = 3, y = 2$
41. 4 ft × 6 ft   **43.** $(6.87, 4.19), (4.19, 6.87)$

Problem Set 9.5, pp. 384–385

1.

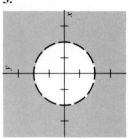

3.

5.

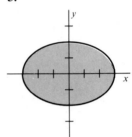

7.

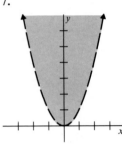

9.

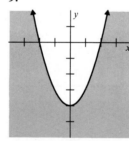

11.

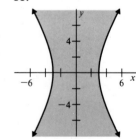

13.

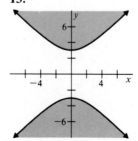

15.

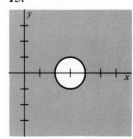

17.

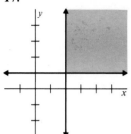

19.

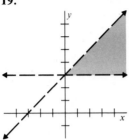

33.

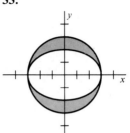

35.

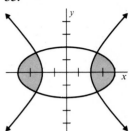

21.

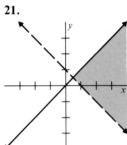

23.

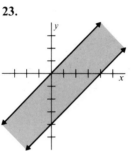

37.

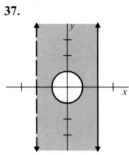

39.

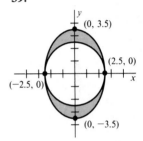

25.

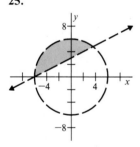

27.

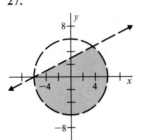

Chapter 9 Review, pp. 385–387

1.

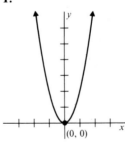

2.

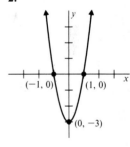

29.

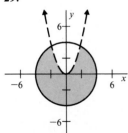

31.

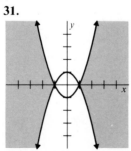

3.

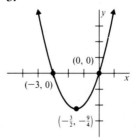

4.

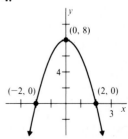

5.

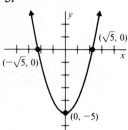

6.

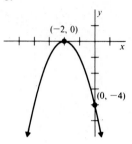

19.

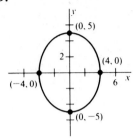

7.

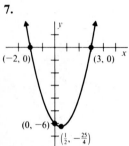

8.

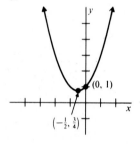

20.

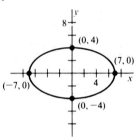

9. 1.5 mi $\times$ 3 mi, maximum area $= 4.5$ mi^2

10. $x^2 + y^2 = 4$ **11.** $(x - 2)^2 + (y + 3)^2 = 9$

12. $C(0, 2)$, $r = 4$ **13.** $C(-4, 5)$, $r = \sqrt{5}$

14.   **15.**   **21.**

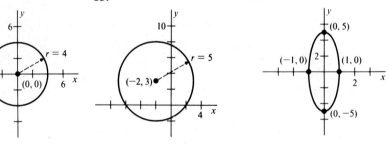

16.   **17.**

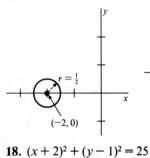

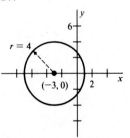

22.

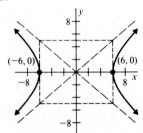

18. $(x + 2)^2 + (y - 1)^2 = 25$

23.

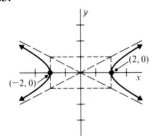

24.

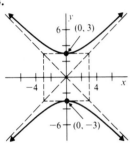

25. Circle   **26.** Parabola   **27.** Hyperbola

28. Ellipse   **29.** $\dfrac{x^2}{4} + \dfrac{y^2}{9} = 1$

30. $\dfrac{x^2}{4} - \dfrac{y^2}{9} = 1$ **31.** $(-1, 1), (3, 9)$

32. $(-3, -4), (3, 4)$ **33.** $(-3, 5), (5, -3)$
34. $(-4, -3), (-4, 3), (4, -3), (4, 3)$
35. $(-3, -1), (-3, 1), \;\;(3, -1), (3, 1)$
36. $(-\sqrt{2}, 3), (\sqrt{2}, 3), (-2, 1), (2, 1)$
37. $(-i, 2), (-i, -2), (i, -2), (i, 2)$
38. 8 in. $\times$ 8 in., 9 in. $\times$ 9 in.

39.

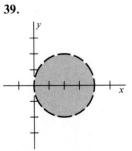

40.

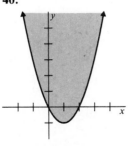

41.

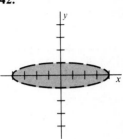

42.

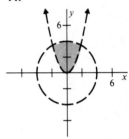

43.

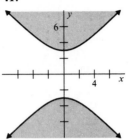

44.

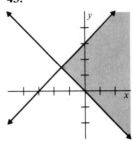

45.

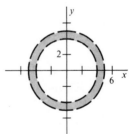

46.

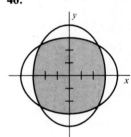

Chapter 9 Test, p. 387

1. $C(6, 8), r = 5$ **2.** $C(-2, 0), r = 2$
3.   **4.**

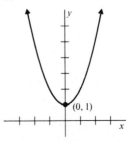

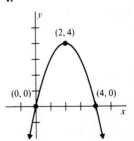

5.

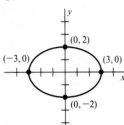

6.

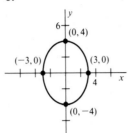

7.

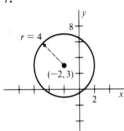

8.

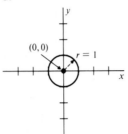

9.

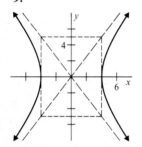

10.

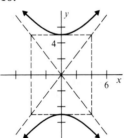

11. $(0, -5)$, $(4, 3)$ **12.** $(5, 1)$, $(-1, -5)$
13. $(18, 3)$, $(\frac{25}{2}, -\frac{5}{2})$
14. $(3, 2)$, $(3, -2)$, $(-3, 2)$, $(-3, -2)$

15.

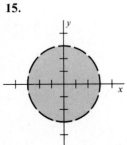

16.

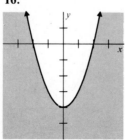

17.

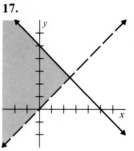

18.

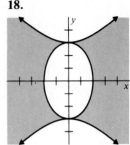

19. $(x + 3)^2 + (y - 2)^2 = 25$
20. 10 ft $\times$ 20 ft, maximum area $= 200$ ft²

Cumulative Review for Chapters 7–9, pp. 388–389

1. 5   **2.** $2\sqrt{13}$

3.

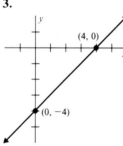

4.

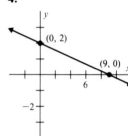

5.

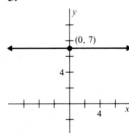

6.

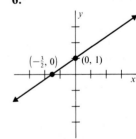

7. $\dfrac{-12}{5}$ **8.** $-\dfrac{3}{2}$ **9.** $y = 5x - 3$

10. $y = \frac{3}{4}x + \frac{9}{2}$ **11.** $m = -2$, $b = -1$

12. $m = \frac{5}{3}$, $b = -5$ **13.** Parallel

14. Perpendicular

15.

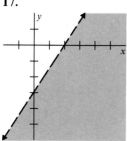

16.

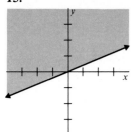

17.

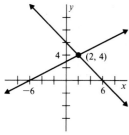

18.

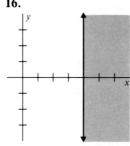

19. $3x - 2y = -4$
20. Independent   **21.** $(1, 1)$

22. $(0, -\frac{1}{2})$ **23.** $(-7, 6)$ **24.** $(8, -4)$
25. 16 and 5   **26.** 13 quarters, 17 nickels
27. $(3, -2, -1)$ **28.** $(2, 6, -1)$ **29.** 23
30. -17 **31.** 16   **32.** 0   **33.** $(4, -1)$
34. $(3, -3, 4)$ **35.** $(1, 5, -2)$
36.

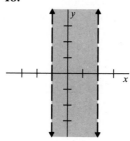

37.

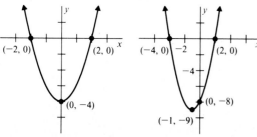

38.

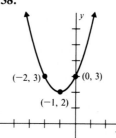

39.

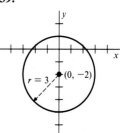

40.

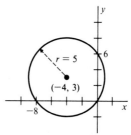

41.

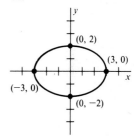

42.

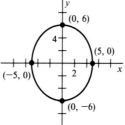

43.

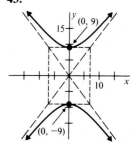

44.

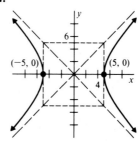

45. $(-2, -7), (7, 2)$
46. $(4, 2), (4, -2), (-4, 2), (-4, -2)$

47.

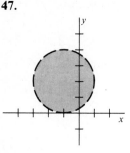

48.

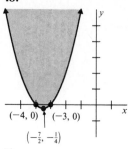

49.

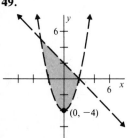

50.

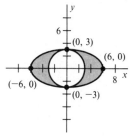

51. $(x + 2)^2 + (y - 5)^2 = 169$ **52.** 16 and 6

CHAPTER 10

Problem Set 10.1, pp. 394–397

1. Domain = {2, −6, 0, 9}, range = {5, 4, 3}; is a function
3. Domain = {3, 1}, range = {9, 7, −8}; not a function
5. Domain = {1, 2, 3}, range = {1, 2, 3}; is a function
7. Domain = {−5, −6, −7}, range = {0}; is a function
9. Domain = {4, 5, 6}, range = {5, 6, 7}; is a function
11. Yes   **13.** Yes   **15.** Yes   **17.** No
19. Yes   **21.** No   **23.** Yes
25. No, one x-value can give many y-values
27. All real numbers   **29.** All real numbers
31. $x \neq 1$ **33.** $x \neq \pm 4$
35. All real numbers   **37.** $x \neq 2, 3$
39. $x \geq 6$ **41.** All real numbers
43. $x \geq -\frac{9}{2}$ **45.** $x < 5$
47. Domain is $-3 \leq x \leq 3$, range is $-2 \leq y \leq 2$; not a function

49. Domain is all real numbers, range is $y \geq 1$; is a function
51. Domain is all real numbers, range is all real numbers; is a function
53. Domain is all real numbers, range is $y = 4$; is a function
55. Domain is all real numbers, range is $y \leq -2$ or $y \geq 2$; not a function
57. Domain is all real numbers, range is all real numbers; is a function
59. Domain is $-3 \leq x \leq 3$, range is $0 \leq y \leq 3$; is a function
61. $A = x^2$, domain is $x > 0$ **63.** $y = 6x$
65. $T = 0.80\,(220 - a)$, domain is $18 \leq a \leq 55$
67. $x \neq \pm 1.75$

Problem Set 10.2, pp. 401–402

1. 11   **3.** −9   **5.** 1   **7.** 4   **9.** $2a + 1$
11. $2a + 3$ **13.** 19   **15.** 35   **17.** $\frac{7}{3}$
19. 5   **21.** $6a^2 + 4a + 3$
23. $6x^2 + 12xh + 6h^2 - 4x - 4h + 3$ **25.** 8
27. 8   **29.** 8   **31.** 8   **33.** 7
35. −24   **37.** −1   **39.** 14   **41.** 116
43. 5   **45.** −27   **47.** 6   **49.** 9
51. 7   **53.** 3   **55.** $t + a$
57. $t^2 + at + a^2$ **59.** 0
61. $f(g(a)) = a^2 + 2a + 1$ $g(f(a)) = a^2 + 1$
63. $C(x) = 4x + 275$; $C(0) = 275$, the cost of producing 0 curtains is \$275; $C(235) = 1215$, the cost of producing 235 curtains is \$1215
65. a) 12.472 **b)** 67,930

Problem Set 10.3, pp. 406–407

1.   **3.**

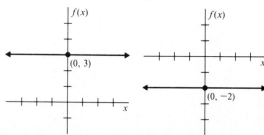

5.

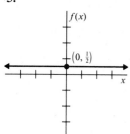

7.

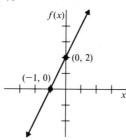

21. Quadratic

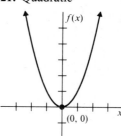

23. Constant

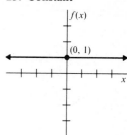

9.

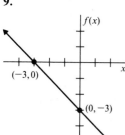

11.

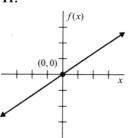

25. Quadratic

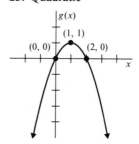

27. Linear

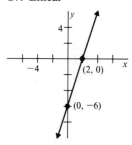

13.

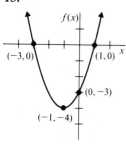

15.

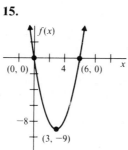

29. Quadratic

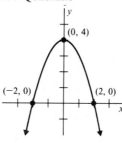

31. Constant

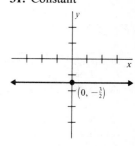

17.

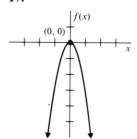

19. Linear

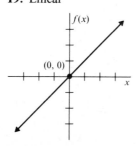

33. Quadratic

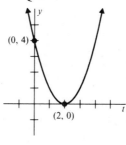

35. Quadratic

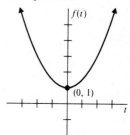

37. 3   **39.** 0   **41.** 5

43.

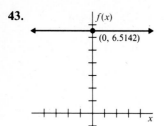

13.

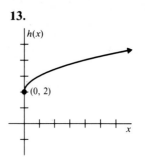

15.

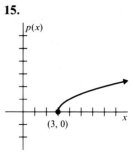

Problem Set 10.4, p. 412

1.

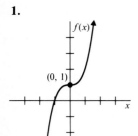

3.

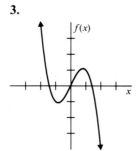

17.

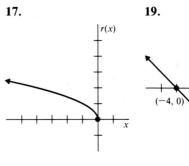

19.

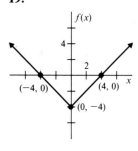

5.

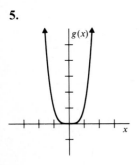

7.

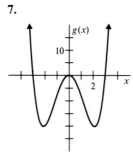

21.

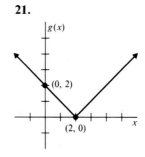

23.

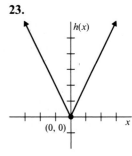

9.

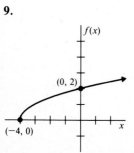

11.

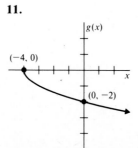

25.

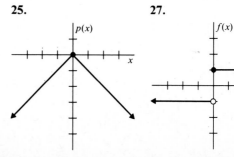

27.

29.

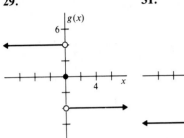

31.

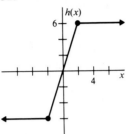

33.

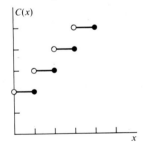

35. 2 hr $< t <$ 7.3 hr

37. $C(x) = \begin{cases} 2 \text{ if } 0 < x \le 1 \\ 3 \text{ if } 1 < x \le 2 \\ 4 \text{ if } 2 < x \le 3 \\ 5 \text{ if } 3 < x \le 4 \end{cases}$

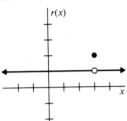

39.

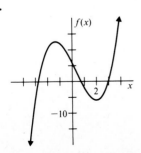

Problem Set 10.5, pp. 416–418

1. $f^{-1} = \{(5, 2), (2, -1), (0, -3)\}$, f^{-1} is a function
3. $f^{-1} = \{(0, 0), (2, 2), (-4, 4)\}$, f^{-1} is a function
5. $f^{-1} = \{(6, 3), (-1, -4), (-2, 5), (6, -7)\}$, f^{-1} is not a function
7. $f^{-1}(x) = x - 2$, f^{-1} is a function

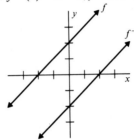

9. $f^{-1}(x) = \frac{1}{2}x$, f^{-1} is a function

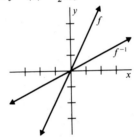

11. $f^{-1}(x) = \frac{1}{3}x + 2$, f^{-1} is a function

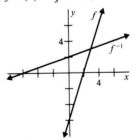

13. $f^{-1}(x) = -x + 3$, f^{-1} is a function

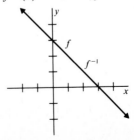

15. $f^{-1}(x) = \pm\sqrt{x-2}$, f^{-1} is not a function

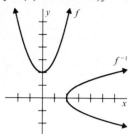

17. $f^{-1}(x) = \sqrt{x-2}$, f^{-1} is a function

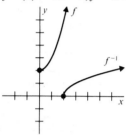

19. f^{-1} is $x = 1$, f^{-1} is not a function

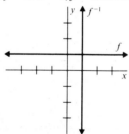

21. $f^{-1}(x) = \pm\sqrt{\frac{1}{2}x + 2}$

23. $g^{-1}(x) = \sqrt[3]{x+1}$ **25.** $h^{-1}(x) = x$

27. $-3x + 2y = 6$

29. a) 2   **b)** 5   **c)** -1 **d)** 6

31. $f^{-1}(x) = \frac{1}{5}x$ **a)** 20   **b)** 4   **c)** a **d)** a

33.

35.

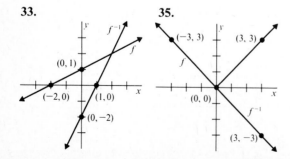

37.

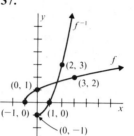

39.

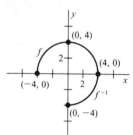

41.

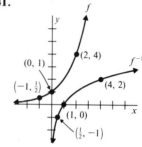

43. $f^{-1}(x) = 4.3625x + 8.85$

Problem Set 10.6, pp. 422–424

1. $p = kq$ **3.** $r = ks^3$ **5.** $u = \dfrac{k}{v}$

7. $w = \dfrac{k}{z^2}$ **9.** $y = kxz^3$ **11.** $T = \dfrac{kr}{s}$

13. $F = \dfrac{km_1 m_2}{d^2}$

15. a) $k = 4$ **b)** $y = 4x$ **c)** $y = 24$

17. a) $k = 3$ **b)** $y = 3x^2$ **c)** $y = 108$

19. a) $k = 40$ **b)** $y = \dfrac{40}{x}$ **c)** $y = \frac{20}{3}$

21. $y = 30$ **23.** $y = 4$

25. \$315 an ounce   **27.** 256 ft   **29.** 18 hr

31. 9 ohms   **33.** 1536 lb

35. 4 times greater   **37.** $y = 126.6325$

Chapter 10 Review, pp. 424–426

1. Domain = {1, 2, -6, 3}, range = {3, 1, 2}; is a function

2. Domain = {0, 4}, range = {4, 0}; not a function

3. Yes   **4.** Yes   **5.** No   **6.** Yes

7. Yes   **8.** All real numbers
9. $x \neq 3, -3$ **10.** $x < 1$
11. All real numbers   **12.** All real numbers
13. Domain = {3}, range is all real numbers; not a function
14. Domain is all real numbers, range is all real numbers; is a function
15. 1   **16.** -2 **17.** $3a + 1$
18. $3a - 2$ **19.** 3   **20.** $-\frac{1}{2}$ **21.** 0
22. 5   **23.** 5   **24.** 3   **25.** $t + a$ **26.** 0
27. $C(x) = 5x + 375$; $C(0) = 375$, the cost of producing 0 fans is $375; $C(100) = 875$, the cost of producing 100 fans is $875

28.

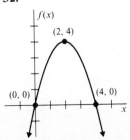

29.

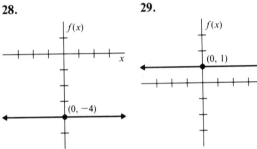

30.

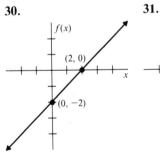

31.

32.

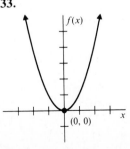

33.

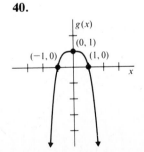

34. Linear

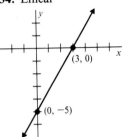

35. Quadratic

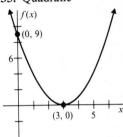

36. Quadratic

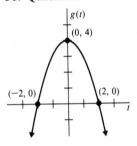

37. Constant

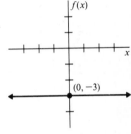

38.

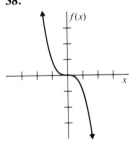

39.

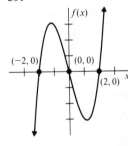

40.

41.

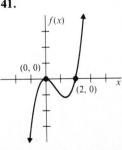

42.

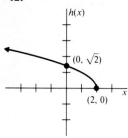

43.

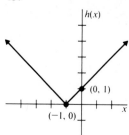

50. $f^{-1}(x) = \frac{1}{2}x - 2, f^{-1}$ is a function

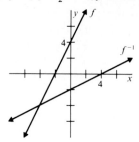

44.

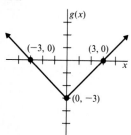

45.

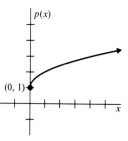

51. $f^{-1}(x) = \pm\sqrt{x - 4}, f^{-1}$ is not a function

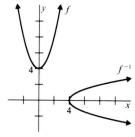

46.

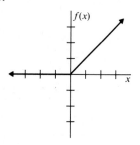

52. $f^{-1}(x) = \sqrt{x - 4}, f^{-1}$ is a function

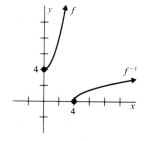

47. $f^{-1} = \{(2, 2), (4, 3), (3, -1)\}, f^{-1}$ is a function
48. $f^{-1} = \{(3, -2), (1, 0), (3, 1), (2, 2)\}, f^{-1}$ is not a function

49. $f^{-1}(x) = -\frac{1}{2}x, f^{-1}$ is a function

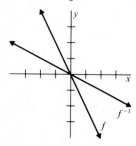

53. $g^{-1}(x) = \sqrt[3]{x}$ **54.** $h^{-1}(x) = \pm\sqrt{\dfrac{x + 4}{3}}$

55. $f^{-1}(x) = \frac{1}{2}x + \frac{1}{2}$ **a)** -1 **b)** $\frac{1}{2}$ **c)** a **d)** -3

56. $s = kt$ **57.** $y = \dfrac{k}{x}$ **58.** $A = \dfrac{k}{x^2}$

59. $C = kr$ **60.** $y = \frac{27}{2}$ **61.** $z = 4$
62. 3 in³

Chapter 10 Test, pp. 426–427

1. a) Yes **b)** Yes   **2. a)** $x \neq 2$ **b)** $x \leq 3$
3. a) $\{0, 1, 2\}$ **b)** $\{3, 4, 2, 1\}$ **c)** No

d) {(3, 0), (4, 1), (2, 2), (1, 1)}

4. a) 3 **b)** −6 **c)** $3 - a^2$

5.

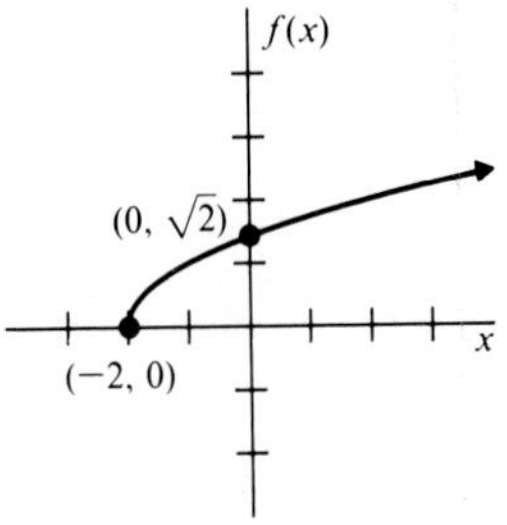

6.

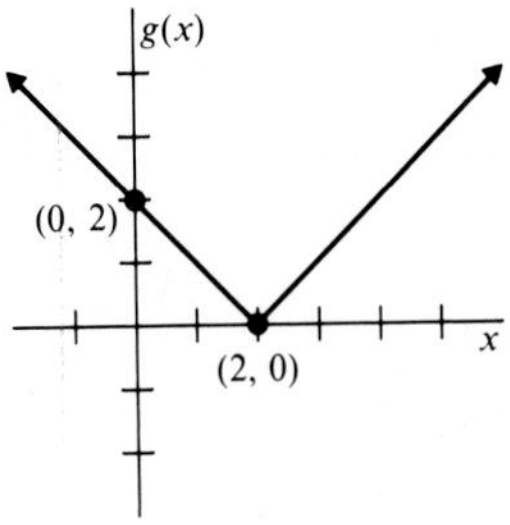

7.

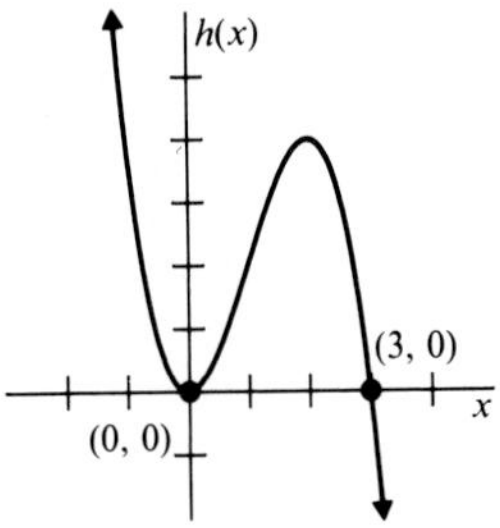

8.

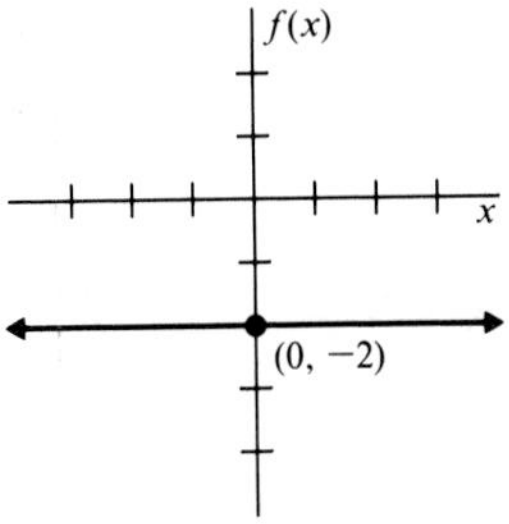

9.

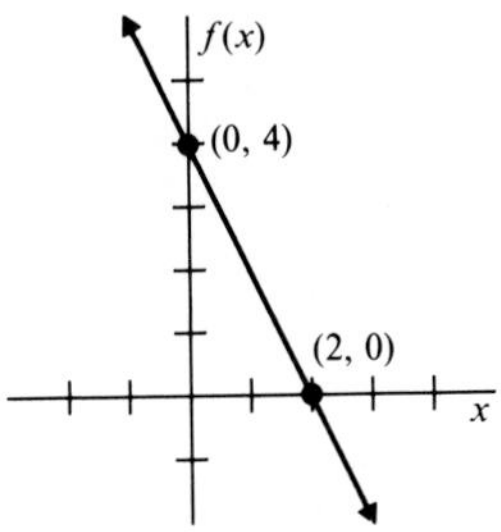

10.

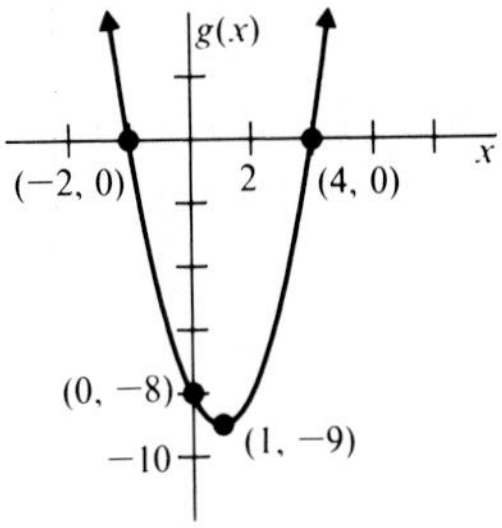

11.

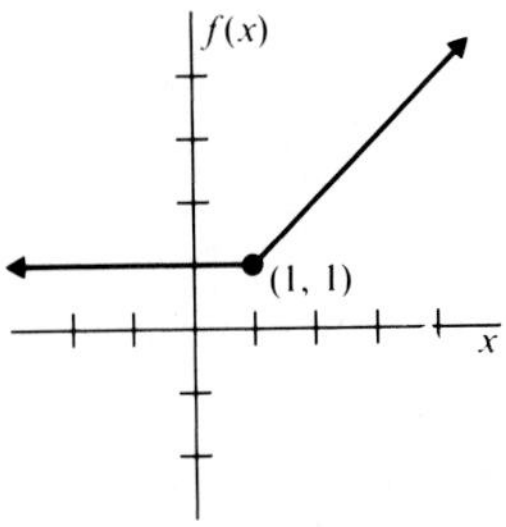

12. $f^{-1}(x) = \frac{1}{3}x - \frac{4}{3}$, f^{-1} is a function

13. $f^{-1}(x) = \pm\sqrt{x+2}$, f^{-1} is not a function

14. a) 1 **b)** −13 **15.** $t + a$

16. $f^{-1}(x) = x - 3$ **a)** −2 **b)** a

17. a) $y = kxz$ **b)** $T = \dfrac{kx}{y}$ **18.** $y = 9$

19. 400 ft **20.** 2 amps

CHAPTER 11

Problem Set 11.1, pp. 431–432

1.

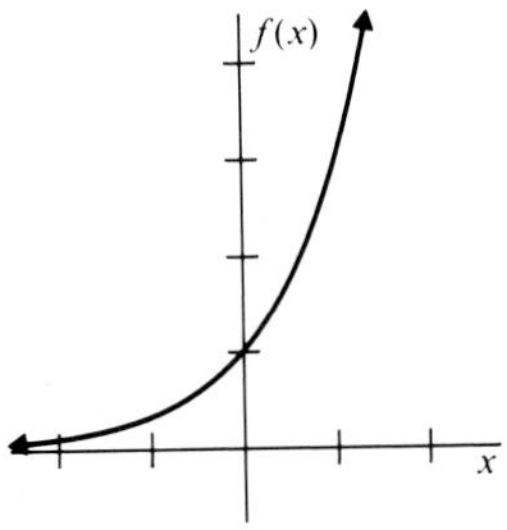

3.

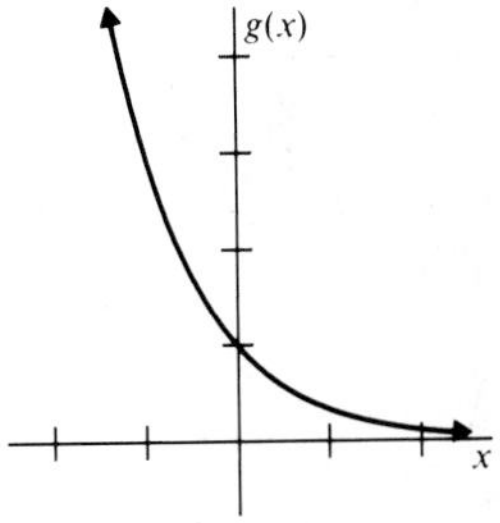

5.

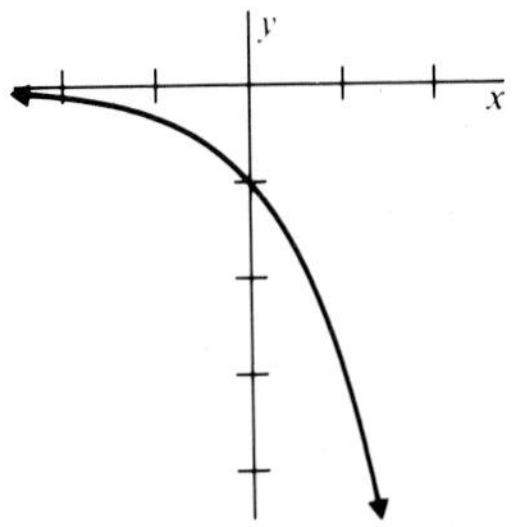

7.

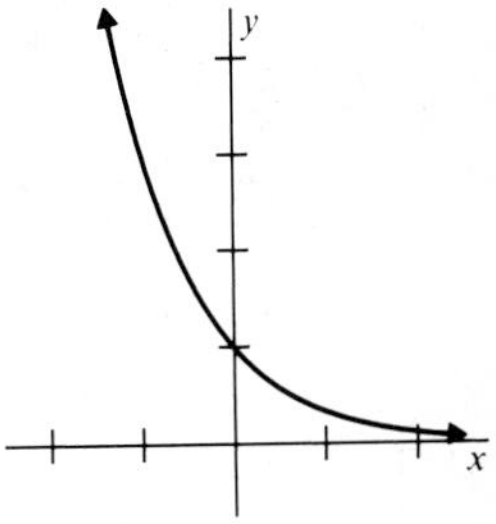

9.

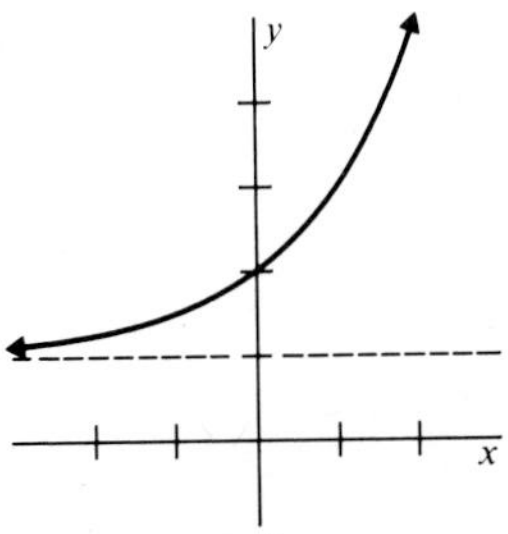

11.

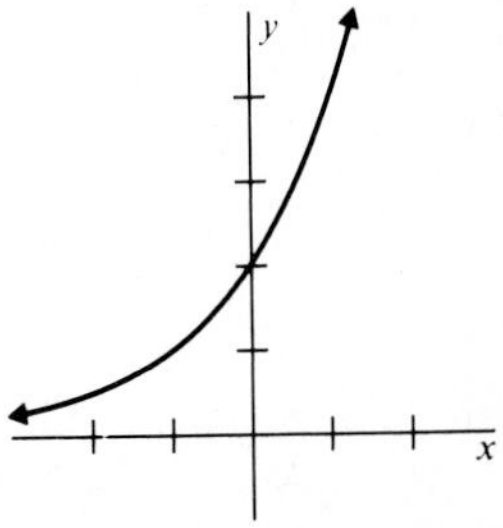

13.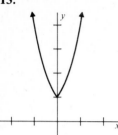

15.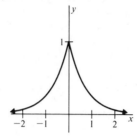

17.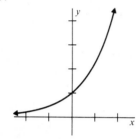

19. a) 5 cells **b)** 10 cells **c)** 20 cells **d)** 40 cells

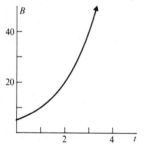

21. a) 100 mg **b)** 50 mg **c)** 25 mg **d)** 12.5 mg

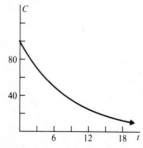

23. Future value $\approx$ \$1,200,000, total investment = \$25,000

Problem Set 11.2, pp. 436–438

1. $\log_5 25 = 2$ **3.** $\log_3 81 = 4$
5. $\log_3 \frac{1}{3} = -1$ **7.** $\log_{10} 100 = 2$
9. $\log_{10} 0.01 = -2$ **11.** $\log_6 6 = 1$
13. $\log_4 2 = \frac{1}{2}$ **15.** $\log_b t = s$
17. $2^4 = 16$ **19.** $5^3 = 125$ **21.** $(\frac{1}{7})^{-2} = 49$
23. $25^{1/2} = 5$ **25.** $(\frac{1}{2})^3 = \frac{1}{8}$ **27.** $(\sqrt{3})^2 = 3$
29. $11^0 = 1$ **31.** $b^r = M$ **33.** $y = 3$
35. $y = -3$ **37.** $y = 4$ **39.** $y = -1$
41. $y = -3$ **43.** $y = 0$ **45.** $y = 1$
47. $y = \frac{1}{2}$ **49.** $y = t$ **51.** $x = 81$
53. $x = 4$ **55.** $x = \frac{1}{32}$ **57.** $a = 4$
59. $a = \frac{1}{4}$ **61.** $a = 16$
63.   **65.**

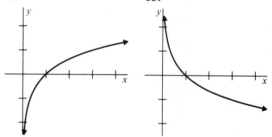

67. If $y = \log_2 0$, then $2^y = 0$, but no value of y makes 2^y zero
69. a) 0 **b)** 6.7 **c)** 8.3
71. $\log_{10} 3 \approx 0.4771$ **73.** $\log_{10} 346 \approx 2.5391$

Problem Set 11.3, pp. 441–443

1. $\log_2 3 + \log_2 7$ **3.** $\log_3 x + \log_3 y$
5. $\log_5 4 + \log_5 x$ **7.** $\log_{10} x + \log_{10}(x+1)$
9. $\log_4 3 - \log_4 7$ **11.** $\log_2 x - \log_2 y$
13. $\log_5 2 - \log_5 x$
15. $\log_{10}(x-4) - \log_{10} x$ **17.** $2\log_6 x$
19. $-3\log_{10} x$ **21.** $\frac{1}{2}\log_2 3$ **23.** $\frac{1}{3}\log_a x$
25. $2 + \log_{10} 3$ **27.** $3 + \log_3 x$
29. $4 - \log_2 3$ **31.** $-\log_4 x$
33. $3\log_2 x + \log_2 y$ **35.** $3 + 4\log_2 x$
37. $2\log_3 x - \log_3 y$ **39.** $5\log_3 x - 2$
41. $\log_4 x + \frac{1}{2}\log_4 y - 2\log_4 z$
43. $\log_5 x - \log_5 y - \log_5 z$
45. $\log_5 x - \log_5 y - 1$ **47.** $\log_5 x^7$

49. $\log_3 \sqrt{x}$ **51.** $\log_{10} \frac{1}{x}$ **53.** $\log_2 rs$

55. $\log_3 \dfrac{r}{s}$ **57.** $\log_5 xyz$

59. $\log_{10} \dfrac{xy}{z}$ **61.** $\log_a \dfrac{\sqrt{x}}{yz}$ **63.** $\log_a \dfrac{x}{y^2 z}$

65. $\log_{10} (x^2 - 1)$ **67.** $\log_{10} (x + 4)$
69. $\log_2 4 \cdot 8 = \log_2 32 = 5$, $(\log_2 4)(\log_2 8) =$
$(2)(3) = 6$

71. $\dfrac{\log_3 9}{\log_3 3} = \dfrac{2}{1} = 2$, $\log_3 9 - \log_3 3 = 2 - 1 = 1$

73. Let $x = \log_a M$ and $y = \log_a N$. Then $a^x = M$
and $a^y = N$. Therefore $\dfrac{M}{N} = \dfrac{a^x}{a^y} = a^{x-y}$. Convert to logarithmic form and get $\log_a \dfrac{M}{N} = x - y$.
Substituting for x and y gives
$\log_a \dfrac{M}{N} = \log_a M - \log_a N$

75. 1.1761   **77.** 1.9084

Problem Set 11.4, p. 447

1. 0.4771   **3.** 0.4150   **5.** 0.8762
7. 1.8129   **9.** 3.8129   **11.** 4.8129
13. -0.2890 **15.** -1.2890 **17.** 10.5740
19. -9.4260 **21.** 5.55   **23.** 55.5
25. 555   **27.** 5,550,000   **29.** 263
31. 6.31×10^{16} **33.** 1.95×10^{54}
35. 0.194   **37.** 0.0194   **39.** 4.13×10^{-16}
41. If $\log 0 = y$, then $10^y = 0$,
but that is impossible
43. 10.5   **45.** 2.5×10^{-6} moles per liter
47. 10 times

Problem Set 11.5, pp. 452–453

1. 2   **3.** 2.32   **5.** 2   **7.** 1.17   **9.** $\frac{1}{2}$
11. 0.95   **13.** -3 **15.** -2.86 **17.** $\frac{1}{3}$
19. $-2, -1$ **21.** 3   **23.** 1   **25.** 5
27. 6   **29.** 32   **31.** 1001   **33.** $-1, 9$
35. 6   **37.** 20   **39.** $\frac{4}{3}$ **41.** 7.3 yr, 7.3 yr
43. 6.6 yr   **45.** 4:35 A.M.

Problem Set 11.6, pp. 455–456

1. 1.58   **3.** 4.01   **5.** 0.68   **7.** -3.19
9. 6.29   **11.** $\log_2 15 = \dfrac{\log_9 15}{\log_9 2}$

13. $\log_7 5 = \dfrac{1}{\log_5 7}$ **15.** 7.1389

17. 4.6052   **19.** 3.8501   **21.** 1.6351
23. -4.9519 **25.** 1   **27.** 3   **29.** -1
31. $\frac{1}{2}$ **33.** 7.39   **35.** 44.7   **37.** 4.57
39. 0.741   **41.** \$2983.65
43. a) 14.7 psi   **b)** 1.8 psi   **45.** 6.9 yr

Chapter 11 Review, pp. 456–458

1.   **2.**

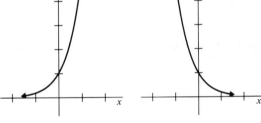

3.   **4.**

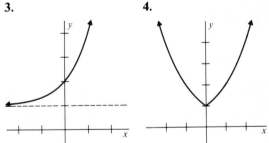

5. a) 10 cells   **b)** 30 cells
 c) 90 cells   **d)** 270 cells
6. $\log_3 243 = 5$ **7.** $\log_3 \frac{1}{27} = -3$
8. $\log_8 4 = \frac{2}{3}$ **9.** $\log_{10} 0.00001 = -5$
10. $(\frac{1}{2})^4 = \frac{1}{16}$ **11.** $7^0 = 1$
12. $4^5 = 1024$ **13.** $32^{2/5} = 4$
14. $y = 5$ **15.** $y = -4$ **16.** $y = 3$
17. $y = -6$ **18.** $x = 625$ **19.** $x = 27$
20. $a = 12$ **21.** $a = \frac{1}{3}$

22.

23.

24. $\log_4 5 + \log_4 7$ **25.** $\log_2 9 - \log_2 x$
26. $\log_6 16 + \log_6 x$
27. $\log_4 x - \log_4 (x + 1)$ **28.** $6 \log_5 x$
29. $\frac{1}{4} \log_4 x$ **30.** $1 - \log_7 x$
31. $\log_4 3 - 3$ **32.** $2 + 2 \log_6 x$
33. $3 + \log_2 x + 2 \log_2 y$
34. $2 + 2 \log_{10} x - 2 \log_{10} y$
35. $2 \log_6 x - \log_6 y - 1$ **36.** $\log_4 x^7$

37. $\log_3 \dfrac{a}{b^4}$ **38.** $\log_a \dfrac{\sqrt{x}}{y^3 z^4}$

39. $\log_{10} (x + 7)$ **40.** 0.8633 **41.** 0.7959
42. 4.5328 **43.** -3.3449 **44.** $x = 3.15$
45. $x = 5,460,000$ **46.** $x = 27.6$
47. $x = 0.402$ **48.** $x = 9.95 \times 10^{-11}$
49. $x = 2.06 \times 10^{-5}$ **50.** $x = 3$
51. $x = 2.86$ **52.** $x = -1, 5$
53. $x = 2.77$ **54.** $x = 4$ **55.** $x = 2$
56. $x = 0$ **57.** $x = 9$ **58.** $t = 9.7$ yr
59. 0.71 **60.** 2.18 **61.** -2.63

62. 4.98 **63.** $\log_3 21 = \dfrac{\log_7 21}{\log_7 3}$

64. 2.3026 **65.** 3.6109 **66.** -6.9078
67. 0.2500 **68.** 11.0 **69.** 0.607
70. $\$7166.65$

Chapter 11 Test, pp. 458–459

1. $5^2 = 25$ **2.** $10^{-2} = 0.01$
3. $\log_6 216 = 3$ **4.** $\log_9 3 = \frac{1}{2}$
5.

6.

7. $a = 2$ **8.** $x = 64$ **9.** $y = 2$
10. $y = -4$ **11.** $1 + \log_2 y$
12. $2 - \log_7 x - \log_7 y$ **13.** $\log x\sqrt{y}$

14. $\log_5 \dfrac{x^3}{y}$ **15.** $x = 2$ **16.** $x = 1$

17. $x = 0.58$ **18.** $x = 2$ **19.** $x = 3$
20. 2.3945 **21.** -3.7696
22. $x = 70,600$ **23.** $x = 0.000444$

24. $\log_2 9 = \dfrac{\log_7 9}{\log_7 2}$ **25.** $\$5399.44$

CHAPTER 12

Problem Set 12.1, pp. 464–466

1. $2, 5, 8, 11, a_{10} = 29$
3. $2, 6, 12, 20, a_{10} = 110$
5. $0, \frac{1}{2}, \frac{2}{3}, \frac{3}{4}, a_{10} = \frac{9}{10}$
7. $2, 2\frac{1}{2}, 3\frac{1}{3}, 4\frac{1}{4}, a_{10} = 10\frac{1}{10}$
9. $6, 6, 6, 6, a_{10} = 6$
11. $-1, \frac{1}{2}, -\frac{1}{3}, \frac{1}{4}, a_{10} = \frac{1}{10}$
13. $2, 4, 8, 16, a_{10} = 1024$ **15.** $a_n = 2n$

17. $a_n = n$ **19.** $a_n = n^3$ **21.** $a_n = \dfrac{1}{n^2}$

23. $a_n = (-1)^n$ **25.** $a_n = (-1)^{n+1}(2n)$
27. $4 + 8 + 12 + 16 + 20 + 24 = 84$
29. $4 + 4 + 4 + 4 + 4 + 4 = 24$
31. $5 + 7 + 9 + 11 = 32$
33. $4 + 6 + 8 + 10 + 1 = 29$
35. $4 + 2 + 2 + 4 + 8 = 20$
37. $0 + 6 + 24 = 30$ **39.** $1 + \frac{1}{2} + \frac{1}{3} + \frac{1}{4} = \frac{25}{12}$
41. $-1 + 4 - 27 = -24$

43. $\dfrac{x^2}{2} + \dfrac{x^3}{3} + \dfrac{x^4}{4} + \dfrac{x^5}{5} + \dfrac{x^6}{6}$

45. $\displaystyle\sum_{n=1}^{6} (3n - 1)$ **47.** $\displaystyle\sum_{n=2}^{7} (3n - 1)$

49. $\displaystyle\sum_{n=1}^{4} 3n$ **51.** $\displaystyle\sum_{n=1}^{4} 3^n$ **53.** $\displaystyle\sum_{n=1}^{4} \dfrac{n}{n + 1}$

55. $\displaystyle\sum_{n=1}^{5} (n^2 + 1)$ **57.** $\displaystyle\sum_{n=0}^{4} x^n$

59. $\displaystyle\sum_{n=1}^{5} n = 1 + 2 + 3 + 4 + 5 = 15$

$\displaystyle\sum_{n=1}^{5} k = 1 + 2 + 3 + 4 + 5 = 15$

61. $10,000, $9000, $8100, $7290, $6561
63. 48
65. $0.94, $1.00, $1.06, $1.12, $1.19, $1.26, $1.34, $1.42, $1.50, $1.59

Problem Set 12.2, pp. 470–471

1. Arithmetic; $d = 1$; 6, 7
3. Arithmetic; $d = 6$; 33, 39
5. Arithmetic; $d = 4$; 14, 18
7. Arithmetic; $d = -5$; -17, -22
9. Arithmetic; $d = \frac{1}{2}$; $\frac{7}{2}$, 4
11. Nonarithmetic **13.** 64 **15.** 658
17. $6n - 2$ **19.** -32 **21.** $\frac{801}{5}$ **23.** 4
25. 2 **27.** 185 **29.** 710 **31.** $\frac{100}{3}$
33. 1278 **35.** 8,006,000 **37.** 17 th
39. 5050 **41.** $2850
43. a) 16 seats **b)** 1426 seats
45. $a_n = 2n + 2$ **47.** 415.438

Problem Set 12.3, pp. 475–476

1. Geometric; $r = 2$; 32, 64
3. Geometric; $r = \frac{1}{3}$; $\frac{1}{81}$, $\frac{1}{243}$
5. Nongeometric
7. Geometric; $r = 5$; -625, -3125
9. Nongeometric
11. Geometric; $r = -1$; -6, 6
13. Geometric; $r = -\frac{1}{2}$; $-\frac{1}{8}$, $\frac{1}{16}$ **15.** 54
17. $2(3)^{n-1}$ **19.** $\frac{1}{8}$ **21.** $-\frac{9}{8}$ **23.** -96
25. 0.2 **27.** 1 **29.** $(-1)^{n-1}$ **31.** 2387
33. $\frac{315}{4}$ **35.** 1094 **37.** $S_{50} = 0$, $S_{51} = 1$
39. $48(1.5^{10} - 1) \approx 2720$ **41.** 204.8 gal
43. $a_n = 15(2)^{n-3}$ **45.** $29,372; $271,521

Problem Set 12.4, pp. 478–479

1. 20 **3.** Does not exist **5.** 3 **7.** 2
9. $\frac{100}{7}$ **11.** $\frac{64}{7}$ **13.** Does not exist
15. Does not exist **17.** 6 **19.** $\frac{5}{6}$ **21.** $\frac{1}{3}$
23. $\frac{4}{11}$ **25.** $\frac{7}{99}$ **27.** $\frac{41}{333}$ **29.** 1
31. $\frac{52}{9}$ **33.** 16 **35.** 40 ft **37.** 37.6

Problem Set 12.5, pp. 483–484

1. $a^7 + 7a^6b + 21a^5b^2 + 35a^4b^3 + 35a^3b^4 + 21a^2b^5 + 7ab^6 + b^7$

3. 15 **5.** 10 **7.** 165 **9.** 8 **11.** 1
13. 15 **15.** $x^3 + 3x^2y + 3xy^2 + y^3$
17. $x^4 + 12x^3 + 54x^2 + 108x + 81$
19. $x^5 - 5x^4y + 10x^3y^2 - 10x^2y^3 + 5xy^4 - y^5$
21. $16x^4 + 32x^3y + 24x^2y^2 + 8xy^3 + y^4$
23. $a^3 - 15a^2 + 75a - 125$
25. $243m^5 - 810m^4 + 1080m^3 - 720m^2 + 240m - 32$
27. $x^4 + 2x^3y + \frac{3}{2}x^2y^2 + \frac{1}{2}xy^3 + \frac{1}{16}y^4$
29. $t^{12} + 6t^{10} + 15t^8 + 20t^6 + 15t^4 + 6t^2 + 1$
31. $a^{20} + 20a^{19}b + 190a^{18}b^2 + \cdots$
33. $x^{15} + 150x^{14} + 10500x^{13} + \cdots$
35. $t^{24} - 12t^{22} + 66t^{20} - \cdots$ **37.** $84a^6b^3$
39. $-64,064p^9q^5$ **41.** $\frac{35}{8}x^4y^4$
43. $x^4 - 10x^3 + 37.5x^2 - 62.5x + 39.0625$

Chapter 12 Review, pp. 484–486

1. 3, 5, 7, 9, $a_{10} = 21$ **2.** $-\frac{1}{3}$, 0, $\frac{1}{5}$, $\frac{1}{3}$, $a_{10} = \frac{2}{3}$
3. -2, 4, -8, 16, $a_{10} = 1024$ **4.** $a_n = 4n$
5. $a_n = \frac{2n}{2n + 1}$ **6.** $a_n = \frac{(-1)^{n+1}}{n^3}$
7. $5 + 10 + 15 + 20 + 25 + 30 = 105$
8. $5 + 4 + 5 + 8 + 13 = 35$
9. $1 + \frac{1}{2} + \frac{1}{4} + \frac{1}{8} + \frac{1}{16} = \frac{31}{16}$ **10.** $\sum\limits_{n=0}^{6} (2n + 1)$
11. $\sum\limits_{n=0}^{3} (5n + 6)$ **12.** $\sum\limits_{n=0}^{4} \left(\frac{1}{4}\right)^n$
13. $d = 4$; 23, 27 **14.** $d = 5$; 19, 24
15. $d = -\frac{1}{2}$; $1\frac{1}{2}$, 1 **16.** 17 **17.** -23
18. -4 **19.** 644 **20.** 336 **21.** 650
22. $660,000 **23.** $r = 4$; 512, 2048
24. $r = -2$; 80, -160 **25.** $r = \frac{2}{3}$; $\frac{32}{243}$, $\frac{64}{729}$
26. $\frac{1}{81}$ **27.** $\frac{80}{81}$ **28.** 2 **29.** 189
30. $\frac{156}{5}$ **31.** $11\frac{23}{27}$ ft **32.** $\frac{27}{2}$ **33.** $\frac{25}{6}$
34. $\frac{64}{3}$ **35.** $\frac{2}{3}$ **36.** 1 **37.** $\frac{2}{3}$ **38.** $\frac{4}{33}$
39. $\frac{25}{9}$ **40.** 140 ft **41.** 35 **42.** 11
43. 1 **44.** 28
45. $x^6 + 6x^5y + 15x^4y^2 + 20x^3y^3 + 15x^2y^4 + 6xy^5 + y^6$
46. $81m^4 + 216m^3 + 216m^2 + 96m + 16$
47. $t^{10} - 10t^8 + 40t^6 - 80t^4 + 80t^2 - 32$
48. $x^3 - \frac{3}{2}x^2y + \frac{3}{4}xy^2 - \frac{1}{8}y^3$
49. $x^{10} + 50x^9 + 1125x^8 + \cdots$
50. $560a^3b^4$

Chapter 12 Test, p. 486

1. a) 6 **b)** 10 **2.** $d = 0.7$; 7.3, 8.0

3. $r = \frac{1}{2}; \frac{5}{2}, \frac{5}{4}$ **4.** $2 + 6 + 12 + 20 = 40$

5. $2 + 7 + 12 + 17 + 22 = 60$ **6.** $\sum\limits_{n=1}^{5} 3n$

7. $\sum\limits_{n=2}^{5} \dfrac{n+1}{n}$ **8.** $a_n = 2n + 3$

9. $a_n = (-1)^{n-1} n^2$ **10.** $\frac{2}{3}, 1, \frac{6}{5}, \frac{4}{3}, a_{10} = \frac{5}{3}$

11. 1, 5, 9, 13, $a_{10} = 37$ **12.** 17 **13.** -4

14. 230 **15.** $\frac{1}{9}$ **16.** $\frac{45}{4}$ **17.** $\frac{8}{33}$

18. $x^4 - 8x^3 + 24x^2 - 32x + 16$

19. $2268a^6b^3$ **20.** 144 in.

Cumulative Review for Chapters 10–12, pp. 487–489

1. Yes **2.** No **3.** $x \neq -2, 3$

4. $x \geq -6$ **5.**

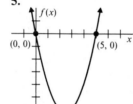

6.

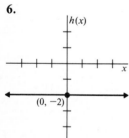

7.

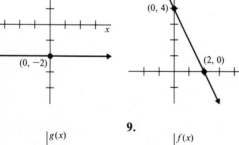

8.

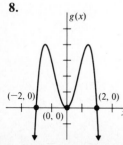

9.

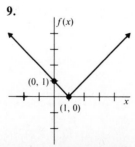

10.

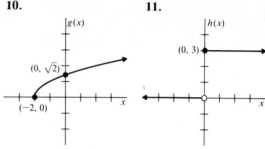

11.

12. a) 1 **b)** 3 **c)** $\sqrt{2(t+h)+1}$
 d) $\sqrt{2\sqrt{2x+1}+1}$

13. $f^{-1}(x) = \dfrac{x+2}{5}$ **14.** $t^2 + at + a^2$

15. $x = \dfrac{ky^2}{z^3}$ **16.** $V = \dfrac{kT}{P}$ **17.** $y = 1$

18. 5120 pounds **19.** $\log_2 32 = 5$

20. $\log_3 100 = t$ **21.** $2^9 = 512$

22. $16^{3/4} = 8$ **23.** $\dfrac{\log_5 8}{\log_5 12}$ **24.** 2

25. $\frac{3}{2}$ **26.** 0 **27.** 6

28.

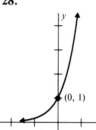

29.

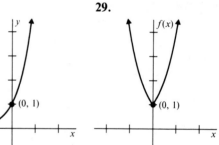

30.

31.

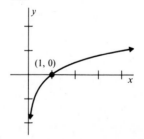
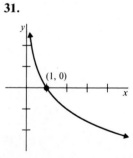

32. $\frac{1}{9}$ **33.** 2 **34.** 2 **35.** ± 4

36. $5 \log x - 3 \log y - 4 \log z$

37. $\ln x^3 \sqrt{x+1}$ **38.** $\log \dfrac{x^2}{y^4}$ **39.** 2.6551

40. -6.0164 **41.** 2.7726   **42.** -3.9120
43. 357   **44.** 0.000691   **45.** 2.59
46. 122   **47.** 4.7 yr   **48.** \$1,019,400,000
49. $4, \frac{9}{2}, \frac{16}{3}, \frac{25}{4}; a_{10} = \frac{121}{10}$

50. $\frac{3}{4}, \frac{6}{5}, \frac{3}{2}, \frac{12}{7}; a_{10} = \frac{30}{13}$ **51.** $a_n = \dfrac{n}{(n+1)^2}$

52. $a_n = \dfrac{(-1)^{n+1}}{2n-1}$ **53.** $\displaystyle\sum_{n=1}^{5} 3^n$

54. $\displaystyle\sum_{n=1}^{4} (4n+3)$ **55.** $d = 4; 18, 22$

56. $d = \frac{1}{2}; 0, \frac{1}{2}$ **57.** $r = \frac{1}{4}; \frac{1}{128}, \frac{1}{512}$
58. $r = -3; 81, -243$ **59.** 770   **60.** $\frac{85}{16}$
61. 2   **62.** 78   **63.** 1
64. $x^8 - 8x^7y + 28x^6y^2 - 56x^5y^3 + 70x^4y^4 - 56x^3y^5 + 28x^2y^6 - 8xy^7 + y^8$
65. $x^5 + 10x^4y + 40x^3y^2 + 80x^2y^3 + 80xy^4 + 32y^5$
66. $364x^6y^{11}$ **67.** $1120a^4b^4$ **68.** 37.5 cm

APPENDIXES

Problem Set A.1, pp. 494–495

1. Finite   **2.** Infinite   **3.** True
4. False   **5.** True   **6.** False
7. $\{1, 2, 3, 4\}$ **8.** $\{8, 9, 10, 11, \ldots\}$
9. $\{2, 3, 4, 5\}$ **10.** $\emptyset$
11. $\{x | x$ is a natural number between 3 and 9$\}$
12. $\{y | y$ is a natural number less than 6$\}$
13. $\{t | t$ is a natural number greater than 9$\}$
14. $\{p | p$ is an odd natural number greater than 6$\}$
15. True   **16.** False   **17.** True
18. True   **19.** False   **20.** True
21. True   **22.** True   **23.** $\emptyset$ **24.** $\emptyset, \{1\}$
25. $\emptyset, \{1\}, \{2\}, \{1, 2\}$
26. $\emptyset, \{1\}, \{2\}, \{3\}, \{1, 2\}, \{1, 3\}, \{2, 3\}, \{1, 2, 3\}$
27. $\{1, 2, 3, 4, 5, 7\}$ **28.** $\{3, 5\}$
29. $\{2, 3, 5, 6, 8\}$ **30.** B **31.** A **32.** A
33. A **34.** $\emptyset$

Problem Set A.2, p. 499

1. 500 cm   **2.** 18,000 mm   **3.** 780 cm
4. 16,000 m   **5.** 5.2 km   **6.** 62 m
7. 0.045 m   **8.** 171 mm   **9.** 0.5 cm
10. 6000 g   **11.** 15 kg   **12.** 1620 g
13. 8000 mℓ **14.** 0.5 ℓ **15.** 37 ℓ
16. 2930 mℓ **17.** 7.315 m   **18.** 198.5 lb
19. 9.843 ft   **20.** 563.2 km   **21.** 311.9 g
22. 0.8119 oz   **23.** 1.143 m   **24.** 60.96 m
25. 17.78 cm   **26.** 4.732 ℓ **27.** 3.937 in.
28. 78.74 in.   **29.** 385.6 kg   **30.** 31.71 qt
31. 14.22 yd   **32.** 124.3 mi
33. 21,240 sq mi   **34.** 439.4 cu in.

Index